Modern International Developments in Data Analysis and Multivariate Statistical Analysis

Sofia D. Anastasiadou
University of Western Macedonia, Greece

Theodore G. Chadjipadelis
Aristotle University of Thessaloniki, Greece

Nikolaos Koutsoupias
University of Macedonia, Greece

Angelos Markos
Democritus University of Thrace, Greece

George Menexes
Aristotle University of Thessaloniki, Greece

IGI Global
Scientific Publishing
Publishing Tomorrow's Research Today

Published in the United States of America by
IGI Global Scientific Publishing
701 East Chocolate Avenue
Hershey, PA, 17033, USA
Tel: 717-533-8845
Fax: 717-533-7115
Website: https://www.igi-global.com E-mail: cust@igi-global.com

Library of Congress Cataloging-in-Publication Data

Names: Anastasiadou, Sofia D., 1968- editor | Chadjipadelis, Theodore editor | Koutsoupias, Nikos, 1969- editor | Markos, Angelos editor | Menexes, Georgios, 1965- editor

Title: Modern international developments in data analysis and multivariate statistical analysis / edited by Sofia Anastasiadou, Theodore Chadjipadelis, Nikos Koutsoupias, Angelos Markos, Georgios (George) Menexes.

Description: Hershey, PA : IGI Global Scientific Publishing, [2026] | Includes bibliographical references and index. | Summary: "The primary objectives of the book are to showcase the latest advancements in analytical techniques, highlight innovative applications across diverse fields, and provide a comprehensive resource for understanding and addressing the complexities of modern data"-- Provided by publisher.

Identifiers: LCCN 2025001267 (print) | LCCN 2025001268 (ebook) | ISBN 9798369394007 hardcover | ISBN 9798369394014 paperback | ISBN 9798369394021 ebook

Subjects: LCSH: Multivariate analysis http://id.loc.gov/authorities/subjects/sh85088390 | Research--Statistical methods http://id.loc.gov/authorities/subjects/sh2010110884 | Social sciences--Statistical methods http://id.loc.gov/authorities/subjects/sh85124018

Classification: LCC QA278 .M648 2025 (print) | LCC QA278 (ebook) | DDC 519.5/35--dc23/eng/20250522

LC record available at https://lccn.loc.gov/2025001267

LC ebook record available at https://lccn.loc.gov/2025001268

British Cataloguing in Publication Data

A Cataloguing in Publication record for this book is available from the British Library.

Modern International Developments in Data Analysis and Multivariate Statistical Analysis

Sofia D. Anastasiadou
University of Western Macedonia, Greece

Theodore G. Chadjipadelis
Aristotle University of Thessaloniki, Greece

Nikolaos Koutsoupias
University of Macedonia, Greece

Angelos Markos
Democritus University of Thrace, Greece

George Menexes
Aristotle University of Thessaloniki, Greece

IGI Global
Scientific Publishing
Publishing Tomorrow's Research Today

Vice President of Editorial	Melissa Wagner
Director of Acquisitions	Mikaela Felty
Director of Book Development	Jocelynn Hessler
Production Manager	Mike Brehm
Cover Design	Jose Rosado

Published in the United States of America by
IGI Global Scientific Publishing
701 East Chocolate Avenue
Hershey, PA, 17033, USA
Tel: 717-533-8845
Fax: 717-533-7115
Website: https://www.igi-global.com E-mail: cust@igi-global.com

Library of Congress Cataloging-in-Publication Data

Names: Anastasiadou, Sofia D., 1968- editor | Chadjipadelis, Theodore editor | Koutsoupias, Nikos, 1969- editor | Markos, Angelos editor | Menexes, Georgios, 1965- editor

Title: Modern international developments in data analysis and multivariate statistical analysis / edited by Sofia Anastasiadou, Theodore Chadjipadelis, Nikos Koutsoupias, Angelos Markos, Georgios (George) Menexes.

Description: Hershey, PA : IGI Global Scientific Publishing, [2026] | Includes bibliographical references and index. | Summary: "The primary objectives of the book are to showcase the latest advancements in analytical techniques, highlight innovative applications across diverse fields, and provide a comprehensive resource for understanding and addressing the complexities of modern data"-- Provided by publisher.

Identifiers: LCCN 2025001267 (print) | LCCN 2025001268 (ebook) | ISBN 9798369394007 hardcover | ISBN 9798369394014 paperback | ISBN 9798369394021 ebook

Subjects: LCSH: Multivariate analysis http://id.loc.gov/authorities/subjects/sh85088390 | Research--Statistical methods http://id.loc.gov/authorities/subjects/sh2010110884 | Social sciences--Statistical methods http://id.loc.gov/authorities/subjects/sh85124018

Classification: LCC QA278 .M648 2025 (print) | LCC QA278 (ebook) | DDC 519.5/35--dc23/eng/20250522

LC record available at https://lccn.loc.gov/2025001267

LC ebook record available at https://lccn.loc.gov/2025001268

British Cataloguing in Publication Data

A Cataloguing in Publication record for this book is available from the British Library.

Table of Contents

Detailed Table of Contents

A multidimensional and multivariate structure with mixed-type data allows researchers to use various statistical approaches and data clustering techniques. The choice of clustering method used can have an impact on the results obtained In this study, the Partitioning Clustering (k-means) and Hierarchical Cluster Analysis methods were compared. The main objective of this study was to apply and compare these methods in segmenting the data set into groups-clusters. The results of the study showed that there are several analysis strategies for clustering mixed-type data, based on data coding and the chosen measurement scale for input variables. Remarkably, Python delivered results over 100% faster than SPSS. In conclusion, it emerged that the results of the classification depend on the coding strategy and the selection of the measurement scale of the variables to be used in the analysis.

Choropleth maps are essential tools for visualizing spatial data and supporting decision-making across fields such as agriculture, environmental science, urban planning, and public health. However, classifying spatial data into distinct categories introduces subjectivity, affecting both interpretation and utility. This study examines how classification methods influence spatial visualization, emphasizing how biases from common techniques—such as equal interval, quantile, and standard deviation—can distort spatial patterns and decisions. We explore alternative approaches, including Natural Breaks, Box Maps, and Local Indicators of Spatial Association (LISA), which may reduce subjectivity and improve map accuracy. A case study shows how combining these methods with spatial autocorrelation can mitigate bias and enhance analytical reliability. The findings underscore the value of context-aware classification and statistical rigor to ensure transparent, reproducible, and data-driven map interpretations.

The 2024 European Elections marked a pivotal post-Brexit moment, influenced by socio-economic challenges like climate change, migration, and the energy crisis. This study examines the evolution of political marketing as a strategic electoral tool, focusing on its role in voter engagement and party positioning at national and supranational levels. By analyzing the dual-level dynamics of political competition, the research explores how European parties employed marketing strategies to shape key political axes and address diverse voter concerns. Using a comparative multivariate analysis, it investigates variations in campaign approaches, alignment with European political groups, and thematic priorities during the 2024 elections.

This chapter investigates the dynamics of political participation and electoral behavior among university students in Thessaloniki, Greece, offering a detailed analysis of the factors shaping engagement within this critical demographic. Using advanced multivariate methodologies, the study identifies distinct respondent clusters based on their voting patterns, political knowledge, interest, mobilization strategies, and issue prioritization. The findings reveal two primary axes of polarization: engagement versus disengagement and progressive versus conservative ideologies. The study also highlights the transformative role of social media as a mobilizing tool for progressive voters, contrasting with the reliance on traditional media among conservative respondents. Issue-based voting patterns further delineate the ideological divides, with pragmatic voters prioritizing safety and economic stability, and progressive voters focusing on societal and cultural reforms. These insights contribute to a deeper understanding of voter segmentation and the factors influencing political behavior.

This study proposes a risk assessment method for the maritime industry to address international threats such as terrorism, piracy, robbery, cyber-attacks, smuggling, and drug trafficking. Due to limited literature on maritime risk acceptance criteria, the method employs the Bow-Tie diagram for Risk Identification, Analysis, Evaluation, and Cost-Benefit Analysis. After identifying the Hazard and Top Event, the risk is calculated based on the defined probability and contribution of each threat, and the effectiveness of each prevention barrier. Similarly, risk for each consequence is computed using the defined risk value and the effectiveness of each mitigation barrier. The total cost of risk treatment is calculated from the cost of all prevention and mitigation measures and compared to the cost in case of the Top Event and consequences. Results are visualized in the Bow-Tie diagram using a color-coded scheme, offering a clear overview of risks and associated costs.

Artificial intelligence has a considerable impact on national and international politics as well on international relations. It challenges geostrategic alliances, introduces new challenges to the international agenda, helps diplomats and negotiators in their work, and raise new questions and opportunities for human rights. Using R language tools, this study explores the digital presence of the scientific field of AI in national and international politics and provides a complete view of its authorship taking into account bibliometrical data including annual production rates, worldwide citation frequencies, word frequencies and word networking dynamics. The acquired datasets undergo additional analysis through the application of Multiple Correspondence Analysis (MCA) in conjunction with Hierarchical Clustering (HC). The goal of this analysis is to discern patterns and correlations among the diverse elements of these digital authorship datasets which can provide a clear semantic overview of the role of Artificial Intelligence in National and International Politics.

This chapter investigates the integration of Sentiment Analysis, Multiple Correspondence Analysis, and Hierarchical Clustering analysis to advance the understanding of emotional dynamics in international relations discourse, with particular emphasis on Sino-American relations. Bibliometric analysis supplements this framework by offering a systematic overview of scholarly output. This integrated methodological approach enables the processing of large-scale textual data, the refinement of categorical distinctions through multiple correspondence analysis, and the identification of emotional patterns via hierarchical clustering. The findings indicate a predominance of positive sentiment and trust, while fear and anger underscore persistent geopolitical anxieties. Furthermore, the bibliometric results identify Australia, the United States, and China as principal contributors, alongside a growing pattern of global academic engagement. The synthesis of these methods provides a multidimensional perspective on emotional valence in geopolitical analysis.

This chapter undertakes a systematic bibliometric analysis of hierarchical clustering applications within the social sciences and humanities, drawing upon publications indexed in the Scopus database from 1975 to 2024. The findings reveal the methodological versatility of clustering techniques in addressing complex research problems across a broad spectrum of academic disciplines. Thematic emphases on social network analysis and data management strategies underscore the centrality of clustering in elucidating intricate sociocultural systems. The United States, China, and the United Kingdom emerge as principal contributors, highlighting the global scope and interdisciplinary nature of this research domain. This analysis provides a comprehensive account of prevailing thematic trends, patterns of scholarly collaboration, and the intellectual architectures that underpin the field. It offers a robust foundation for both the methodological advancement of clustering techniques and their expanded application to pressing societal and academic challenges

This article examines the impact of work stress on substitute secondary school teachers in Greece, a group facing unique challenges and uncertainties. It analyzes the relationship between job stress and factors such as job performance and burnout. A questionnaire was designed with five factors: (1) job stress, (2) job performance, (3) burnout, (4) job satisfaction, and (5) coping strategies. Participants responded using a Likert scale, and data will be analyzed through regression to identify correlations and effects of stress. The study also explores strategies teachers use to manage stress and how these affect satisfaction and performance. Preliminary results suggest stress significantly impacts burnout and satisfaction. Coping strategies mitigate stress, while performance varies based on stress levels and strategies. This study highlights the need for interventions to support substitute teachers, aiming to improve mental health and performance. Findings could guide programs to enhance resilience and satisfaction in this critical educational sector.

Gender-based violence in the workplace is a complex and multidimensional issue that affects employees' psychological and professional lives daily. This phenomenon can manifest in various forms, such as sexual harassment, verbal abuse, psychological abuse, and bullying. The aim of this research is to identify the forms of gender-based violence in the workplace, analyze the risk factors, and evaluate the effectiveness of existing policies and practices aimed at addressing it. The research, conducted through the collection of electronic questionnaires, seeks to document the experiences of employees across various sectors and levels regarding gender-based violence in the workplace. Understanding and mapping the factors that increase or reduce the risk of gender-based violence will help develop more effective prevention and response strategies. The research will employ a quantitative approach using questionnaires distributed to employees from various sectors and different professional levels within the Region of Eastern Macedonia and Thrace.

Osteoporosis is a skeletal disease, characterised by reduced bone density, disrupted bone microarchitecture and susceptibility to fracture. Its epidemiology and pathophysiology justify the tendency to be considered as a women's disease. The study aims at determining the level of knowledge of Greek men on osteoporosis. Data were collected with the FOOQ and the MOKQ. It comprises 20 statements, assessing the knowledge on bone development, osteoporosis risk factors, prevention and treatment. 330 men participated in the study. SIA' results showed that the hardest statement of the FOOQ was the one referring to whether postmenopausal hormone replacement therapy can slow down bone loss. The easiest statement refers to the presentation of an osteoporotic fracture in 20% of women older than 50, if no preventative measures are taken. The most difficult statement of the MOKQ is the one that assesses whether hormonal treatment for prostate cancer does not reduce the risk of osteoporosis, whereas the easiest statement assesses whether bone loss increases in men after the age of 70.

Preface

In recent decades, the world of data analysis has undergone a seismic transformation. The proliferation of large and complex datasets—often multivariate, mixed-type, and high-dimensional—has demanded equally sophisticated methods for analysis, interpretation, and decision-making. *Modern International Developments in Data Analysis and Multivariate Statistical Analysis* emerges from this dynamic landscape as a scholarly response to the growing need for advanced statistical approaches that can address intricate research questions across diverse disciplines. This edited volume brings together cutting-edge methodologies and applications, emphasizing comparative frameworks and empirical rigor.

Each chapter in this book represents a unique contribution to the evolving field of multivariate data analysis. Drawing on a variety of statistical tools such as PCA, MCA, CATPCA, clustering algorithms, and sentiment analysis techniques, the contributors offer insights that are both methodologically innovative and practically relevant. Whether applied to political science, environmental mapping, machine learning, artificial intelligence ethics, or maritime security, the studies herein highlight how multivariate methods can transform raw data into meaningful patterns, actionable intelligence, and informed decisions.

Our contributors include statisticians, data scientists, and domain experts from leading Greek universities, each of whom brings both theoretical knowledge and empirical insight to their work. The diversity of topics—from spatial data visualization to electoral behavior and international political sentiment—reflects the wide applicability of multivariate analysis in solving complex problems. This volume is designed not only to showcase academic excellence but also to serve as a practical resource for researchers, educators, and professionals across data-intensive fields.

This volume is dedicated to the Greek professor Ioannis Papadimitriou, an influential figure in the fields of mathematics, statistics, and data analysis. By applying methods derived from the French school of Data Analysis, Professor Papadimitriou has provided innovative solutions to complex problems in diverse fields such as finance, literature, economics, applied informatics, politics, and education. His outstanding academic career is marked by substantial contributions to scientific research and exceptional mentorship. Inspired by his dedication and passion, his students continue today to advance in the fields of Data Analysis and Multivariate Statistical Analysis. We express our deepest gratitude and admiration to this exceptional scholar and scientist.

CHAPTER OVERVIEW

Chapter 1: Dimensionality Reduction Techniques for Multivariate Categorical Data

This chapter presents a comparative analysis of dimensionality reduction methods— SVD with PCA, MCA, CATPCA, FAMD, MFA, and NLCCA—applied to high-dimensional categorical data using the "Forest Cover Type" dataset. The study finds that no single method consistently outperforms the others, highlighting that the choice of technique must align with the data structure and research goals. It also explores hybrid strategies combining multiple methods, demonstrating the value of diverse, objective-driven approaches to reduce dimensionality and enhance data interpretation.

Chapter 2: Dimensionality Reduction in Data Cubes with Independent Slices

Focusing on three-dimensional data cubes with independent "slices", this chapter evaluates PCA-based strategies using two data cubes with varying internal structures. Some strategies focus on decomposing the total variability of the data cube into between- and within-"slices" variability. Other strategies either ignore the effect of "slices" or consider it on the overall data cube structure. The findings emphasize the need for context-sensitive strategies tailored to the cube's composition and analytical goals, providing new insights into dimensionality reduction in data cubes-multidimensional datasets.

Chapter 3: Clustering Mixed-Type Data: A Comparative Study Using Python and SPSS

This chapter investigates clustering strategies for mixed-type datasets, comparing k-means and hierarchical cluster analysis implemented in Python and SPSS. The study demonstrates how data coding and variable measurement scales influence clustering outcomes. Notably, Python outperforms SPSS in processing speed, completing analyses over 100% faster. The results underscore the importance of preprocessing decisions in clustering and advocate for efficient tool selection based on data characteristics and analytical needs.

Chapter 4: Reducing Subjectivity in Spatial Data Classification for Decision-Making

This chapter analyses the inherent subjectivity of choropleth map classification and its implications for decision-making. It compares commonly used classification methods and explores alternatives such as Natural Breaks, Box Maps, and LISA. A case study demonstrates how incorporating spatial autocorrelation and statistical rigor can mitigate bias. The proposed framework aims to support more accurate and transparent visualizations in domains such as agriculture, urban planning, and public health.

Chapter 5: Predictive Classification and Feature Selection in Forest Cover Type Data

Using the UCI "Forest Covertype" dataset, this chapter compares the performance of multiple machine learning algorithms in predictive classification and feature selection both in the original and standardized data. Algorithms including K-NN, SVC, Logistic Regression, SGD, Decision Trees, Gaussian Naïve Bayes, Neural Networks, Random Forest, and Extra Trees are evaluated with optimization techniques like GridSearch and AdaBoost. Results show that Random Forest and Extra Trees provide the highest accuracy, particularly on standardized data. Feature importance analysis identifies elevation, soil type, and wilderness area as key predictors, offering practical insights for environmental modeling.

Chapter 6: Analyzing European Elections: A Multivariate Perspective (2019–2024)

This chapter employs Hierarchical Cluster Analysis and Factorial Correspondence Analysis to compare the 2019 and 2024 European Elections. It uncovers growing political polarization and ideological consolidation, particularly among nationalist and Eurosceptic parties like PfE and ECR. Regional clustering based on voting patterns highlights evolving voter preferences and provides a multivariate framework to interpret EU-wide electoral dynamics.

Chapter 7: Political Marketing Strategies in the 2024 European Elections

Examining the role of political marketing in the 2024 European Elections, this chapter analyzes how parties across the EU crafted campaign messages in response to climate, migration, and energy challenges. A multivariate comparative analysis reveals strategic positioning across national and EU contexts, illustrating how parties tailored their messaging to ideological trends and voter priorities.

Chapter 8: Political Participation among Greek University Students

This chapter investigates electoral behavior among university students in Thessaloniki using multivariate techniques. It identifies key clusters based on engagement levels, ideological alignment, and media usage. The study reveals significant generational divides in political mobilization, with progressive students favoring social media and conservatives relying on traditional media. The results contribute to a nuanced understanding of youth political engagement and segmentation.

Chapter 9: Maritime Security Risk Assessment Using the Bow-Tie Method

Addressing global maritime threats, this chapter introduces a risk assessment framework based on the Bow-Tie diagram. It quantifies risks from threats like piracy and cyberattacks through probability-weighted models and evaluates the effectiveness of mitigation measures. A color-coded visualization system enhances interpretability, offering a practical decision-making tool for maritime security planning and cost-benefit analysis.

Chapter 10: Mapping AI's Influence in Politics through Digital Authorship Analysis

Using multivariate statistical analysis, this chapter explores AI's presence in national and international politics by analyzing digital authorship patterns. Bibliometric data are processed with MCA and hierarchical clustering to identify global publication trends, authorship clusters, and thematic dynamics. The study offers a comprehensive semantic map of AI's role in reshaping geopolitical discourse and policy.

Chapter 11: Sentiment and Geopolitical Analysis in Sino-American Relations

This chapter integrates sentiment analysis, MCA, and hierarchical clustering to examine emotional narratives in Sino-American discourse. Results highlight dominant sentiments of trust and anxiety, while bibliometric insights reveal global scholarly engagement. The chapter presents a novel multidimensional framework for understanding emotional dynamics in international relations.

Chapter 12: Hierarchical Clustering in Social Sciences and Humanities: A Review

This systematic review analyzes the application of hierarchical clustering in social sciences and humanities research from 1975 to 2024. Drawing on Scopus-indexed literature, it identifies dominant themes such as social network analysis and interdisciplinary methodologies. The review maps out evolving research trends and highlights methodological advances, supporting future development in clustering applications.

Chapter 13: Work-Related Stress among Substitute Teachers in Greece

Focusing on substitute secondary school teachers, this chapter explores the effects of work-related stress on job performance and burnout. Using survey-based analysis across five dimensions—including job stress, satisfaction, and burnout—it highlights the precarious conditions faced by substitutes and their implications for educational effectiveness. The findings underscore the need for targeted interventions and systemic reforms to support teacher well-being.

Chapter 14: Gender-Based Violence in the Workplace: Psychological and Professional Impacts

This chapter explores the multifaceted phenomenon of gender-based violence (GBV) in the workplace, including sexual harassment, verbal abuse, psychological aggression, and bullying. Using quantitative data collected via electronic questionnaires from employees across sectors in Eastern Macedonia and Thrace, the study identifies common forms of GBV, risk factors, and the effectiveness of current mitigation policies. The findings contribute to the development of more targeted prevention strategies and inform institutional practices aimed at fostering safer, more inclusive work environments.

Chapter 15: Greek Men's Knowledge of Osteoporosis

This chapter assesses the awareness and understanding of osteoporosis among Greek men using two validated instruments: the FOOQ and the MOKQ. Based on responses from 330 participants, the study identifies key gaps in knowledge, particularly regarding hormonal treatment and fracture risk. The findings challenge the perception of osteoporosis as a predominantly female condition and underscore the need for public health campaigns targeting male populations to improve preventative care and education.

Chapter 16: Exploring AI Ethics Through Factorial Correspondence Analysis

This chapter examines students' awareness of key ethical issues in artificial intelligence, such as privacy, bias, autonomy, and justice. Based on responses from 239 participants and analyzed through factorial correspondence analysis, the study identifies conceptual groupings around core ethical constructs. The findings suggest a high level of student awareness regarding AI ethics and support the need for ethical literacy as a foundational component of AI education and policy development.

We extend our deepest gratitude to the contributors whose scholarly dedication has shaped this book, and to the institutions that supported their research. The high quality of the contributions reflects the strength of the Greek academic community in the field of data science and statistical methodology. Special thanks are due to our respective universities for fostering an environment of innovation and interdisciplinary exchange that made this volume possible.

To our readers—students, academics, analysts, and policy-makers—we offer this volume as both a reference and an inspiration. Whether you seek to deepen your methodological knowledge or apply new techniques to your field of inquiry, we hope that the insights presented here will prove both enlightening and actionable. The future of data analysis lies in continual adaptation, cross-disciplinary integration, and a commitment to rigor—and we are proud to contribute to that future with this book.

Finally, we reaffirm our belief that data, when carefully analyzed and responsibly interpreted, has the power to inform better decisions, solve complex problems, and contribute meaningfully to society. We trust that *Modern International Developments in Data Analysis and Multivariate Statistical Analysis* will serve as a valuable guide and companion in that journey.

Chapter 1
Dimensionality Reduction Techniques for Multivariate Categorical Data

Nikolaos Papafilippou
https://orcid.org/0009-0003-3148-7229
Aristotle University of Thessaloniki, Greece

Zacharenia Kyrana
https://orcid.org/0000-0001-9269-0675
Aristotle University of Thessaloniki, Greece

Emmanouil D. Pratsinakis
https://orcid.org/0000-0002-3725-3525
Aristotle University of Thessaloniki, Greece

Christos Dordas
https://orcid.org/0000-0002-7027-474X
Aristotle University of Thessaloniki, Greece

Georgios C. Menexes
https://orcid.org/0000-0002-1034-7345
Aristotle University of Thessaloniki, Greece

ABSTRACT

High-dimensional categorical data challenges Data Science, Machine Learning, and Statistics in analyzing variability and interpretation. Using the "Forest Cover Type" dataset (n=581,012), this paper explores dimensionality reduction methods: PCA,

DOI: 10.4018/979-8-3693-9400-7.ch001

MCA, CATPCA, FAMD, MFA, and NLCCA. Results show no method is universally best; the choice depends on data and goals. Advanced strategies like combining PCA, MCA, and CATPCA offer unique insights, emphasizing diverse approaches tailored to study objectives.

INTRODUCTION

Categorical data, representing "qualitative information" and characterized by discrete values, is frequently encountered in various research and scientific fields. These data are divided into two main categories: nominal data, which include categories without any inherent order, for example forest cover type or color, and ordinal data, where the categories have a natural or hierarchical order, such as educational level or agreement levels in statements (Agresti, 2012; Keller, 2018). Analyzing categorical data is often more challenging than numerical data, as they do not have a natural numerical scale, their categories do not have a specific order or consistent distances between them. For this reason, specialized methods and dimensionality reduction techniques are often required to extract meaningful insights from the data. The challenges that arise when analyzing categorical data concern both the structure of the data and the interpretability of the results (Agresti, 2018; Flowerdew & Martin, 2013; Upton, 2016).

The present study focuses on the above challenges by analyzing and evaluating various methods for dimensionality reduction of categorical multivariate/multidimensional data. At the same time, it proposes new approaches regarding dimensionality reduction strategies, aiming to better capture the variability (inertia) of the data, highlight the importance of selecting different analytical strategies for studying their structure and characteristics, and improve the interpretation of the results.

METHODS FOR ANALYZING MULTIVARIATE/MULTIDIMENSIONAL CATEGORICAL DATA

The methods of Multivariate and Multidimensional Statistical Data Analysis represent an alternative methodological and philosophical approach to statistical inference, and include three main families of methods (Benzécri, 1992; Menexes, 2006): a) Correspondence Analysis (CA and MCA) for bivariate and multivariate data respectively, b) Principal Component Analysis (PCA), and c) Hierarchical Cluster Analysis (HCA). A distinctive feature of these methods is the symmetrical treatment of variables, where no distinction is made between dependent and independent variables. Their goal is to reveal and describe latent structures that may exist in multidimensional data tables through processes of transformation and reduction of the dimensions of the original mathematical space in which the phenomenon under examination can be described. The new dimensions, typically constructed from complex relationships among variables, are ultimately interpreted as new composite variables or factors (or components). Another key characteristic of these methods is that they do not require a priori assumptions about the existence of a theoretical distribution or hypotheses regarding the parameters of the population under study. One of the main disadvantages of dimensionality reduction methods is the "curse of dimensionality" (Oseledets & Tyrtyshnikov, 2009), referring to the difficulty of determining the appropriate number of significant dimensions. Moreover, these methods demand substantial computational resources and present challenges in managing missing values, addressing outliers, and visualizing results (Hair et al., 2010; Menexes, 2006). The evolution of these techniques includes machine learning (Bishop, 2016; Song et al., 2018), which focuses on the analysis of multidimensional data for decision-making and the development of predictive models.

For categorical data analysis, following the "French School of Data Analysis" (Benzécri, 1992), methods such as CA and MCA treat all data as categorical, measured at the nominal scale (including quantitative data, if any, through appropriate transformations – categorization of their values). Within the methodological and philosophical frame of "Dutch School of Data Analysis" the user defines the scale of the variables used in an analysis depending on prior knowledge, experience and the research's conceptual and theoretical frame. The Categorical data are quantified using optimal scaling methods (such as CATPCA), assigning optimal values to the variables' categories and subsequently applying methods suitable for quantitative data (Gifi, 1990; Menexes, 2006; Di Franco, 2015). For the Dutch authors, there are no *a priori* intrinsic properties of a variable that necessarily define the level of scaling which must be chosen for this variable (nominal, ordinal or scale). For example, consider a variable that categorizes individuals based on their age. While age categories are often considered as the values of an ordinal variable, there may be scenarios where relationships with other variables suggest nominal treatment of the variable age (see Gifi 1990, Menexes 2006). For instance, a driving safety index might show a positive linear association with younger ages (under 30 yrs), no association with ages between 30 and 40, a non – linear association with ages 40-60 and a negative monotonic association with the age of older individuals (over 60 yrs). In such cases, it would be more appropriate to treat age as a categorical variable measured at the nominal scale for better reflecting the inherent relationship with the safety index data. Additionally, there are methods that simultaneously handle variables with different measurement scales, either by treating them as a unified set or by dividing them into groups. The main methods are presented below.

CORRESPONDENCE ANALYSIS

Correspondence Analysis (CA) (Benzécri, 1992; Greenacre, 1991; Le Roux & Rouanet 2004) is a dimensionality reduction method used to identify patterns and relationships among categorical variables without a priori assumptions or prerequisites. The purpose of CA is to highlight and visualize the intrinsic structure of the data, which is not immediately apparent, revealing how variables are connected and how their categories interact. In its simplest form, the case of two variables X and Y with k and l categories respectively, CA is applied to the "correspondence" table P, which results from the contingency table, where the categories of the variables cross (Table 1), by dividing the frequencies f_{ij}, i=1, 2, …, k and j=1, 2, …, l of each cell of the contingency table, by the total n, that is $p_{ij} = \frac{f_{ij}}{n}$,

Table 1. Contingency table of categorical variables X and Y

	Categories of variable Y				
Categories of variable X	Y_1	Y_2	…	Y_l	Row sum (total)
X_1	f_{11}	f_{12}	…	f_{1l}	$n_{1.}$
X_2	f_{21}	f_{22}	…	f_{2l}	$n_{2.}$
⋮	⋮	⋮	…	⋮	⋮
X_k	f_{k1}	f_{k2}	…	f_{kl}	$n_{k.}$
Columns sum (total)	$n_{.1}$	$n_{.2}$	…	$n_{.l}$	n

where f_{ij} is the cell frequency of row i and column j, $n_{i.}$ is the sum of frequencies of i row, $n_{.j}$ is the sum of frequencies of j column, and n is the total of the observations. To measure the relationship between the rows (or columns) of the table, the chi-squared distance proposed by Benzécri is used (Benzécri, 1992):

$$d^2_{x^2}(i, i') = \sum_{j=1}^{\lambda} \frac{1}{c_j}\left(\frac{p_{ij}}{r_i} - \frac{p_{i'j}}{r_{i'}}\right)^2, \text{ distance for rows i and i' or}$$

$$d^2_{x^2}(j, j') = \sum_{i=1}^{\kappa} \frac{1}{r_i}\left(\frac{p_{ij}}{c_j} - \frac{p_{ij'}}{c_{j'}}\right)^2, \text{ distance for columns j and j'}$$

where $r_i = n_{i.}/n$, $c_j = n_{.j}/n$ the mass of i row and j column respectively.

Using the chi-squared distance and the row (or column) masses, the inertia of the contingency table is calculated. Inertia can be considered a measure of the dispersion of the profile points in the multidimensional space to which they belong. It has been shown (see Benzécri, 1992; Greenacre, 2006; Menexes, 2006) that inertia is equal to $I=Q/n$, where Q is the chi-squared statistic calculated during the test of independence between the two variables. Inertia can also be considered as the information embedded in the data of the contingency table. The main idea of CA is to simultaneously project the point clouds of the rows and columns into a space, usually two-dimensional (Biplot) or three-dimensional, so that the new optimal representation retains as much of the original information as possible. With CA, the main axes of inertia are identified, and the optimal subspace is defined by the first factorial axes in order. The maximum number of axes is p=min $\{k-1, l-1\}$, where k and l are the categories of variables X and Y, respectively. The main axes are determined by the eigenvectors resulting from the application of Singular Value Decomposition (SVD) to the matrix of the standardized residuals $S=(s_{ij})=\frac{p_{ij}-r_i c_j}{\sqrt{r_i c_j}}$, where p_{ij} is the (i, j) element of the correspondence table, and r_i, c_j are the masses of the i-th row and j-th column, respectively. The inertia Is of each axis s, where s=1, 2, …, p and p= min $\{k-1, l-1\}$, is equal to $Is=ds^2==\lambda s$, where ds is the singular value and λs is the corresponding eigenvalue. The proportion of the explained inertia for each axis is equal to $\frac{\lambda_s}{I} \times 100$. Additionally, if each element of the matrix S is divided by the total inertia I of the contingency table, the resulting new matrix expresses the contribution of each cell as a percentage of the total inertia.

Each row or column point (Benzécri, 1992; Greenacre, 2006) projected onto the s-axis contributes its partial inertia to the axis inertia. The ratio of the partial inertia of a row (or column) point projected onto an axis s to the total inertia of the axis is denoted as *CTR(s)*. This ratio allows for identifying the points that contribute most to the axis taking the specific orientation in the space. The ratio of the partial inertia of a row (or column) point projected onto an axis s to the total inertia of the point is denoted as *COR*(s) and measures the quality of representation of the point on the given axis. These two indices (*CTR* and *COR*) are used in the selection of significant points that require interpretation on each axis (Benzécri, 1992; Greenacre, 2006). Based on the significant points, an effort is made to assign a physical interpretation and identity to the factorial axes.

In the case of Multiple Correspondence Analysis (MCA) (Benzécri, 1992; Greenacre, 2006; Abdi & Valentin, 2007), we have a table in the form of (*objects*) x (*variables*). Initially, the categorical variables are encoded as binary (0/1) variables. Each category of a variable is represented as a separate binary variable, indicating the presence or absence of the specific category. The data are summarized in a Z_{0-1} table (logical table), where each row corresponds to an observation-object and each column corresponds to a level of a categorical variable. The Z_{0-1} table is decomposed using Singular Value Decomposition (SVD), extracting the principal axes that explain the maximum inertia in the data. MCA also produces plots (biplots) that display both the objects and the categories, as well as their relationships. Similar results can be obtained by applying SVD to the Burt table $B=Z^T Z$, where Z^T is the transpose of the (Z_{0-1}) table. Analyzing the B table saves storage space and computational time but loses the information about the objects, and the original data table cannot be reconstructed. However, there is a methodology that makes it possible to retrieve information about the objects even when the method is applied to the Burt table (Greenacre, 1984).

Non-Linear Categorical Principal Components Analysis (CATPCA)

CATPCA is a method for analyzing categorical data (Gifi, 1990; De Leeuw & Mair, 2009; Linting et al., 2007; Meulman et al., 2004), similar to MCA, which can simultaneously process categorical variables (nominal, ordinal) and scale variables. Categorical variables are transformed by assigning optimal values to their categories, creating variables with numeric values. This is typically achieved using iterative techniques such as the Alternating Least Squares (ALS) method. The transformations can be non-monotonic or monotonic. Monotonic transformations are generally suitable for ordinal categorical or continuous data, whereas non-monotonic functions are used for nominal data. In fact, non-monotonic transformations can also be applied to numeric and ordinal variables when non-linear relationships between variables are assumed. CATPCA finds the optimal quantifications of the categories in the sense that the total variance (inertia) explained in the transformed variables is maximized (Meulman et al., 2004; De Leeuw & Mair, 2009). During the process of optimal scaling, the information contained in the original categorical data is preserved in the optimal quantifications. The optimal scaling method generates numerical scores for categorical and mixed data, making it possible to use these variables in analyses traditionally designed for quantitative data. By transforming nominal and ordinal variables into optimal values, this method ensures compatibility with techniques such as PCA, regression, and discriminant analysis. The resulting scores preserve the inherent structure of the original data for instance, maintaining category order for ordinal variables or assigning independent scores to nominal categories, while optimizing variance explained or correlations. This adaptability enables researchers to include mixed data types in a unified analytical framework, broadening the application of quantitative methods to complex datasets. CATPCA is particularly useful when working with large datasets that include many categorical and quantitative variables, as it can help identify the most important features and reduce noise in the data. It is also beneficial for visualizing the phenomenon under investigation in a two- or three-dimensional space.

Principal Component Analysis (PCA)

PCA (Hair et.al., 2010; Jolliffe, 2002; Jolliffe & Cadima, 2016), although primarily designed for dimensionality reduction of quantitative data, can also be adapted for categorical data analysis under specific conditions, either on the factor scores obtained from their transformation using optimal scaling methods or on the original data (in the case of ordinal variables), if treated as 'quasi' quantitative (Robitzsch, 2020). PCA can also be applied to Spearman's *rho* and Kendall's *tau* correlation matrices for ordinal variables, to the polychoric correlation matrix for nominal and ordinal variables, or to the tetrachoric correlation matrix in the case of dichotomous (binary) variables (Kolenikov & Angeles, 2004; Mitra & Zhang, 2014; Watanabe et al., 2024; Ziyanak & Yagci, 2024). It is a method used to identify patterns and relationships in multidimensional datasets. Its goal is to transform a set of potentially correlated variables into a smaller set of variables, which could be uncorrelated if orthogonal rotation of the axes is applied, known as principal components, while retaining as much of the original information as possible. PCA is based on the computation of eigenvalues and eigenvectors of the data's correlation or covariance matrix, with the principal components being defined as linear combinations of the original variables. This technique is particularly useful for visualizing data in two- or three-dimensional space, understanding latent structures, and preprocessing in machine learning models (Bishop, 2016; Song et al., 2018). PCA relies on certain assumptions that are often unmet, such as all variables being numeric (interval, ratio, or discrete) in scale and the relationships between variables being linear. To overcome these limitations, nonlinear PCA, known as CATPCA, is used (Gifi, 1990; Michailidis & De Leeuw, 1998).

Factor Analysis for Mixed Data (FAMD)

FAMD is a method designed for analyzing datasets that include both quantitative and qualitative variables. It is particularly useful when dealing with mixed data types, extending the principles of PCA and MCA to accommodate this combination (Pagès, 2014c; Chavent et al., 2014; Ran, 2019; Markos et al., 2020; Ptak et al., 2021). FAMD begins by standardizing the quantitative variables so that they have a mean of zero and a variance of one. The qualitative variables are represented using indicator coding (0/1), also known as binary or dummy variables, and each category is treated as a separate variable. A qualitative variable with k categories, it is transformed into $k-1$ binary variables. Qualitative variables are also standardized by subtracting each i raw values (0/1) of indicator matrix G_{ij} with the proportion of observations in category j, and dividing with the square root of this proportion:

$$G^*_{ij} = \frac{G_{ij} - p_j}{\sqrt{p_j}}$$

whereG^*_{ij} is the standardized element of indicator matrix in i raw and j category, G_{ij} is the element of indicator matrix, which value is 1 if the observation i belongs to category j, and 0 otherwise, and p_j is the proportion of observations in category j. This approach balances the contribution of continuous and categorical features, allowing them to be compared on the same variance scale (Pagès, 2014c). FAMD uses Singular Value Decomposition (SVD) on a matrix that includes standardized quantitative variables and standardized indicator variables for qualitative data. The decomposition identifies the principal components (axes) that explain the greatest amount of combined variability in the dataset. Like PCA, FAMD aims to reduce the dimensionality of the dataset while retaining as much variance (inertia) as possible. It identifies principal components that explain the maximum variance (inertia) in the data. The resulting principal components are used to project the original data into a lower-dimensional space. This projection facilitates the visualization and interpretation of relationships between variables and objects. Quantitative variables contribute to the definition of the components in a manner similar to PCA, while qualitative variables contribute in a way similar to MCA. FAMD provides factor scores for the objects and loadings for the variables. The factor scores represent the coordinates of the observations/objects in the reduced space, while the loadings indicate the contribution of each variable to the principal components.

Non-Linear Canonical Correlation Analysis (NLCCA)

NLCCA is a method (Gifi, 1990; Michaeli, 2016) used to explore and quantify the relationships between two or more sets or blocks of variables of the same dataset (or objects), when these relationships are nonlinear. Several scale types can be handled (discrete, numerical, ordinal, and nominal). For ordinal variables, each monotone ascending transformation gives the same information about the objects as the original coding, while for nominal variables each isomorphic transformation and for numerical variables each linear transformation contains the same information (Van der Burg et al., 1994). In contrast to traditional Canonical Correlation Analysis (CCA), which is limited to linear relationships, NLCCA (Kramer, 1991; Dauxois & Nkiet, 1998) leverages nonlinear transformations to capture more complex associations. NLCCA applies nonlinear transformations to the data to maximize the canonical correlations between the transformed sets of variables. This process often involves iterative techniques, such as the Alternating Least Squares (ALS) method. After the nonlinear transformation, we compute the eigenvalues and eigenvectors of a specific matrix that describes the relationships between the transformed data sets.

This matrix could be a cross-covariance matrix that captures the relationship between the transformed data sets. The eigenvalues represent the strength of the canonical correlations, such as the amount of shared variance between the transformed sets of variables. The eigenvectors represent the directions 'axes' in the transformed space, where the linear combinations of the variables in each set are most strongly correlated with each other. The performance of the model is evaluated using three indicators (Van der Burg et al., 1994; Theodosiou et al., 2007): Multiple Fit, Single Fit, and Single Loss. The Multiple Fit index measures the overall percentage of variance in the data sets explained by the model. The Single Fit index indicates how well each individual variable is represented by the model. The Single Loss index expresses the percentage of variance in each variable that is not explained by the model.

Multiple Factor Analysis (MFA)

MFA (Bécue-Bertaut & Pagès, 2008; Pagès, 2014c) is an extension of factor analysis methods designed for the analysis of data structured into groups (sets or blocks) of variables. The number and/or nature of the variables used to describe the observations can vary from one set of variables to the other, but the observations should be the same in all the data sets (Abdi et al., 2013). It is particularly useful when the data include categorical variables (nominal and ordinal), as it allows for the simultaneous analysis of multiple sets of variables, combining quantitative and categorical data while preserving the structure of the variable groups. Suppose we have K groups of variables, each with n observations. Initially, each group of variables is standardized. For the quantitative variables, this is done by subtracting the mean and dividing by the standard deviation and for the categorical variables as it was described above (see FAMD). Then, a block matrix X is constructed, by concatenating all the data tables, where each block corresponds to a group of variables. This matrix has dimensions $n \times m$, where m is the total number of variables across all the groups. Next, weights are assigned to each block by dividing the values of X by the first (max) eigenvalue that results from applying PCA to each block consisting of quantitative variables, and the first (max) eigenvalue that results from applying MCA to each block of categorical variables, in order to balance the influence of each group in the analysis. Then, generalized GSVD (Abdi et al., 2013) is performed on the unified matrix X. This analysis identifies common factors that explain the maximum variance across all groups. The method reduces the dimensionality of the dataset by projecting it onto a lower-dimensional space defined by the principal components. These components summarize the most important information shared by the groups. The extracted factors are used to create plots that visualize the relationships between variables and objects within and between the groups. This helps in identifying patterns and associations that span multiple groups of variables.

Dimensionality reduction is a fundamental aspect of analyzing high-dimensional datasets, including those containing categorical variables. Several methods have been developed to address the unique challenges of categorical data, each with specific advantages and applications. Correspondence Analysis (CA) and its multivariate extension, Multiple Correspondence Analysis (MCA), are among the most established methods (Benzécri, 1992; Greenacre, 1991; Greenacre & Blasious, 2006). These methods utilize chi-square distance to reveal associations between categorical variables, providing interpretable, low-dimensional representations. The computational foundation of these methods is Singular Value Decomposition (SVD), which enables efficient dimensionality reduction while preserving key relational structures in the data (Markos et al., 2009). Categorical Principal Component Analysis (CATPCA) is another widely-used approach (Gifi 1990; De Leeuw & Mair, 2009; Linting et al., 2007; Meulman et al., 2004) extending PCA by incorporating optimal scaling transformations. This method transforms categorical variables into numerical values that maintain their qualitative properties, making it particularly effective for datasets containing ordinal variables. Optimal scaling techniques, such as those employed in CATPCA, allow categorical data to be analyzed using numerical methods (Di Franco, 2015; Donaires et al., 2023). Additionally, distance metric Hellinger has been used in comparison to chi-square (Cuadras et al., 2006). Correlation-based measures, such as Spearman's rho, Kendall's tau, and polychoric correlations also used to quantify relationships among variables, forming the basis for methods like Canonical Correlation Analysis (CCA) and PCA on transformed data (Kolenikov & Angeles, 2004; Mitra & Zhang, 2014; Watanabe et al., 2024; Ziyanak & Yagci, 2024). Furthermore, graphical tools such as biplots enhance interpretability by representing variables and observations in reduced-dimensional spaces. These visualizations are instrumental in uncovering patterns, clusters, and relationships within categorical data (Jacoby, 1998; Blasious & Greenacre, 1998). In recent years, the emergence of machine learning has introduced and compared innovative applications for dimensionality reduction with categorical data combining with tasks such as clustering, classification, and similarity search, highlighting their potential in practical scenarios. These approaches address both computational efficiency and the increasing complexity of real-world datasets (Meng et al., 2016; Nguyen & Holmes, 2019; Bera et al., 2021; Armstrong et al., 2022).

Building on these established methodologies, our proposed strategies integrate methods such as MCA, CATPCA, PCA after optimal scaling transformations, and advanced hybrid approaches like FAMD, NLCCA and MFA, in a comparative frame. By systematically combining these methods, our approach identifies the strategy that retains the most information while offering superior interpretability. This novel framework bridges theoretical advancements and practical applications, contributing a comprehensive and flexible toolkit for the analysis of multivariate categorical data.

MATERIALS AND METHODS

The multivariate data analysis methods applied and evaluated in this study were: Multiple Correspondence Analysis (MCA), Nonlinear Categorical Principal Component Analysis with Optimal Scaling, Factor Analysis for Mixed Data (FAMD), Nonlinear Canonical Correlation Analysis (NLCCA), Multiple Factor Analysis (MFA), and finally, after appropriate data transformation, Principal Component Analysis (PCA). In addition, the square root of the maximum eigenvalue and the optimal scores resulting from the application of MCA and CATPCA were used in proposed strategies for transformation and replacement of the original values.

Analysis Strategies

The choice of method and the proposed analysis strategy depends on the measurement scale of the data and whether we consider them as one or more groups of variables (in this study variables are considered synonymous with features). The proposed analysis strategies based on the study of variability (inertia), its structure and the interpretability of the results, which is the key point of Statistical - Data Analysis in biological sciences and especially in agricultural ones (Kitchen & Clay, 2016). The proposed analysis strategies are summarized as follows (see also Figure 1).

Strategy 1 *(S1)*

If all the data are nominal, then we apply Multiple Correspondence Analysis (MCA).

Strategy 2 *(S2)*

In case all data are ordinal we can apply MCA treating them as nominal (*S*2.1), or CATPCA (*S*2.2), or PCA on the Spearman's *rho* correlation matrix (*S*2.3), or PCA on the Kendall's *tau* correlation matrix (*S*2.4), or PCA on the polychoric correlation matrix (*S*2.5), or PCA on the original data treating them as ‘quasi’ quantitative (quasi scale) with equal distance intervals between categories (*S*2.6).

Strategy 3 *(S3)*

If the data are both nominal and ordinal, then we can apply MCA treating them all as nominal (*S3*.1) or apply CATPCA (*S*3.2).

Strategy 4 *(S4)*

If the data are both nominal and ordinal, then we apply MCA to the nominal data and CATPCA or PCA to the ordinal data (see strategies *S*2.3 – *S*2.6). We divide the 0-1 values of the logical matrix (*Z0-1*) of the nominal variables by the square root of the maximum eigenvalue $\sqrt{\lambda_{max}}$ resulting from the application of MCA, and the initial coded values of the ordinal variables by the square root of the maximum eigenvalue $\sqrt{\lambda_{max}}$ resulting from the application of CATPCA or PCA, in order to balance the contribution of nominal and ordinal variables, such as in methods FAMD and MFA. Then, we apply PCA to all the transformed data.

Strategy 5 *(S5)*

If the data are both nominal and ordinal, then we apply MCA to the nominal data and CATPCA to the ordinal data, replacing their original values with the optimal quantification values of the first factor axis from MCA and CATPCA, respectively. Then, we apply PCA to all the transformed data.

Strategy 6 *(S6)*

If the data are both nominal and ordinal, then we either apply FAMD (*S*6.1) or apply NLCCA (*S*6.2) or MFA (*S*6.3) considering them as two separate groups of variables.

The dataset to which the above methods and strategies were applied was the Forest Cover type from the UCI Machine Learning Repository (Blackard, 1998). This dataset includes information on tree type, shadow coverage, distance to nearby landmarks, soil type, and local topography. It consists of thirteen (13) variables, ten (10) quantitative and three (3) qualitative, and contains 581,012 rows. The quantitative variables included Elevation measured in meters, Horizontal_ Distance_to_Hydrology (H_D_To_H) and Vertical_Distance_to_Hydrology (V_D_To_H) which are measure in meters the distance to the nearest surface water features, Horizontal_Distance_to_Roadways (H_D_To_R), and Horizontal_Distance_to_Fire_Points (H_D_To_FP), which are measure in meters the distance to nearest roadway and to nearest wildfire ignition points respectively, Aspect, which is measured in degrees 'azimuth', Slope, which is measured in degrees, and the degree of Hillshade during the summer solstice at 9 a.m., noon, and 3 p.m. (Hillshade 9am, HS_9am; Hillshade Noon, HS_Noon; Hillshade 3pm, HS_3pm), which are measured with an index from 0 to 255, where 0 indicates complete shadow with no sunlight reaches the terrain and 255 represents full illumination, where the terrain receives the maximum light. The qualitative variables (or features) included Wilderness Area (W_A) with four (4) categories, Soil Type (S_T) with forty (40) categories, and Cover Type (Cover_Type) with seven (7) categories. For the purposes of this work, the quantitative variables were transformed into ordinal variables with 4 categories each (Visual Binning with 3 cutpoints).

Figure 1. Flowcharts for the proposed analysis strategies

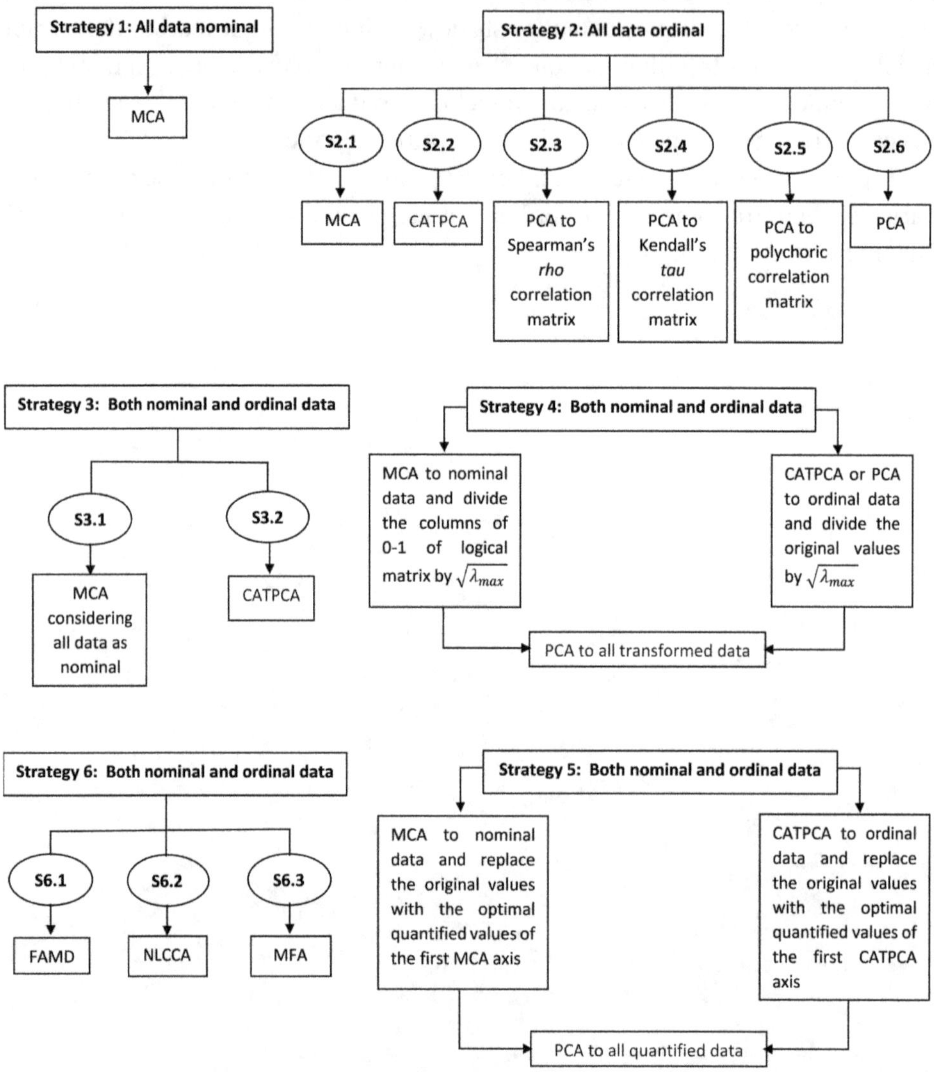

MCA: Multiple Correspondence Analysis, CATPCA: Categorical Principle Component Analysis, PCA: Principle Component Analysis, FAMD: Factor Analysis for Mixed Data, NLCCA: Non Linear Canonical Correlation, MFA: Multiple Factor Analysis, λ_{max}: the maximum eigenvalue results from MCA and CATPCA

The data analyses were performed using the statistical software R v4.3.2 (S6.1, S6.3), IBM SPSS Statistics v26 and v29 (S1, S2.1, S2.5, S3.1, S2.2, S3.2, S4, S5, S6.3), Python 3.10 (S1, S2.1, S3.1) via the Anaconda platform, and Jupyter notebook

6.4.5. For the visualization of the results, we considered the first two principal component axes of the analyses. To identify the variables (or features) with the highest contribution to the principal axes, we set a loading threshold of 0.50.

RESULTS

In strategies S1, S2.1, and S3.1, where we applied MCA considering all data as nominal, the first two factorial axes explained 8.85% of the total inertia (without correction). The variables S_T, W_A, and Cover_Type, related to soil type, wilderness area, and forest cover type, contributed most to the 1st factorial axis. In contrast, the variables HS_3pm, Aspect, and HS_9am, concerning orientation and shading in the morning and afternoon, had the highest contributions to the 2nd factorial axis. Notably, the first thirty-three (33) axes explained 61.14% of the total inertia (without correction), while 40 axes explained approximately 70% of the total inertia (without correction) (Figure 2, Table 2).

Figure 2. Factorial plane 1x2 from the application of MCA

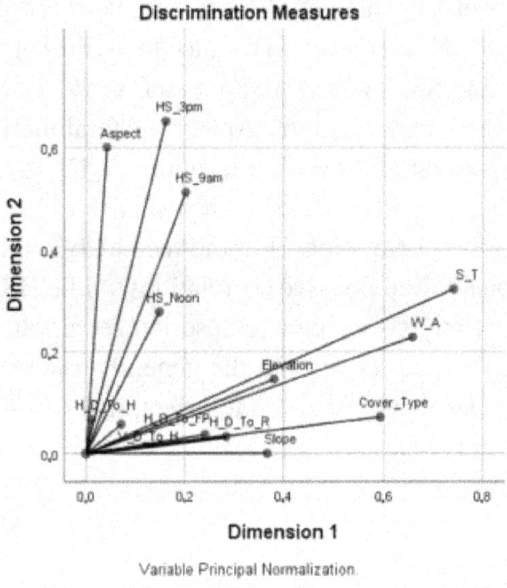

Table 2. Discrimination measures of the variables on the first two axes

W_A	**.658**	**.228**
S_T	**.741**	.323
Elevation	.380	.146
Aspect	.042	**.601**
Slope	.367	.000
H_D_To_H	.011	.067
V_D_To_H	.071	.058
H_D_To_R	.284	.033
H_D_To_FP	,242	.037
HS_9am	.202	**.512**
HS_Noon	.148	.277
HS_3pm	.162	**.651**
Cover_Type	**.593**	.072

With **bold** variables with higher contribution to axes S_T: Soil Type, W_A: Wilderness Area, HS: Hillshade, H_D_To_R: Horizontal Distance to Roadways, H_D_To_F_P: Horizontal Distance to Fire Points, V_D_To_H: Vertical Distance to Hydrology, H_D_To_H: Horizontal Distance to Hydrology

In strategies S2.2 and S3.2, where we applied CATPCA to all the data, four (4) significant axes emerged, explaining 62.26% of the total inertia, with the first two explaining 39.41%. The variables S_T, Elevation, H_D_To_R, and H_D_To_FP contributed most to the first factorial axis with positive correlations, while W_A and Cover_Type had negative correlations. For the second axis, the variables HS_3pm, Aspect, and HS_Noon showed positive correlations, while HS_9am showed a negative correlation. The first axis appears to describe geographical or environmental characteristics related to land cover and the general structure of the areas, while the second axis describes features related to orientation (Aspect) and the temporal distribution of sunlight (Figure 3, Table 3).

Figure 3. Factorial plane 1x2 from the application of CATPCA

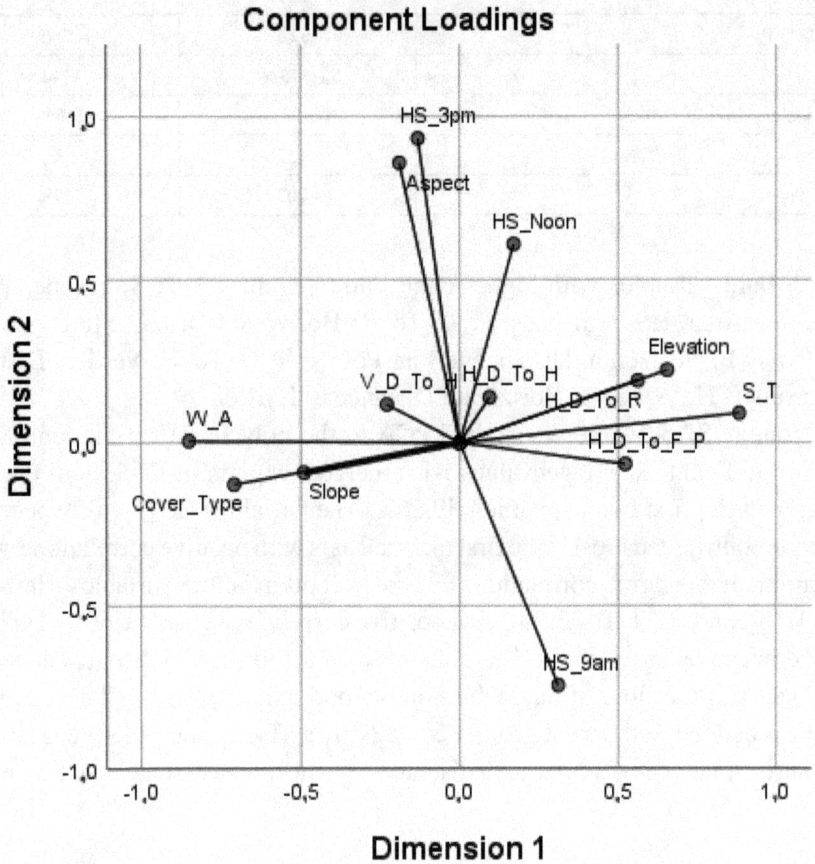

Table 3. Variables loadings to the first two factorial axes

W_A	**-.849**	**.005**
S_T	**.881**	.090
Elevation	**.655**	.222
Aspect	.189	**.858**
Slope	.490	-.091
H_D_To_H	.095	.138
V_D_To_H	-.230	.117
H_D_To_R	**.561**	.190

continued on following page

Table 3. Continued

W_A	**-.849**	**.005**
H_D_To_FP	**.522**	-.066
HS_9am	.313	**-.744**
HS_Noon	.171	**.608**
HS_3pm	-.132	**.933**
Cover_Type	**-.707**	-.128

With **bold** variables with higher contribution to axes S_T: Soil Type, W_A: Wilderness Area, HS: Hillshade, H_D_To_R: Horizontal Distance to Roadways, H_D_To_F_P: Horizontal Distance to Fire Points, V_D_To_H: Vertical Distance to Hydrology, H_D_To_H: Horizontal Distance to Hydrology

In strategy S2.5, where we applied PCA to the polychoric correlation matrix, five (5) significant axes (eigenvalues >1) emerged, explaining 71.15% of the total inertia, with the first two explaining 39.26%. The variables HS_3pm, Aspect, and HS_Noon contributed most to the first factorial axis with positive correlations, while HS_9am had a negative correlation. For the second axis, the variables Elevation, H_D_To_R, and H_D_To_FP showed positive correlations, while Cover_Type and S_T had negative correlations. The first axis appears to explain differences related to sunlight and the time of day, while the second axis explains differences more closely associated with forest ground cover (Cover_Type) and other geographical information (elevation, horizontal distances to the nearest road and fire point) (Figure 4, Table 4).

Figure 4. Factorial plane 1x2 from the application of PCA to the polychoric correlation matrix

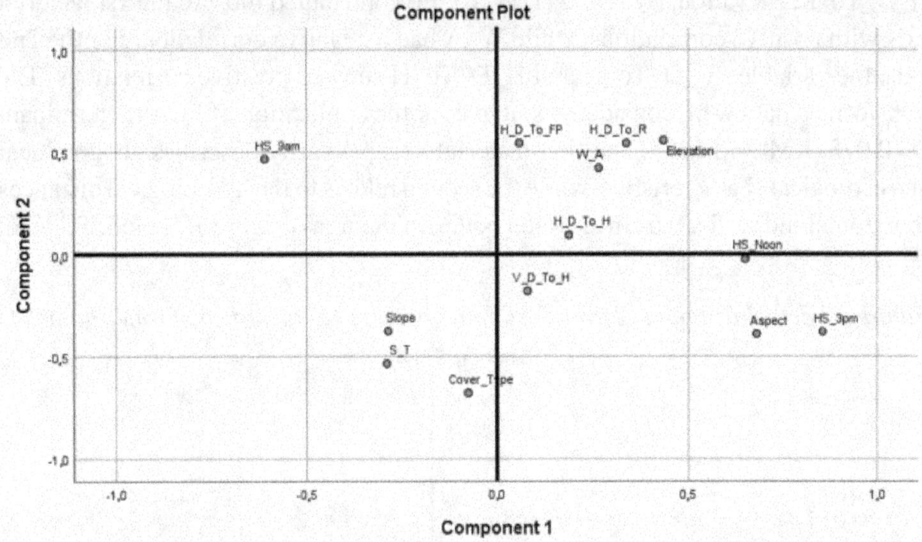

Table 4. Variable loadings (>0.50) on the first two factorial axes

HS_3pm	**.857**	
Aspect	.679	
HS_Noon	.649	
HS_9am	-.609	
H_D_To_H		
Cover_Type		-.679
Elevation		.557
H_D_To_FP		.543
H_D_To_R		.542
S_T		-.535
W_A		
Slope		
V_D_To_H		

S_T: Soil Type, W_A: Wilderness Area, HS: Hillshade, H_D_To_R: Horizontal Distance to Roadways, H_D_To_F_P: Horizontal Distance to Fire Points, V_D_To_H: Vertical Distance to Hydrology, H_D_To_H: Horizontal Distance to Hydrology

From the application of strategy S4, four (4) significant axes emerged, explaining 60.43% of the total inertia, with the first two explaining 37.09%. The variables H_D_To_R, Elevation, S_T, and H_D_To_FP contributed most to the 1st factorial axis with positive correlations, while W_A had a negative correlation. For the 2nd axis, the variables V_D_To_H and H_D_To_H showed positive correlations. The transformed data were considered suitable for the application of PCA (determinant D=0.073, KMO=0.598). The first factorial axis primarily describes geographical and ecological characteristics, while the second relates to the hydrological distances (horizontal and vertical) from the water bodies of the areas (Figure 5, Table 5, Table 6).

Figure 5. Factorial plane 1x2 from the application of PCA to the transformed data (S4)

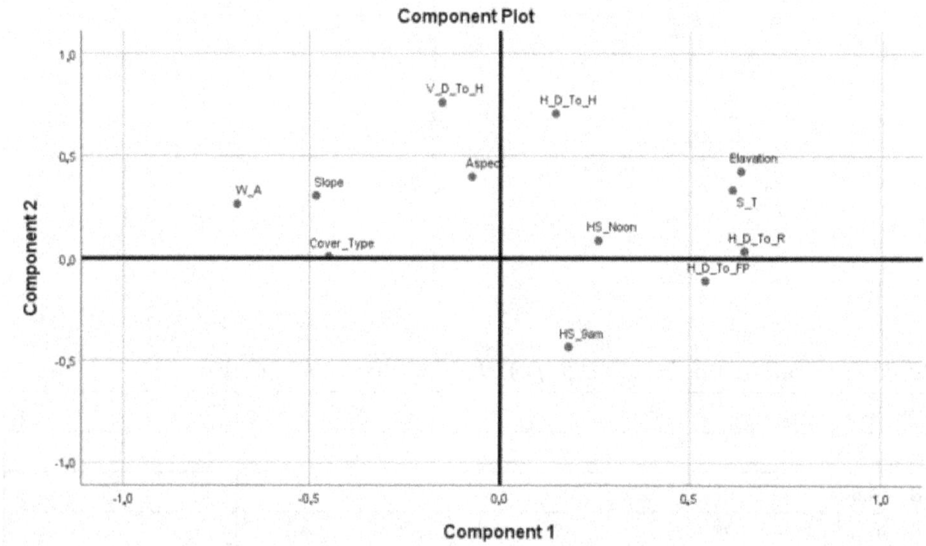

Table 5. Variable loadings (>0.50) on the first two factorial axes

W_A	**-.694**	
H_D_To_R	.641	
Elevation	.633	
S_T	.611	
H_D_To_FP	.540	
Slope		
Cover_Type		

continued on following page

Table 5. Continued

W_A	**-.694**	
HS_Noon		
V_D_To_H		.762
H_D_To_H		.708
HS_9am		
Aspect		

S_T: Soil Type, W_A: Wilderness Area, HS: Hillshade, H_D_To_R: Horizontal Distance to Roadways, H_D_To_F_P: Horizontal Distance to Fire Points, V_D_To_H: Vertical Distance to Hydrology, H_D_To_H: Horizontal Distance to Hydrology

Table 6. Explained inertia by the first factorial axes (S4)

Axes	Eigenvalues	% Percentage	Cumulative %
1	2.543	21.192	21.192
2	1.907	15.893	37.085
3	1.696	14.136	51.221
4	1.106	9.213	60.434

From the application of strategy S5, four (4) significant axes also emerged, explaining 67.99% of the total inertia, with the first two explaining 46.89%. The variables S_T, W_A, Cover_Type, Elevation, and H_D_To_R contributed most to the first factorial axis with positive correlations, while the variables HS_3pm, Aspect, and HS_Noon showed positive correlations, and HS_9am showed a negative correlation for the second axis. The quantified data were considered suitable for the application of PCA (determinant D=0.003, KMO=0.722). The first factorial axis primarily interprets topographical variation, such as soil type, forest cover, and wildness of the area, as well as the relationship with anthropogenic features like distance to roads. The second factorial axis interprets solar parameters (Hillshade) and their relationship with geographic orientation (Aspect) (Figure 6, Table 7, Table 8).

Figure 6. Factorial plane 1x2 from the application of PCA to the quantified data (S5)

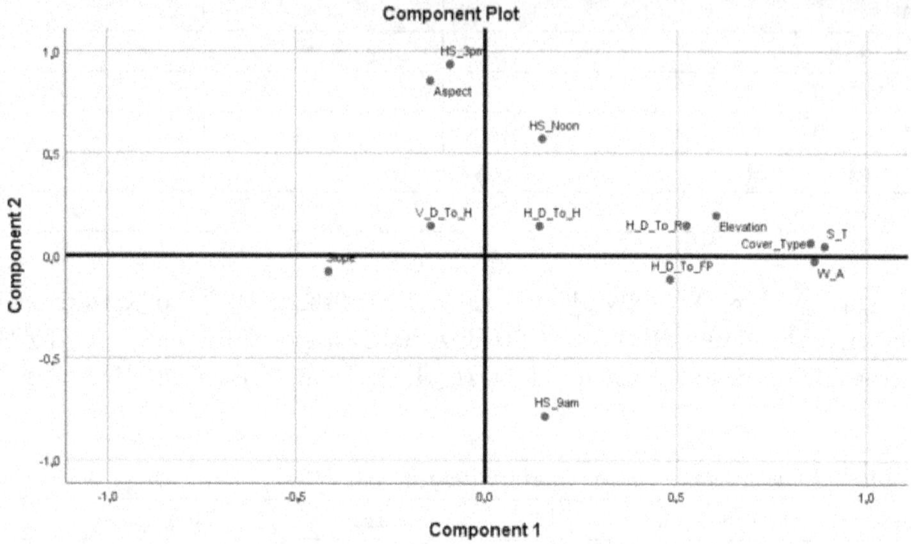

Table 7. Variable loadings (>0.50) on the first two factorial axes

S_T	**.889**	
W_A	.862	
Cover_Type	.852	
Elevation	.603	
H_D_To_R	.524	
H_D_To_FP		
Slope		
HS_3pm		.936
Aspect		.855
HS_9am		-.785
HS_Noon		.572
V_D_To_H		
H_D_To_H		

S_T: Soil Type, W_A: Wilderness Area, HS: Hillshade, H_D_To_R: Horizontal Distance to Roadways, H_D_To_F_P: Horizontal Distance to Fire Points, V_D_To_H: Vertical Distance to Hydrology, H_D_To_H: Horizontal Distance to Hydrology

Table 8. Explained inertia by the first factorial axes (S5)

Axes	Eigenvalues	% Percentage	Cumulative %
1	3.415	26.271	26.271
2	2.680	20.615	46.886
3	1.699	13.070	59.956
4	1.044	8.034	67.990

The application of strategy S6.2 (NLCCA) resulted in zero eigenvalues and zero fit, indicating either a lack of differentiation or that the model is not effectively capturing the relationships within the data (Table 9). This unexpected result is likely due to the large number of rows in the dataset. When tests were conducted using approximately 50% of the rows, the method produced meaningful results. Additionally, when the analysis was attempted using Python, an error message indicated that the large dataset could not be processed. This suggests that the size of the dataset exceeds the computational capacity of the method or the software, leading to the observed discrepancies. Reducing the dataset size appears to alleviate this issue and enables the method to provide interpretable outcomes.

Table 9. Summary of analysis

		Dimension		Sum
		1	2	
Loss	Set 1	1.000	1.000	2.000
	Set 2	1.000	1.000	2.000
	Mean	1.000	1.000	2.000
Eigenvalue		.000	.000	
Fit				.000

From the application of strategy S6.1 (FAMD), the first two factorial axes explained 10.94% of the total inertia, with the variables S_T, W_A, and Cover_Type contributing most to the first factorial axis, while the variables HS_3pm, Aspect, and HS_9am contributed most to the second axis. The first axis is related to environmental and geographical characteristics (soil and forest cover type, wildness of the area, and elevation), while the second is related to solar radiation and orientation (Figure 7, Figure 8).

Figure 7. Factorial plane 1x2 in the first two dimensions from the application of FAMD

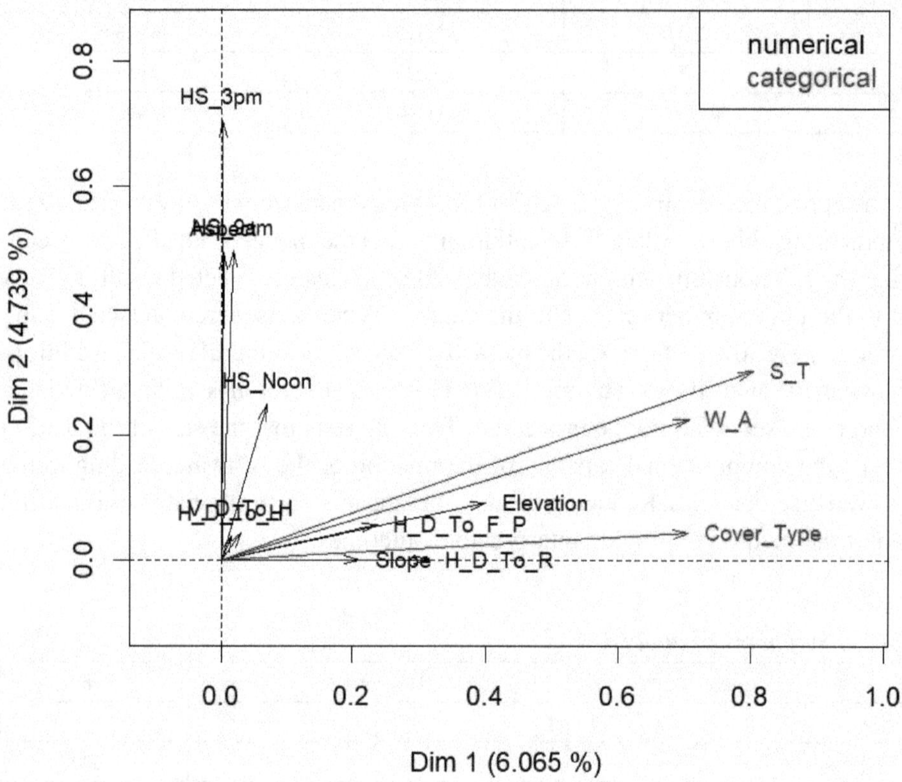

S_T: Soil Type, W_A: Wilderness Area, HS: Hillshade, H_D_To_R: Horizontal Distance to Roadways, H_D_To_F_P: Horizontal Distance to Fire Points, V_D_To_H: Vertical Distance to Hydrology

Figure 8. Contribution of the features to the first two dimensions

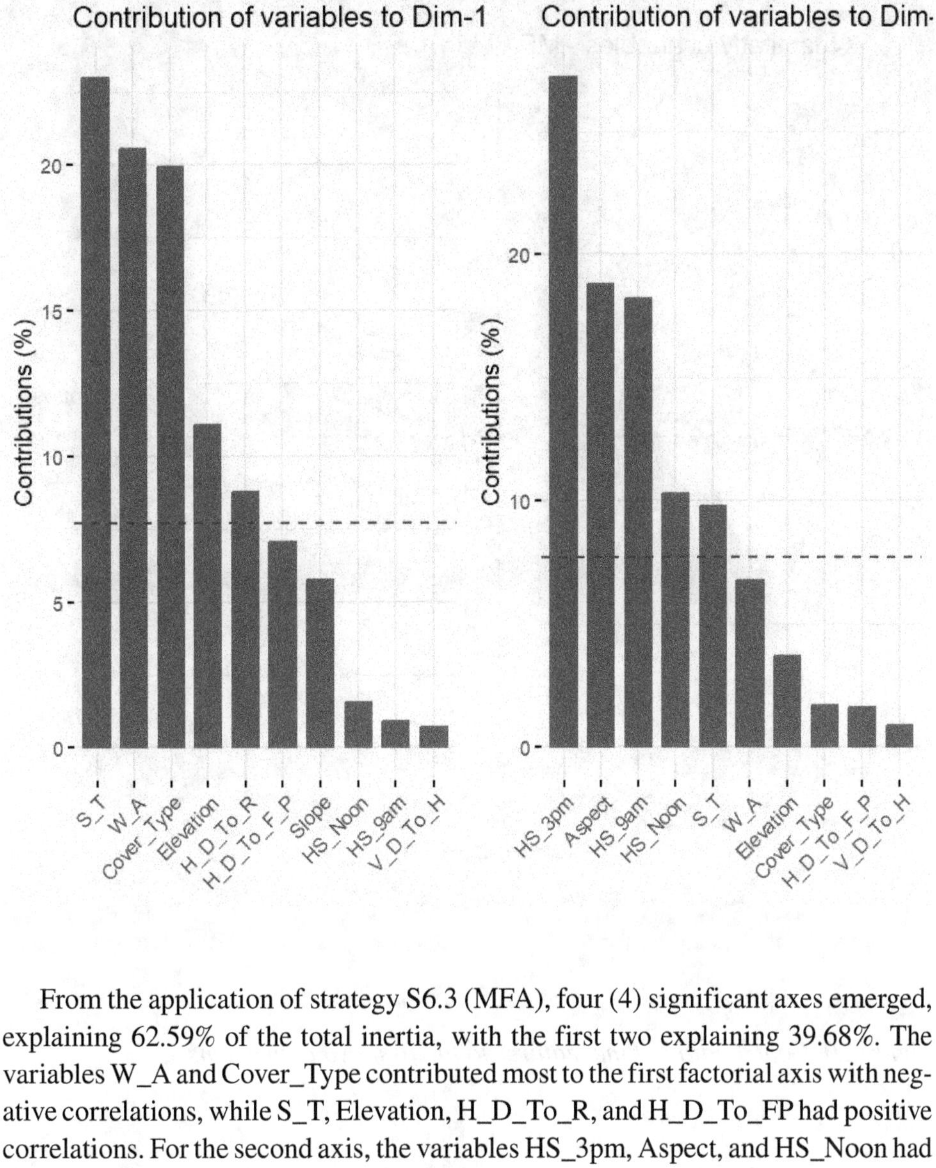

From the application of strategy S6.3 (MFA), four (4) significant axes emerged, explaining 62.59% of the total inertia, with the first two explaining 39.68%. The variables W_A and Cover_Type contributed most to the first factorial axis with negative correlations, while S_T, Elevation, H_D_To_R, and H_D_To_FP had positive correlations. For the second axis, the variables HS_3pm, Aspect, and HS_Noon had positive correlations, while HS_9am had a negative correlation (Figure 9, Figure 10, Table 10).

Figure 9. Factorial plane 1x2 in the first two dimensions from the application of MFA

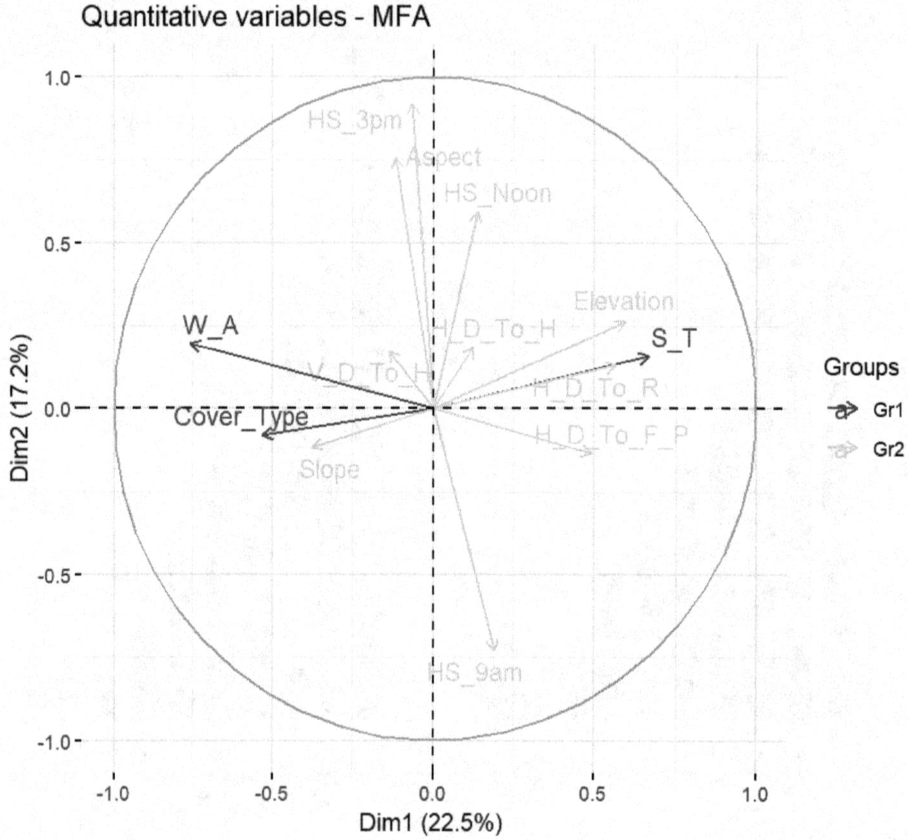

Figure 10. Contribution of the features to the first two dimensions

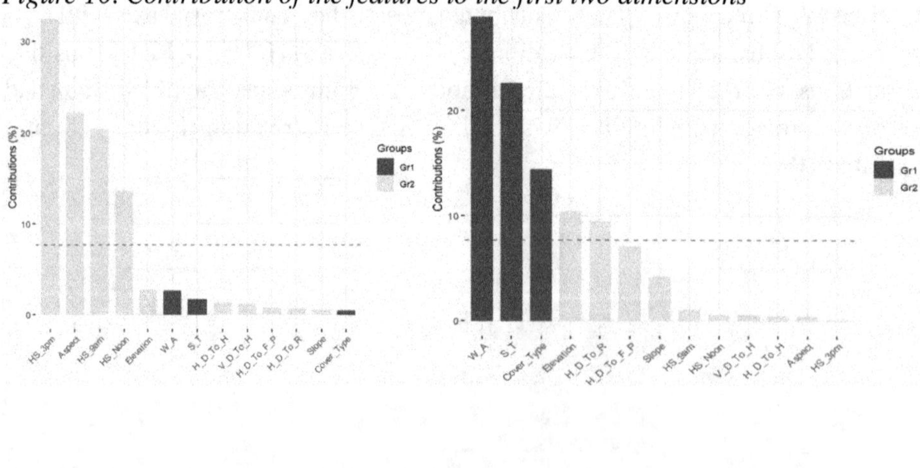

S_T: Soil Type, W_A: Wilderness Area, HS: Hillshade, H_D_To_R: Horizontal Distance to Roadways, H_D_To_F_P: Horizontal Distance to Fire Points, V_D_To_H: Vertical Distance to Hydrology, H_D_To_H: Horizontal Distance to Hydrology

Table 10. Explained inertia by the first factorial axes (S6.3)

Dimension	Eigenvalue	% Percentage	Cumulative %
1	1.3387	22.50	22.50
2	1.0221	17.18	39.68
3	0.7954	13.37	53.05
4	0.5676	9.54	62.59
5	0.4889	8.21	70.80

DISCUSSION

The application of different strategies to the dataset yielded distinct insights regarding the explained inertia and variables contributions. These results underscore the importance of selecting a strategy aligned with the dataset's characteristics and research objectives. Each strategy applied provided unique insights into the relationships between the variables, with varying approach in explaining the total inertia and in the way they identified meaningful patterns. Also, revealed significant insights into the relationships between environmental, geographical, and temporal factors, as well as their contributions to the overall inertia. The results highlight the complexity of the data and the variety of relationships that can be captured by different statistical methods and strategies. To the best of our knowledge, there are no comparative studies between these analysis strategies. This fact makes this study innovative and useful to the scientific community.

The choice of strategy heavily depends on how users wish to address the measurement scales of the variables. Each method assumes specific properties for the data, whether treating ordinal data as nominal, as quasi-quantitative, or using specialized transformations like optimal scaling. These decisions can significantly influence the interpretation of results and the balance between information retention and interpretability. Consequently, users must carefully consider the nature of their variables and the objectives of their analysis to select the most appropriate strategy. By aligning methodological choices with the scales of measurement, users can ensure more accurate and meaningful insights from their data.

While most of the strategies successfully identified meaningful patterns in the data, there were differences in the proportion of inertia explained by the axes across the different methods and strategies. For instance, methods like CATPCA (S2.2, S3.2) and PCA (S2.5, S.4 and S.5) on polychoric correlations matrix, and on transformed and quantified data, provided robust explanations, with the first two factorial axes explaining over 60% of the inertia in some cases. These strategies captured significant geographical, environmental, and temporal components, demonstrating the high explanatory power of the selected variables. On the other hand, the NLCCA (S6.2) model failed to identify any meaningful correlation, as indicated by the zero eigenvalues and zero fit. This suggests that NLCCA may not be suitable due to the large number of rows in the dataset. Furthermore, strategies like MFA (S6.2) and FAMD, which integrate multiple data types, provided a different view of the data by highlighting how variables from distinct categories such as geographical, temporal, and environmental, interact. These methods proved particularly useful in identifying cross-variable relationships, such as the negative correlation between Cover_Type and W_A (Wilderness Area) on the first axis, and the positive correlations with S_T (Soil Type) and Elevation. This provides a comprehensive understanding of how ecological characteristics relate to geographical and temporal factors. In addition, MCA's treatment of all data as nominal (S1, S2.1, and S3.1) have led to limited performance, indicating that it may not be ideal for datasets with a mix of nominal and continuous variables.

One main finding across most strategies is the significant role of soil type (S_T), wilderness area (W_A), and forest cover type (Cover_Type) in explaining the first factorial axis. These variables, often tied to the ecological and geographical characteristics of the study area, describe broad distinctions in the landscape. For example, the proximity to certain environmental features, such as roads and fire points, influenced the distribution of these variables, showing a strong relationship between land cover types and geographical features like elevation and distances to key infrastructure, such as roads, and fire points.

Another consistent result across strategies was the contribution of sunlight exposure variables, such as HS_3pm, Aspect, and HS_9am (Hillshade), which were strongly associated with the second factorial axis. This axis often represented temporal aspects, such as daily solar exposure, and its relationship with geographical orientation (Aspect). These variables provide insight into the directional and shading effects that influence the distribution of ecological and environmental features over time. It suggests that the spatial arrangement and the time of day play crucial roles in shaping the ecological landscape, with areas receiving different amounts of sunlight throughout the day showing varied characteristics.

Interestingly, the distance to roads (H_D_To_R) and Elevation had also significant contribution in several strategies, particularly in explaining the first axis.

These variables were correlated with the geographical structure of the study area, indicating that proximity to roads and elevation are significant factors in the spatial distribution of the environmental variables. Similarly, horizontal distance to the nearest fire point (H_D_To_FP) showed a strong relationship with this axis, underlining the importance of human-made and natural features in shaping the environment.

The findings also reinforce the importance of temporal factors in ecological studies, particularly the role of sunlight exposure and its effect on various environmental processes. The positive correlations between Hillshade HS_3pm, HS_Noon, and Aspect indicate that solar radiation and the time of day significantly influence the landscape, possibly affecting factors like vegetation growth, temperature distribution, and wildlife behavior. This insight is valuable for understanding how different environmental features respond to solar radiation patterns, which could inform land management strategies or ecological predictions.

Overall, the findings of this study underscore the importance of employing multiple analytical strategies to capture the diverse and complex relationships inherent in ecological and geographical datasets. While different methods varied in their ability to explain the inertia and the structure of data, the results consistently pointed to the importance of both geographical and temporal factors in shaping the environment. The challenges encountered with NLCCA (S6.2) suggest that this method may not be optimal for datasets with extensive rows due to computational constraints. This highlights the importance of carefully selecting analytical approaches that align with the specific characteristics of the dataset. Nonetheless, the results from the other strategies provided valuable insights into the dynamics of land cover, elevation, human influence, and solar exposure, contributing to a deeper understanding of the study area's ecological characteristics.

While the proposed strategies demonstrate robust performance in integrating various dimensionality reduction techniques, several limitations must be acknowledged. The interpretability of results may vary depending on the chosen strategy, as different methods prioritize distinct aspects of the data structure. Strategies involving MCA, CATPCA, or PCA transformations may lead to information loss, particularly when assumptions about variable scales or correlations are not entirely appropriate. Furthermore, the computational demands of hybrid methods like FAMD and MFA can pose challenges for datasets with high dimensionality or a large number of observations. Finally, the selection of axes for interpretation remains subjective and may not fully capture complex interactions among variables, highlighting the need for careful evaluation of each strategy's assumptions and outcomes.

CONCLUSIONS

In this study, we investigated different strategies and methods of analysis on a large dataset with nominal and ordinal variables, evaluating their effectiveness in explaining the proportion of inertia and identifying the most significant features contributing to the formation of factorial axes. The analysis demonstrated that different strategies significantly influence the proportion of explained inertia and the structure of the features highlighted.

Strategies employing Multiple Correspondence Analysis (MCA) and CATPCA revealed clear contributions of soil type (S_T), wilderness area (W_A), and forest cover type (Cover_Type) to overall variability. PCA approaches, whether applied to polychoric correlation matrix or to transformed data, captured distinctions related to sunlight dynamics, topographical variation, and hydrological distances. Notably, strategies using PCA with quantified and transformed data (S4 and S5) explained a larger proportion of inertia, indicating their ability to provide deeper insights into feature relationships. On the other hand, NLCCA had unexpected results likely due to the large number of rows in the dataset.

The results emphasize the value of combining feature quantification, dimensionality reduction, and exploratory analysis to better understand complex and voluminous datasets. Future research could build upon these methods, incorporating additional or more refined features to further improve analytical outcomes and support ecological decision-making.

Generally, the choice of the appropriate dimensionality reduction strategy depends largely on the nature of the data and the research objectives. The analysis showed that there is no one 'universally' best method, but rather that using different approaches provides more and often different information about the structure of variability in the data. More generally, it is wiser to talk about analysis strategies and not about a 'universal' dimensionality reduction method.

Future research could focus on developing hybrid models that integrate dimensionality reduction with machine learning techniques to enhance scalability and predictive accuracy. Investigating the application of these strategies across different domains, such as bioinformatics or social sciences, could provide valuable insights into their generalizability. Additionally, studies could focus into the selection of optimal strategies based on data characteristics, leveraging advancements in artificial intelligence. These directions have the potential to enhance the analytical toolkit for multivariate categorical data, contributing to both theoretical advancements and practical applications.

ACKNOWLEDGEMENT

Competing Interests:
The authors of this publication declare there are no competing interests.
Funding:
This research received no specific grant from any funding agency in the public, commercial, or not-for-profit sectors. Funding for this research was covered by the author(s) of the article.

REFERENCES

Abdi, H., & Valentin, D. (2007). Multiple correspondence analysis. *Encyclopedia of measurement and statistics, 2*(4), 651-657.

Abdi, H., Williams, L. J., & Valentin, D. (2013). Multiple factor analysis: Principal component analysis for multitable and multiblock data sets. *Wiley Interdisciplinary Reviews: Computational Statistics*, *5*(2), 149–179. DOI: 10.1002/wics.1246

Agresti, A. (2012). *Categorical data analysis* (Vol. 792). John Wiley & Sons.

Agresti, A. (2018). *An introduction to categorical data analysis*. John Wiley & Sons.

Armstrong, G., Rahman, G., Martino, C., McDonald, D., Gonzalez, A., Mishne, G., & Knight, R. (2022). Applications and comparison of dimensionality reduction methods for microbiome data. *Frontiers in Bioinformatics*, *2*, 821861. DOI: 10.3389/fbinf.2022.821861 PMID: 36304280

Bécue-Bertaut, M., & Pagès, J. (2008). Multiple factor analysis and clustering of a mixture of quantitative, categorical and frequency data. *Computational Statistics & Data Analysis*, *52*(6), 3255–3268. DOI: 10.1016/j.csda.2007.09.023

Benzécri, J.-P. (1992). *Correspondence Analysis Handbook*. Marcel Dekker, Inc. DOI: 10.1201/9780585363035

Bishop, C. M. (2016). *Pattern recognition and machine learning*. Springer.

Blackard, J. (1998). Covertype [Dataset]. UCI Machine Learning Repository. DOI: 10.24432/C50K5N

Blasius, J., & Greenacre, M. L. (1998). Visualization of categorical data. In *Elsevier eBooks*. DOI: 10.1016/B978-0-12-299045-8.X5000-1

Bond, J., & Michailidis, G. (1996). Homogeneity Analysis in Xlisp-Stat. *Journal of Statistical Software*, *1*(2). DOI: 10.18637/jss.v001.i02

Chavent, M., Kuentz-Simonet, V., Labenne, A., & Saracco, J. (2014). Multivariate analysis of mixed data: The R package PCAmixdata. *arXiv preprint arXiv:1411.4911*. https://cran.auckland.ac.nz/web/packages/PCAmixdata/index.html

Cuadras, C. M., Cuadras, D., & Greenacre, M. J. (2006). A comparison of different methods for representing categorical data. *Communications in Statistics. Simulation and Computation*, *35*(2), 447–459. DOI: 10.1080/03610910600591875

Dauxois, J., & Nkiet, G. M. (1998). Nonlinear canonical analysis and independence tests. *Annals of Statistics*, *26*(4). DOI: 10.1214/aos/1024691242

De Leeuw, J., & Mair, P. (2009). GIFI Methods for Optimal Scaling INR: The Package Homals. *Journal of Statistical Software*, *31*(4). DOI: 10.18637/jss.v031.i04

Di Franco, G. (2015). Multiple correspondence analysis: One only or several techniques? *Quality & Quantity*, *50*(3), 1299–1315. DOI: 10.1007/s11135-015-0206-0

Donaires, O. S., Cezarino, L. O., Liboni, L. B., Ribeiro, E. M. S., & Martins, F. P. (2023). Multivariate data analysis of categorical data: Taking advantage of the rhetorical power of numbers in qualitative research. *Quality & Quantity*, *57*(6), 5283–5312. DOI: 10.1007/s11135-022-01589-1

Flowerdew, R. T. N., & Martin, D. (2013). Methods in human Geography. In *Routledge eBooks*. DOI: 10.4324/9781315837277

Gifi, A. (1990). *Nonlinear multivariate analysis* (Heiser, W., Meulman, J. J., & van den Berg, G., Eds.). Wiley.

Greenacre, M. (1984). *Theory and Applications of Correspondence Analysis*. Academic Press.

Greenacre, M. (1991). Interpreting Multiple Correspondence Analysis. *Applied Stochastic Models and Data Analysis*, *7*(2), 195–210. DOI: 10.1002/asm.3150070208

Greenacre, M., & Blasius, J. (2006). Multiple correspondence analysis and related methods. *In Chapman and Hall/CRC eBooks*. DOI: 10.1201/9781420011319

Greenacre, M., Groenen, P. J. F., Hastie, T., D'Enza, A. I., Markos, A., & Tuzhilina, E. (2022). Principal component analysis. *Nature Reviews. Methods Primers*, *2*(1), 100. DOI: 10.1038/s43586-022-00184-w

Hair, J. F., Black, W. C., Babin, B. J., & Anderson, R. E. (2010). *Multivariate Data Analysis: A Global Perspective* (7th ed.). Pearson Education, Inc.

Jacoby, W. (1998). Statistical Graphics for visualizing multivariate data. In *SAGE Publications, Inc. eBooks*. DOI: 10.4135/9781412985970

Jolliffe, I. T. (2002). *Principal component analysis* (2nd ed.). Springer.

Jolliffe, I. T., & Cadima, J. (2016). Principal component analysis: A review and recent developments. *Philosophical Transactions. Series A, Mathematical, Physical, and Engineering Sciences, 374*(2065), 20150202. DOI: 10.1098/rsta.2015.0202 PMID: 26953178

Kassambara, A. (2017). *Practical guide to principal component methods in R: PCA, M (CA), FAMD, MFA, HCPC, factoextra* (Vol. 2). Sthda.

Keller, G. (2018). *Statistics for management and economics*. Cengage Learning.

Kitchen, N. R., & Clay, S. A. (2016). Understanding and Identifying Variability. *Agronomy, Horticulture and Plant Science Faculty Publications. 69.* https://openprairie.sdstate.edu/plant_faculty_pubs/69

Kolenikov, S., & Angeles, G. (2004). The use of discrete data in PCA: theory, simulations, and applications to socioeconomic indices. *Chapel Hill: Carolina Population Center, University of North Carolina, 20*, 1-59.

Kramer, M. A. (1991). Nonlinear principal component analysis using autoassociative neural networks. *AIChE Journal. American Institute of Chemical Engineers, 37*(2), 233–243. DOI: 10.1002/aic.690370209

Le Roux, B., & Rouanet, H. (2004). *Geometric data analysis: from correspondence analysis to structured data analysis*. Springer Science & Business Media.

Linting, M., Meulman, J. J., Groenen, P. J., & Van der Koojj, A. J. (2007). Nonlinear principal components analysis: Introduction and application. *Psychological Methods, 12*(3), 336–358. DOI: 10.1037/1082-989X.12.3.336 PMID: 17784798

Markos, A., Menexes, G., & Papadimitriou, T. (2009b). Multiple correspondence analysis for "tall" data sets. *Intelligent Data Analysis, 13*(6), 873–885. DOI: 10.3233/IDA-2009-0398

Menexes, G. (2006). *Experimental Designs in Data Analysis*. Doctoral dissertation, University of Macedonia, Thessaloniki, Greece (in Greek).

Meng, C., Zeleznik, O. A., Thallinger, G. G., Kuster, B., Gholami, A. M., & Culhane, A. C. (2016). Dimension reduction techniques for the integrative analysis of multi-omics data. *Briefings in Bioinformatics, 17*(4), 628–641. DOI: 10.1093/bib/bbv108 PMID: 26969681

Meulman, J., Van Der Kooij, A., & Heiser, W. (2004). Principal components Analysis with nonlinear optimal scaling transformations for ordinal and nominal data. In *SAGE Publications, Inc. eBooks* (pp. 50–71). DOI: 10.4135/9781412986311.n3

Michaeli, T., Wang, W., & Livescu, K. (2016). Nonparametric canonical correlation analysis. *In International conference on Machine Learning* (pp. 1967-1976). PMLR. http://proceedings.mlr.press/v48/michaeli16.pdf

Michailidis, G., & De Leeuw, J. (1998). The Gifi System of Descriptive Multivariate Analysis. *Statistical Science*, *13*(4), 307–336. DOI: 10.1214/ss/1028905828

Mitra, R., & Zhang, C. H. (2014). Multivariate analysis of nonparametric estimates of large correlation matrices. *arXiv preprint arXiv:1403.6195*.

Nanga, S., Bawah, A. T., Acquaye, B. A., Billa, M. I., Baeta, F. D., Odai, N. A., Obeng, S. K., & Nsiah, A. D. (2021). Review of dimension reduction methods. *Journal of Data Analysis and Information Processing*, *9*(3), 189–231. DOI: 10.4236/jdaip.2021.93013

Nevill, A. M., Atkinson, G., Hughes, M. D., & Cooper, S. M. (2002). Statistical methods for analyzing discrete and categorical data recorded in performance analysis. *Journal of Sports Sciences*, *20*(10), 829–844. DOI: 10.1080/026404102320675666 PMID: 12363298

Oseledets, I. V., & Tyrtyshnikov, E. E. (2009). Breaking the curse of dimensionality, or how to use SVD in many dimensions. *SIAM Journal on Scientific Computing*, *31*(5), 3744–3759. DOI: 10.1137/090748330

Pagès, J. (2014c). Multiple factor analysis by example using R. In *Chapman and Hall/CRC eBooks*. DOI: 10.1201/b17700

Painsky, A., Feder, M., & Tishby, N. (2020). Nonlinear canonical correlation analysis: A compressed representation approach. *Entropy (Basel, Switzerland)*, *22*(2), 208. DOI: 10.3390/e22020208 PMID: 33285982

Ptak, S. H., Errico, M., & Christensen, K. V. (2021). Extracting activity patterns: Exploratory data analysis on a fucoidan extract data set with mixed variables. *Algal Research*, *54*, 102220. DOI: 10.1016/j.algal.2021.102220

Ran, X. (2019). *Using the dimension reduction method FAMD in the data preprocessing step for risk prediction and for unsupervised clustering* (Doctoral dissertation, University of Pittsburgh). Doctoral dissertation, University of Pittsburgh, USA. http://d-scholarship.pitt.edu/36564/

Robitzsch, A. (2020). Why ordinal variables can (Almost) always be treated as continuous variables: Clarifying assumptions of robust continuous and ordinal factor analysis estimation methods. *Frontiers in Education*, *5*, 589965. DOI: 10.3389/feduc.2020.589965

Song, Y., Li, Y., & Qu, J. (2018). A New Approach for Supervised Dimensionality Reduction. [IJDWM]. *International Journal of Data Warehousing and Mining*, *14*(4), 20–37. DOI: 10.4018/IJDWM.2018100102

Theodosiou, T., Angelis, L., & Vakali, A. (2007). Non-linear correlation of content and metadata information extracted from biomedical article datasets. *Journal of Biomedical Informatics*, *41*(1), 202–216. DOI: 10.1016/j.jbi.2007.06.004 PMID: 17643352

Upton, G. J. (2016). *Categorical data analysis by example*. John Wiley & Sons., DOI: 10.1002/9781119450382

Van der Burg, E., De Leeuw, J., & Dijksterhuis, G. (1994). OVERALS: Nonlinear canonical correlation with k sets of variables. *Computational Statistics & Data Analysis*, *18*(1), 141–163. DOI: 10.1016/0167-9473(94)90136-8

Watanabe, K., Naito, K., & Koch, I. (2024). Robustness of Principal Component Analysis with Spearman's Rank Matrix. *Journal of Statistical Theory and Practice*, *18*(1), 6. DOI: 10.1007/s42519-023-00358-z

Ziyanak, S., & Yagci, J. (2024). Examining retention methods in factor analysis: A comparison of Polychoric and Pearson correlations for categorical and continuous data. *Journal of Educational Impact.*, *1*(1), 13–34. DOI: 10.70617/tx7v8v28

KEY TERMS AND DEFINITIONS

Categorical Variables/Data: Variables that represent categories or groups rather than numerical values.

Nominal Variables: Categorical variables without a natural order, such as names or labels (e.g., gender, country).

Ordinal Variables: Categorical variables with a meaningful order but no consistent difference between values (e.g., rankings like "low," "medium," "high").

Dimensionality Reduction: A process used in data analysis to reduce the number of variables under consideration, simplifying the dataset while preserving essential information.

Multiple Correspondence Analysis (MCA): A statistical method used to analyze categorical data by summarizing relationships between multiple variables in a low-dimensional space.

Nonlinear Categorical Principal Components Analysis (CATPCA): A dimensionality reduction method designed for categorical data, which uses optimal scaling to transform variables into numerical values for principal component analysis.

Principal Component Analysis (PCA): A method for reducing the dimensionality of continuous data by transforming variables into a set of linearly uncorrelated components.

Factor Analysis for Mixed Data (FAMD): A method that handles datasets with both categorical and continuous variables, combining features of PCA and MCA to analyze mixed data effectively.

Multiple Factor Analysis (MFA): A method for analyzing datasets organized into groups of variables, allowing comparisons across multiple variable sets while accounting for their structure and relationships.

Nonlinear Canonical Correlation Analysis (NLCCA): An advanced method for exploring nonlinear relationships between two or more sets of variables by finding canonical components that maximize correlations.

Singular Value Decomposition (SVD): A mathematical method for factorizing a matrix into components, often used in dimensionality reduction to identify patterns in data.

Chapter 2
Dimensionality Reduction in Data Cubes With Independent Slices

Zacharenia Kyrana
https://orcid.org/0000-0001-9269-0675
Aristotle University of Thessaloniki, Greece

Alexandra M. Michaelidou
https://orcid.org/0000-0002-2471-5655
Aristotle University of Thessaloniki, Greece

Emmanouil D. Pratsinakis
https://orcid.org/0000-0002-3725-3525
Aristotle University of Thessaloniki, Greece

Efstratios Kiranas
https://orcid.org/0009-0002-6358-2071
International Hellenic University, Greece

Nikolaos Papafilippou
https://orcid.org/0009-0003-3148-7229
Aristotle University of Thessaloniki, Greece

Georgios C. Menexes
https://orcid.org/0000-0002-1034-7345
Aristotle University of Thessaloniki, Greece

ABSTRACT

Data cubes are p-dimensional data structures, where $p \geq 3$. The application of different dimensionality reduction strategies based on PCA to data cubes was investigated. Some strategies focused on decomposing the total variability of the data cube into between-"slices" and within-"slices" variability. Other strategies either ignored the effect of "slices" or considered it on the overall data cube structure. The primary aim was to provide new insights into dimensionality reduction strategies on data cubes with independent "slices". For the implementation and comparison of the proposed strategies, two three-dimensional data cubes with independent "slices"

DOI: 10.4018/979-8-3693-9400-7.ch002

and differing structures in terms of the number of objects per "slice" were examined. The proposed strategies highlighted the importance of applying different analysis strategies for dimensionality reduction according to the research objectives.

INTRODUCTION

The world, sciences, and most experimental procedures are characterized by complex, multidimensional data, while capturing scientific interest for their observation, study, and comprehension. Advancements in computing power and the prevalence of statistical and programming software packages have made managing large volumes of multivariate and multidimensional data (big data) more accessible.

In Mathematics, according to the sixth edition of the *Concise Oxford Dictionary of Mathematics*, "The cube is one of the Platonic solids, a polyhedron with six square faces, eight vertices, and 12 edges" (Earl & Nicholson, 2021), while according to MathWorld, "A data cube is a three-dimensional dataset consisting of stacked two-dimensional data "slices" as a function of a third coordinate" (Weisstein, 2024).

In Computer Science and especially in the field of Data Warehousing and Data Mining, according to Han et al. (2012), "A data warehouse is usually modeled by a multidimensional data structure, called a data cube".

In Statistics, according to the dictionary of the Organization for Economic Co-operation and Development (OECD), "The data cube is a multidimensional structure for storing statistical information" (OECD, 2008). Furthermore, based on Pavoine et al. (2007), "The data cube is a three-dimensional dataset consisting of stacked two-dimensional data "slices" as a function of a third coordinate", while Gilbert et al. (2018) explain that "The data cube is related to the concept of database alignment, where multiple databases are aligned in different dimensions under certain conditions".

According to the above definitions, in this study data cubes are p-dimensional data structures, where $p \geq 3$. p can take any integer value greater than or equal to three, depending on the structure of the data. While p could be a prime number in some cases (such as $p = 3$), there is no inherent constraint preventing p from being any integer that reflects the number of dimensions in the data cube. For example, in a three-dimensional data cube, the first dimension consists of the objects (e.g., different plants), the second dimension consists of the variables (e.g., four quantitative variables), and the third dimension consists of the variable of "slices" (e.g., three different species). If an additional fourth dimension was added (e.g., two different locations), the data cube would become four-dimensional.

Data cubes are used to explore and analyze multidimensional data from various perspectives, answer complex questions, and extract more information (Suh &

Chung, 2016). According to Kroonenberg (2008, p. 17), the main research question that can be answered when processing and analyzing a data cube is "Which groups of objects behave differently, in which variables, and under which conditions?".

In international scientific literature, a data cube appears under different names, according to its structural characteristics (e.g., multi-way, multi-block, multi-group, etc.) (Kroonenberg, 2008; Tenenhaus & Tenenhaus, 2014). Therefore, a data cube can consist of the same or different objects per "slice", an equal or unequal number of objects per "slice", the same or different variables per "slice", quantitative or categorical "slices", and dependent or independent "slices". In most scientific fields, data cubes consist of dependent "slices" (i.e., repeated measurements of the same objects per "slice"). These data cubes are commonly called multi-way data (Acar & Yener, 2009; Kolda & Bader, 2009; Kroonenberg, 2008). In contrast, in most biological sciences, including fields such as nutrition and agricultural experiments, the objects are the same or different and equal or unequal in number per "slice", the variables are the same per "slice", and the "slices" are categorical and dependent or independent (Kroonenberg, 2008; Pavoine et al., 2007). In this study, the objects were different and either equal or unequal in number per "slice", the variables were the same per "slice", and the "slices" were categorical and independent. According to Tenenhaus & Tenenhaus (2014) and Codesido et al. (2021), such a data cube is referred to as a multi-group data.

Data Cubes With Dependent "Slices" and the Same Objects per "Slice"

Regarding the mathematical background of data cubes with dependent "slices" (with same objects per "slice"), assume $\underline{\mathbf{Y}}$ is a three-dimensional $I \times J \times K$ data cube (consequently $p = 3$), with y_{ijk} elements, where $i = 1, 2, \ldots, I, j = 1, 2, \ldots, J$ and $k = 1, 2, \ldots, K$. The first dimension I denotes the rows or else the objects (experimental or sampling units) of the data cube. The second dimension J denotes the columns or else the variables (or features) of the data cube. The third dimension K denotes the different conditions or else the "depth" of the data cube. Each condition of the third dimension is called a "slice". In other words, the data cube $\underline{\mathbf{Y}}$ can be seen as a collection of two-dimensional matrices $\mathbf{Y}_k$, where each matrix represents a "slice" of the data cube (De Roover et al., 2013; Kroonenberg, 2008; Pavoine et al., 2007).

The most common way to manage a data cube is to transform it into a two-dimensional data matrix. If the data cube consists of dependent "slices", this procedure is called unfolding (or matricization or stringing out) into tall or wide format, and it can be done in six different ways depending on the type of data cube and the objectives of the study. In terms of mathematical background, data cube $\underline{\mathbf{Y}}$ transformed into a "stacked" matrix $\mathbf{Y}$ with dimensions $IK \times J$, which is the most common and computational convenient unfolding way (Burgas et al., 2018). The "stacked" matrix contains elements denoted by y_{ijk}, where $i = 1, 2, \ldots, I$, $j = 1, 2, \ldots, J$ and $k = 1, 2, \ldots, K$ (Burgas et al., 2018; Kroonenberg, 2008) (see Figure 1).

Figure 1. Transformation of a data cube with dependent "slices" into a "stacked" data matrix through unfolding into tall format

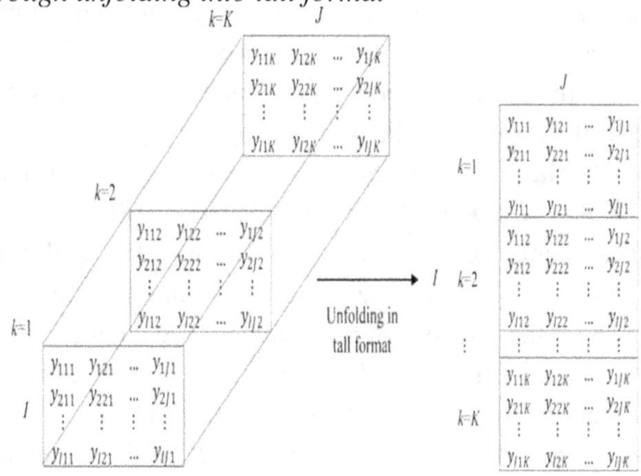

Data Cubes With Independent "Slices" and With Different and Equal Number of Objects per "Slice"

Regarding the mathematical background of data cubes with independent "slices" (with different and equal numbers of objects per "slice"), assume $\underline{\mathbf{X}}$ is a three-dimensional $I \times J \times K$ data cube (consequently $p = 3$), with x_{ijk} elements, where $i = 1, 2, \ldots, I$, $j = 1, 2, \ldots, J$ and $k = 1, 2, \ldots, K$. The first dimension I denotes the rows, or the objects of the data cube, which are different and equal in number per "slice". The second dimension J denotes the columns or else the variables of the data cube. The third dimension K denotes the independent "slices" of the data cube. In other words, the data cube $\underline{\mathbf{X}}$ can be seen as a collection of two-dimensional matrices $\mathbf{X}_k$, where each matrix represents a "slice" of the data cube (Rotari et al., 2024).

The most classic way to manage a data cube is to transform it into a two-dimensional data matrix. If the data cube consists of independent "slices" with different objects per "slice", this procedure is called vertical concatenation. Regarding the mathematical background, data cube $\underline{\mathbf{X}}$ transformed into a "stacked" matrix $\mathbf{X}$ with dimensions $IK \times J$ and x_{ijk} elements, where $i = 1, 2, \ldots, I$, $j = 1, 2, \ldots, J$ and $k = 1, 2, \ldots, K$ (Codesido et al., 2021; De Roover et al., 2013; Rotari et al., 2024) (see Figure 2).

Figure 2. Transformation of a data cube with independent "slices" and with different and equal number of objects per "slice" into a "stacked" data matrix through vertical concatenation

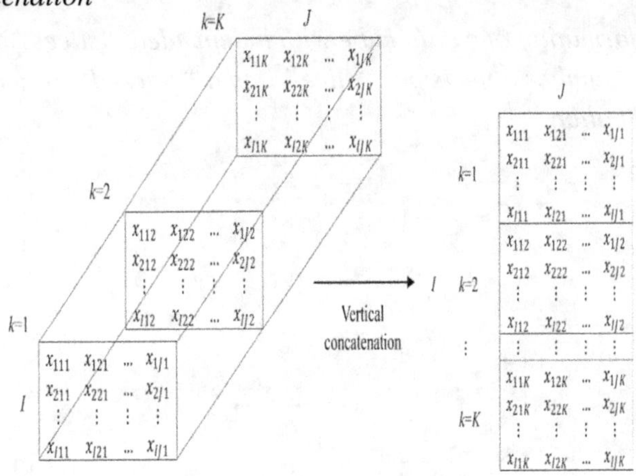

Data Cubes With Independent "Slices" and With Different and Unequal Number of Objects per "Slice"

When a data cube with independent "slices" consisted of different and unequal number of objects per "slice", then its structure isn't a true data cube but can still be referred to as such in a generalized sense. Regarding the mathematical background, assume $\underline{\mathbf{D}} = [\mathbf{D}_1^{\mathrm{T}}, \mathbf{D}_2^{\mathrm{T}}, \ldots, \mathbf{D}_K^{\mathrm{T}}]^{\mathrm{T}}$ is a generalized three-dimensional data cube (consequently $p = 3$) with $\mathbf{D}_k$ two-dimensional unbalanced "slices". Each "slice" $\mathbf{D}_k$ has dimensions $n_k \times J$ and contains d_{ijk} elements, where $i = 1, 2, \ldots, n_k$, $j = 1, 2, \ldots, J$ and $k = 1, 2, \ldots, K$. The first dimension n_k denotes the number of objects in the kth "slice". The second dimension J denotes the columns or else the variables of the data cube. The third dimension K denotes the independent "slices" of the data cube. In other words, the generalized data cube $\underline{\mathbf{D}}$ can be seen as a collection of two-dimensional matrices $\mathbf{D}_k$, where each matrix represents a "slice" of the data cube (Tenenhaus & Tenenhaus, 2014).

The classic way to manage a data cube is to transform it into a two-dimensional data matrix. If the data cube consists of independent "slices" with different and unequal number of objects per "slice", this procedure is called vertical concatenation (De Roover et al., 2013). Specifically, the generalized data cube $\underline{\mathbf{D}}$ transformed into a "stacked" data matrix $\mathbf{D}$ with dimensions $N \times J$ and d_{ijk} elements, where $i = 1, 2, \ldots, n_k$, $j = 1, 2, \ldots, J$, $k = 1, 2, \ldots, K$, and N is the total number of objects in the entire data cube ($\sum_{k=1}^{K} n_k = N$) (Codesido et al., 2021; De Roover et al., 2013; Tenenhaus & Tenenhaus, 2014) (see Figure 3).

Figure 3. Transformation of a data cube with independent "slices" and with different and unequal number objects per "slice" into a "stacked" data matrix through vertical concatenation

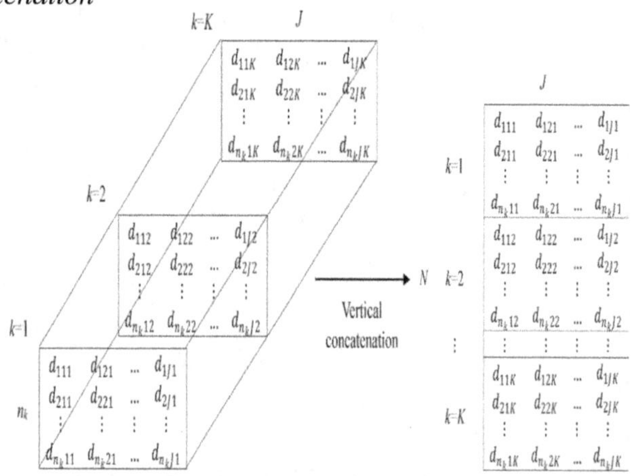

Dimensionality reduction is an ever-growing field of study. This has led to the development of a variety of methods and their variants, which suggests the great complexity of dimensionality reduction techniques (Cunningham & Ghahramani, 2015).

Dimensionality reduction methods aim to provide a compressed representation of the original dataset, smaller in "volume" but maintaining critical and informative content (Dash et al., 1997; Sharma, 1996). However, these methods also come with inherent disadvantages, such as the "curse of dimensionality" (Bellman, 1961; Kyrana et al., in press; Nguyen & Holmes, 2019), challenges in determining the number of important dimensions, high computational requirements (Kyrana et al., in press), handling missing values, outliers, and extreme values, as well as challenges in visualizing the results (Hair et al., 2010; Menexes, 2006).

In international scientific literature, there are several dimensionality reduction methods for either data cubes or data matrices. The most widely used methods for data matrices are Principal Components Analysis (PCA) (Anderson, 1984; Hair et al., 2010; Jolliffe, 2002; Jolliffe & Cadima, 2016; Kaiser, 1958; Song et al., 2018), Factor Analysis (FA) (Anderson, 1984; Hair et al., 2010; Jolliffe, 2002), Multiple Correspondence Analysis (MCA) (Greenacre & Blasius, 2006; Menexes, 2006; Messaoud et al., 2007), Categorical Principal Components Analysis with optimal scaling (CATPCA) (Gifi, 1990; Linting et al., 2007; Markos et al., 2020; Tabachnick

& Fidell, 2007) and Factor Analysis for Mixed Data (FAMD) (Kassambara, 2017; Markos et al., 2020; Mothobi et al., 2021; Pagès, 2014). The best-known methods for data cubes with dependent "slices" are Tucker decomposition (Acar & Yener, 2009; Kolda & Bader, 2009; Kroonenberg et al., 2009; Li et al., 2015; Sun & Braatz, 2020) and CP (CANDECOMP/PARAFAC) decomposition, which consider the data cube as a tensor (Acar & Yener, 2009; Kolda & Bader, 2009; Kroonenberg et al., 2009; Li et al., 2015; Sun & Braatz, 2020). In contrast, the best-known methods for data cubes with independent "slices" are Simultaneous Components Analysis (SCA) (Smilde et al., 2022; Tenenhaus & Tenenhaus, 2014; Timmerman & Kiers, 2003), Multi-group PCA (Eslami et al., 2013; Rotari et al., 2024; Tenenhaus & Tenenhaus, 2014) and Common PCA (Eslami et al., 2013; Rotari et al., 2024; Tenenhaus & Tenenhaus, 2014).

However, the application of the above "traditional" dimensionality reduction methods depends on the scientific field, the objectives of the study, the transformation of the data cube, the observed data and the type of data cube. Therefore, it is more appropriate to discuss analysis strategies rather than a "global" dimensionality reduction method suitable for every case, in order to increase the range of applications for each method and extract more usable results (Kyrana et al., in press).

There have been several comparative studies among different analysis strategies and methods for data cubes with independent "slices". For example, in recent literature, Eslami et al. (2013) proposed and compared Multi-group PCA, Dual Multiple Factor Analysis, and Dual STATIS, which are used to determine a common variance-covariance matrix. In their study, between-group comparisons and Multi-group PCA were used for stepwise determination of common loading vectors, as well as Common PCA and Dual Generalized Procrustes Analysis. The methods' comparison in this study aimed to show how multi-block methods can be adapted to the multi-group context, to identify common vectors of loadings across groups, and to highlight the similarities and differences among the methods, in terms of how similar the common vectors of loadings obtained by the various methods are, and in terms of the total variances in the different groups recovered by the successive components, as well as computational structure and interpretability.

Tenenhaus & Tenenhaus (2014) compared Regularized Generalized Canonical Correlation Analysis (RGCCA) and Clusterwise SCA with Equal Average Cross-Products (Clusterwise SCA-ECP), in order to highlight the extension of RGCCA to the multi-group context and investigate both the common relationships among variables across groups and the specific patterns within each group, illustrating RGCCA's ability to uncover shared structures while accounting for group-specific variability.

Codesido et al. (2020) compared PCA and Network PCA (NetPCA) on a multi-group dataset to investigate the influence of each group on the global model and demonstrate NetPCA's ability to account for the links between groups.

Rotari et al. (2024) compared PARAFAC model and an extension of it on multi-group data, in order to identify the common structure of variability of all groups by capturing the common and explanatory loadings of the data (even when the variances of the groups were unequal).

However, to the best of our knowledge, there are no comparative studies that explore the components and structure of variability of a data cube with independent "slices", focusing on the different management and coding of the "slices" variable, and the integration of different analysis methods, leading finally to the application of PCA. In general, investigating the effect of data cube "slices" on the overall structure of the data cube is crucial, as it can result in either simpler or more complex data cube management and statistical analysis practices.

The aims of this study were to propose some new insights, beyond Analysis of Variance (ANOVA), into analysis strategies for dimensionality reduction on data cubes with independent "slices", and to highlight the importance of applying different analysis strategies in the interpretation of results.

The outline of the study is as follows: The Materials and Methods section introduces the two real datasets used for the analyses, the 13 proposed analysis strategies, and the mathematical framework for decomposing total variability into between-"slices" and within-"slices" variability. The Results section presents the outcomes (PCA loading plots) of the analysis strategies for both the "Iris Plants" and "ADONUT" data cubes. The Discussion section provides an in-depth analysis of the results for both data cubes and compares the findings. The Conclusions section summarizes the study's findings, highlights its originality and contribution to the international scientific community, and discusses its limitations and recommendations for future research.

MATERIALS AND METHODS

The analysis strategies used in this study were based on the PCA method without rotation. Alternatively, the FA method can also be used. It is important to note that PCA is the best-known dimensionality reduction method when the data lies on Euclidean spaces (Greenacre et al., 2022; Jolliffe & Cadima, 2016). In other cases, such as data on non-Euclidean spaces like manifolds, other dimensionality reduction methods are more suitable, including t-distributed Stochastic Neighbor Embedding (t-SNE) (Draganov & Dohn, 2023; Hinton & Roweis, 2002; McInnes et al., 2018), Uniform Manifold Approximation and Projection (UMAP), (Draganov & Dohn,

2023; McInnes et al., 2018) and Isometric Mapping (ISOMAP) (Draganov & Dohn, 2023; Hinton & Roweis, 2002; McInnes et al., 2018). Both PCA and FA aim to reduce the mathematical dimensions of a data matrix. However, PCA assumes that the common variance equals the total variance, while FA posits that the total variance is divided into common and unique variance (Fabrigar et al., 1999; Kyrana et al., in press). Both PCA and FA methods can utilize Singular Value Decomposition (SVD) as a core computational tool on the appropriate data matrix (Greenacre et al., 2022; Jolliffe & Cadima, 2016; Smilde et al., 2022).

For practical purposes, including visualization and interpretation, only the first two principal components were selected. The statistical software packages used for the analyses were IBM SPSS Statistics v29.0 (SPSS), R v4.3.2 (R) and XLSTAT v2014.5.03 (XLSTAT).

Dataset 1

The agricultural dataset "Iris Plants" from the UCI Machine Learning Repository (Fisher, 1936; UCI Machine Learning Repository) was employed in this study. The dataset was treated as a three-dimensional data cube $\underline{\mathbf{X}}(50 \times 4 \times 3)$ with independent "slices" which transformed into a "stacked" data matrix $\mathbf{X}(150 \times 4)$. The first dimension consisted of different and equal number of objects (plants) per "slice" ($I = 50$). The second dimension consisted of the same four quantitative variables per "slice" ($J = 4$). Specifically, the variables were sepal length (cm), sepal width (cm), petal length (cm), and petal width (cm) of the plants. The third dimension was a categorical variable denoting three independent "slices" ($K = 3$) corresponding to three iris plant species (iris-setosa, iris-versicolor, iris-virginica).

Dataset 2

The nutritional dataset "Adolescent Nutrition (ADONUT)", from a Panhellenic study on the nutritional habits of adolescents conducted during the years 2010-2012, was also employed in this study (Grammatikopoulou et al., 2014). The study was conducted by the Nutritional Sciences and Dietetics Department at the International Hellenic University. Permissions for data collection were conducted from the Directorate of Secondary Education, the Institute's Research Committee, the Greek Ministry of Education, as well as from the parents or/and guardians of the adolescents. The study was fully funded by the Educational Institute's Research Committee. A total number of 42,061 adolescents were selected via proportionate stratified random sampling. The dataset was considered a three-dimensional data cube $\underline{\mathbf{D}}$ with independent "slices" which transformed into a "stacked" data matrix $\mathbf{D}(42{,}061 \times 13)$. The first dimension consisted of different and unequal number of objects (adolescents) per "slice" (n_1 = 5,549, n_2 = 26,875, n_3 = 9,637). The second dimension consisted of the same 13 quantitative variables per "slice" (J = 13). Specifically, the initial dataset consisted of the weekly frequency consumption of 140 foods. These foods were categorized into 13 main food categories according to their nutritional content, culinary use, and biological origin. The 13 main food categories (Dairy and products, Fats and Oils, Grains, Meat, Poultry, Seafood, Eggs, Vegetables, Legumes, Fruits, Sauces, Sweets and Snacks, and Beverages) served as the final quantitative variables of the data cube. The third dimension was a categorical variable denoting three independent "slices" (K = 3) corresponding to three different types of areas (urban, suburban and rural), from which adolescents came from.

ANALYSIS STRATEGIES

The proposed analysis strategies based on the study of variability (inertia) which is the key point of Statistical-Data Analysis in biological sciences and especially in agricultural ones (Kitchen & Clay, 2016). Specifically, some strategies were based on the idea of decomposing the total variability of the data cube into between-"slices" variability and within-"slices" variability (Eslami et al., 2013). According to Rencher and Christensen (2012), total variability can be decomposed into between and within variability. Specifically, based on the mathematical background of a "stacked" data matrix **X** with different and equal number of objects per "slice":

Total variability = Between-"slices" variability + Within-"slices" variability $\Longleftrightarrow$

$$\Longleftrightarrow \mathbf{X}_{\mathrm{T}} = \mathbf{X}_{\mathrm{B}} + \mathbf{X}_{\mathrm{W}}$$

Assume a "stacked" data matrix **X** with dimensions $IK \times J$ and x_{ijk} elements, where $i = 1, 2, \ldots, I$, $j = 1, 2, \ldots, J$ and $k = 1, 2, \ldots, K$.

Let $\mathbf{x}_{ik} = \begin{bmatrix} x_{i1k} \\ x_{i2k} \\ \vdots \\ x_{iJk} \end{bmatrix}$ a J-dimensional vector of variables for object i in "slice" k.

The J-dimensional mean vector for the kth "slice" is

$\overline{\mathbf{x}}_{.k} = \frac{1}{I}\sum_{i=1}^{I}\mathbf{x}_{ik} = \begin{bmatrix} \overline{x}_{.1k} \\ \overline{x}_{.2k} \\ \vdots \\ \overline{x}_{.Jk} \end{bmatrix}$. The J-dimensional mean vector for the entire

"stacked" matrix **X** is $\overline{\mathbf{x}}_{..} = \frac{1}{IK}\sum_{k=1}^{K}\sum_{i=1}^{I}\mathbf{x}_{ik} = \begin{bmatrix} \overline{x}_{.1.} \\ \overline{x}_{.2.} \\ \vdots \\ \overline{x}_{.J.} \end{bmatrix}$.

Total variability is defined as a $J \times J$ data matrix $\mathbf{X_T} = \sum_{k=1}^{K}\sum_{i=1}^{I}(\mathbf{x}_{ik} - \overline{\mathbf{x}}_{..})(\mathbf{x}_{ik} - \overline{\mathbf{x}}_{..})^{\mathrm{T}}$. Between-"slices" variability is defined as a $J \times J$ data matrix $\mathbf{X_B} = \sum_{k=1}^{K} I(\overline{\mathbf{x}}_{.k} - \overline{\mathbf{x}}_{..})(\overline{\mathbf{x}}_{.k} - \overline{\mathbf{x}}_{..})^{\mathrm{T}}$. Within-"slices" variability is defined as a $J \times J$ data matrix $\mathbf{X_W} = \sum_{k=1}^{K}\sum_{i=1}^{I}(\mathbf{x}_{ik} - \overline{\mathbf{x}}_{.k})(\mathbf{x}_{ik} - \overline{\mathbf{x}}_{.k})^{\mathrm{T}}$. It is important to note that $\mathbf{A}^{\mathrm{T}}$ is the transpose of matrix $\mathbf{A}$.

Based on the mathematical background of a "stacked" data matrix $\mathbf{D}$ with different and unequal number of objects per "slice":

Total variability = Between-"slices" variability + Within-"slices" variability $\Longleftrightarrow$

$$\Longleftrightarrow \mathbf{D_T} = \mathbf{D_B} + \mathbf{D_W}$$

Assume a "stacked" data matrix $\mathbf{D}$ with dimensions $N \times J$ and d_{ijk} elements, where $i = 1, 2, \ldots, n_k$, $j = 1, 2, \ldots, J$, $k = 1, 2, \ldots, K$, and N is the total number of objects in the entire data cube ($\sum_{k=1}^{K} n_k = N$).

Let $\mathbf{d}_{ik} = \begin{bmatrix} d_{i1k} \\ d_{i2k} \\ \vdots \\ d_{iJk} \end{bmatrix}$ a J-dimensional vector of variables for object i in "slice" k.

The J-dimensional mean vector for the kth "slice" is

$\overline{\mathbf{d}}_{.k} = \frac{1}{n_k}\sum_{i=1}^{n_k}\mathbf{d}_{ik} = \begin{bmatrix} \overline{d}_{.1k} \\ \overline{d}_{.2k} \\ \vdots \\ \overline{d}_{.Jk} \end{bmatrix}$. The J-dimensional mean vector for the entire "stacked" matrix $\mathbf{D}$ is $\overline{\mathbf{d}}_{..} = \frac{1}{N}\sum_{k=1}^{K}\sum_{i=1}^{N}\mathbf{d}_{ik} = \begin{bmatrix} \overline{d}_{.1.} \\ \overline{d}_{.2.} \\ \vdots \\ \overline{d}_{.J.} \end{bmatrix}$.

Total variability is defined as a $J \times J$ data matrix $\mathbf{D_T} = \sum_{k=1}^{K}\sum_{i=1}^{n_k}(\mathbf{d}_{ik} - \overline{\mathbf{d}}_{..})(\mathbf{d}_{ik} - \overline{\mathbf{d}}_{..})^{\mathrm{T}}$. Between-"slices" variability is defined as a $J \times J$ data matrix $\mathbf{D_B} = \sum_{k=1}^{K} n_k(\overline{\mathbf{d}}_{.k} - \overline{\mathbf{d}}_{..})(\overline{\mathbf{d}}_{.k} - \overline{\mathbf{d}}_{..})^{\mathrm{T}}$. Within-"slices" variability is defined as a $J \times J$ data matrix $\mathbf{D_W} = \sum_{k=1}^{K}\sum_{i=1}^{n_k}(\mathbf{d}_{ik} - \overline{\mathbf{d}}_{.k})(\mathbf{d}_{ik} - \overline{\mathbf{d}}_{.k})^{\mathrm{T}}$.

Other strategies either initially ignored the effect of "slices" on the structure and components of the variability and introduced it at a second level, or completely ignored the effect of the data cube "slices" on the structure and components of the variability. Some others took into account the effect of the data cube "slices" on the structure and components of variability through different coding of the "slices" variable. The proposed analysis strategies are presented below for both "stacked" data matrices $\mathbf{X}$ and $\mathbf{D}$.

Strategy 1 (S1)

The effect of "slices" was not of scientific interest and the study focused on the overall structure of the observed variability. Specifically, PCA was applied to both data matrices **X**(150 × 4) and **D**(42,061 × 13), while the variable of "slices" was not considered, respectively.

Strategy 1.1 (S1.1)

The effect of "slices" was not of particular scientific interest and the study focused on the overall structure of the observed variability after balancing the effect of "slices". Specifically, SVD was applied to each "slice's" correlation matrix **R** and **R***, respectively: $\mathbf{R}_1(4 \times 4)$, $\mathbf{R}_2(4 \times 4)$, $\mathbf{R}_3(4 \times 4)$, and $\mathbf{R}^*_1(13 \times 13)$, $\mathbf{R}^*_2(13 \times 13)$, $\mathbf{R}^*_3(13 \times 13)$. The largest eigenvalue of each "slice" ($\lambda_{max(1)}, \lambda_{max(2)}, \lambda_{max(3)}$) was found, and later the values of the variables in each "slice" were divided by the corresponding square root of the largest eigenvalue ($\sqrt{\lambda_{max}}$). This balancing procedure is also applied to the Multiple Factor Analysis (MFA) method (Pagès, 2014). Finally, PCA was applied to the "new" data matrices $\mathbf{X}_T(150 \times 4)$ and $\mathbf{D}_T(42{,}061 \times 13)$, respectively.

Strategy 2 (S2)

The effect of "slices" was of scientific interest and the study focused on the structure of the variability between-"slices". Specifically, the matrices $\mathbf{X}_B(150 \times 4)$ and $\mathbf{D}_B(42{,}061 \times 13)$ were constructed capturing the variability between the "slices", to which PCA was later applied.

Strategy 2.1 (S2.1)

The effect of "slices" was of scientific interest and the study focused on the structure of the variability between-"slices", after balancing the effect of the "slices". Specifically, the matrices $\mathbf{X}_B(150 \times 4)$ and $\mathbf{D}_B(42{,}061 \times 13)$ capturing the variability between the "slices" were constructed and SVD was applied to each "slice's" correlation matrix **R** and **R***, respectively: $\mathbf{R}_{B1}(4 \times 4)$, $\mathbf{R}_{B2}(4 \times 4)$, $\mathbf{R}_{B3}(4 \times 4)$, and $\mathbf{R}^*_{B1}(13 \times 13)$, $\mathbf{R}^*_{B2}(13 \times 13)$, $\mathbf{R}^*_{B3}(13 \times 13)$. The largest eigenvalue of each "slice" ($\lambda_{max(1)}, \lambda_{max(2)}, \lambda_{max(3)}$) was found, and later the values of the variables in each "slice" were divided by the corresponding square root of the largest eigenvalue ($\sqrt{\lambda_{max}}$). Finally, PCA was applied to the "new" data matrices $\mathbf{X}_{BT}(150 \times 4)$ and $\mathbf{D}_{BT}(42{,}061 \times 13)$, respectively.

Strategy 3 (S3)

The effect of "slices" was not of scientific interest and the study focused on the structure of the variability within-"slices", with the "linear" effect of "slices" removed. Specifically, the matrices $\mathbf{X}_W(150 \times 4)$ and $\mathbf{D}_W(42{,}061 \times 13)$ were constructed to represent the variability within the "slices", to which PCA was later applied.

Strategy 3.1 (S3.1)

The effect of "slices" was not of scientific interest and the study focused on the structure of the variability within-"slices", with the "linear" or non-"linear" effect of "slices" removed. Specifically, Categorical Regression with optimal scaling (CATREG) (Van der Kooij et al., 2006) was applied to the data matrices $\mathbf{X}(150 \times 5)$ and $\mathbf{D}(42{,}061 \times 14)$, where each of the four and 13 quantitative variables, respectively, were entered as dependent variables, while the categorical variable of "slices" was entered as independent variable, respectively. The residuals from each regression analysis were calculated, which were then used to construct the matrices $\mathbf{X}_{W*}(150 \times 4)$ and $\mathbf{D}_{W*}(42{,}061 \times 13)$, capturing the variability within-"slices". Finally, PCA was applied to the data matrices $\mathbf{X}_{W*}$ and $\mathbf{D}_{W*}$.

Strategy 4 (S4)

The effect of "slices" was of scientific interest, but it was not desirable for this to influence the structure and components of the overall variability. The study focused on the overall structure of the variability. Specifically, PCA was applied to the data matrices $\mathbf{X}(150 \times 4)$ and $\mathbf{D}(42{,}061 \times 13)$. The effect of "slices" was introduced at a second level through scatter plots, with different colors and markers representing each "slice".

Strategy 5 (S5)

The effect of "slices" was of scientific interest, but it was not desirable for this to influence the structure and components of the overall variability. The study focused on the overall structure of the variability. Specifically, PCA was applied to the data matrices $\mathbf{X}(150 \times 4)$ and $\mathbf{D}(42{,}061 \times 13)$, respectively, and the effect of "slices" was introduced at a second level through scatter plots with the centroids of each "slice" as supplementary points. An alternative visualization of the results was performed through the corresponding biplots.

Strategy 6 (S6)

The effect of "slices" was not of scientific interest and the study focused on the separate analysis of the data cube "slices". Specifically, PCA was applied to the each "slice": $\mathbf{X}_1(50 \times 4)$, $\mathbf{X}_2(50 \times 4)$, $\mathbf{X}_3(50 \times 4)$, and $\mathbf{D}_1(5{,}549 \times 13)$, $\mathbf{D}_2(26{,}875 \times 13)$, $\mathbf{D}_3(9{,}637 \times 13)$, while the variable of "slices" was not considered. Finally, a conceptual comparison of the results could be made.

Strategy 7 (S7)

The effect of "slices" was of scientific interest and the study focused on the overall structure of the variability. The variable of "slices" was introduced in the analysis (PCA) as a quantitative variable after the application of CATREG. Specifically, CATREG was applied to the data matrices $\mathbf{X}(150 \times 5)$ and $\mathbf{D}(42{,}061 \times 14)$, where the four and 13 quantitative variables, respectively, were entered as independent variables and the categorical variable of "slices" was entered as dependent variable, respectively. Then, the optimal quantification of the "slices" variable was found and introduced into both data matrices **X** and **D**, respectively, replacing the categorical variable of "slices". Finally, PCA was applied to the "new" data matrices $\mathbf{X}(150 \times 5)$ and $\mathbf{D}(42{,}061 \times 14)$.

Strategy 8 (S8)

The effect of "slices" was of scientific interest and the study focused on the overall structure of the variability. The variable of "slices" was introduced into the analysis as a categorical variable. Specifically, CATPCA was applied to both mixed type datasets $\mathbf{X}(150 \times 5)$ and $\mathbf{D}(42{,}061 \times 14)$, with four and 13 quantitative variables, respectively, and the variable of "slices" as a categorical variable. It is worth noting that CATPCA is a popular method for reducing the dimensionality of mixed type data (Hendrickson, 2014) measured on quantitative, nominal, binary and/or ordinal scales (Vichi et al., 2019).

Strategy 9 (S9)

The effect of "slices" was of scientific interest and the study focused on the overall structure of the variability. The variable of "slices" was introduced into the analysis (PCA) as a quantitative variable after applying FAMD. Specifically, FAMD was applied to both data matrices $\mathbf{X}(150 \times 5)$ and $\mathbf{D}(42{,}061 \times 14)$, with four and 13 quantitative variables, respectively, and the variable of "slices" as a categorical variable. Then, the optimal quantification of the "slices" variable was found, and introduced into both data matrices **X** and **D**, respectively, replacing the categorical variable of "slices". Finally, PCA was applied to the "new" datasets **X**(150 × 5) and **D**(42,061 × 14).

Strategy 10 (S10)

The effect of "slices" was of scientific interest and the study focused on the overall structure of the variability. The variable of "slices" was introduced into the analysis as a categorical variable. Specifically, FAMD was applied to both mixed type datasets **X**(150 × 5) and **D**(42,061 × 14), with four and 13 quantitative variables, respectively, and the variable of "slices" as a categorical variable. It is worth noting that FAMD is also a popular method for reducing the dimensionality of mixed type data (Costa et al., 2022).

Strategy 11 (S11)

The effect of "slices" was of scientific interest and the study focused on the overall structure of the variability, as the components of the variability of "slices" were made comparable through Generalized Procrustes Analysis (GPA) (Gower, 1975). Specifically, PCA was applied to each "slice": $\mathbf{X}_1$(50 × 4), $\mathbf{X}_2$(50 × 4), $\mathbf{X}_3$(50 × 4), and $\mathbf{D}_1$(5,549 × 13), $\mathbf{D}_2$(26,875 × 13), $\mathbf{D}_3$(9,637 × 13). Later, the six loading matrices of the variables in each "slice" were constructed (three loading matrices from **X**, with dimensions 4 × 2 each, and three loading matrices from **D**, with dimensions 13 × 2 each). Finally, GPA was applied to both united loading matrices $\mathbf{L}_\mathbf{X}$(4 × 6) and $\mathbf{L}_\mathbf{D}$(13 × 6), with the variables' loadings in each "slice".

Strategy 12 (S12)

The effect of "slices" was of scientific interest and the study focused on the overall structure of the variability. The "slices" were introduced as supplementary pseudo-variables (dummy coded as 0-1) in the analysis, through loading plots that projected their position in relation to the other variables. Specifically, the variable of "slices" was transformed into three binary pseudo-variables. Finally, PCA was applied to both data matrices **X**(150 × 4) and **D**(42,061 × 13), and the corresponding three pseudo-variables introduced as supplementary variables into the analysis (the "slices" were not desirable to affect the structure and components of the variability of the other variables).

Strategy 13 (S13)

The effect of "slices" was of scientific interest and the study focused on the overall structure of the variability. The "slices" were introduced as pseudo-variables (dummy coded as 0-1) in the analysis. Specifically, the variable of "slices" was transformed into three binary pseudo-variables. Finally, PCA was applied to both data matrices **X**(150 × 7) and **D**(42,061 × 16).

Strategy 13.1 (S13.1)

The effect of "slices" was of scientific interest and the study focused on the overall structure of the variability. The "slices" were introduced as pseudo-variables (dummy coded as 0-1) in the analysis after ensuring that the variability of the pseudo-variables and the quantitative variables were comparable, allowing for a balanced contribution to the analysis. This procedure is also applied to MCA and MFA (Husson & Josse, 2013; Pagès, 2004). Specifically, the variable of "slices" was transformed into three binary pseudo-variables. The percentage p of units of each pseudo-variable was calculated, and each pseudo-variable was later divided by the corresponding $\sqrt{p}$. Finally, PCA was applied to both data matrices **X***(150 × 7) and **D***(42,061 × 16), respectively.

RESULTS

Results From the "Iris Plants" Dataset

The PCA loading diagrams for all strategies applied to the data cube $\underline{\mathbf{X}}$ are presented below (Figures 4-9). Specifically, Figures 4.a and 4.b present the results of S1 and S1.1, Figures 4.c and 4.d present the results of S2 and S2.1, and Figures 4.e and 4.f present the results of S3 and S3.1, respectively.

Figure 4. PCA loading plots of strategies S1 (a), S1.1 (b), S2 (c), S2.1 (d), S3 (e), S3.1 (f). The axes represent the first two principal components (principal component 1-Dim 1, principal component 2-Dim 2) with the corresponding percentages of the variance explained by each principal component and the vectors (arrows) represent the loadings of each variable on the principal components (analysis and visualization from R)

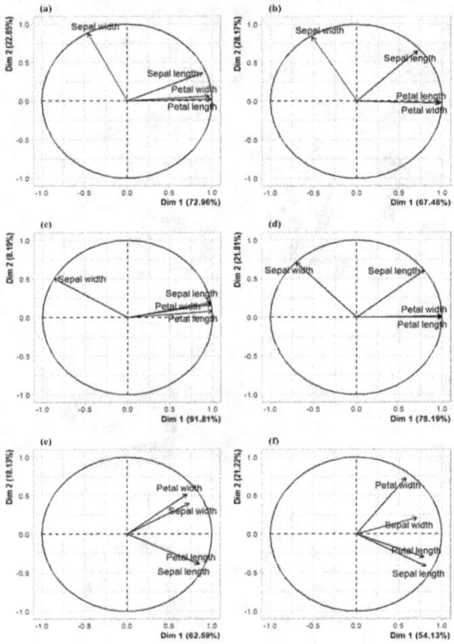

Figures 5.a and 5.b present the results of S4 and S5, respectively.

Figure 5. Scatter plot of strategies S4 (a) and S5(b). The axes represent the first two principal components (principal component 1-Dim 1, principal component 2-Dim 2) with the corresponding percentages of the variance explained by each principal component. Each point represents an object (experimental unit), the different colors and markers represent the "slice" to which each object belongs, and the bolded points in plot (b) represent the centroids of each "slice" (analysis and visualization from R)

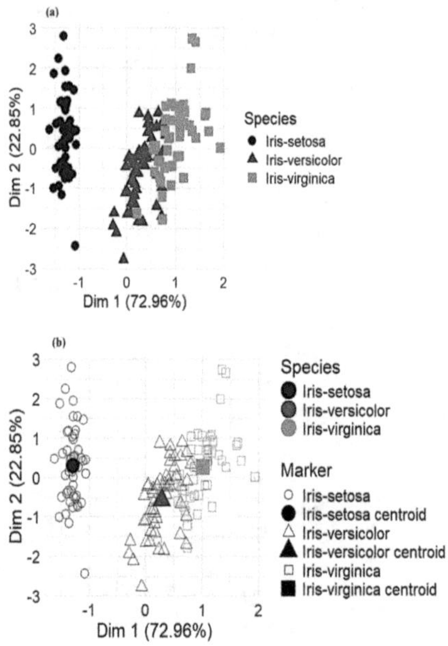

Figure 6 presents the results of S6. More specifically, Figure 6.a corresponds to the first "slice" (iris-setosa), Figure 6.b corresponds to the second "slice" (iris-versicolor) and Figure 6.c corresponds to the third "slice" (iris-virginica) of the data cube.

Figure 6. PCA loading plots of strategy S6. Plot (a) corresponds to the first "slice" (iris-setosa), plot (b) corresponds to the second "slice" (iris-versicolor) and plot (c) corresponds to the third "slice" (iris-virginica). The axes represent the first two principal components (principal component 1-Dim 1, principal component 2-Dim 2) with the corresponding percentages of the variance explained by each principal component and the vectors (arrows) represent the loadings of each variable on the principal components (analysis and visualization from R)

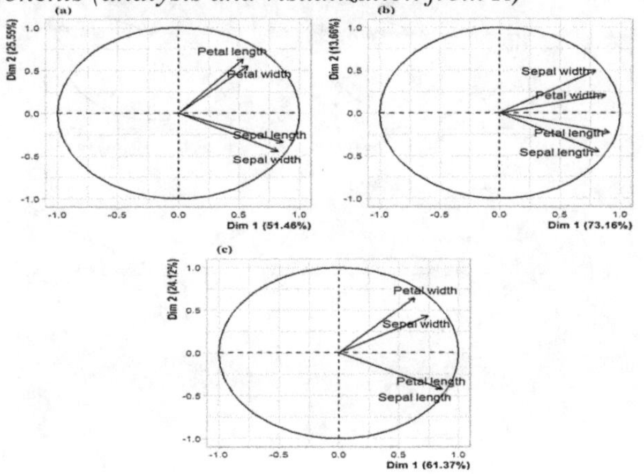

Figure 7.a presents the results of S7, Figure 7.b presents the results of S8, Figure 7.c presents the results of S9, and Figure 7.d presents the results of S10, respectively.

Figure 7. PCA loading plots of strategies S7 (a), S8 (b), S9 (c) and S10 (d). The axes represent the first two principal components (principal component 1-Dim 1, principal component 2-Dim 2) with the corresponding percentages of the variance explained by each principal component and the vectors (arrows) represent the loadings of each variable on the principal components (analysis and visualization from R)

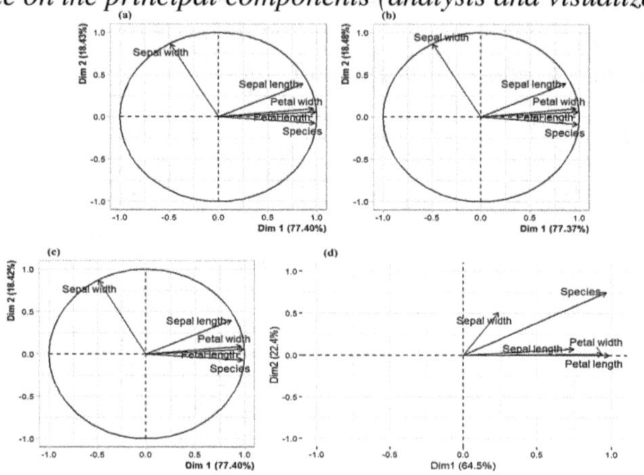

Figure 8 presents the results of S11.

Figure 8. PCA loading plot of strategy S11. The axes represent the first two principal components (principal component 1-Dim 1, principal component 2-Dim 2) with the corresponding percentages of the variance explained by each principal component and the vectors (arrows) represent the loadings of each variable per "slice" and their "mergers" (consensus) in Dim 1 and Dim 2 (analysis and visualization from XLSTAT)

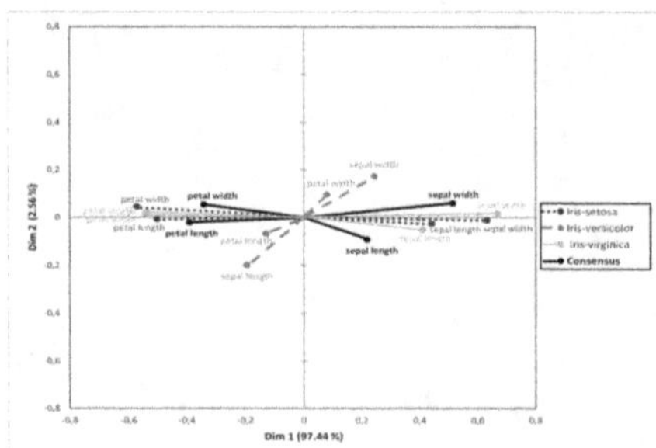

Figures 9.a and 9.b present the results of S12. In Figure 9.a, the variable of "slices" (Species) is presented as a supplementary variable, while in Figure 9.b, the three "slices" (iris-setosa, iris-versicolor, and iris-virginica) are shown as three

supplementary points. Figures 9.c and 9.d present the results of S13 and S13.1, respectively.

Figure 9. PCA loading plots of strategies S12, S13 and S13.1. In plot (a) the variable of "slices" is presented as a supplementary variable, while in plot (b) all the three supplementary "slices" are presented. The axes represent the first two principal components (principal component 1-Dim 1, principal component 2-Dim 2) with the corresponding percentages of the variance explained by each principal component and the vectors (arrows) represent the loadings of each variable on the principal components (analysis and visualization from R)

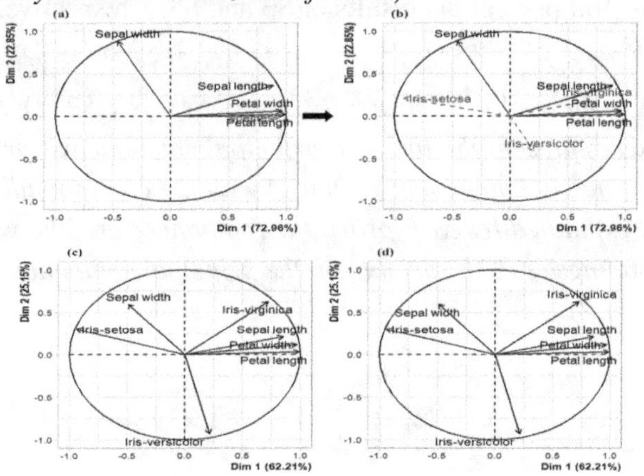

The results of S4 and S5 showed that the three "slices" of the "Iris Plants" data cube were well-separated with each other.

The percentages of total inertia explained by the first two principal components for S1, S1.1 were 95.81% and 95,65%, respectively, while for S2 and S2.1 was 100%. For S3 and S3.1 the percentages of total inertia explained by the first two principal components were 80.72% and 75.35%, respectively. For the first, second and third "slice" of the "Iris Plants" data cube, the corresponding percentages were 77.03%, 73.16% and 85.49%. Also, for S7, S8, S9 and S10 the percentages of total inertia explained by the first two principal components were 95.83%, 95.85%, 95.82% and 86.90%, respectively. For S11 the percentage of total inertia explained by the first two principal components was 100%, while for S13 and S13.1 the corresponding percentage was 87.36%.

Regarding the correlations between the variables, S1, S2 and S2.1, as well as S7, S8 and S9 showed a common structure between them. Additionally, S3, S3.1 and all three "slices" also shared a common structure. Also, S13 and S13.1 showed a common structure, while the remaining strategies differed.

Regarding the contribution of variables in the first two principal components, S1, S1.1 and S2.1, as well as S7, S8 and S9 showed a common structure, respectively. In addition, S2 and S3 also shared a common structure. Also, S13 and S13.1 showed a common structure. In contrast, the remaining strategies differed.

Results of the "ADONUT" Dataset

The PCA loading diagrams for all strategies applied to the data cube **D** are presented below (Figures 10-15). Specifically, Figures 10.a and 10.b present the results of S1 and S1.1, Figures 10.c and 10.d present the results of S2 and S2.1, and Figures 10.e and 10.f present the results of S3 and S3.1, respectively.

Figure 10. PCA loading plots of strategies S1 (a), S1.1 (b), S2 (c), S2.1 (d), S3 (e), S3.1 (f). The axes represent the first two principal components (principal component 1-Dim 1, principal component 2-Dim 2) with the corresponding percentages of the variance explained by each principal component and the vectors (arrows) represent the loadings of each variable on the principal components (analysis and visualization from R)

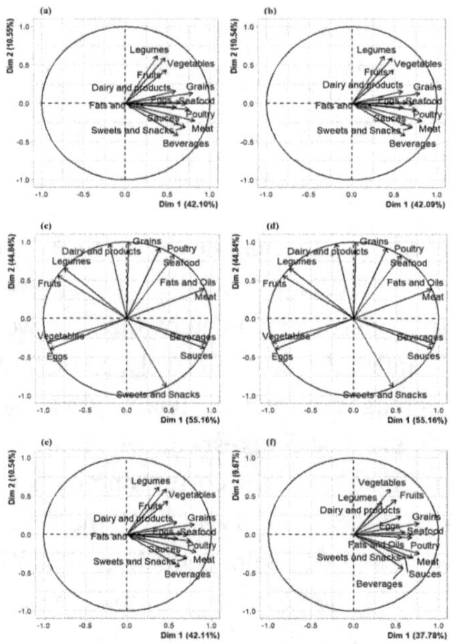

Figures 11.a and 11.b present the results of S4 and S5, respectively.

Figure 11. Scatter plot of strategies S4 (a) and S5(b). The axes represent the first two principal components (principal component 1-Dim 1, principal component 2-Dim 2) with the corresponding percentages of the variance explained by each principal component. Each point represents an object (experimental unit), the different colors and markers represent the "slice" to which each object belongs, and the bolded points in plot (b) represent the centroids of each "slice" (analysis and visualization from R)

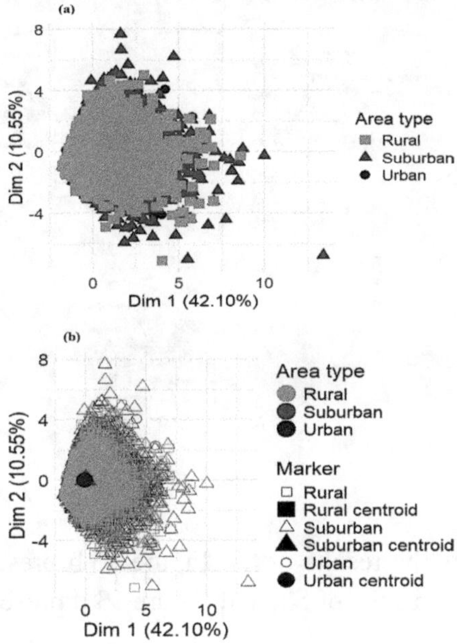

Figure 12 presents the results of S6. More specifically, Figure 12.a corresponds to the first "slice" (urban), Figure 12.b corresponds to the second "slice" (suburban) and Figure 12.c corresponds to the third "slice" (rural) of the data cube.

Figure 12. PCA loading plots of strategy S6. Plot (a) corresponds to the first "slice" (urban), plot (b) corresponds to the second "slice" (suburban) and plot (c) corresponds to the third "slice" (rural). The axes represent the first two principal components (principal component 1-Dim 1, principal component 2-Dim 2) with the corresponding percentages of the variance explained by each principal component and the vectors (arrows) represent the loadings of each variable on the principal components (analysis and visualization from R)

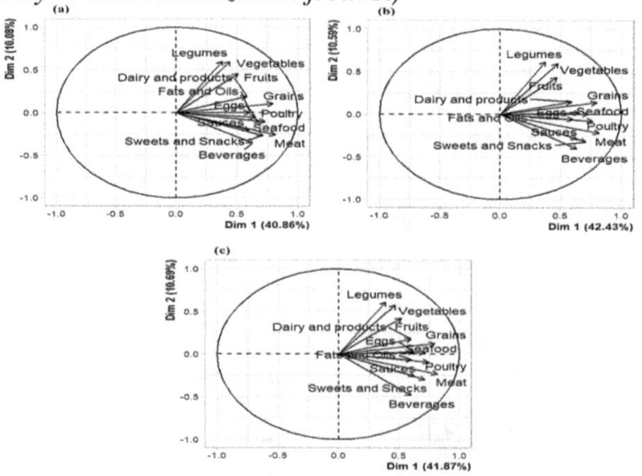

Figure 13.a presents the results of S7, Figure 13.b presents the results of S8, Figure 13.c presents the results of S9, and Figure 13.d presents the results of S10, respectively.

Figure 13. PCA loading plots of strategies S7 (a), S8 (b), S9 (c) and S10 (d). The axes represent the first two principal components (principal component 1-Dim 1, principal component 2-Dim 2) with the corresponding percentages of the variance explained by each principal component and the vectors (arrows) represent the loadings of each variable on the principal components (analysis and visualization from R)

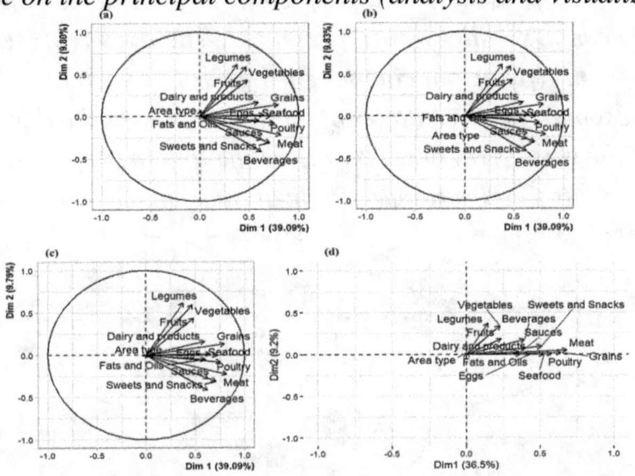

Figure 14 presents the results of S11.

Figure 14. PCA loading plots of strategy S11. The axes represent the first two principal components (principal component 1-Dim 1, principal component 2-Dim 2) with the corresponding percentages of the variance explained by each principal component and the vectors (arrows) represent the loadings of each variable per "slice" and their "mergers" (consensus) in Dim 1 and Dim 2 (analysis and visualization from XLSTAT)

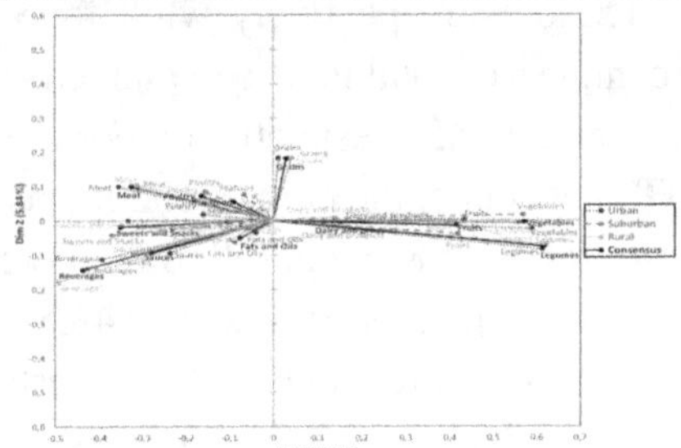

Figures 15.a and 15.b present the results of S12. In Figure 15.a, the variable of "slices" (Area type) is presented as a supplementary variable, while in Figure 15.b,

the three “slices” (urban, suburban, and rural) are shown as three supplementary points. Figures 15.c and 15.d present the results of S13 and S13.1, respectively.

Figure 15. PCA loading plots of strategies S12, S13 and S13.1. In plot (a) the variable of “slices” is presented as a supplementary variable, while in plot (b) all the three supplementary “slices” are presented. The axes represent the first two principal components (principal component 1-Dim 1, principal component 2-Dim 2) with the corresponding percentages of the variance explained by each principal component and the vectors (arrows) represent the loadings of each variable on the principal components (analysis and visualization from R)

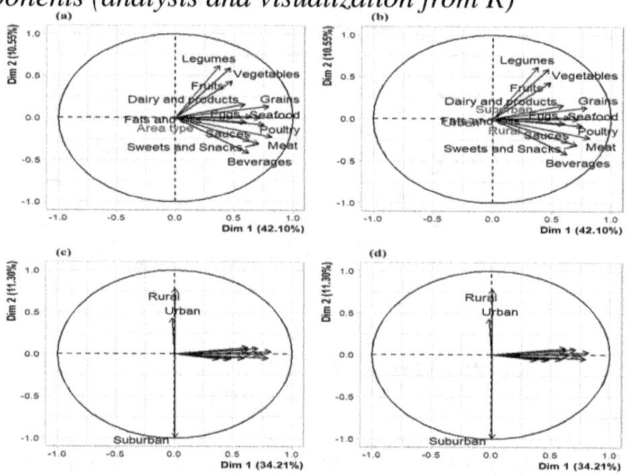

The results of S4 and S5 showed that the three “slices” of the “ADONUT” data cube were overlapping with each other.

The percentages of total inertia explained by the first two principal components for S1, S1.1 were 52.65% and 52.63%, respectively, while for S2 and S2.1 it was 100%. For S3 and S3.1 the percentages of total inertia explained by the first two principal components were 52.65% and 47.45%respectively. For the first, second and third “slice” of the “ADONUT” data cube, the corresponding percentages were 50.94%, 53.02% and 52.56%. Also, for S7, S8, S9 and S10 the percentages of total inertia explained by the first two principal components were 48.89%, 48.92%, 48.88% and 45.66%, respectively. For S11 the percentage of total inertia explained by the first two principal components was 100%, while for S13 and S13.1 the corresponding percentage was 47.51%.

Regarding the correlations between the variables, S1, S1.1, S3, and S3.1, as well as the first and third “slices” displayed a common structure. Additionally, S2 and S2.1 showed a common structure between them, while S13, S13.1, and the first "slice" also shared a common structure. In contrast, the remaining strategies differed in terms of their correlation structures.

Regarding the contribution of variables in the first two principal components, S1, S1.1, S3, S3.1 and the three "slices" of the "ADONUT" data cube showed a common structure. S2 and S2.1 also shared a common structure. Similarly, S13 and S13.1 showed a common structure. In addition, strategies S7, S8, S9 and S10 showed a common structure between them, while the structure of strategy S11 differed.

DISCUSSION

Traditionally, the two datasets could be analyzed with the ANOVA (multiple one-way ANOVAs) or with the Multivariate ANOVA (MANOVA) method. Specifically, ANOVA or MANOVA examine the mean differences between "slices" across one or more dependent variables, as well as their interactions. While both ANOVA and MANOVA provide valuable information about how "slices" differ across the variables, they focus primarily on comparing the "slices" means and detecting overall multivariate differences. In contrast, the proposed strategies provide deeper insights into the structure and components of variability through visualizations in lower dimensions resulting from the application of PCA. Furthermore, the proposed strategies can be used complementary with the ANOVA (multiple one-way ANOVAs) or MANOVA methods.

The two datasets, "Iris Plants" and "ADONUT", differed significantly in both data cube structure and sample size, which led to notable differences in the results of the proposed analysis strategies. Specifically, regarding the total inertia explained by the first two principal components, all strategies for the "Iris Plants" data cube showed higher percentages of total inertia, except of S2, S2.1, and S11. In both "Iris Plants" and "ADONUT" data cube, these strategies displayed a total inertia percentage of 100%. Regarding the correlations between variables and their contributions to the first two principal components, notable differences emerged between the "Iris Plants" and "ADONUT" data cubes. In addition, in the "Iris Plants" data cube, the "slices" were well-separated, whereas in the "ADONUT" data cube, the "slices" were completely overlapped.

Particularly, based on the results of "Iris Plants" dataset it seemed that the structure and components of the total variability and the variability between-"slices" differed slightly after balancing the effect of "slices" (S1-S1.1 and S2-S2.1). The structure and components of the variability within-"slices" differed slightly between the removal of the "linear" and non-"linear" effect of "slices" (S3-S3.1). Furthermore, removing the effect of "slices" affected the structure and components of total variability (S1, S1.1-S3, S3.1), while the structure and components of total variability and variability between-"slices" remained almost constant (S1, S1.1-S2, S2.1). The second and third "slice" showed a common structure and components of variability, unlike the first "slice" (S6). Also, there was a clear separation of the first "slice" from the other two "slices" along the first principal component, indicating that it has distinct characteristics (S12). The inclusion of "slices" in the analysis as pseudo-variables did not affect the overall structure and components of the variability, each "slice" was correlated with distinct patterns across the measured variables, indicating that "slices" were correlated with specific combinations of petal and sepal dimensions (S13).

Based on the results of "ADONUT" dataset it seemed that the structure and components of the total variability and the variability between-"slices" remained constant, even after balancing the effect of "slices" (S1-S1.1 and S2-S2.1). The structure and components of the variability within-"slices" remained constant, either after removing the "linear" or non-"linear" effect of "slices" (S3-S3.1). Furthermore, removing the effect of "slices" did not affect the structure and components of total variability (S1, S1.1-S3, S3.1), while the structure and components of total inertia and inertia between-"slices" differed (S1, S1.1-S2, S2.1). All three "slices" showed a common structure and components of variability (S6). Also, the three "slices" were very close to the origin (low contribution to the principal components) and uncorrelated with each other and with all variables (S12). The inclusion of "slices" in the analysis as pseudo-variables affected the overall structure and components of the variability, while "slices" were uncorrelated with all variables (S13).

For both datasets, the inclusion of "slices" variable in the analysis as a quantitative variable (S7, S9) and as a categorical variable by applying CATPCA (S8) did not affect the overall structure and components of variability. On the contrary, the inclusion of "slices" variable in the analysis as a categorical variable with the application of FAMD, affected the overall structure and components of variability (S10). In addition, the application of GPA changed the structure and components of variability of the data cubes (S11) by redistributing variability among the principal components, shifting the contribution of some variables to the principal components, changing the groupings among some variables, and increasing the positive correlations of the grouped variables. Balancing the contribution of pseudo-variables to the analysis did not play a role (S13-S13.1).

CONCLUSIONS

Based on the above, it can be concluded that in "ADONUT" dataset, the data cube "slices" do not significantly affect the overall structure of the data cube, while in "Iris Plants" dataset, the data cube "slices" significantly affect the overall structure of the data cube. Therefore, in "ADONUT" dataset and generally in similar data cubes, where "slices" do not significantly affect the overall structure of the data cube, PCA can be applied to the overall dataset without considering the variable of "slices".

The proposed strategies allow for the exploration of the overall data structure, capturing patterns of variability across all variables simultaneously. In addition, some of these strategies capture total variability, as well as within-"slices" variability and between-"slices" variability, in a more integrated manner by handling the variable of "slices" in different ways, depending on the specific aims of each study. This integrated approach allows for a more comprehensive understanding of the data's underlying structure, going beyond simple comparisons of "slices" and revealing complex relationships between variables and data cube "slices".

The originality of this study lies in the comparison of 13 different dimensionality reduction strategies applied to two distinct types of real (not simulated) data cubes with independent "slices". The study also focuses on the different management and coding of the "slices" variable, and the integration of different analysis methods, leading finally to the application of PCA.

The contribution of this study to the existing literature lies in highlighting the importance of applying a variety of analysis strategies tailored to the study's aims. Furthermore, the application of various analysis strategies as opposed to applying individual dimensionality reduction methods broadens the range of potential applications of each method and, by extension, extracts more interesting results.

For future research, we propose the development of new and innovative analysis strategies on data cubes with independent "slices" for dimensions $p \geq 4$. These strategies can be applied on both real and simulated data cubes with independent "slices".

This study applied and compared 13 dimensionality reduction strategies to two specific datasets. Both the datasets and the strategies used were limited to the scope of this study. Therefore, caution should be exercised when generalizing the findings to other contexts or datasets.

ACKNOWLEDGEMENTS

Competing Interests:

The authors of this publication declare there are no competing interests.

Funding:

This research received no specific grant from any funding agency in the public, commercial, or not-for-profit sectors. Funding for this research was covered by the author(s) of the article.

REFERENCES

Acar, E., & Yener, B. (2009). Unsupervised Multiway Data Analysis: A Literature Survey. *IEEE Transactions on Knowledge and Data Engineering*, *21*(1), 6–20. DOI: 10.1109/TKDE.2008.112

Anderson, T. W. (1984). *An Introduction to Multivariate Statistical Analysis* (2nd ed.). John Willey and Sons Inc.

Bellman, R. E. (1961). *Adaptive Control Processes: A Guided Tour*. Princeton University Press. DOI: 10.1515/9781400874668

Burgas, L., Melendez, J., Colomer, J., Massana, J., & Pous, C. (2018). N-dimensional extension of unfold-PCA for granular systems monitoring. *Engineering Applications of Artificial Intelligence*, *71*, 113–124. DOI: 10.1016/j.engappai.2018.02.013

Codesido, S., Hanafi, M., Gagnebin, Y., González-Ruiz, V., Rudaz, S., & Boccard, J. (2021). Network principal component analysis: A versatile tool for the investigation of multigroup and multiblock datasets. *Bioinformatics (Oxford, England)*, *37*(9), 1297–1303. DOI: 10.1093/bioinformatics/btaa954 PMID: 33165510

Costa, E., Papatsouma, I., & Markos, A. (2022). Benchmarking distance-based partitioning methods for mixed-type data. *Advances in Data Analysis and Classification*, *17*(3), 701–724. DOI: 10.1007/s11634-022-00521-7

Cunningham, J. P., & Ghahramani, Z. (2015). Linear dimensionality reduction: Survey, insights, and generalizations. *Journal of Machine Learning Research*, *16*(89), 2859–2900. DOI: 10.48550/arXiv.1406.0873

Dash, M., Liu, H., & Yao, J. (1997). Dimensionality reduction of unsupervised data. In Anon (Ed.), *Proceedings of the International Conference on Tools with Artificial Intelligence* (pp. 532-539). IEEE. https://doi.org/DOI: 10.1109/TAI.1997.632300

De Roover, K., Ceulemans, E., Timmerman, M. E., Nezlek, J. B., & Onghena, P. (2013). Modeling differences in the dimensionality of multiblock data by means of clusterwise simultaneous component analysis. *Psychometrika*, *78*(4), 648–668. DOI: 10.1007/s11336-013-9318-4 PMID: 24092482

Draganov, A., & Dohn, S. (2023). Unexplainable Explanations: Towards Interpreting tSNE and UMAP Embeddings. *arXiv (Cornell University)*. https://doi.org//arxiv.2306.11898DOI: 10.48550

Earl, R., & Nicholson, J. (2021). *The Concise Oxford Dictionary of Mathematics*. Oxford University Press. DOI: 10.1093/acref/9780198845355.001.0001

Eslami, A., Qannari, E. M., Kohler, A., and Bougeard, S. (2013). General overview of methods of analysis of multi-group datasets. *Revue des Nouvelles Technologies de l'Information. 25*.

Fabrigar, L. R., Wegener, D. T., MacCallum, R. C., & Strahan, E. J. (1999). Evaluating the use of exploratory factor analysis in psychological research. *Psychological Methods*, *4*(3), 272–299. DOI: 10.1037/1082-989X.4.3.272

Fisher, R. A. (1936). The use of multiple measurements in taxonomic problems. *Annals of Eugenics*, *7*(2), 179–188. DOI: 10.1111/j.1469-1809.1936.tb02137.x

Gifi, A. (1990). *Nonlinear Multivariate Analysis*. John Willey and Sons Inc.

Gilbert, R., Lafferty, R., Hagger-Johnson, G., Harron, K., Zhang, L. C., Smith, P., Dibben, C., & Goldstein, H. (2018). GUILD: Guidance for information about linking data sets. *Journal of Public Health (Oxford, England)*, *40*(1), 191–198. DOI: 10.1093/pubmed/fdx037 PMID: 28369581

Gower, J. C. (1975). Generalized procrustes analysis. *Psychometrika*, *40*(1), 33–51. DOI: 10.1007/BF02291478

Grammatikopoulou, M. G., Poulimeneas, D., Gounitsioti, I. S., Gerothanasi, K., Tsigga, M., & Kiranas, E.ADONUT Study Group. (2014). Prevalence of simple and abdominal obesity in Greek adolescents: The ADONUT study. *Clinical Obesity*, *4*(6), 303–308. DOI: 10.1111/cob.12070 PMID: 25826159

Greenacre, M., Groenen, P. J. F., Hastie, T., Iodice D'Enza, A., Markos, A., & Tuzhilina, E. (2022). Principal component analysis. *Nature Reviews. Methods Primers*, *2*(1), 100. DOI: 10.1038/s43586-022-00184-w

Greenacre, M. J., & Blasius, J. (2006). Multiple Correspondence Analysis and Related Methods. In *Chapman and Hall/CRC eBooks*. https://doi.org/DOI: 10.1201/9781420011319

Hair, J. F., Black, W. C., Babin, B. J., & Anderson, R. E. (2010). *Multivariate Data Analysis: A Global Perspective* (7th ed.). Pearson Education Inc.

Han, J. W., Kamber, M., & Pei, J. (2012). *Data Mining Concepts and Techniques* (3rd ed.). Morgan Kaufmann Publishers.

Hendrickson, J. L. (2014). *Methods for Clustering Mixed Data*. Doctoral dissertation, University of South Carolina, Columbia.

Hinton, G. E., & Roweis, S. T. (2002). Stochastic neighbor embedding. *Advances in neural information processing systems, 15*, 857-864. https://papers.neurips.cc/paper/2002/file/6150ccc6069bea6b5716254057a194ef-Paper.pdf

Husson, F., & Josse, J. (2013). Handling missing values in multiple factor analysis. *Food Quality and Preference, 30*(2), 77–85. DOI: 10.1016/j.foodqual.2013.04.013

Jolliffe, I. T. (2002). *Principal component analysis* (2nd ed.). Springer.

Jolliffe, I. T., & Cadima, J. (2016). Principal component analysis: A review and recent developments. *Philosophical Transactions. Series A, Mathematical, Physical, and Engineering Sciences, 374*(2065), 20150202. DOI: 10.1098/rsta.2015.0202 PMID: 26953178

Kaiser, H. F. (1958). The varimax criterion for analytic rotation in factor analysis. *Psychometrika, 23*(3), 187–200. DOI: 10.1007/BF02289233

Kassambara, A. (2017). *Practical Guide to Cluster Analysis in R: Unsupervised Machine Learning* (Vol. 1). STHDA.

Kitchen, N. R., & Clay, S. A. (2016). Understanding and Identifying Variability. *Agronomy, Horticulture and Plant Science Faculty Publications. 69*. https://openprairie.sdstate.edu/plant_faculty_pubs/69

Kolda, T. G., & Bader, B. W. (2009). Tensor Decompositions and Applications. *SIAM Review, 51*(3), 455–500. DOI: 10.1137/07070111X

Kroonenberg, P., Harshman, R. A., & Murakami, T. (2009). Analysing Three-way Profile Data Using the Parafac and Tucker3 Models Illustrated with Views on Parenting. *Applied Multivariate Research, 13*(1), 5–41. DOI: 10.22329/amr.v13i1.2833

Kroonenberg, P. M. (2008). *Applied Multiway Data Analysis*. Wiley Series in Probability and Statistics. DOI: 10.1002/9780470238004

Kyrana, Z., Pratsinakis, E. D., Papafilippou, N., Markos, A., Dordas, C., & Menexes, G. C. (in press). Evaluating dimensionality reduction strategies on mixed-type datasets: A comparative analysis using Python, R and SPSS. *International Journal of Sustainable Agricultural Management and Informatics*.

Li, Z., Cao, J., & Gu, Q. (2015). Temporal-Aware QoS-Based Service Recommendation using Tensor Decomposition. [IJWSR]. *International Journal of Web Services Research*, *12*(1), 62–74. DOI: 10.4018/IJWSR.2015010105

Linting, M., Meulman, J. J., Groenen, P. J., & Van der Koojj, A. J. (2007). Nonlinear principal components analysis: Introduction and application. *Psychological Methods*, *12*(3), 336–358. DOI: 10.1037/1082-989X.12.3.336 PMID: 17784798

Machine Learning Repository, U. C. I. *Covertype Data Set*. Available online: https://archive.ics.uci.edu/dataset/53/iris (accessed on 11 November 2024).

Markos, A., Moschidis, O., & Chadjipantelis, T. (2020). Sequential dimension reduction and clustering of mixed-type data. *Int. J. Data Analysis Techniques and Strategies*, *12*(3), 228–246. DOI: 10.1504/IJDATS.2020.108043

McInnes, L., & Healy, J. (2018). UMAP: uniform manifold approximation and projection for dimension reduction. *arXiv (Cornell University)*. https://doi.org//arxiv.1802.03426DOI: 10.48550

Menexes, G. (2006). *Experimental Designs in Data Analysis*. Doctoral dissertation, University of Macedonia, Thessaloniki, Greece (in Greek).

Messaoud, R. B., Boussaïd, O., & Loudcher-Rabaseda, S. (2007). A Multiple Correspondence Analysis to Organize Data Cubes. *Frontiers in Artificial Intelligence and Applications*, *155*(1), 133–146. https://halshs.archives-ouvertes.fr/halshs-00476483

Mothobi, N., Van Ryneveld, L., & Graham, M. A. (2021). A Cross-Country Comparison of Mathematics Teachers' Beliefs About Technology in Education. [IJICTE]. *International Journal of Information and Communication Technology Education*, *17*(4), 1–13. DOI: 10.4018/IJICTE.20211001.oa13

Nguyen, L. H., & Holmes, S. (2019). Ten quick tips for effective dimensionality reduction. *PLoS Computational Biology*, *15*(6), e1006907. DOI: 10.1371/journal.pcbi.1006907 PMID: 31220072

OECD. (2008). *OECD Glossary of Statistical Terms*. OECD Publishing.

Pagès, J. (2004). Analyse Factorielle de Données Mixtes. *Revue de Statistique Appliquee*, *52*(4), 93–111. http://archive.numdam.org/item/RSA_2004__52_4_93_0.pdf

Pagès, J. (2014). *Multiple Factor Analysis by Example Using R*. CRC Press. DOI: 10.1201/b17700

Pavoine, S., Blondel, J., Baguette, M., & Chessel, D. (2007). A new technique for ordering asymmetrical three-dimensional data sets in ecology. *Ecology*, *88*(2), 512–523. DOI: 10.1890/05-1806 PMID: 17479768

Rencher, A. C., & Christensen, W. F. (2012). *Methods of Multivariate Analysis* (3rd ed.). John Willey and Sons Inc. DOI: 10.1002/9781118391686

Rotari, M., Diaz, V. F., De Ketelaere, B., & Kulahci, M. (2024). An extension of PARAFAC to analyze multi-group three-way data. *Chemometrics and Intelligent Laboratory Systems*, *246*, 105089. DOI: 10.1016/j.chemolab.2024.105089

Sharma, S. (1996). *Applied Multivariate Techniques*. John Willey and Sons Inc.

Smilde, B. J., Botman, E., de Ruiter, R. D., Smit, J. M., Teunissen, B. P., Lubbers, W. D., Schwarte, L. A., Schober, P., & Eekhoff, E. M. W. (2022). Monitoring and Management of Fibrodysplasia Ossificans Progressiva: Current Perspectives. *Orthopedic Research and Reviews*, *14*, 113–120. DOI: 10.2147/ORR.S337491 PMID: 35480068

Song, Y., Li, Y., & Qu, J. (2018). A new approach for supervised dimensionality reduction. *International Journal of Data Warehousing and Mining*, *14*(4), 20–37. DOI: 10.4018/IJDWM.2018100102

Suh, I., & Chung, Y. D. (2016). A Workload Assignment Strategy for Efficient ROLAP Data Cube Computation in Distributed Systems. *International Journal of Data Warehousing and Mining*, *12*(3), 51–71. DOI: 10.4018/IJDWM.2016070104

Sun, W., & Braatz, R. D. (2020). Opportunities in tensorial data analytics for chemical and biological manufacturing processes. *Computers & Chemical Engineering*, *143*, 107099. DOI: 10.1016/j.compchemeng.2020.107099

Tabachnick, B. G., & Fidell, L. S. (2007). *Using Multivariate Statistics* (5th ed.). Allyn and Bacon/Pearson Education.

Tenenhaus, A., & Tenenhaus, M. (2014). Regularized generalized canonical correlation analysis for multiblock or multigroup data analysis. *European Journal of Operational Research*, *238*(2), 391–403. DOI: 10.1016/j.ejor.2014.01.008

Timmerman, M. E., & Kiers, H. A. L. (2003). Four simultaneous component models for the analysis of multivariate time series from more than one subject to model intraindividual and interindividual differences. *Psychometrika*, *68*(1), 105–121. DOI: 10.1007/BF02296656

Van der Kooij, A. J., Meulman, J. J., & Heiser, W. J. (2006). Local minima in categorical multiple regression. *Computational Statistics & Data Analysis*, *50*(2), 446–462. DOI: 10.1016/j.csda.2004.08.009

Vichi, M., Vicari, D., & Kiers, H. A. L. (2019). Clustering and dimension reduction for mixed variables. *Behaviormetrika*, *46*(2), 243–269. DOI: 10.1007/s41237-018-0068-6

Weisstein, E. W. (2024). Data Cube. From MathWorld-A Wolfram Web Resource. https://mathworld.wolfram.com/DataCube.html

KEY TERMS AND DEFINITIONS

Multidimensional: A dataset with multiple dimensions.

Data Cube: A p-dimensional data structure, where $p \geq 3$.

"Slice": A submatrix of a data cube.

Multi-Block: A data cube with independent "slices", where each "slice" shares the same set of objects on different blocks of variables.

Multi-Group: A data cube with independent "slices", where each "slice" shares the same set of variables on different groups of objects.

Unfolding: The transformation of data cubes with dependent "slices" into data matrices.

Concatenation: The transformation of data cubes with independent "slices" into data matrices.

Variability: Heterogeneity of the data.

Chapter 3
Clustering Mixed-Type Data: A Comparative Study Using Python and SPSS

Emmanouil D. Pratsinakis
https://orcid.org/0000-0002-3725-3525
Aristotle University of Thessaloniki, Greece

Lefkothea Karapetsi
https://orcid.org/0000-0001-5222-141X
Institute of Applied Biosciences, Thessaloniki, Greece

Zacharenia Kyrana
https://orcid.org/0000-0001-9269-0675
Aristotle University of Thessaloniki, Greece

Christos Dordas
https://orcid.org/0000-0002-7027-474X
Aristotle University of Thessaloniki, Greece

Nikolaos Papafilippou
https://orcid.org/0009-0003-3148-7229
Aristotle University of Thessaloniki, Greece

Angelos Markos
https://orcid.org/0000-0002-4204-3573
Democritus University of Thrace, Greece

ABSTRACT

A multidimensional and multivariate structure with mixed-type data allows researchers to use various statistical approaches and data clustering techniques. The choice of clustering method used can have an impact on the results obtained In this study, the Partitioning Clustering (k-means) and Hierarchical Cluster Analysis methods were compared. The main objective of this study was to apply and compare these methods in segmenting the data set into groups-clusters. The results of the study showed that there are several analysis strategies for clustering mixed-type data, based on data coding and the chosen measurement scale for input variables. Remarkably, Python

DOI: 10.4018/979-8-3693-9400-7.ch003

delivered results over 100% faster than SPSS. In conclusion, it emerged that the results of the classification depend on the coding strategy and the selection of the measurement scale of the variables to be used in the analysis.

INTRODUCTION

Mixed-type data sets include variables measured on nominal, ordinal, and/or quantitative scales, whether continuous or discrete (Florou, 1997; Menexes, 2006). Such data is used across diverse scientific disciplines, including Biology, Social Sciences, Economics, and Health Sciences (Ahmad & Dey, 2007; Ahmad & Khan, 2019; Hennig & Liao, 2013; Morlini & Zani, 2010).

Despite the widespread use of mixed-type data, integrating them seamlessly into the same multivariate-multidimensional analysis poses challenges. To overcome this problem, one can either convert the categorical data into quantitative using optimal scaling methods, as proposed by the Dutch School of Data Analysis-Gifi System, (Menexes, 2006; Moschidis et al., 2023; van Rijckevorsel & de Leeuw, 1988; Young, 1981), or transform quantitative data into categorical data as suggested by the French School of Data Analysis (Menexes, 2006; Moschidis et al., 2023; Papadimitriou, 1994). Both approaches aim to achieve dataset homogeneity (Moschidis et al., 2023; Pagès, 2004). However, these data set conversions may lead to a new dataset with reduced information relative to the original data set (Menexes, 2006).

Clustering methods are used to group entities (like plants, animals, people, events, microorganisms, and others) into clusters, in a way that individuals with common characteristics are grouped together, while individuals with different characteristics are grouped into different clusters (Ahmad & Khan, 2019; Jain and Dubes, 1988; Markos et al., 2020). However, conventional clustering methods often grapple with datasets comprising both quantitative and categorical data (Ahmad & Khan, 2019; Witten & Frank, 2005). This highlights the pressing need to devise innovative analytical methods or strategies tailored to efficiently classify mixed-type data (Ahmad & Khan, 2019).

Notable drawbacks of clustering methods are that the results depend on the distance metric and the aggregation method used, in the sense that different combinations of these parameters can lead to different results (Pratsinakis et al., 2020, 2021). Furthermore, they may not always be effective for large data sets, handling missing values (Hair et al., 2010) and managing extreme values-outliers (Hair et al., 2010; Menexes, 2006). Additionally, the visualization and presentation of results might not always be "useful" to the end user (Menexes, 2006). In many cases, the outputs (be it numerical or diagrammatic) from clustering methods are not straightforward for end users to interpret.

The objectives of this study were to compare the results of two clustering methods, Hierarchical and Partitioning Clustering (or *k*-means), considering various analysis strategies regarding the coding and the measurement scale of input variables, to investigate the application of these methods to multidimensional mixed-type data and to compare the efficiency of two statistical software in terms of result extraction time.

MATERIALS AND METHODS

The clustering techniques employed in this study include the Hierarchical and the Partitioning (or *k*-means) Clustering, implemented with diverse analytical strategies. Specifically, analyses were conducted using: a) the raw data, b) the data derived from the factorial axes after applying the Principal Components Analysis (PCA), c) the data derived from the factorial axes after applying Multiple Correspondence Analysis (MCA), and finally d) the data derived from the factorial axes after applying Non-Linear Categorical Principal Components Analysis with Optimal Scaling Scaling (CATPCA). Using dimensional reduction methods as data preprocessing step prior to cluster analysis is a common strategy, therefore the results of cluster analysis depend on the dimensional reduction method used (Cuell & Bonsal, 2009; Markos et al., 2020).

MATHEMATICAL BACKGROUND

The mathematical foundations of the two clustering methods are detailed below.

Hierarchical Cluster Analysis

The main goal of cluster analysis is to categorize observations into clusters in a way that maximizes intra-cluster similarity based on the defined clustering variables (Everitt, 2011; Hair et al., 2010; Sharma, 1996). The initial stage involves choosing a similarity or dissimilarity measure (examples include squared Euclidean distance, Minkowski, Canberra and others). Following this, a suitable clustering aggregation method aligned with the selected metric is determined (such as single linkage, complete linkage, Ward's method and others) (Everitt, 2011; Hair et al., 2010; Sharma, 1996). Subsequently, a decision is made regarding the appropriate number of clusters (Everitt, 2011; Hair et al., 2010; Sharma, 1996). Lastly, the interpretation of the cluster solution concludes the process.

The squared Euclidean distance is given by the following formula (Sharma, 1996):

$$D_{ij}^2 = \sum_{k=1}^{p} \left(x_{ik} - x_{jk} \right)^2$$

where D_{ij}^2 represents the squared Euclidean distance between subjects i and j, x_{ik} is the value of the kth variable for the ith subject, x_{jk} is the value of the kth variable for the jth subject, and p is the number of variables. As the similarity between subjects increases, their mutual distance decreases, and the reverse holds for subjects that are more dissimilar.

One of the most widely used distance metrics for mixed-type variables (D'Orazio, 2021) (nominal and scale types) is obtained by Gower's similarity coefficient (Gower, 1971). This coefficient involves calculating distances on a variable-by-variable basis and then averaging them, with each distance scaled to span a range from 0 (indicating minimum distance) to 1 (maximum distance). Gower's distance is attractive due to its ability to handle missing values and the flexibility to incorporate user-defined weighting schemes when averaging distances for individual variables (D'Orazio, 2021; Gower, 1971).

Ward's method (Sharma, 1996; Ward, 1963) deviates from the typical approach of computing distances between clusters. Instead, it focuses on maximizing within-cluster homogeneity. This is achieved by using the within-group sum of squares as the measure of homogeneity. In essence, the goal of Ward's method is to minimize the total within-group or within-cluster sums of squares. At each step, clusters are formed in a way that leads to the smallest within-cluster sums of squares, which is also referred to as the error sums of squares (ESS).

In this study, the squared Euclidean distance was used as a similarity measure, and the Ward's method as the aggregation method. Given that our research primarily focused on contrasting the algorithms shared by SPSS and Python, Gower's distance wasn't examined. This decision stemmed from SPSS's lack of support for this particular distance and its associated clustering algorithm.

Partitioning Cluster Analysis

According to Sharma (Sharma, 1996), in nonhierarchical clustering (termed Partitioning Cluster Analysis), the dataset is divided into k partitions or groups, each representing a distinct cluster. Unlike hierarchical clustering, the number of clusters needs to be known in advance. Partitioning clustering techniques generally involve the following steps:

- Choose k initial cluster centroids or seeds, where k represents the desired number of clusters
- Assign each observation to the cluster with the closest centroid

- Reassign or reallocate each observation to one of the *k* clusters based on a predetermined stopping rule
- Terminate the process if no further reallocation of data points occurs or if the reassignment meets the criteria specified by the stopping rule. Otherwise, return to Step 2 and repeat the process

Computer System Specifications

All the described methods were implemented and compared using a PC equipped with an AMD® Ryzen™9 5900X processor @ 3.7GHz (24 CPUs), ~4.8 GHz, and Windows 11 Pro 64-bit as the operating system. The PC used DDR4 technology RAM with a speed of 3200 MHz, with a memory capacity of 64 GB. The hard drive used was an M.2 NVMe PCI Express 4.0 technology with a capacity of 1 TB and read speed of 7000 MB/s and a write speed of 5000 MB/s. The statistical software packages included Python 3.10 via the Anaconda platform, Jupyter Notebook 6.4.5, and IBM SPSS Statistics v26.0. For the sake of brevity, statistical software packages will be referred to as Python, and SPSS, respectively, in the remainder of the text.

Dataset

The "Forest Cover Type" dataset from the UCI Machine Learning Repository (Blackard, 1998) was employed in this study. This dataset encompasses the forest cover type of four wilderness areas within the Roosevelt National Forest in northern Colorado, USA. These four wilderness areas (Rawah, Neota, Comanche Peak, and Cache la Poudre) represent minimally disturbed forests, indicating that the seven forest cover types (Spruce/Fir, Lodgepole Pine, Ponderosa Pine, Cottonwood/Willow, Aspen, Douglas-fir, and Krummholz) present are the product of ecological processes rather than forestry management practices.

The dataset consists of 581,012 objects, obtained from 30m × 30m forest sections and 55 mixed-type variables. Specifically, "Cover Type" is a nominal variable composed of seven categories. "Wilderness Area" and "Soil Type" are nominal variables, with four and 40 categories, respectively, appearing in the dataset as independent binary pseudo-variables (coded as 0 or 1). "Elevation" (m) of the forests and the four variables related to the "Distances of forests from nearby landmarks" (hydrology, roadways, and fire points) (m) are scale variables. The three variables of "Hillshade" at 9 am, noon, and 3 pm used an index ranging from 0 to 255 as a unit of measurement and considered scale variables. Finally, the "Aspect" and "Slope" of the forests are scale variables representing circular data, as their unit of measurement is in degrees (xo) and converted into radians (rad), to facilitate the analysis as scale variables. The conversion from degrees to radians was performed according to the following equation:

$$rad = x^o\ (\pi\ /\ 180^o),\ \pi \cong 3.14$$

RESULTS

Hierarchical Cluster Analysis

Hierarchical Cluster Analysis (using squared Euclidean distance along with Ward's aggregation method) was applied to the ten quantitative variables (more specifically, elevation, slope, vertical and horizontal distance to hydrology, horizontal distance to roadways, hillshade at 9 am, noon and 3 pm, horizontal distance to fire points, and orientation) and to the three significant factor axes resulting from PCA (explaining 64.8% of the total variance). In both cases, as mentioned before, the squared Euclidean distance was used as a measure of dissimilarity between objects, and Ward's criterion was used as the aggregation method. In both analyses Hierarchical Clustering could not be executed in either software platforms, SPSS and Python, due to the large amount of data (mainly due to the large number of objects), as shown in Figure 1.

Figure 1. Warning messages for the inability to perform Hierarchical Clustering in SPSS (a) and Python (b).

Warnings
Insufficient memory for Cluster
Execution of this command stops.

(a)

MemoryError: Unable to allocate 1.23 TiB for an array with shape (168787181566,) and data type float64

(b)

K-Means Clustering

K-means was performed with eight different analysis strategies.

The first strategy was to apply *k*-means to the ten quantitative variables (more specifically. elevation, slope, vertical and the horizontal distance to hydrology, horizontal distance to roadways, hillshade at 9 am, noon and 3 pm, horizontal distance to fire points, and orientation). Using the Elbow Score criterion (Tripathi et al., 2018), as shown in Figure 2, six clusters were identified (C1-C6) that included 24.8%, 3.7%, 17.1%, 12. 5%, 34.3% and 7.5% of the items respectively.

Figure 2. Elbow score diagram for the determination of the number of clusters in k-means, created using Python.

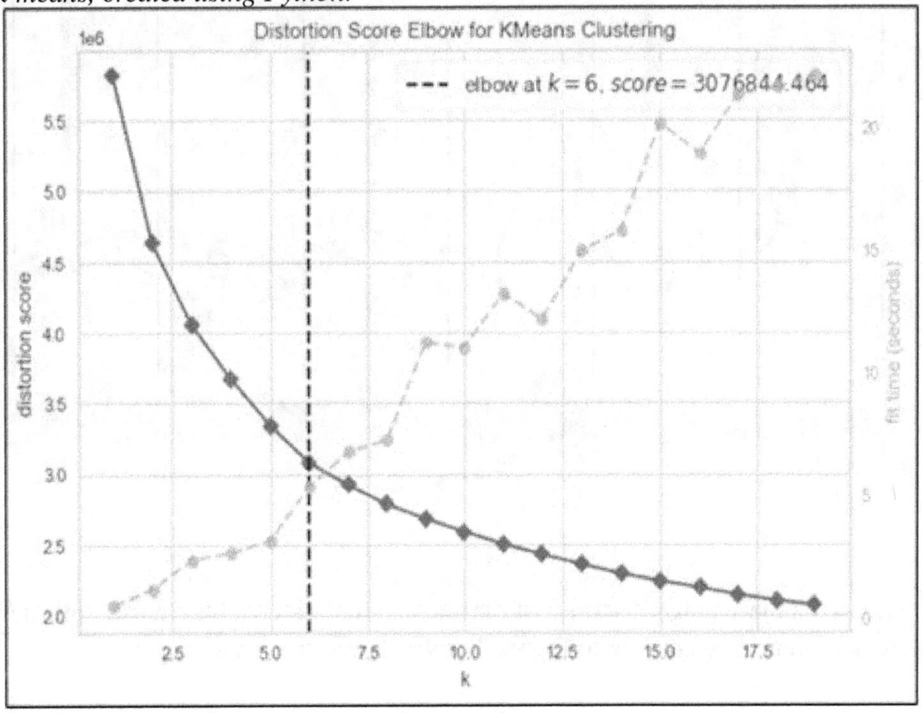

Figure 3. Cluster profiles of the six clusters that resulted from k-means based on the average Z-scores of the ten quantitative variables.

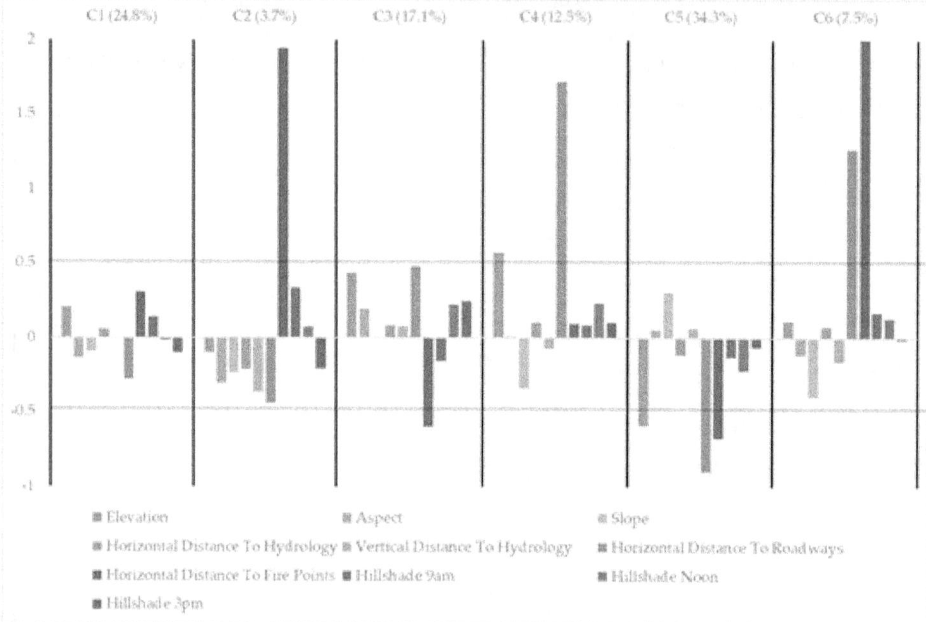

Based on the average Z-scores, of the ten quantitative variables, cluster profiles were determined. According to Figure 3, it was found that on average:

- The ten quantitative variables related to the forest surfaces belonging to cluster C1 have averages that did not differ significantly from the averages in the whole sample
- The forest surfaces belonging to cluster C2 was characterized by a long horizontal distance from the fire points compared to the forest surfaces of the other clusters
- The forest surfaces belonging to cluster C3 was characterized by a short horizontal distance from the fire points compared to the forest surfaces of the other clusters
- The forest surfaces belonging to cluster C4 was characterized by a long horizontal distance to roadways, and low elevation compared to the forest surfaces of the other clusters
- The forest surfaces belonging to cluster C5 was characterized by a short horizontal distance to roadways, a short horizontal distance from the fire points and great elevation compared to the forest surfaces of the other clusters

- The forest surfaces belonging to cluster C6 was characterized by a long horizontal distance to roadways, and a long horizontal distance from the fire points compared to the forest surfaces of the other clusters

Based on the eta square (η^2) coefficient (Michos et al., 2012), as shown in Figure 4, greater contribution to the formation of the clusters had the horizontal distance from the roadways (η^2=0.831) and the horizontal distance to the fire points (η^2=0.776). The runtime in SPSS was 5.7 sec while in Python was 1.2 sec.

Figure 4. Contributions based on the η^2 coefficient of the ten quantitative variables to the formation of the six clusters resulted from k-means.

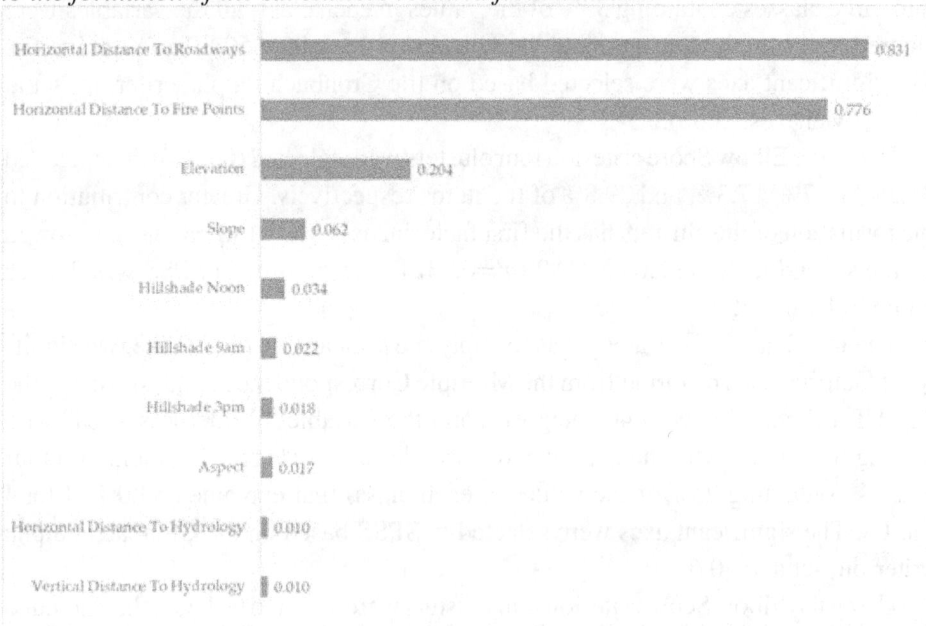

The second analysis strategy was to apply the *k*-means method to the three significant factor axes resulting from the Principal Component Analysis-PCA accounting for 64.8% of total variance. Initially, the analysis was applied to the whole dataset (the ten quantitative variables and the 40-1=39 binary pseudo-variables for "Soil Type" and 4-1=3 binary pseudo-variables for "Wilderness Area"). However, multiple crashes occurred, and then the analysis was applied only to the ten quantitative variables. The significant factor axes were determined by the following selection criteria:

- Selecting factorial axes with a total variance percentage of 60% or more

- Using eigenvalues' diagram (Scree-plot)

Using the Elbow Score criterion five clusters were selected (C1-C5) that included 8.7%, 13.6%, 29.1%, 28.5%, and 20.2% of the items respectively. Greater contribution to the formation of the clusters has the first factorial axis PCA1 (η^2=0.663), followed by the second factorial axis PCA2 (η^2=0.598) and the third factor axe PCA3 (η^2=0.534). The runtime in SPSS was 2.2 sec while in Python was 0.9 sec.

The third analysis strategy was to apply the *k*-means method to the two significant factorial axes resulting from the Multiple Correspondence Analysis-MCA (using the "Soil Type" variable with 40 categories and the variable "Wilderness Area" with 4 categories along with the quantitative variables after categorizing them into three classes containing 33% of the values in each class, all the variables were entered in the analysis as nominal variables) that explained 50.8% of total inertia. The significant axes were selected based on the Cronbach's alpha criterion, set at $a \geq 0.6$ (Menexes, 2006).

Using the Elbow Score criterion four clusters were selected (C1-C4) that included 35.2%, 17.7%, 17.3%, and 29.8% of the items respectively. Greater contribution to the formation of the clusters has the first factorial axis MCA1 (η^2=0.889), followed by the second factorial axis MCA2 (η^2=0.732). The runtime in SPSS was 2.8 sec while in Python it was 0.8 sec.

The fourth analysis strategy was to apply the *k*-means method to the two significant factorial axes resulting from the Multiple Correspondence Analysis (using the "Soil Type" variable with 40 categories and the variable "Wilderness Area" with 4 categories along with the quantitative variables after categorizing them into four classes containing 25% of the values in each class) that explained 54.0% of total inertia. The significant axes were selected in SPSS based on the Cronbach's alpha criterion, set at $a \geq 0.6$.

Using the Elbow Score criterion four clusters were selected (C1-C4) that included 27.3%, 27.2%, 23.8%, and 21.7% of the items respectively. Greater contribution to the formation of the clusters has the first factorial axis MCA1 (η^2=0.851), followed by the second factorial axis MCA2 (η^2=0.738). The runtime in SPSS was 2.8 sec while in Python it was 0.8 sec.

The fifth analysis strategy was to apply the *k*-means method to the two significant factorial axes resulting from the Categorical Principal Components Analysis-CATPCA (using the "Soil Type" variable with 40 categories and the variable "Wilderness Area" with 4 categories, as nominal, along with the quantitative variables, measured as ordinal, after categorizing them into three classes containing 33% of the values in each class) that explained 46.5% of total variance. The significant axes were selected based on the Cronbach's alpha criterion, set at $a \geq 0.6$.

Using the Elbow Score criterion four clusters were selected (C1-C4) that included 28.9%, 19.2%, 14.5%, and 37.4% of the items respectively. Greater contribution to the formation of the clusters has the first factorial axis CATPCA1 (η^2=0.892), followed by the second factorial axis CATPCA2 (η^2=0.702). The runtime in SPSS was 1.8 sec while in Python was 0.9 sec.

The sixth analysis strategy was to apply the *k*-means method to the two significant factorial axes resulting from the CATPCA (using the "Soil Type" variable with 40 categories and the variable "Wilderness Area" with 4 categories, as nominal, along with the quantitative variables, measured as ordinal, after categorizing them into four classes containing 25% of the values in each class) that explained 46.7% of total variance. The significant axes were selected based on the Cronbach's alpha criterion, set at $a \geq 0.6$.

Using the Elbow Score criterion four clusters were selected (C1-C4) that included 37.0%, 25.5%, 16.7%, and 20.9% of the items respectively. Greater contribution to the formation of the clusters has the first factorial axis CATPCA1 (η^2=0.836), followed by the second factorial axis CATPCA2 (η^2=0.687). The runtime in SPSS was 1.8 sec while in Python was 0.9 sec.

The seventh analysis strategy was to apply the *k*-means method to the two significant factorial axes resulting from the CATPCA (using the "Soil Type" variable with 40 categories and the variable "Wilderness Area" with 4 categories, as nominal, along with the quantitative variables after categorizing them into three classes containing 33% of the values in each class) that explained 46.5% of total variance. For the ten quantitative variables, five were considered as been measured in ordinal (after categorized them into classes) and five in nominal scale, the selection was random. The significant axes were selected based on the Cronbach's alpha criterion, set at $a \geq 0.6$.

Using the Elbow Score criterion four clusters were identified (C1-C4) that included 29.6%, 14.4%, 36.5%, and 19.2% of the items respectively. Greater contribution to the formation of the clusters has the first factorial axis CATPCA1 (η^2=0.892), followed by the second factorial axis CATPCA2 (η^2=0.701). The runtime in SPSS was 1.8 sec while in Python was 0.9 sec.

Finally, the eighth analysis strategy was to apply the *k*-means method to the two significant factorial axes resulting from the CATPCA (using the "Soil Type" variable with 40 categories and the variable "Wilderness Area" with 4 categories, as nominal, along with the quantitative variables after categorizing them into four classes containing 25% of the values in each class) that explained 46.7% of total variance. For the ten quantitative variables, five were considered as been measured in ordinal (after categorized them into classes) and five in nominal scale, the selection was random. The significant axes were selected based on the Cronbach's alpha criterion, set at $a \geq 0.6$.

Using the Elbow Score criterion four clusters were selected (C1-C4) that included 38.5%, 15.5%, 20.1%, and 26.1% of the items respectively. Greater contribution to the formation of the clusters has the first factorial axis CATPCA1 (η^2=0.841), followed by the second factorial axis CATPCA2 (η^2=0.689). The runtime in SPSS was 1.8 sec while in Python was 0.9 sec.

Figures 5a and 5b provide an exhaustive presentation of results derived from the eight analytical strategies.

Figure 5. Overall presentation and results of k-means analysis across eight strategies [types of variables involved in the analysis, number of selected clusters based on the Elbow Score criterion, statistical software packages used, execution times and number of objects in each cluster]. Analyses were performed on (a) quantitative variables, three PCA factorial axes, two MCA factorial axes with two different strategies and (b) two CATPCA factorial axes with four different strategies.

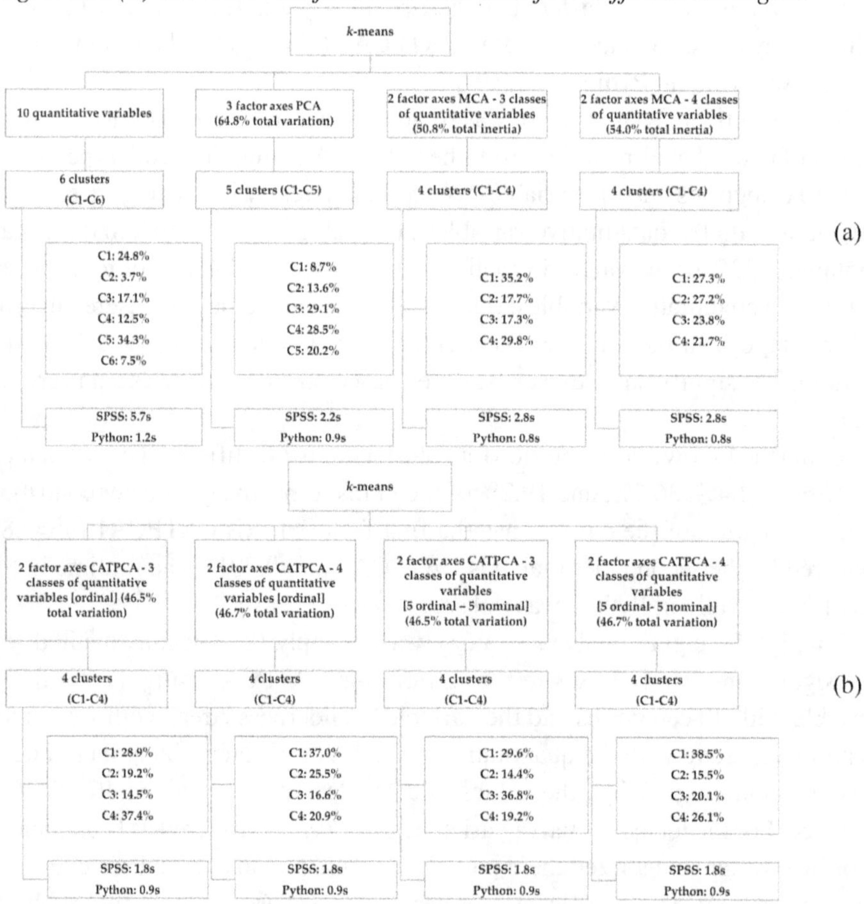

DISCUSSION

The analyses conducted highlight the existence of various strategies for clustering mixed type data sets, regarding the coding and measurement scale of the input variables. However, the composition and the number of clusters were different according to the analysis strategy. Furthermore, the chosen variables, whether raw data or factor axes, lead to different number of clusters each time, different compositions of the clusters, in terms of the items in each cluster, and with different profile according to the average (*Z*-score).

In terms of computational efficiency, Python consistently outperformed SPSS, often executing tasks over 100% faster. It is worth mentioning, however, that the computational system encountered challenges and lags while running methods on both software platforms. Despite this, the outcomes from both software were largely consistent, with only minor discrepancies attributed to numerical rounding.

Concerning the drawbacks of Hierarchical Clustering, it appears that this approach was not viable for application on the particular dataset or datasets of similar size (especially concerning the number of objects) and with the specific PC composition, as both software packages experienced crashes. This limitation poses challenges, especially when visualizing "tall" data through dendrogram construction. Nonetheless, it offers the advantage of estimating the number of clusters by combining different distances and aggregation methods. On the other hand, the main limitation of *k*-means is the inability for users to specify a particular distance and clustering approach.

The challenges highlighted by this study's results, which warrant attention in future research, include the development of novel clustering methods specifically tailored for distance measures suitable for mixed type data (binary, nominal, ordinal, scale, and circular). Another aspect involves comparing clustering methods based on machine learning algorithms, such as the *K*-Neighbors method. Lastly, there's a need for developing tree diagram construction methods that consider only the top nodes of the Hierarchical Clustering when applied to "tall" datasets.

In our efforts to conduct analyses—incorporating both dimensionality reduction techniques and clustering—on a dataset featuring numerous binary variables (at a minimum of 40 variables, notably the 39 "Soil Type" pseudo-variables and the three "Wilderness Area" pseudo-variables), we often encountered computational failures. Moreover, the results did not consistently yield coherent interpretations. The results' variability was also observed depending on the selected reference binary variable. Thus, exploring methods capable of analyzing such datasets without these complications would be beneficial.

CONCLUSIONS

In terms of execution times, Python exhibited a distinct advantage over SPSS. This investigation revealed that at least eight diverse analytical strategies can be adopted for a single data set. These strategies revolve around data coding and the selection of variable measurement scales. Additionally, outcomes were influenced by the chosen dimensionality reduction technique, a prevalent preparatory step in data processing before cluster analysis. This research underscores the importance of employing a range of analytical approaches until an optimal clustering solution is achieved or the robustness of the results is affirmed.

ACKNOWLEDGEMENTS

Competing Interests:

The authors of this publication declare there are no competing interests.

Funding:

This research received no specific grant from any funding agency in the public, commercial, or not-for-profit sectors. Funding for this research was covered by the author(s) of the article.

REFERENCES

Ahmad, A., & Dey, L. (2007). A *k*-mean clustering algorithm for mixed numeric and categorical data. *Data & Knowledge Engineering*, *63*(2), 503–527. DOI: 10.1016/j.datak.2007.03.016

Ahmad, A., & Khan, S. S. (2019). Survey of state-of-the-art mixed data clustering algorithms. *IEEE Access : Practical Innovations, Open Solutions*, *7*, 31883–31902. DOI: 10.1109/ACCESS.2019.2903568

Blackard, J. (1998). Covertype. UCI Machine Learning Repository. DOI: 10.24432/C50K5N

Cuell, C., & Bonsal, B. (2009). An assessment of climatological synoptic typing by principal component analysis and *k*-means clustering. *Theoretical and Applied Climatology*, *98*(3–4), 361–373. DOI: 10.1007/s00704-009-0119-8

D'Orazio, M. (2021). Distances with mixed type variables some modified Gower's coefficients. *arXiv preprint arXiv:2101.02481*.

Everitt, B. (Ed.). (2011). *Cluster analysis* (5th ed.). Wiley. DOI: 10.1002/9780470977811

Florou, G. (1997). *Best distance and best way grouping data determination for various hierarchical clustering agglomerative methods* [Doctoral dissertation, University of Macedonia].

Gower, J. C. (1971). A general coefficient of similarity and some of its properties. *Biometrics*, *27*(4), 857. DOI: 10.2307/2528823

Hair, J. F., Black, W. C., Babin, B. J., & Anderson, R. E. (2010). *Multivariate data analysis: A global perspective* (7th ed.). Pearson.

Hennig, C., & Liao, T. F. (2013). How to find an appropriate clustering for mixed-type variables with application to socio-economic stratification. *Applied Statistics*, *62*(3), 309–369. DOI: 10.1111/j.1467-9876.2012.01066.x

Jain, A. K., & Dubes, R. C. (1988). *Algorithms for clustering data*. Prentice Hall. (Prentice Hall advanced reference series).

Markos, A., Moschidis, O., & Chadjipantelis, T. (2020). Sequential dimension reduction and clustering of mixed-type data. *International Journal of Data Analysis Techniques and Strategies*, *12*(3), 228. DOI: 10.1504/IJDATS.2020.108043

Menexes, G. (2006). *Experimental designs in data analysis* [Doctoral dissertation, University of Macedonia].

Michos, M. C., Menexes, G., Tsadilas, C. D., & Tsatsarelis, C. A. (2012). Energy inputs, outputs and greenhouse gas emissions in organic, integrated and conventional peach orchards. *Ecological Indicators*, *13*(1), 22–28. DOI: 10.1016/j.ecolind.2011.05.002

Morlini, I., & Zani, S. (2009, November). Comparing approaches for clustering mixed mode data: An application in marketing research. In *Data Analysis and Classification: Proceedings of the 6th Conference of the Classification and Data Analysis Group of the SocietàItaliana di Statistica* (pp. 49-57). Berlin, Heidelberg: Springer Berlin Heidelberg.

Moschidis, O., Markos, A., & Chadjipadelis, T. (2023). Hierarchical clustering of mixed-type data based on barycentric coding. *Behaviormetrika*, *50*(1), 465–489. DOI: 10.1007/s41237-022-00165-z

Pagès, J. (2004). Analyse factorielle de données mixtes. *Revue de Statistique Appliquee*, *52*(4), 93–111.

Papadimitriou, G. (1994). *Data analysis methods: University lectures*. University of Macedonia Economics and Social Sciences.

Pratsinakis, E. D., Karapetsi, L., Ntoanidou, S., Markos, A., Madesis, P., Eleftherohorinos, I., & Menexes, G. (2021). Comparison of hierarchical clustering methods for binary data from SSR and ISSR molecular markers. In *Data analysis and rationality in a complex world 16* (pp. 233–241). Springer International Publishing., DOI: 10.1007/978-3-030-60104-1_26

Pratsinakis, E. D., Ntoanidou, S., Polidoros, A., Dordas, C., Madesis, P., Eleftherohorinos, I., & Menexes, G. (2020). Comparison of hierarchical clustering methods for binary data from molecular markers. *International Journal of Data Analysis Techniques and Strategies*, *12*(3), 190–212. DOI: 10.1504/IJDATS.2020.108036

Sharma, S. (1996). *Applied multivariate techniques*. John Wiley & Sons.

Tripathi, S., Bhardwaj, A., & Poovammal, E. (2018). Approaches to clustering in customer segmentation. *International Journal of Engineering & Technology, 7*(3.12), 802. DOI: 10.14419/ijet.v7i3.12.16505

van Rijckevorsel, J. L. A., & de Leeuw, J. (1988). *Component and correspondence analysis: Dimension reduction by functional approximation*. Wiley.

Ward, J. H.Jr. (1963). Hierarchical grouping to optimize an objective function. *Journal of the American Statistical Association, 58*(301), 236–244. DOI: 10.1080/01621459.1963.10500845

Witten, I. H., & Frank, E. (2005). *Data mining: Practical machine learning tools and techniques* (2nd ed.). Morgan Kaufmann.

Young, F. W. (1981). Quantitative analysis of qualitative data. *Psychometrika, 46*(4), 357–388. DOI: 10.1007/BF02293796

Chapter 4
Reducing Subjectivity in Spatial Data Classification for Decision-Making

Thomas M. Koutsos
https://orcid.org/0000-0003-4035-7971
Aristotle University of Thessaloniki, Greece

Thomas Alexandridis
Aristotle University of Thessaloniki, Greece

ABSTRACT

Choropleth maps are essential tools for visualizing spatial data and supporting decision-making across fields such as agriculture, environmental science, urban planning, and public health. However, classifying spatial data into distinct categories introduces subjectivity, affecting both interpretation and utility. This study examines how classification methods influence spatial visualization, emphasizing how biases from common techniques—such as equal interval, quantile, and standard deviation—can distort spatial patterns and decisions. We explore alternative approaches, including Natural Breaks, Box Maps, and Local Indicators of Spatial Association (LISA), which may reduce subjectivity and improve map accuracy. A case study shows how combining these methods with spatial autocorrelation can mitigate bias and enhance analytical reliability. The findings underscore the value of context-aware classification and statistical rigor to ensure transparent, reproducible, and data-driven map interpretations.

DOI: 10.4018/979-8-3693-9400-7.ch004

1. INTRODUCTION

Choropleth maps serve as critical instruments for synthesizing spatial data and informing decision-making processes in fields ranging from urban planning and environmental science to social studies and medical research. While displaying a single parameter, the use of spatial data classifications has already been proved essential across various fields, including agriculture, soil science, geography, environmental, epidemiological and social sciences (Anselin, 2021; Brewer & Pickle, 2002; Sobjak et al., 2023; Viscarra Rossel et al., 2015). They enable effective visualization and analysis of data patterns, supporting targeted interventions, resource allocation, and strategic planning in diverse disciplines (Brewer & Pickle, 2002; Słomska-Przech & Gołębiowska Małgorzata I., 2021; Zhang & MacIejewski, 2017). Therefore, data visualization is critical because it translates numerical values into visual representations, simplifying complex patterns and enabling stakeholders to identify trends and allocate resources effectively (Lavanya et al., 2023). However, the transformation of raw data into classified maps is not an objective endeavor (Fang et al., 2022; Oxoli et al., 2020; Prestby, 2023).

In addition, although it has been clearly noted so far that the choice of classification method impacts the visual interpretation of the underlying pattern of spatial distribution (Coulson, 1987; Jenks & Caspall, 1971; Słomska-Przech & Gołębiowska Małgorzata I., 2021), the subjective biases introduced by classification choices remain an underexplored area. The selection of classification methods and procedures that categorize data into distinct groups introduces inherent subjectivity, shaping both the map's visual narrative and its interpretive conclusions (Cromley & McLafferty, 2011; Mu & Tong, 2020; Slocum et al., 2022; Zhang & MacIejewski, 2017). For instance, when creating a choropleth map, the primary concern is often optimizing the number of classes rather than considering the subjectivity of the chosen classification method (Al-Ghamdi, 2014). As a result, and despite its profound implications, this subjectivity is frequently underestimated, potentially leading to skewed interpretations that undermine evidence-based policymaking (Słomska-Przech & Gołębiowska Małgorzata I., 2021). Thus, reducing subjectivity in spatial classification is important to ensure that data-driven decision-making is influenced as little as possible by methodological biases.

Common classification techniques for creating choropleth maps, such as equal interval, quantile, and standard deviation, are often favored for their algorithmic simplicity (Mu & Tong, 2020). Choropleth maps can also be combined into a bivariate choropleth map, allowing users to visualize two or more variables on a single map to reveal their relationships (Alyaqout & Anzah, 2025). Yet each method imposes assumptions about data distribution and spatial structure, which can alter the story a map conveys. For example, quantile classification prioritizes relative rankings but may obscure extreme values, while equal interval preserves data range at the expense of masking clusters (Fang et al., 2022; Oxoli et al., 2020). These discrepancies raise concerns about reproducibility, particularly when maps guide high-stakes decisions like disaster preparedness or public health funding. Recent research highlights that method selection is rarely driven solely by data characteristics; instead, it reflects cartographers' biases, institutional norms, or analytical priorities and this is why especially for public health certain cartographic guidelines must be followed (Centers for Disease Control and Prevention (U.S.), 2012; Stieb et al., 2019). Such findings underscore the necessity of context-driven classification strategies or the existence of specific mapping guidelines.

Given the potential for classification subjectivity to distort spatial patterns and misinform decision-making, exploring alternative methods that minimize bias and enhance representational accuracy becomes essential. Less conventional approaches, including Natural Breaks (Jenks) (Coulson, 1987; Jenks, 1967), Box Maps (Anselin et al., 2006), and Local Indicators of Spatial Association (LISA) cluster maps (Anselin, 1988, 1995, 2019, 2021), offer alternatives to mitigate distortions caused by traditional methods. Natural Breaks identify class boundaries by minimizing within-group variance, adapting flexibly to data structure to reveal inherent groupings (Ke et al., 2023; Sujithlal et al., 2024). Box Maps, grounded in exploratory data analysis, employ median-based thresholds to reduce outlier influence, enhancing robustness in skewed datasets (Atanaka-Santos et al., 2007). LISA cluster maps integrate spatial autocorrelation, explicitly mapping clusters of high or low values while accounting for neighborhood relationships (Anselin, 1988, 2019, 2021; Oxoli et al., 2020). Despite their potential, these methods remain underutilized, partly due to unfamiliarity or perceived complexity.

The subjectivity of classification is further complicated by the interplay between analytical objectives and data properties, which is another issue not fully mentioned to the extent it should be. A policymaker mapping infrastructure needs might prioritize methods emphasizing spatial outliers, while an ecologist tracking gradual habitat changes could favor techniques preserving gradient continuity. Additionally, user expertise influences method adoption: novices may rely on software defaults, while specialists often advocate for approaches aligned with disciplinary conventions or develop their own rules based on design perception, a process susceptible to cognitive bias (Jimenez et al., 2024). This variability challenges efforts to standardize practices, prompting critical questions about balancing methodological rigor with interpretive flexibility.

While previous studies (Al-Ghamdi, 2014; Jiménez-Carvajal et al., 2016; Zhang & MacIejewski, 2017) have acknowledged the impact of classification choices on spatial interpretation of a single parameter, they have primarily focused on comparing methods rather than addressing the underlying subjectivity that shapes these choices. This manuscript builds upon that foundation by critically examining the biases introduced by traditional classification techniques and exploring alternative methods that offer greater objectivity and interpretive accuracy. Therefore, the aim is not just about comparing different classification methods and discussing their pros and cons, an already well-covered topic in literature. Rather, it investigates the subjectivity challenges inherent in map classification and evaluates underused less subjective methods that also enhance representational accuracy.

The objectives of the current study are as follows: (1) to expose how subjective classification choices shape map outputs and decision-making, (2) to assess the strengths and limitations of other alternative data clustering techniques regarding subjectivity such as Natural Breaks, Box Maps, and LISA cluster maps, and (3) to propose strategies for more transparent methods aligned with analytical goals, with a primary focus on combining results from different classifications to reduce subjectivity issues. By contrasting traditional and alternative techniques, this work seeks to empower practitioners to create maps that are statistically rigorous and interpretively faithful. Using a case study, this study examines how to select classification techniques to improve the reliability of spatial data analysis, minimize bias, and guide more effective decision-making across various disciplines. Special attention is given to spatial autocorrelation, particularly through Local Indicators of Spatial Association (LISA), which minimize subjectivity by incorporating spatial relationships between data points into the clustering process. This approach reveals high-value and low-value clusters, as well as spatial outliers. The spatial autocorrelation method demonstrated significantly reduced subjectivity in supporting decision-making by ensuring that the identified patterns are statistically significant and spatially coherent, rather than artifacts of the classification method.

1.1 Quantile and Equal Intervals Maps: The Traditional Classification Methods

Quantile maps and Equal Intervals maps represent two distinct approaches in thematic cartography, each with specific mathematical foundations and use cases (Zongfan et al., 2022). Quantile mapping employs a data-driven classification method that arranges observations in ascending order and divides them into four classes based on their position in cumulative frequency distribution. This approach is statistically robust as it accounts for the actual distribution of the data, making it particularly valuable for comparing different geographical areas or time periods. The break points, or class boundaries, are determined by identifying the values at the 25th, 50th, and 75th percentiles of the dataset. This method has proven especially effective in epidemiological and socioeconomic mapping where data often exhibits significant skewness (Brewer & Pickle, 2002; Slocum et al., 2022; Stieb et al., 2019).

Equal Intervals mapping, alternatively, employs a mathematically straightforward approach by dividing the data range into classes of equal amplitude. The class interval is calculated as (maximum value - minimum value) / number of desired classes, resulting in breaks that are equidistant across the data range (Al-Ghamdi, 2014; R. G. Cromley & Cromley, 2009; Jenks & Caspall, 1971; Zhang & MacIejewski, 2017). While this method offers simplicity in interpretation and calculation, its effectiveness can be significantly impacted by data distribution characteristics. In datasets with clustered values or extreme outliers, Equal Intervals may produce empty classes or classes with vastly different frequencies of observations. However, this method excels in representing continuous phenomena like elevation, temperature, or rainfall, where the absolute differences between values carry meaningful information (MacEachren, 2008). Recent studies have shown that combining these methods with modern visualization techniques can enhance their utility, particularly in web-based mapping applications where interactive features allow users to switch between classification methods dynamically (Alyaqout & Anzah, 2025; Jimenez et al., 2024; Ke et al., 2023; Slocum et al., 2022).

1.2. The Box Map Classification Approach

A *Box Map* is a classification mapping method that provides a clear visualization overview of the spatial distribution of data (Anselin, 1996; Peterman et al., 2022) and uses statistics from the well-known box plot (median, quartiles, whiskers) to classify data, where "classes" are usually the following: "low outlier," "lower quartile," "median," "upper quartile," and "high outlier". The method is based on Tukey's box plot logic, defining outliers as values beyond 1.5× (or 3× depending on the software and how the outliers are considered) the interquartile range (IQR). The main strengths of using a Box Map are as follows (Anselin, 1996; Anselin et al., 2006): a) *Outlier Detection*: Explicitly identifies extremes; b) *Robust to Skewness*: Focuses on central 50% of data (IQR), reducing sensitivity to outliers; c) *Intuitive for Non-Experts*: Mimics box plots, a familiar visualization tool. However, there are biases that must be considered such as a) *Arbitrary Outlier Threshold*: The 1.5×IQR rule is conventional but not universally applicable; b) *Masking Spatial Clusters:* Outliers may be scattered, obscuring regional trends; c) *Class Imbalance:* The "median" class can dominate, underemphasizing variation.

Compared to other data clustering classifications when producing maps for decision making Box Map can be more useful during the stage of exploratory analysis as it focuses on distributional extremes and central tendency. Compared to methods that consider spatial dependencies, such as Spatial Autocorrelation, Box Map can highlight lower or higher outliers but not spatial clusters or patterns (Anselin, 1996, 2021; Anselin et al., 2006; Peterman et al., 2022). Therefore, Box Map can be more useful to identify outliers and central tendency in spatial data, while SA is more efficient as it focuses on spatial dependencies, and it can be better for formal hypothesis testing.

On the other hand, Box maps (Anselin et al., 2006), a not traditional but unfortunately underused classification, typically employ box-plot-like visualizations to help users to better understand the distribution of data, while Equal Intervals maps often include graduated legend bars that emphasize uniform class sizes. Research has demonstrated that the choice between these methods can significantly impact map interpretation and decision-making processes, particularly in public health and urban planning contexts (Brewer & Pickle, 2002; R. G. Cromley & Cromley, 2009). Modern GIS software packages incorporate both methods, often with automated optimization algorithms that can suggest the most appropriate settings for classification method based on data characteristics and intended use.

1.3 The Jenks Natural Breaks (JNB) Classification Method

The Jenks Natural Breaks Optimization (JNB) method is a data classification technique commonly used in geographic information systems (GIS) to partition data into distinct classes based on natural groupings present in the dataset (Jenks, 1967; Jenks & Caspall, 1971; Simela et al., 2025; Slocum et al., 2022). This method is widely applied in cartography and spatial analysis to uncover meaningful patterns and distributions within datasets. Mathematically, the Jenks Natural Breaks algorithm minimizes the sum of squared deviations (SSD) within each class and maximizes the SSD between classes. The total variance of the dataset is decomposed as follows (Brewer & Pickle, 2002; Jenks, 1967; Jenks & Caspall, 1971):

$$\textit{Total variance} = \sum_{k=1}^{m} \sum_{i \in C_k} (x_i - \bar{x}_k)^2 + \sum_{k=1}^{m} n_k (x_i - \bar{x}_k)^2$$

where x_i represents individual data values, $\bar{x}_k$ is the mean, k is the number of measurements

The goal of the algorithm is to minimize the within-class variance while maximizing between-class variance. The Jenks Natural Breaks method is an iterative optimization process that typically uses dynamic programming techniques to search for the optimal arrangement of class boundaries (Haining, 2012; Slocum et al., 2022; Zongfan et al., 2022). The first step involves sorting the data in ascending order, followed by the calculation of potential breakpoints. Each potential classification scheme is evaluated based on an objective function that seeks to minimize within-class variance. While the method is computationally intensive for large datasets, it provides robust results when applied to datasets that exhibit natural groupings. A key advantage of the Jenks Natural Breaks classification is its ability to produce intuitively meaningful class intervals that align well with naturally occurring patterns in the data (Jenks, 1967; Jenks & Caspall, 1971; Slocum et al., 2022). This makes it especially valuable in environmental studies, urban planning, and demographic analysis, where data often exhibit skewed distributions and outliers. However, a drawback of the method is its sensitivity to the number of classes chosen by the user, which can influence the interpretability of the results (Al-Ghamdi, 2014; Jenks & Caspall, 1971; Zhang & MacIejewski, 2017). Selecting an inappropriate number of classes may lead to overfitting or loss of meaningful detail.

Thus, the Jenks Natural Breaks method offers an effective way for classifying both spatial and non-spatial data by uncovering inherent patterns through variance optimization (R. G. Cromley & Cromley, 2009; Slocum et al., 2022). Despite its computational complexity and sensitivity to parameter selection, it remains a widely used technique due to its ability to highlight meaningful structure within datasets. Its applications in geospatial analysis, demographic segmentation, and environmental monitoring demonstrate its versatility across various fields requiring data-driven decision-making.

1.4 The Spatial Autocorrelation Approach

Spatial autocorrelation (SA) measures the extent to which similar values cluster (positive SA) or disperse (negative SA) in space. While SA analysis can mitigate biases in traditional classification methods, it is not entirely free of bias, as its effectiveness depends on the context, methodology, and assumptions (Anselin, 2021; Sergio J. Rey & Luc Anselin, 2010). Global SA is assessed using Moran's *I* and Geary's *C*, while local SA is described by the Local Indicators of Spatial Association (LISA) and the local G_i and G_i^* statistics. Global Moran's *I* is an inferential statistic, interpreted within the null hypothesis that values are randomly distributed across the study area (Anselin, 2019). Global Moran's *I* can be calculated as follows (Anselin, 1995, 2021; Sergio J. Rey & Luc Anselin, 2010) using Eq. (1):

$$I = \frac{n}{\sum_{i=1}^{n}\sum_{j=1}^{n} w_{ij}} \frac{\sum_{i=1}^{n}\sum_{j=1}^{n} w_{ij}(x_i - \bar{x})(x_j - \bar{x})}{\sum_{i=1}^{n}(x_i - \bar{x})^2} \quad (1)$$

where n equals the number of observations; w_{ij} the weight between locations i and j; x_i and x_j are the values at locations i and j; $\bar{x}$ is the average over all locations of the variable.

The local indicator of spatial association (LISA) can be described (Anselin, 1995) using Eq. (2):

$$I_i = \frac{x_i - \bar{x}}{\sum_i (x_i - \bar{x})^2} \sum_j w_{ij}\left(x_j - \bar{x}\right) \tag{2}$$

whereas w_{ij} the weight between locations i and j; x_i and x_j are the values at locations i and j; $\bar{x}$ is the average over all locations of the variable. As shown above in Eqs. (1) and (2), a spatial weight matrix $\boldsymbol{W}$ (consisted of several w_{ij} pairs) is needed for the calculation of the spatial autocorrelation. Each weight element w_{ij}, as element of this normalized neighborhood matrix, corresponds to a pair of observations at locations i and j. Non-zero values reflect the potential spatial interaction between two observations, while zero values indicate a lack of spatial interaction (Anselin, 1988).

For calculating spatial weights, two are the most used methods (Anselin, 2021; Anselin et al., 2006): a) Rook's, where w_{ij} is 1 if two units share a common edge and 0 otherwise, and b) Queen's, where w_{ij} is 1 if they share either an edge or a vertex. By convention, w_{ii} is defined as zero. The weight matrix $\boldsymbol{W}$ can also be based on actual distance, inverse distance with powers from 1 to 5, or the k-nearest points method, often relying on Euclidean distances between observations, as shown in Eq. (3) (Anselin, 1988, 2021):

$$d_{ij} = \sqrt{\left(x_i - x_j\right)^2 + \left(y_i - y_j\right)^2} \tag{3}$$

where i and j any two positions in the given area, with (x_i, y_i) and (x_j, y_j) their coordinates, respectively. Once d_{ij} is obtained from Eq. (3), then it can be used to calculate weights as inverse distance weights:

$$w_{ij} = \frac{1}{d_{ij}^m} \tag{4}$$

where m is the power parameter, which controls the influence of distance in weighting.

For k-nearest weight matrices, distances between point pairs are calculated, and the k-nearest points (typically between 4 and 10) are selected and retained in the matrix (Anselin, 1988, 1996; Sergio J. Rey & Luc Anselin, 2010). Row-standardization is usually applied to simplify the calculation of spatial autocorrelation statistics. The spatial weight matrix $\boldsymbol{W}$ determines each observation's contribution to global spatial autocorrelation, as Moran's I is calculated by summing the products of weights and deviations from the mean for neighboring observations (Anselin, 2021). Concerning the calculated variance for global Moran's I, this can be obtained from Eq. (5) (Anselin, 1988, 2021; Anselin et al., 2006):

$$var_N(I) = \frac{1}{(n-1)(n+1)\left(\sum_{i=1}^{n}\sum_{j=1}^{n} w_{ij}\right)^2}\left[n^2 S_1 - nS_2 + 3\left(\sum_{i=1}^{n}\sum_{j=1}^{n} w_{ij}\right)^2\right] - \frac{1}{(n-1)^2} \quad (5)$$

where:

$$S_1 = \frac{1}{2}\sum_{i=1}^{n}\sum_{j=1}^{n}\left(w_{ij} + w_{ji}\right)^2 \quad (6)$$

$$S_2 = \frac{1}{2}\sum_{i=1}^{n}\sum_{j=1}^{n}\left(w_{i.} + w_{.i}\right)^2 \quad (7)$$

Where:

$var_N(I)$: the variance of Moran's I under normality assumptions.
n: the number of spatial units (observations) in the dataset.
w_{ij}: the spatial weight between observation i and observation j, indicating the strength of their spatial relationship.
S_1: the sum of squared spatial weights, considering both directions.
S_2: row and column sums of the spatial weight matrix, considering all spatial interactions.

The expected Moran's I is calculated by using the following equation (8) (Anselin, 1988, 1995, 2021):

$$E_N(I) = -(n-1)^{-1} \quad (8)$$

The expected significance of Global Moran's I can be tested on their z-scores (standard deviations) by using the equation (9) (Anselin, 2021; Sergio J. Rey & Luc Anselin, 2010)

$$z - score = \frac{I - E(I)}{\sqrt{var(I)}} \quad (9)$$

In the case of a normal distribution of data, the threshold of 1.96 can be applied to test the significance level of z. If the z-value is greater than 1.96 or smaller than −1.96, this implies that the spatial autocorrelation is significant

Compared to other spatial data classifications, Spatial Autocorrelation can mitigate certain biases because(Anselin, 2021; Brus et al., 1996; Jenks & Caspall, 1971): a) avoids arbitrary class breaks: unlike choropleth maps or cluster classifications (e.g., Jenks, quantiles), SA does not rely on discretizing data into classes, reducing bias from subjective classification choices; b) quantifies spatial structure in data: SA metrics (e.g., Moran's *I*, Geary's *C*, Local Indicators of Spatial Association (LISA)) explicitly model spatial dependencies, helping identify true clusters or outliers rather than artifacts of classification; c) assumption of stationarity: by accounting for spatial relationships, SA can mitigate errors from assuming homogeneity within aggregated units (e.g., census tracts).

Despite its strengths, SA has its own limitations and biases that should be considered when used in decision making. A key concept is the researcher degrees of freedom, which refers to the many choices made regarding data inclusion, hypothesis selection, and method application—often described as the "garden of forking paths" (Gelman & Loken, 2014). In SA using the software GeoDa, critical decisions include spatial scale selection (MAUP) (Openshaw, 1983), handling outliers, inputting missing values, defining spatial weights (Queen or Rook calculation method), addressing unconnected observations, and adjusting tuning parameters (Anselin et al., 2006). These choices can impact data exploration outcomes, making sensitivity analysis essential for mitigating bias. Therefore, it should be noted that SA results can vary with the scale or zoning of spatial units (e.g., counties vs. neighborhoods), like the commonly used classification methods. Another issue might be the "edge" effect because SA metrics at study area boundaries may have unreliable SA values due to missing neighbor data. In addition, SA heavily depends on how spatial relationships are defined (e.g., distance thresholds, contiguity). Mis-specified weights matrices (e.g., ignoring barriers like rivers) can distort results. Also, scale dependency should be considered too, as SA can detect clusters at specific scales but miss patterns at others (e.g., hotspots in cities but not neighborhoods) (Anselin, 1996, 2021; Anselin et al., 2006).

However, despite the potential limitations and biases mentioned above, we can deduce that SA may be less biased than commonly used classification methods if: a) the spatial weights matrix accurately reflects real-world interactions; b) the scale of analysis matches the underlying process (e.g., disease spread at the neighborhood level); non-stationarity is addressed (e.g., using geographically weighted regression); c) results are validated with domain knowledge or ground truthing.

2. THE CHALLENGE OF SUBJECTIVITY IN SPATIAL DATA CLASSIFICATION: A CASE STUDY

The subjective nature of traditional classification methods arises from the arbitrary choices analysts make when setting classification thresholds, breakpoints, and the number of clusters (Al-Ghamdi, 2014; Charles et al., 2024). This subjectivity often leads to inconsistencies in decision-making, resulting in varying conclusions drawn from the same dataset. Common spatial data classification methods, including equal interval and quantile maps, aim to simplify data interpretation but inherently introduce subjectivity. Quantile maps group data into bins containing equal numbers of data points, while equal interval maps divide the data range into intervals of equal size, irrespective of the number of data points in each bin (Zhang & MacIejewski, 2017). Figure 1 depicts the clustering logic underlying each classification.

Figure 1. Conceptual representation of the clustering logic underlying each classification (illustration by the authors).

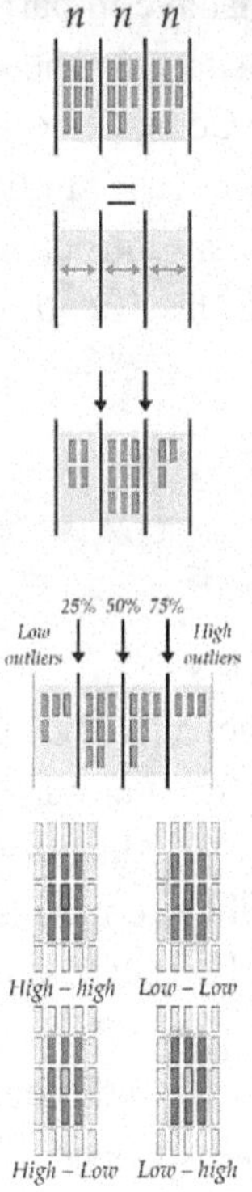

Quantile
Data are distributed equally across class intervals, resulting in unequal class widths but maintaining the same frequency of observations per class.

Equal Interval
The data range for each class is identical, producing equal class widths while allowing for varying frequencies of observations per class.

Natural Breaks (Jenks)
Data are analyzed to account for non-uniform distributions, leading to unequal class widths with varying frequencies of observations per class. This method optimizes classification by minimizing within-class variance and maximizing between-class variance.

Box Map
Data are classified according to the following scheme: "low outliers", "< 25%", "25% – 50%", "50% – 75%", ">75%", and "high outliers."
This method is based on Tukey's box plot logic.

Spatial Autocorrelation - LISA Cluster maps
Data are compared with neighboring values and classified into the following clusters:
"*High – high*": high values surrounded by similarly high values.
"*Low – low*": low values surrounded by similarly low values.
"*High – low*": high values surrounded by low values (outliers).
"*Low – high*": low values surrounded by high values (outliers).

While these methods help users compare values across different ranges, their outcomes are heavily influenced by user-defined parameters, such as the number of clusters and breakpoint selection(Al-Ghamdi, 2014; Sergio J. Rey & Luc Anselin, 2010). As a result, analysts applying the same method to identical datasets may arrive at different visual representations and conclusions. Box Map is also used for classifying data into quartiles and identifying outliers. The Natural Breaks (Jenks) method offers an alternative by grouping values based on natural groupings within the data. For instance, this approach may highlight a cluster of high values in a particular region, whereas a quantile map might disperse these values across multiple categories, leading to divergent interpretations. This variability becomes particularly problematic when maps inform policy or management decisions, as subjective methods can undermine the robustness, reproducibility, and objectivity of conclusions (Openshaw, 1983). These inconsistencies can have significant implications, particularly in policymaking, where spatial data informs resource allocation, urban planning, and public health interventions. To address these limitations, alternative classification methods have been developed, offering improved objectivity and adaptability to different data structures.

A case study as a theoretical hypothesis – scenario was conducted to identify potential subjectivity issues in each classification method through visual comparisons, considering their specific characteristics in addressing the question of where attainment levels in a megacity are high or low based on the available Census tracts, to support local authorities in making informed decisions. Within the case study, changes in spatial patterns resulting from data partitioning, which may introduce subjectivity, were assessed using Spatial Autocorrelation for each classification method.

3. MATERIALS AND METHODS: A CASE STUDY ON DIFFERENT MAP CLASSIFICATIONS

To examine how spatial data classification methods introduce subjectivity and influence result interpretation, particularly in identifying areas with high and low attainment levels, we analyzed data from the GeoDa Data Center on NYC education and socio-demographic characteristics (2,166 observations, 113 variables, covering the years 2008–2012; Source: American Community Survey 2008–2012, US Census Bureau) (Murugesan M., 2017). The study area was New York City (Figure 2). The focus was on the variable “master,” representing the population aged 25 and older with at least a master's degree. Since we are focusing on the variable “master”, it was considered useful to normalize this variable for meaningful spatial analysis. Therefore, we created a new variable, “mst2poptot,” which represents “master's degree holders as a proportion of the total population.” It is defined as the ratio master/

poptot, where poptot represents the total population in each area. The software used includes ArcGIS Pro v.3.4, GeoDa v.1.22.0.4, QGIS Desktop v.3.36.1.

Three classification methods—quantile maps, equal intervals, and natural breaks—were applied to visualize the data and assess their impact on interpretation. These methods were then compared with a box map using the same dataset. Finally, LISA cluster maps were constructed to illustrate the spatial autocorrelation of the "master" variable. Visual comparisons were the primary method used to evaluate the results of each classification method.

Figure 2. Study area in NYC using Education and socio-demographic characteristics for New York City Census tracts (Source: American Community Survey 2008-2012, US Census Bureau).

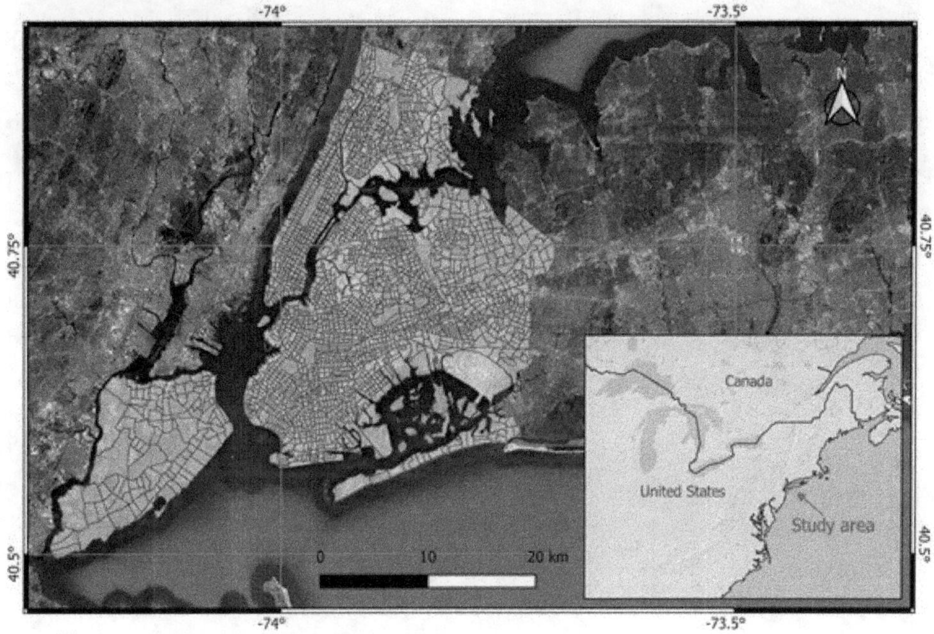

The descriptive statistics for both the "master" and "Mst2poptot" variables were calculated using the data engineering module in ArcGIS (Table 1). The analysis of the "master" variable reveals a highly skewed distribution with substantial variability (Table 1). With a mean of 245.19 and median of 141.50, along with a standard deviation of 326.48 and range of 2,729, the data shows strong right-skewness (skewness=3.30) and heavy tails (kurtosis=17.11).

Table 1. Descriptive statistics for variable "master" and the normalized variable "mst2poptot".

Variable	Min	Max	Mean	SD	Median	Count	Outliers	Range	IQR	CV (%)	Skewness	Kurtosis
master	0	2729	245.19	326.48	141.50	2166	174	2729	225	1.33	3.30	17.11
Mst2poptot	0	0.478	0.061	0.05	0.045	2166	136	0.478	0.061	0.92	1.61	6.62

where: *SD*: Standard Deviation, *IQR*: Interquartile, *CV*: Coefficient of Variation

The presence of 174 outliers (as calculated via data engineering option in ArcGIS) in variable "master" and 136 outliers in "mst2poptot" among 2,166 observations, coupled with a low coefficient of variation (1.33% in "master" and 0.92% in "mstpoptot"), suggests that while there is strong asymmetry in the distribution, the relative variability of the data is quite small (Figure 3).

Figure 3. Data distribution: Histogram and boxplot of the "mst2poptot" variable.

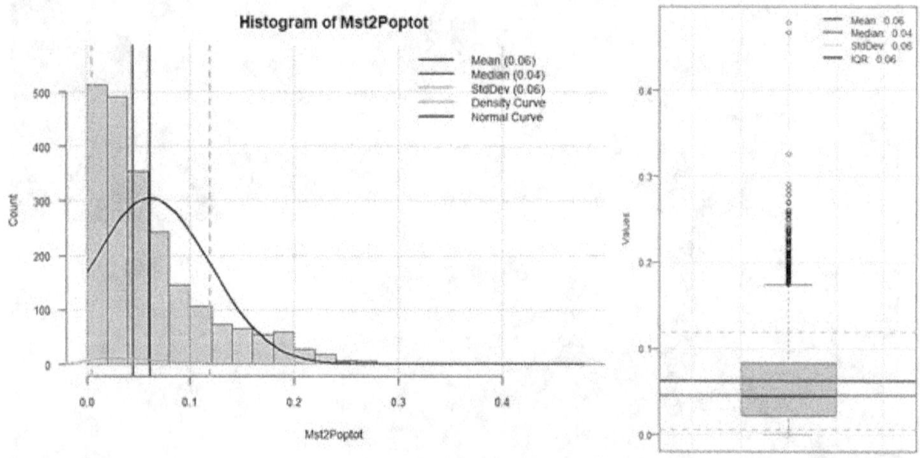

These characteristics indicate that values are tightly clustered, despite the presence of some high-value outliers affecting the mean.

4. RESULTS

4.1 Analyzing Visualizations of the Classification Methods

Figure 4 presents clustering results using six classification methods: (a) Quantile, (b) Equal Intervals, (c) Natural Breaks, (d) Box Map, (e) LISA cluster map, and (f)

Significance map (as part of the Spatial Autocorrelation analysis). These methods were evaluated to determine optimal representation of areas with high concentrations of people aged 25+ with master's degrees, while assessing potential bias in each approach:

- The *Quantile map* (Figure 4a) reveals spatial distribution patterns of educational attainment. Notable clusters of higher educational attainment appear in Manhattan, western Brooklyn, and parts of Queens, indicated by areas with similar color patterns. These clusters suggest neighborhoods with comparable educational characteristics. The clustering patterns remain somewhat subtle, lacking definitive separation between areas of higher and lower educational attainment, which may necessitate additional analytical approaches to fully identify trends.
- The *Equal Interval map* (Figure 4b) emphasizes stark contrasts, with Manhattan's Upper East and West Sides emerging as prominent high-value clusters shown in darker shades, while most of the city appears in lighter shades. This method provides clearer visual distinction between high and low concentration areas compared to the quantile approach. However, this classification potentially obscures subtle variations in lower educational attainment levels, as the lighter shade spectrum may compress differences in these regions.
- The *Natural Breaks map* (Figure 4c) employs statistical grouping to organize data based on inherent divisions within the dataset. This method effectively displays educational attainment gradients, particularly as they radiate outward from Manhattan. The approach highlights transition zones corresponding to neighborhood boundaries and socioeconomic divisions, providing detailed visualization of educational attainment variation. While offering more accurate representation of both higher and lower educational attainment levels, this method introduces potential subjectivity in the selection of class numbers and in defining meaningful breaks.
- The *Box Map* (Figure 4d) demonstrates pronounced spatial contrasts, particularly highlighting the north-south divide in Manhattan and stratification in western Brooklyn. This method effectively captures spatial patterns and variations, clearly delineating areas with distinct characteristics. Additionally, it successfully identifies outliers, providing a more comprehensive visualization of the educational landscape compared to several other classification methods.
- *Spatial Autocorrelation analysis* (Figure 4e-f) includes the LISA cluster map showing high-value clusters (red areas) and spatial outliers (pink and light blue regions), alongside a significance map highlighting statistically signifi-

cant clusters. By incorporating spatial relationships between data points, this approach facilitates identification of meaningful patterns while enhancing analytical robustness. Methodological concerns include the subjective selection of spatial weights matrix calculation method (Queen or Rook), which may influence results based on data structure. Furthermore, while the significance map appropriately focuses on statistically significant values, it might overlook meaningful patterns that fall below statistical thresholds, potentially underrepresenting practical significance. Compared to other classification methods, this approach provides a clear overview of statistically significant areas where attainment levels are high (high-high cluster) or low (low-low cluster). However, there might be a potential concern since LISA cluster maps focus on identifying hot-spots or cold-spots of data rather than presenting a range of data or quantitative information. Therefore, the results from these classifications might not be directly comparable.

Figure 4. Applying different classifications to the same data: (a) Quantile map; (b) Equal Intervals map, (c) Natural Breaks (Jenks) classification; (d) Box Map with hinge 3.0; (e) Spatial Autocorrelation – LISA cluster map; (f) Spatial Autocorrelation – significance map. In each map, the index provides the classes, with the number of records in each class shown in parentheses.

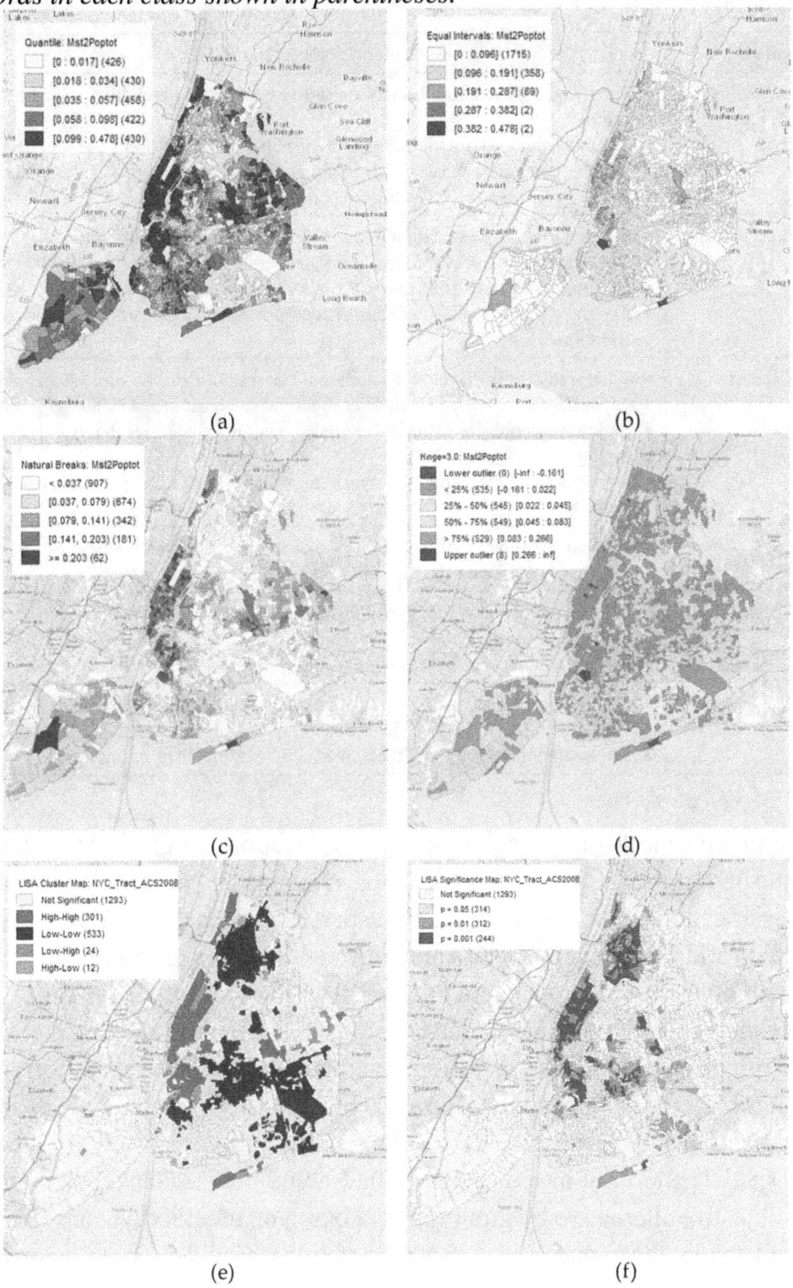

After comparing the visualizations of the classification methods under consideration, we identified potential subjectivity and bias issues (Table 2).

Table 2. Potential subjectivity and bias issues in using different classification mapping methods.

Classification	Potential subjectivity/Bias
Quantile	• Forces equal number of observations into each class, potentially masking true data patterns • May underrepresent extreme values by grouping them with more moderate values • Creates artificial boundaries that might not reflect natural breaks in the data • Could suggest gradual transitions where sharp distinctions exist
Equal Intervals	• May underrepresent variation in the middle ranges of the distribution • Can create empty classes if data is not uniformly distributed • Overemphasizes extreme values, potentially exaggerating disparities • May show fewer regions in darker shades, potentially misrepresenting the overall distribution.
Natural Breaks (Jenks)	• While statistically driven, the number of classes must still be subjectively chosen. • May reflect researcher bias in determining what constitutes a meaningful break • Could overemphasize historical or socioeconomic boundaries that may be changing. • Interpretation of "natural" groupings might be influenced by preexisting assumptions.
Box Map	• Bivariate choropleth visualization may oversimplify complex relationships • The use of quartiles as boundaries may affect or limit interpretation.
LISA Cluster Map	• Selection of spatial weights matrix is subjective (Queen or Rook) and may affect results. • May overemphasize cluster boundaries that align with expected patterns.
Significance Map	• May overlook meaningful patterns that don't meet statistical thresholds. • Focusing on statistical significance might overshadow practical significance. • Areas without coloring might be incorrectly interpreted as unimportant.

In addressing the theoretical hypothesis on the location of high and low attainment levels to support local authorities in decision-making, we can conclude that while each classification method provides a unique perspective on the data, only the LISA cluster map offers statistically significant information on where attainment levels are high (high-high clusters) and where they are low (low-low clusters). These results and potential subjectivity issues underscore the importance of selecting meticulously the appropriate spatial data classification method to minimize subjectivity and enhance the reliability of spatial analyses. While traditional methods such as equal intervals and quantile maps are useful for specific purposes, they often obscure critical spatial patterns or introduce bias. The Natural Breaks (Jenks) method offers a less subjective alternative by grouping data based on natural divisions. However, advanced spatial techniques like LISA cluster maps provide a more rigorous and objective means of analyzing spatial data by incorporating spatial relationships and statistical significance.

4.2 Quantifying Changes in Spatial Patterns as a Measure of Induced Subjectivity Using Different Classification Approaches

Spatial autocorrelation analysis, using both global Moran's *I* and local Moran's *I* (LISA cluster maps), was conducted to compare patterns in the original dataset with those observed after applying the following spatial data classification methods: Quartile, Equal Intervals, Natural Breaks, and Box Map with a hinge of 3.0 and 1.5 (Figure 5). The objective was to determine which classification approach best preserved the spatial structure of the unclassified data. A greater divergence from the initial spatial autocorrelation measures would indicate a higher degree of information loss and increased subjectivity in data representation.

Figure 5. Comparing the results of spatial data classification using Spatial Autocorrelation (SA) with the same dataset under different conditions: (a) Baseline – SA on the original data; (b) SA on Quantile-classified data; (c) SA on Equal Intervals; (d) SA on Natural Breaks; (e) SA on Box Map with a hinge of 3.0; (f) SA on Box Map with a hinge of 1.5 (using the Queen contiguity weight calculation method with 999 permutations). In each map, the index provides the classes, with the number of records in each class shown in parentheses.

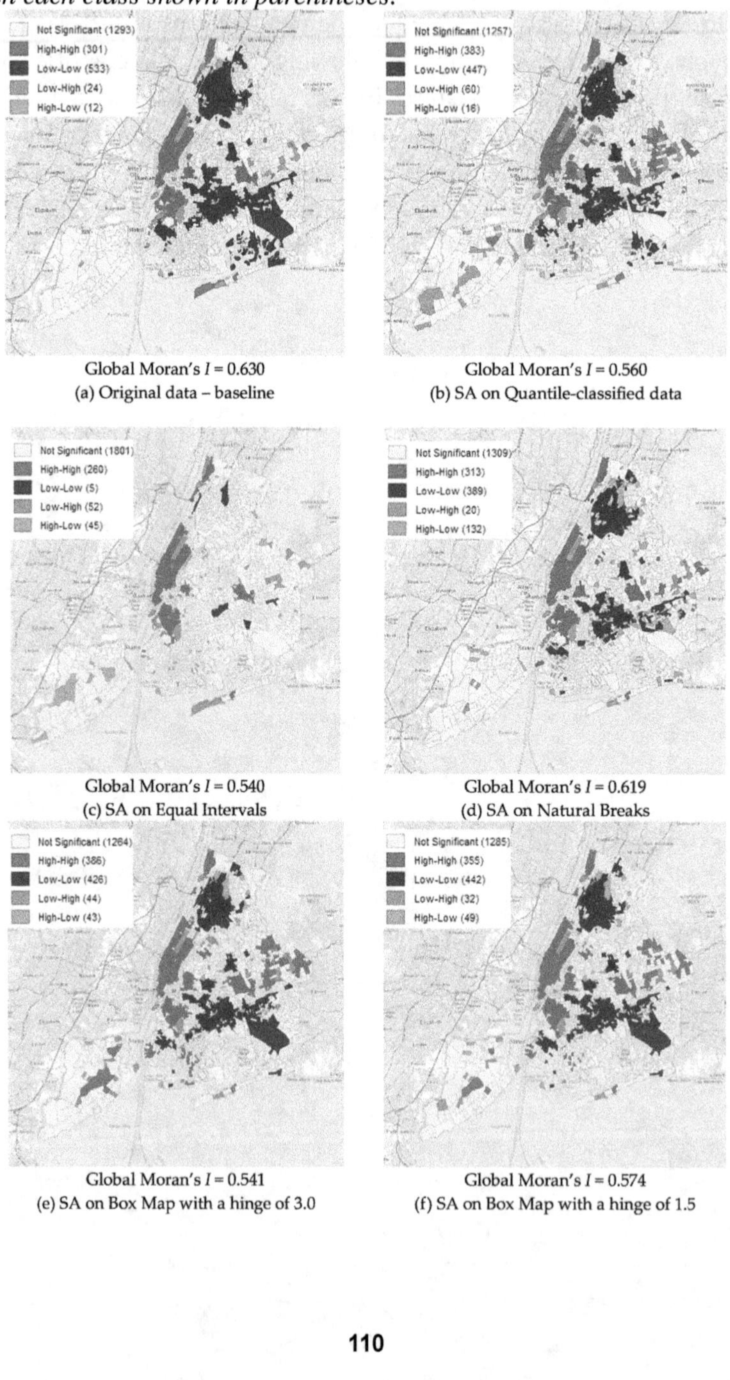

Global Moran's $I = 0.630$
(a) Original data – baseline

Global Moran's $I = 0.560$
(b) SA on Quantile-classified data

Global Moran's $I = 0.540$
(c) SA on Equal Intervals

Global Moran's $I = 0.619$
(d) SA on Natural Breaks

Global Moran's $I = 0.541$
(e) SA on Box Map with a hinge of 3.0

Global Moran's $I = 0.574$
(f) SA on Box Map with a hinge of 1.5

The analysis started by calculating Global Moran's I for the original dataset to establish a baseline measure of spatial dependence. The dataset was then classified using Quantile, Equal Intervals, Natural Breaks, and Box Map (with hinges 1.5 and 3.0), and spatial autocorrelation was recalculated for each classification. By comparing these values with those derived from the original data, we assessed the impact of each classification method on spatial patterns. Differences in Moran's I values provided insight into how classification methods altered spatial relationships, potentially leading to distortions in data interpretation. However, using Global Moran's I averages the spatial patterns, which could be considered as a limitation of this method. This is why LISA cluster maps were also constructed and compared not only visually but also based on the number of records identified in each cluster and the number of outliers (high-low' and 'low-high' clusters). Based on the comparison of individual LISA cluster maps (Figure 5), the Equal Intervals method showed the most discrepancies.

Table 3 presents the effects of different classification methods on spatial autocorrelation, measured by Global Moran's I. The baseline spatial autocorrelation of the original data ($I = 0.630$) shows moderate positive autocorrelation, meaning the spatial data are moderately clustered. Among the classification methods, Natural Breaks ($I =$ 0.619, $\Delta I =$ -0.01 or -1.7%) shows the least loss in spatial autocorrelation, with only a very slight decrease in Moran's I. This indicates that Natural Breaks preserves the original spatial structure almost entirely, with minimal disruption to the clustering of data points. On the other hand, methods such as Quantile ($I = 0.560$, $\Delta I =$ -0.07 or -11.1%) and Equal Intervals ($I = 0.540$, $\Delta I =$ -0.09 or -14.2%) show a moderate reduction in spatial autocorrelation, with Equal Intervals causing a slight reduction in spatial clustering, yet still largely preserving the original structure. The Box Map (Hinge 3.0, $I = 0.541$, $\Delta I =$ -0.09 or -14.1%) causes a moderate loss of spatial continuity, but less severely compared to what might have been expected. Similarly, Box Map (Hinge 1.5, $I = 0.574$, $\Delta I =$ -0.06 or -8.9%) also shows a smaller decrease, with spatial patterns remaining closer to the original data. Overall, the results highlight that Natural Breaks introduces the least disruption to spatial patterns, making it the most effective in retaining the original spatial autocorrelation (Table 3).

Table 3. Difference between the Spatial Autocorrelation (SA) on the original data (variable: mst2poptot) and the data after employing SA on the spatial data of the classification methods under consideration.

Classification Method	Global Moran's I	Difference from Original Data (ΔI)	Evaluation of Spatial Autocorrelation (SA)	Interpretation/Notes
Baseline: SA on the Original Data	0.630	-	Moderate positive autocorrelation: Spatial structure is somewhat disrupted due to equal distribution of data values.	Spatial data are moderately dispersed.
SA on Quantile Data	0.560	-0.07 (-11.1%)	Moderate positive autocorrelation: Spatial structure is largely preserved, like the original data.	A slight decrease in Moran's I compared to the baseline, indicating moderate reduction in spatial autocorrelation.
SA on Equal Intervals	0.540	0.09 (-14.2%)	Moderate positive autocorrelation: Slight reduction in spatial clustering, but the method retains much of the original structure.	Very little change in Moran's I, suggesting the spatial pattern is nearly unchanged.
SA on Natural Breaks	0.619	-0.01 (-1.7%)	Low positive autocorrelation: Significant disruption of spatial patterns due to stricter class boundaries.	A noticeable decrease in Moran's I, indicating significant loss in spatial autocorrelation.
SA on Box Map (Hinge 3.0)	0.541	-0.09 (-14.1%)	Moderate positive autocorrelation: Loss of spatial continuity, but less severe compared to Hinge 3.0.	A moderate decrease in Moran's I, reflecting noticeable but less severe disruption in spatial continuity than Hinge 3.0.
SA on Box Map (Hinge 1.5)	0.574	-0.06 (-8.9%)	Moderate positive autocorrelation: Spatial structure is somewhat disrupted due to equal distribution of data values.	The decrease in Moran's I indicates a slight loss of spatial autocorrelation.

These findings highlight the importance of carefully selecting a classification method that minimizes distortion and preserves spatial integrity in thematic mapping, particularly when the final product is intended for decision-making in critical agricultural, social, or epidemiological contexts.

5. DISCUSSION ON SUBJECTIVITY ISSUES

5.1 The Challenge of Subjectivity in Spatial Data Classification

Traditional spatial data classification methods, such as quantile and equal interval mapping, have long been used to simplify complex datasets and facilitate interpretation. However, they overlook outliers by incorporating them into clusters. Despite their widespread use, these methods introduce an element of subjectivity, as the chosen classification scheme and the number of classes can substantially influence the outcome. On the other hand, Natural Breaks attempts to identify clusters within data by grouping values where significant differences occur, but still the optimal number of clusters might affect the final interpretation. These are critical points that can lead to discrepancies in spatial data interpretation and inconsistencies in subsequent decision-making processes.

The reliance on user-defined parameters introduces variability in results, making it difficult to achieve consistency and reproducibility. For instance, an analyst using a natural breaks method might emphasize a particular cluster of high values in one region, whereas another analyst employing a quantile classification might disperse the same values across multiple categories, leading to divergent conclusions. Such inconsistencies pose a significant challenge when spatial data is used to inform critical decisions in fields such as environmental management, urban planning, and agricultural development. Outlier detection plays a crucial role in spatial data analysis, and different classification techniques offer varying levels of effectiveness in identifying and handling these extreme values. Equal interval classification performs poorly as it tends to mask the presence of outliers by lumping them into broad categories. For example, income data may place moderately high earners and millionaires in the same class, obscuring meaningful distinctions. Quantile classification offers moderate effectiveness by ensuring an even distribution of values across categories, but it often groups extreme values with moderately high ones, making outliers less distinguishable. Natural breaks classification improves on this by potentially isolating outliers when significant deviations occur, although its effectiveness is heavily influenced by the number of breaks chosen.

Furthermore, the spatial heterogeneity inherent in many datasets exacerbates the limitations of traditional classification methods. Variations in spatial patterns, scale effects, and boundary delineations can all influence the results, further complicating the decision-making process. Addressing these challenges requires the adoption of more objective methods, such as spatial autocorrelation, which can mitigate the influence of subjective decision-making by incorporating the spatial structure of data directly into the analysis.

5.2 A Critical Comparison of the Classification Methods to the Spatial Autocorrelation Approach

Considering the visual comparison of the different classification methods and their potential to preserve or distort spatial patterns, we conducted a critical appraisal of the subjectivity each classification approach may introduce, as outlined in Table 4. A more detailed assessment of the classification methods with respect to subjectivity issues is provided below:

- *Quantile*: The results indicate that the quantile classification method enforces equal-sized groups, which distorts the natural distribution of the data. This distortion leads to artificial spatial patterns, as actual value differences are ignored. The analysis further reveals that spatial relationships are entirely disregarded. Interpretability remains moderate due to the familiarity of quantiles, though they obscure the presence of outliers. In fact, outlier detection is constrained, as extreme values are grouped with non-extreme ones. User control over classification is moderate, requiring the selection of class numbers, while subjectivity is notably high, as the method imposes equal groupings irrespective of the data's natural structure.
- *Equal Interval*: The results show that the Equal Interval classification method applies uniform bin widths, which can misrepresent skewed data by grouping extreme values into broad categories. While it performs well for uniformly distributed data, it distorts spatial trends in datasets with non-uniform distributions. The analysis confirms that spatial context is ignored, reducing its effectiveness for detecting spatially dependent patterns. Interpretability is high due to its similarity to histograms; however, outlier detection is poor, as the method fails to distinguish extreme values from regular data points. User control is moderate, requiring the selection of interval widths, while subjectivity remains high since the assumption of an even data distribution frequently leads to the misrepresentation of actual patterns.
- *Box Map*: The results indicate that the Box Map classification method is effective for outlier detection but is influenced by statistical thresholds that impact classification outcomes. However, spatial context is not considered, as classification is based solely on statistical thresholds. Interpretability is high due to its similarity to box plots, making it a familiar method for data analysis. Outlier detection is excellent, as extreme values are explicitly separated. User control is minimal, as the method adheres to predefined statistical rules, while subjectivity is moderate – high, depending on the chosen threshold values that define outliers.

- *Natural Breaks*: The results indicate that the Natural Breaks classification method optimizes class divisions to minimize variance within groups, though it still requires subjective decisions regarding the number of classes. However, spatial relationships are only partially accounted for, limiting the applicability for spatial analysis. Interpretability is moderate since the class break calculations are not always intuitive. Outlier detection is also moderate, depending on how effectively class breaks separate extreme values. User control is moderate, as it involves selecting the number of classes, while subjectivity remains moderate the statistical optimization helps reduce bias but does not eliminate the influence of user choices.
- *Spatial Autocorrelation*: The results show that the Spatial Autocorrelation classification method effectively models spatial relationships but is influenced by methodological choices, including weight matrices and significance thresholds. However, interpretability is moderate, as it requires statistical expertise to understand the results. Outlier detection is also moderate, with spatial outliers identified separately from general patterns. User control is moderate, involving the selection of a weight matrix, while subjectivity remains moderate although the method is grounded in statistical principles, methodological decisions still influence classification outcomes.

Table 4. Subjectivity assessment of different classification approaches compared to spatial autocorrelation

Aspect	Quantile	Equal Interval	Box Map	Natural Breaks	Spatial Autocorrelation
Bias Source	Artificial class homogeneity	Skewed data, outliers	Arbitrary outlier thresholds	Subjectivity in class breaks	MAUP, edge effects, weights matrix
Use Case	Ranking/ percentiles	Uniform distributions	Outlier detection, skewed data	Skewed/ clustered data	Hypothesis-driven cluster detection
Spatial Context	Ignored	Ignored	Ignored	Partially addressed	Explicitly modeled
Interpretability	Moderate	High (familiar histogram logic)	High (familiar box plot logic)	Moderate	Moderate (requires statistical knowledge)
Outlier detection	Moderate (outliers may be grouped with high values)	Poor (outliers grouped into wide intervals)	Excellent (explicitly identifies outliers)	Moderate (depending on class breaks)	Moderate (spatial outliers are detected as separate High-Low and Low-High clusters)

continued on following page

Table 4. Continued

Aspect	Quantile	Equal Interval	Box Map	Natural Breaks	Spatial Autocorrelation
User choices	Moderate (selection of the number of clusters)	Moderate (selection of the number of clusters)	Low to none (no user intervention)	Moderate (selection of the number of clusters)	Moderate (selection of the method for calculating weights)
Preservation of spatial patterns*	Moderate	Moderate	Moderate/Low	Moderate	High
Subjectivity/ Bias level	Very High	High	Moderate-High	Moderate	Moderate
Justification for subjectivity ranking	Highest subjectivity due to forced equal grouping regardless of actual data distribution; creates artificial patterns	Highly biased when data is not uniformly distributed; can severely misrepresent spatial patterns in skewed distributions	Shows clear outlier patterns but threshold choices and interpretation can be subjective	More objective due to statistical optimization, but still requires subjective decisions on number of classes	Based on statistical methods but subject to methodological choices (weights matrix, significance levels)

*The assessment was based on the results of Spatial Autocorrelation using the data of each classification method.

Although the results from choropleth maps are not directly comparable to those of LISA cluster maps, the above assessment suggests that Spatial Autocorrelation is a superior classification method. Unlike other approaches that either ignore or only partially address spatial context, it explicitly models spatial relationships. By incorporating spatial dependencies through a weight matrix, it provides a statistically robust framework for identifying meaningful clusters and spatial outliers. Unlike Quantile and Equal Interval methods, which impose artificial class breaks and distort natural patterns, Spatial Autocorrelation detects spatial structures based on real geographic relationships. While Box Map effectively identifies outliers, it relies on arbitrary statistical thresholds, whereas Spatial Autocorrelation distinguishes between random and significant spatial patterns. Natural Breaks optimizes class divisions but remains subjective due to user-defined class numbers, whereas Spatial Autocorrelation minimizes bias by using hypothesis-driven analysis. Although it requires statistical knowledge, Spatial Autocorrelation preserves spatial patterns more effectively than other methods, making it the best choice for studies that rely on accurate spatial pattern detection and hypothesis testing.

5.3 Benefits of Using Spatial Autocorrelation in Decision-Making

Spatial autocorrelation offers a statistically rigorous alternative to traditional classification methods by quantifying the degree of spatial dependency between observations. Unlike arbitrary classification schemes, spatial autocorrelation measures whether similar values are clustered together or dispersed across the study area based on spatial proximity. The primary advantage of employing spatial autocorrelation in decision-making is its ability to minimize subjectivity and enhance the objectivity of spatial analyses. By integrating spatial relationships into the classification process, spatial autocorrelation methods allow for a more accurate representation of spatial patterns, leading to better-informed decisions across various domains. In demographic analysis, for instance, spatial autocorrelation can identify areas with high concentrations of specific populations, such as elderly or low-income groups, enabling targeted social programs and resource allocation. In medical research, particularly with Covid-19 data, spatial autocorrelation can detect clusters of high infection rates, helping public health authorities focus interventions on the most affected regions. In public health monitoring, identifying spatial clusters of disease outbreaks can inform the deployment of healthcare resources and mitigation strategies more effectively. In agricultural management, spatial autocorrelation can identify productivity hotspots and underperforming areas, enabling targeted interventions that optimize resource allocation and improve crop yields. In environmental monitoring, the detection of spatial clusters can aid in identifying pollution sources and assessing the effectiveness of mitigation strategies. In urban planning, spatial autocorrelation facilitates the identification of socio-economic disparities, allowing policymakers to allocate resources more effectively and design interventions that address specific community needs.

Furthermore, the statistical rigor of the spatial autocorrelation method (supported also by the significance map as part of the spatial autocorrelation analysis) enhances the reproducibility and transparency of spatial analyses. The ability to identify statistically significant clusters and outliers ensures that decision-makers can rely on the results with confidence, reducing the risk of misinterpretation and improving the overall quality of spatial decision-making.

5.4 Recommendations for Reduced Subjectivity

Building on the potential subjectivity of classification methods, and going beyond the application of spatial autocorrelation (SA), the following recommendations are proposed to achieve less biased results:

- *Combine SA results with other classification methods*: Spatial autocorrelation (SA) can be used effectively to validate clusters identified through classification methods like Natural Breaks or outliers identified by Box Maps. The traditional classification techniques (like quantile and equal intervals) group data based on statistical properties, but they overlook spatial relationships between the observations. By integrating SA, which accounts for spatial dependencies, the analysis becomes more robust and less objective. SA helps confirm whether the identified clusters reflect true patterns or are merely artifacts of the classification method. Using the statistically significant results from spatial autocorrelation (SA) while also considering the natural groupings provided by Natural Breaks and the outliers identified by Box Maps can strengthen the conclusions, ensuring that the clusters are not merely the result of arbitrary classification thresholds but are indeed spatially significant. Based on the data our case study, an example of considering results from two different classification methods (Spatial Autocorrelation – LISA cluster map and Natural Breaks map) is given in Figure 6.
- *Contextualize Results*: While spatial analysis provides valuable insights, it is essential to interpret the results within the proper context. Domain knowledge plays a critical role in understanding the significance of spatial patterns. Without domain knowledge, one could misinterpret the results and draw incorrect conclusions. Understanding the cause-and-effect relationship behind the observed spatial patterns ensures that the analysis is used effectively to inform decision-making. For instance, in epidemiology, recognizing that a high concentration of healthcare facilities might overlap with areas of higher disease incidence can help guide public health interventions without misattributing outbreaks.

Figure 6. Considering the results of two spatial data classifications using both Spatial Autocorrelation (SA) and the lowest and highest class of the Natural Breaks classification (5 classes).

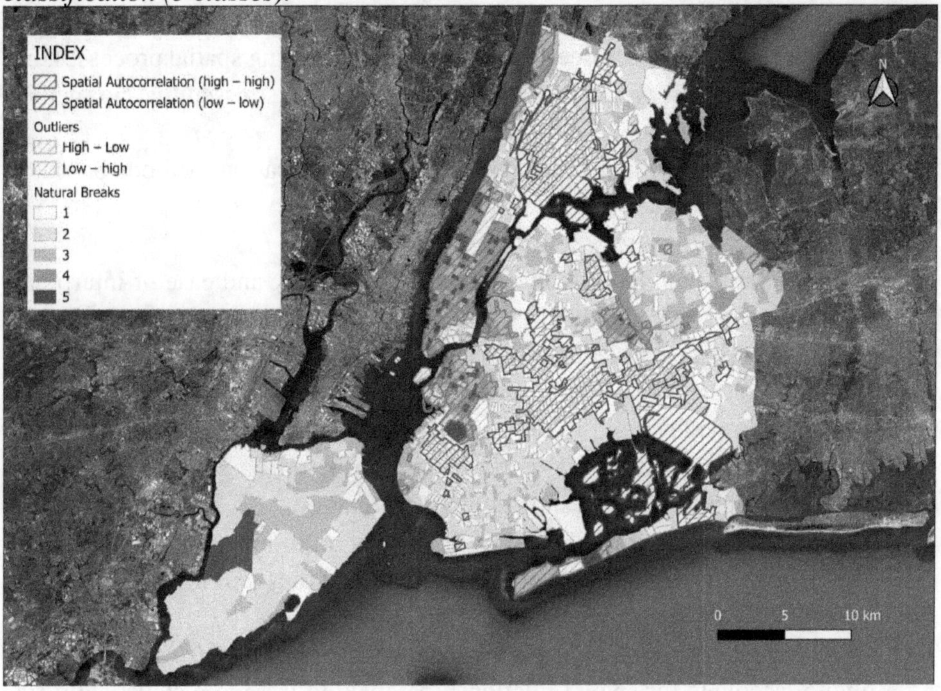

In the case presented, combining the results from the LISA cluster map (hot spots, cold spots, outliers) with clustering using Natural Breaks can provide a more informative spatial overview and help limit bias and potential subjectivity, offering crucial and less misleading insights for decision-making.

6. CONCLUSIONS: ADVANCING OBJECTIVITY IN SPATIAL DECISION-MAKING

The analysis of spatial data classification methods reveals important considerations regarding subjectivity in spatial analysis. While methods such as quantile and equal intervals have been widely adopted for their accessibility and interpretability, they inevitably introduce subjectivity through user-defined parameters and classification schemes. This subjectivity can lead to inconsistent interpretations and potentially biased decision-making when analyzing complex spatial datasets. Spatial autocorrelation (e.g. Local Indicators of Spatial Association – LISA) offers a more rigorous alternative by quantifying spatial dependencies and identifying statistically significant clusters. The results of this method can complement results from other classification techniques that preserve spatial patterns, such as Natural Breaks. These approaches reduce the influence of arbitrary decisions and provide more consistent, reproducible results. The ability to identify high-high, low-low, and spatial outlier clusters enables deeper insights into underlying spatial processes that might otherwise remain obscured. However, it should be stressed that no method is universally unbiased.

The conclusions regarding the comparison of classification methods based on potential subjectivity issues are as follows:

- *Equal interval and quantile methods* offer simplicity and ease of interpretation for non-technical audiences, but they fail to account for spatial relationships and outliers. They might introduce subjectivity due to forced grouping, regardless of the actual data distribution, which can create artificial patterns.
- *Box maps* provide strong statistical visualization by emphasizing outlier detection, yet they lack spatial context and are based on statistical threshold choices, which can lead to subjective interpretation.
- *Natural Breaks* optimize class divisions by minimizing variance within groups, making them useful for skewed data, but they still involve subjective decisions about the number of classes.
- *Spatial autocorrelation* reduces dependence on arbitrary class breaks and explicitly accounts for spatial relationships, making it a more statistically rigorous approach for cluster detection, with minor subjectivity issues related to the calculation of the weight matrix.

Key conclusions drawn from this analysis:

- *Combining spatial autocorrelation with other classification methods* enhances validation and mitigates subjectivity. The choice of the classification methods should be driven by research objectives, data structure, and the target audience.
- *The optimal classification approach* depends on the specific research context (simplicity and ease of interpretation, outlier detection, optimized class divisions etc.).

As spatial data availability expands, objective analytical methods become increasingly important. Future research should focus on advancing spatial clustering methods and integrating them with emerging technologies such as advanced machine learning techniques to improve decision-making across fields such as urban planning and public health. These developments will strengthen data-driven policy formulation and interventions while minimizing subjective biases

REFERENCES

Al-Ghamdi, A. M. (2014). Optimising the Selection of a Number of Choropleth Map Classes. In *Lecture Notes in Geoinformation and Cartography* (Vol. 0, pp. 61–77). Springer., DOI: 10.1007/978-3-319-08180-9_6

Alyaqout, A., & Anzah, F. (2025). Application of bivariate mapping to assess geodiversity and its geomorphic constraints: A case study in Kuwait. *International Journal of Geoheritage and Parks*, *13*(1), 17–30. DOI: 10.1016/j.ijgeop.2025.01.002

Anselin, L. (1988). Spatial Econometrics. *Methods and Models.*, *4*. DOI: 10.1007/978-94-015-7799-1

Anselin, L. (1995). Local Indicators of Spatial Association—LISA. *Geographical Analysis*, *27*(2), 93–115. DOI: 10.1111/j.1538-4632.1995.tb00338.x

Anselin, L. (1996). Interactive Techniques and Exploratory Spatial Data Analysis. *Regional Research Institute Working Papers, 200*. https://researchrepository.wvu.edu/rri_pubs

Anselin, L. (2019). A Local Indicator of Multivariate Spatial Association: Extending Geary's c. *Geographical Analysis*, *51*(2), 133–150. DOI: 10.1111/gean.12164

Anselin, L. (2021). *Spatial Models in Econometric Research*. DOI: 10.1093/acrefore/9780190625979.013.643

Anselin, L., Syabri, I., & Kho, Y. (2006). GeoDa: An introduction to spatial data analysis. In *Geographical Analysis* (Vol. 38, Issue 1, pp. 5–22). DOI: 10.1111/j.0016-7363.2005.00671.x

Atanaka-Santos, M., Souza-Santos, R., & Czeresnia, D. (2007). Spatial analysis for stratification of priority malaria control areas, Mato Grosso State, Brazil. *Cadernos de Saude Publica*, *23*(5), 1099–1112. DOI: 10.1590/S0102-311X2007000500012 PMID: 17486233

Brewer, C. A., & Pickle, L. (2002). Evaluation of methods for classifying epidemiological data on choropleth maps in series. In *Annals of the Association of American Geographers* (Vol. 92, Issue 4, pp. 662–681). Routledge. DOI: 10.1111/1467-8306.00310

Brus, D. J., De Gruijter, J. J., Marsman, B. A., Visschers, R., Bregt, A. K., Breeuwsma, A., & Bouma, J. (1996). The performance of spatial interpolation methods and choropleth maps to estimate properties at points: A soil survey case study. *Environmetrics*, *7*(1), 1–16. DOI: 10.1002/(SICI)1099-095X(199601)7:1<1::AID-ENV157>3.0.CO;2-Y

Centers for Disease Control and Prevention (U.S.). G. and G. S. W. G. (GeoSWG). (2012). *Cartographic guidelines for public health*. https://stacks.cdc.gov/view/cdc/13359

Charles, A. C., Armstrong, A., Nnamdi, O. C., Innocent, M. T., Obiageri, N. J., Begianpuye, A. F., & Timothy, E. E. (2024). Review of Spatial Analysis as a Geographic Information Management Tool. *American Journal of Engineering and Technology Management*. DOI: 10.11648/j.ajetm.20240901.12

Coulson, M. R. C. (1987). In The Matter Of Class Intervals For Choropleth Maps: With Particular Reference To The Work Of George F Jenks. *Cartographica: The International Journal for Geographic Information and Geovisualization, 24*(2), 16–39. DOI: 10.3138/U7X0-1836-5715-3546

Cromley, E. K., & McLafferty, S. L. (2011). GIS and Public Health. In *GIS and Public Health*. Guilford Press.

Cromley, R. G., & Cromley, E. K. (2009). Choropleth map legend design for visualizing community health disparities. *International Journal of Health Geographics*, *8*(1), 52. DOI: 10.1186/1476-072X-8-52 PMID: 19778435

Fang, H., Xin, S., Pang, H., Xu, F., Gui, Y., Sun, Y., & Yang, N. (2022). Evaluating the effectiveness and efficiency of risk communication for maps depicting the hazard of COVID-19. *Transactions in GIS*, *26*(3), 1158–1181. DOI: 10.1111/tgis.12814 PMID: 34512105

Gelman, A., & Loken, E. (2014). *The Statistical Crisis in Science. 102*, 160. https://api.semanticscholar.org/CorpusID:146855235

Haining, R. (2012). GIS and public health, by E.K. Cromley and S.L. McLafferty. *International Journal of Geographical Information Science, April 2013*, 1–2. https://www.guilford.com/books/GIS-and-Public-Health/Cromley-McLafferty/9781609187507

Jenks, G. F., & Caspall, F. C. (1971). Error on Choroplethic Maps: Definition, Measurement, Reduction. *Annals of the Association of American Geographers*, *61*(2), 217–244. DOI: 10.1111/j.1467-8306.1971.tb00779.x

Jenks. (1967). *The Data Model Concept in Statistical Mapping* (Vol. 7). International Yearbook of Cartography.

Jimenez, S. H., Godot, X., Petronijevic, J., Lassagne, M., & Daille-Lefevre, B. (2024). Considering cognitive biases in design: An integrated approach. *Procedia Computer Science*, *232*, 2800–2809. DOI: 10.1016/j.procs.2024.02.097

Jiménez-Carvajal, C., Ruiz-Peñalver, L., Vera-Repullo, J. A., Jiménez-Buendía, M., Antolino-Merino, A., & Molina-Martínez, J. M. (2016). Weighing lysimetric system for the determination of the water balance during irrigation in potted plants. *Agricultural Water Management*, *183*, 6–13. DOI: 10.1016/j.agwat.2016.10.006

Ke, C., He, S., & Qin, Y. (2023). Comparison of natural breaks method and frequency ratio dividing attribute intervals for landslide susceptibility mapping. *Bulletin of Engineering Geology and the Environment*, *82*(10), 384. DOI: 10.1007/s10064-023-03392-0

Lavanya, A., Sindhuja, S., Gaurav, L., & Ali, W. (2023). A Comprehensive Review of Data Visualization Tools: Features, Strengths, and Weaknesses. *International Journal of Computer Engineering in Research Trends*, *10*(1), 10–20. DOI: 10.22362/ijcert/2023/v10/i01/v10i0102

MacEachren, A. M. (2008). *How maps work: representation, visualization, and design* (softcover 2004). Guilford Press.

Mu, W., & Tong, D. (2020). Mapping uncertain geographical attributes: Incorporating robustness into choropleth classification design. *International Journal of Geographical Information Science*, *34*(11), 2204–2224. DOI: 10.1080/13658816.2020.1726921

Murugesan, M. (2017, July 10). *NYC Education + Socio-Demographics*. https://geodacenter.github.io/data-and-lab/NYC_Tract_ACS2008_12/

Openshaw, S. (1983). *The modifiable areal unit problem: Spacial analysis-Concepts and Techniques in Modern Geography (CATMOG 38)* (Geo abstracts University of Anglia, Ed.). Geo abstracts University of Anglia.

Oxoli, D., Sabri, S., Rajabifard, A., & Brovelli, M. A. (2020). A classification technique for local multivariate clusters and outliers of spatial association. *Transactions in GIS*, *24*(5), 1227–1247. DOI: 10.1111/tgis.12639

Peterman, N. J., Palsgaard, P., Vashi, A., Vashi, T., Kaptur, B. D., Yeo, E., & Mccauley, W. (2022). Demographic and Geospatial Analysis of Buprenorphine and Methadone Prescription Rates. *Cureus*. DOI: 10.7759/cureus.25477 PMID: 35800815

Prestby, T. J. (2023). *Trust in maps: what we know and what we need to know*. DOI: 10.1080/15230406.2023.2281306

Rey, S. J., & Anselin, L. (2010). Handbook of Applied Spatial Analysis. In *Handbook of Applied Spatial Analysis*. Springer Berlin Heidelberg., DOI: 10.1007/978-3-642-03647-7

Simela, S. R., Kelepile, M., & Sebobi, T. I. (2025). Spatial analysis and associated risk factors of HIV prevalence in Botswana: Insights from the 2021 Botswana AIDS Impact Survey (BAIS V). *BMC Infectious Diseases*, *25*(1), 69. DOI: 10.1186/s12879-025-10464-x PMID: 39815215

Slocum, T. A., McMaster, R. B., Kessler, F. C., & Howard, H. H. (2022). *Thematic Cartography and Geovisualization* (4th ed.). Thematic Cartography and Geovisualization., DOI: 10.1201/9781003150527

Słomska-Przech, K., & Gołębiowska Małgorzata, I. (2021). Do different map types support map reading equally? Comparing choropleth, graduated symbols, and isoline maps for map use tasks. *ISPRS International Journal of Geo-Information*, *10*(2), 69. DOI: 10.3390/ijgi10020069

Sobjak, R., de Souza, E. G., Bazzi, C. L., Opazo, M. A. U., Mercante, E., & Aikes, J.Junior. (2023). Process improvement of selecting the best interpolator and its parameters to create thematic maps. *Precision Agriculture*, *24*(4), 1461–1496. DOI: 10.1007/s11119-023-09998-4

Stieb, D. M., Huang, A., Hocking, R., Crouse, D. L., Osornio-Vargas, A. R., & Villeneuve, P. J. (2019). Using maps to communicate environmental exposures and health risks: Review and best-practice recommendations. In *Environmental Research* (Vol. 176). Academic Press Inc., DOI: 10.1016/j.envres.2019.05.049

Sujithlal, S. P., Ahana, K. K., Satheesan, K., & Kottayil, A. (2024). Identification of the Tropopause Using the Jenks Natural Breaks Classification From 205-MHz Stratosphere-Troposphere Wind Profiler Radar. *IEEE Transactions on Geoscience and Remote Sensing*, *62*, 1–6. DOI: 10.1109/TGRS.2024.3386564

Viscarra Rossel, R. A., Chen, C., Grundy, M. J., Searle, R., Clifford, D., & Campbell, P. H. (2015). The Australian three-dimensional soil grid: Australia's contribution to the GlobalSoilMap project. *Soil Research (Collingwood, Vic.)*, *53*(8), 845–864. DOI: 10.1071/SR14366

Zhang, Y., & MacIejewski, R. (2017). Quantifying the Visual Impact of Classification Boundaries in Choropleth Maps. *IEEE Transactions on Visualization and Computer Graphics*, *23*(1), 371–380. DOI: 10.1109/TVCG.2016.2598541 PMID: 27875153

Zongfan, B., Ling, H., Xuhai, J., Ming, L., Liangzhi, L., Huiqun, L., & Jiaxin, L. (2022). Spatiotemporal evolution of desertification based on integrated remote sensing indices in Duolun County, Inner Mongolia. *Ecological Informatics*, *70*, 101750. DOI: 10.1016/j.ecoinf.2022.101750

Chapter 5
Predictive Classification and Feature Selection in Forest Cover Type Data

Nikolaos Papafilippou
https://orcid.org/0009-0003-3148-7229
Aristotle University of Thessaloniki, Greece

Zacharenia Kyrana
https://orcid.org/0000-0001-9269-0675
Aristotle University of Thessaloniki, Greece

Emmanouil D. Pratsinakis
https://orcid.org/0000-0002-3725-3525
Aristotle University of Thessaloniki, Greece

Christos Dordas
https://orcid.org/0000-0002-7027-474X
Aristotle University of Thessaloniki, Greece

Angelos Markos
https://orcid.org/0000-0002-4204-3573
Democritus University of Thrace, Greece

ABSTRACT

This study assessed the predictive accuracy and key features of Machine Learning algorithms for classifying forest cover types using the UCI "Forest Covertype" dataset (n=581.012). Algorithms included SVC, Logistic Regression, SGD, Decision Trees, K-NN, Gaussian Naïve Bayes, Neural Networks, Random Forest, and extra Trees. Methods such as GridSearch, Adaboosting, and cross-validation were applied

DOI: 10.4018/979-8-3693-9400-7.ch005

for optimization. Results indicated that K-NN achieved the highest accuracy on the original data, while random Forest and Extra Trees outperform others on both original and standardized data. Standardization improved the accuracy of SVC, Neural Networks, and SGD, had no effect on decision Trees, Random Forest, and Extra Trees, but reduced accuracy for K-NN and Gaussian Naïve Bayes. Feature importance analysis highlighted Elevation, Soil type, and Wilderness area as major contributors. Furthermore, prediction of a random data vector was consistent across algorithms on standardized data but varied for K-NN and Extra Trees when using the original data.

INTRODUCTION

Machine Learning has gained a significant position in data science and serves as a powerful tool for extracting valuable insights and developing predictive models across various fields (Hassan & Alsamhi, 2024). With the increasing availability of large volumes of data, the need for efficient and accurate prediction and classification algorithms is greater than ever (Punia et al., 2021). When selecting the appropriate algorithm for a specific application, comparing the performance of different Machine Learning algorithms becomes a necessary step. Of particular interest is understanding the features of the data that contribute most to the predictive capability of the models (Joshi et al., 2023) and, consequently, to improving the accuracy of the results.

One of the fields where machine learning has applied is forest cover type classification, a critical task for ecological monitoring and sustainable resource management (Kamilaris & Prenafeta-Boldú, 2018; Aziz et al., 2024). Forests play a vital role in maintaining ecological balance, acting as carbon sinks, regulating water cycles, and providing habitats for diverse species. Accurately classifying forest cover types aids in managing resources, preserving biodiversity, and supporting environmental planning. Recent advancements in machine learning have significantly improved the accuracy and efficiency of this process (Sjöqvist et al., 2020). Various studies have explored the application of algorithms such as Random Forest (RF) (Kumar & Sinha, 2020), RF, Support Vector Machines (SVM), Naïve Bayes (NB) (Sjöqvist et al., 2020), RF and Convolutional Neural Networks (CNNs) (Aziz et al., 2024), Logistic Regression (LR) (Kumar et al., 2014), CNN and Linear Discriminant Analysis (LDA) (Blackard & Dean, 1999) or have compared different classification algorithms (Kishore et al., 2016) or focused on feature engineering and extraction (Joshi et al., 2023). RF has been widely favored due to its robustness against overfitting and ability to handle high-dimensional datasets, SVM has shown strong performance in scenarios with limited data but requires careful tuning of parameters for optimal results, while comparative analyses often highlight the superior accuracy of ensemble methods like Extra Trees or Adaboosting, and emerging approaches, including deep learning models such as CNNs, are also being explored for their ability to automatically extract spatial features from remote sensing data (Kumar & Sinha, 2020; Sjöqvist et al., 2020; Aziz et al., 2024; Kishore et al., 2016). Despite these advancements, challenges such as imbalanced datasets, computational costs, and the need for higher accuracy models remain active areas of research.

This study focuses on evaluating and comparing Machine Learning algorithms in predictive classification and analyzing their significant features. Specifically, using the Forest Covertype dataset from the UCI repository, the aim of this study was to compare Machine Learning algorithms regarding their predictive ability to classify data into one of the forest cover types (Cover Type) based on certain features. Through this analysis, the study highlights the performance of each algorithm either on original or on standardized data, the advantages of data standardization, and the features that significantly contribute to classification accuracy, aiming to draw useful conclusions for future applications in the field of Machine Learning.

MACHINE LEARNING AND ALGORITHMS IN SUPERVISED LEARNING

Machine Learning (ML) refers to the field of computer science that studies the creation of algorithms capable of "learning" from collected data, utilizing prior knowledge and experience, without being explicitly programmed with specific rules (Mitchell, 1997). The aim is to discover patterns and relationships to make predictions or decisions (Sarker, 2021). There are three main types of Machine Learning (Sarker, 2021): a) Supervised Learning, where the algorithm builds a function that maps given inputs (training set) to known desired outputs, with the goal of generalizing the function to inputs with unknown outputs, b) Unsupervised Learning, where the algorithm constructs a model for a set of inputs in the form of observations without knowing the desired outputs and c) Reinforcement Learning, where the algorithm learns a strategy of actions through direct interaction with the environment. The first type is used for problems such as classification, prediction, and interpretation, the second for problems like association analysis, clustering, and dimensionality reduction, while the third is applied to planning problems, such as controlling the movement of a robot (Mahesh, 2020; Sen et al., 2020).

Supervised Learning (Johri et al., 2021), as previously mentioned, is a type of Machine Learning algorithm where the system is trained using data that includes both inputs and their corresponding desired outputs. The goal for the algorithm is to learn to associate inputs with correct outputs so that it can make accurate predictions when provided with new, unknown inputs. This process involves two stages: training, where the model learns from known data, and testing, where its ability to generalize to new data is evaluated. Classification and regression are two fundamental techniques in supervised learning, each with different objectives (Murphy, 2012). In classification, the model is trained to categorize data into predefined categories, producing discrete outputs, such as spam detection in emails, disease diagnosis based on symptoms, or land cover type classification (Murphy, 2012). In regression, the goal is to predict continuous numerical values, estimating within a range, such as predicting temperature or the price of a property (Bishop, 2016; Sen et al., 2020).

The main supervised learning algorithms for classification, used depending on the type of problem and the available data, are summarized below.

Logistic Regression

Despite its name, logistic regression is used for classification, as it estimates the probability that a sample belongs to a category. In its simplest form, it is used for binary classification, where the goal is to predict a binary output variable, such as whether an event will occur (yes or no), based on one or more input features (Bishop, 2016; Bisong, 2019). The fundamental function used is the sigmoid function, which predicts the probability of a category and is defined as:

$$\sigma(z) = \frac{1}{1+e^{-z}},$$

where z is the linear combination of input variables x_i and their corresponding weights w_i $z=w_1x_1+w_2x_2+\ldots+w_nx_n$. The goal of the model training is to find the set of weights $\boldsymbol{w}$ that minimizes the cross-entropy loss function:

$$L(w) =-1m\sum\sum yi\,jlogkj=1mi=1P(y=j|xi)$$

where m is the number of training samples, $\boldsymbol{x}(i)$ which could be continues, categorical or binary are the input variables and $\boldsymbol{y}(i)$ (0 or 1) is the output variable for the *i-th* sample, and $h(\boldsymbol{x}(i))=\sigma(\boldsymbol{w}^T$ ⍰$\boldsymbol{x}(i))$ is the predicted probability that the output variable for the *i-th* sample belongs to category $y(i)$.

In the case of multiclass classification (Jiang et al., 2016), where the goal is to predict an output variable with more than two categories (k categories), such as the type of forest cover in an area or the genre of an article (e.g., political, sports, technology, health), multinomial logistic regression (softmax) is used. The softmax function computes the probability for each category k:

$$P(y = k|x) = \frac{e^{z_k}}{\sum_{j=1}^{k} e^{z_j}},$$

where $z_k = w_1x_1 + w_2x_2 + \ldots + w_nx_n$ for category k. The softmax function ensures that the probabilities of all categories sum to 1. The goal of the model training is to find the set of weights $\boldsymbol{w}$ that minimizes the cross-entropy loss function:

$$L(w) =-1m\sum\sum yijlogkj=1mi=1P(y=j|xi)$$

where m is the number of samples, k is the number of categories, y_{ij} is a binary variable that equals 1 if sample i belongs to category j and 0 otherwise, and $P(y=j|x_i$) is the predicted probability that sample i belongs to category j. The logistic regression algorithm is simple and fast, especially for large datasets, and performs well on linearly separable data. However, it is not suitable for non-linearly separable data.

Support Vector Machine (SVM)

The SVM algorithm (Dong, 2021; Zhang, 2012) uses the maximum-margin hyperplane to separate data into categories with the largest possible margin between classes. The margin is defined as the distance between the two closest points from different categories. These points are called support vectors. Suppose we have a dataset that is linearly separable, (x_i, y_i) with $i=1, ..., n$ are feature vectors and $y_i \in \{-1, 1\}$ is the label of the class to which the x_i sample belongs (in the case of two categories). The SVM aims to find a hyperplane that separates the samples with $y_i=1$ from those with $y_i=-1$. A hyperplane can be described by the equation:

$$\mathbf{w} \cdot \mathbf{x} + \mathrm{b} = 0$$

where w is the weight vector, x is the feature vector, b is the bias term, and $w{\cdot}x$ is the dot product of the vectors $\boldsymbol{w}$ and $\boldsymbol{x}$. The margin is given by $M=\frac{2}{w}$, where $||w||$ is the measure of the weight vector. Essentially, maximizing the margin is equivalent to minimizing $||w||$. The mathematical formulation of the SVM problem is to minimize the value of the function min $\frac{1}{2}w^2$ subject to the constraint $y_i(w \cdot x_i + b) \geq 1$ for each i, ensuring that the points are correctly classified according to their labels. The problem is solved using the Lagrange function:

$$\mathrm{L}(\mathrm{w},\mathrm{b},\alpha) = \frac{1}{2}w^2 - \sum_{i=1}^{n} a_i\left[y_i(w \cdot x_i + b) - 1\right]$$

where α_i are the Lagrange multipliers, which satisfy the condition $\alpha_i \geq 0$. The values of α_i obtained from solving the problem determine which points are the support vectors. To predict the category of a new point $\boldsymbol{x}$, we use the function:

$$f(x) = sign \sum_{i=1}^{n} a_i y_i \left(x_i \square x + b \right)$$

where sign denotes the sign (positive or negative) of the expression.

Figure 1. Graphical representation of finding the optimal separating hyperplane. In García-Gonzalo et al. (2016), Hard-rock stability analysis for span design in entry-type excavations with learning classifiers. Materials, 9(7), 531, (p. 7).

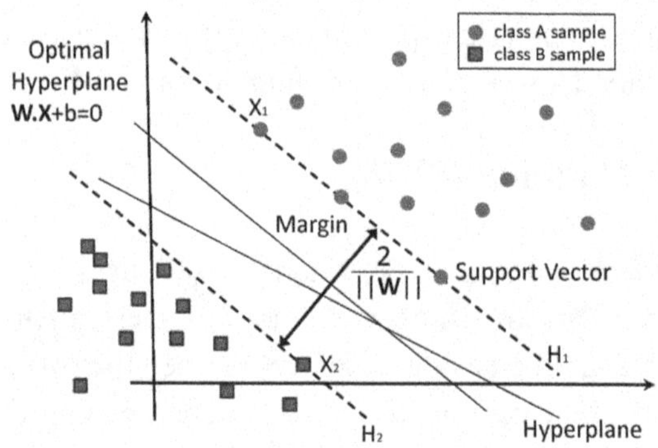

In the case of non-linearly separable data, kernel functions (Murphy, 2012) are used to project the data into a higher-dimensional space where they become linearly separable. The kernel functions are:

· The linear kernel $K(x_i, x_j) = x_i \cdot x_j$, where $x_i \cdot x_j$ is the dot product of two vectors.
· The polynomial kernel $K(x_i, x_j) = (x_i \cdot x_j + c)^d$, where d is the degree of the polynomial.
· The RBF (Radial Basis Function - Gaussian) kernel $K(x_i, x_j) = exp(-\gamma x_i - x_j^{2})$, where $x_i - x_j$ is the distance between x_i, x_j, and γ is a parameter that determines the separation of classes based on how "close" or "far" from each training point the model considers other points.

When γ is large, the Gaussian function becomes narrower (Murphy, 2012). This means that each training example has a more localized influence. Essentially, the model will focus on creating decision boundaries that are tighter around the training points, leading to high complexity. This can result in overfitting because the model captures noise as part of the decision boundary. When γ is small, the Gaussian function becomes broader. Training points have a wider influence, and the model creates smoother decision boundaries. This can lead to underfitting, as the model might fail to capture sufficient detail to distinguish between classes. Parameter γ works in conjunction with the C parameter (Zhang, 2012). The C parameter is a regularization parameter which balances the trade-off between achieving a low error on training data and maintaining a simple decision boundary. The value of C determines how strictly the SVM algorithm enforces correct classification. A high C value is suitable for datasets where misclassification is costly, for example a medical diagnosis, while a low C value is appropriate for noisy datasets, where overfitting to training data may be problematic. C and γ parameters are selected using methods like GridSearch or cross-validation.

Figure 2. Projection of data into a higher-dimensional space using the Kernel function.

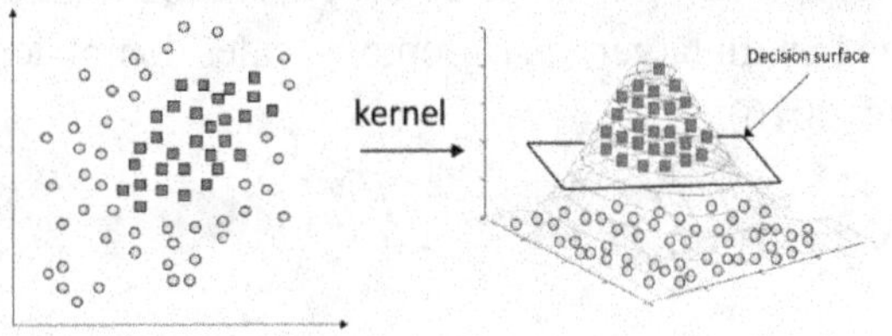

Jain A. (2024). Available at https://medium.com/@abhishekjainindore24/svm-kernels-and-its-type-dfc3d5f2dcd8

The SVM algorithm (Zhang, 2012) can also be extended to multi-class classification problems using two methods: One-vs-One and One-vs-Rest. In the first approach, a separate binary classifier is trained for each possible pair of classes, creating classifiers for every pair combination. Thus, if there are k classes, $\binom{k}{2} = \frac{k \cdot (k-1)}{2}$ classifiers need to be trained. During prediction, each classifier "votes" for one of the two classes it has been trained on, and the class with the most votes is selected as the final prediction. In the second approach, a separate binary classifier is trained for each class, treating that class as positive and all other classes combined as negative. For k classes, k binary classifiers need to be trained. During prediction, each classifier produces a score or probability, and the class with the highest score or probability is chosen as the final prediction.

K-Nearest Neighbors

K-Nearest Neighbors algorithm (Zhang, 2016; Suyal & Goyal, 2022) is a simple and effective machine learning algorithm for classification and regression. The basic idea of K-NN is that a new sample is classified or estimated based on the characteristics of its K nearest neighbors, that is, its K closest points in the feature space.

We define a value for K, which is the number of nearest neighbors to be considered for the decision, and calculate the distance of the new sample from each point in the training dataset. Usually, the Euclidean distance is used:

$d(x,y)=\sqrt{}\sum(xi-yi)2ni=1,$

where $\boldsymbol{x}, \boldsymbol{y}$ are two points in the feature space, $i=1, 2, \ldots, n$, where n is the space dimension. The K closest data points to the new sample are selected based on the distance. For classification, the category of the new sample is determined by the majority class of the K nearest neighbors, while for regression, the value of the new sample is determined by the average of the values of the K nearest neighbors. The choice of K is critical for the performance of K-NN algorithm, and the optimal value is usually selected using hyperparameter optimization methods such as Grid-Search. Since K-NN algorithm is distance-based, feature normalization is required to ensure that features with larger measurement scales do not dominate the distance calculation (Zhang, 2016).

Gaussian Naïve Bayes

Gaussian Naïve Bayes algorithm (Jahromi & Taheri, 2017; Kamel et al., 2019) is mainly used for classification problems with continuous data. The Naive Bayes algorithm is based on Bayes' theorem and the "naive" assumption that features are independent of each other. Gaussian Naive Bayes assumes that each feature follows a normal (Gaussian) distribution, which simplifies the calculation of probabilities for continuous data. Bayes' theorem states that the probability of a sample $\boldsymbol{x}$ belonging to a class C_k can be calculated as:

$$P(C_k \mid x) = \frac{P(x \mid C_k) \cdot P(C_k)}{P(x)}$$

where $P(C_k| x)$ is the posterior probability, that is the probability that x belongs to class C_k

$P(x| C_k)$ is the likelihood, that is the probability of x given that it belongs to class C_k, $P(C_k)$ is the prior probability of class C_k, and $P(x)$ is the total probability of sample $\boldsymbol{x}$. We calculate the probability $P(C_k)$ based on the training data and use the normal distribution to compute $P(x_j| C_k)$ for each feature x_j and each class C_k. We combine the probabilities of the features and the prior probabilities to compute $P(C_k| x)$ for each class. The sample is then assigned to the class with the highest posterior probability (Jahromi & Taheri, 2017; Kamel et al., 2019).

Decision Trees

Decision Trees algorithm (Charbuty & Abdulazeez, 2021; Liu, 2020) is used for both classification and regression. This is a rule-based method that creates a tree-like decision model by analyzing the data based on their features. A decision tree consists of the root (the first split point), nodes (points where the data is split into two or more categories), branches (connections between nodes representing decisions), and leaves (final nodes representing predictions or outcomes). The algorithm selects the feature that best separates the data using appropriate measures such as the Gini index, Entropy, and Information Gain (IG) for classification, and mean squared error (MSE) for regression. The data is split into subsets based on the feature's value, and this process continues iteratively for each subset until a stopping criterion is met, such as the maximum depth of the tree (max_depth) or the minimum number of samples in a leaf node.

The Gini index is defined as $Gini=1-\sum_{i=1}^{n} p_i$, where p_i is the probability of a sample belonging to class i at node t, and n is the number of different classes (Tangirala, 2020). It is used to estimate the "purity" of a node. The lower the Gini value, the "purer" the node, meaning that it contains samples from only one class. For each feature, the algorithm calculates the Gini index before and after splitting the data and selects the split that minimizes the Gini value.

Entropy, defined as $S=-\sum_{i=1}^{n} p_i \cdot log_2 p_i$, is used to evaluate the quality of splits and identify features that reduce uncertainty, leading to "cleaner" categories. The entropy value ranges from 0 (when all samples belong to the same class) to $log_2 n$, where n is the number of classes. For each feature A, the algorithm computes the entropy before and after the data split and then calculates the reduction in entropy (Information Gain)

$$IG(S,A)=S-\sum|Sv||S|\bullet(Sv)v\in values(A) ,$$

where S is the entropy of the original set (node), and Sv is the entropy of the subset based on the ν values of the feature A. The feature selected for the split is the one that maximizes the Information Gain (Tangirala, 2020).

Decision trees are an easily interpretable algorithm that essentially represents a sequence of decisions (Liu et al., 2020). They can handle complex, non-linear datasets and do not require normalization. However, deep trees may lead to overfitting and fail to generalize, while small changes in the data can significantly alter the structure of the tree. To address the limitations of decision trees, there are boosting methods (Parmar et al., 2019) such as Random Forests and Extra Trees.

Random Forests (Breiman, 2001; Schonlau & Zou, 2020) is an ensemble learning method used for both classification and regression tasks. It builds multiple decision trees during training and combines their outputs to make more accurate and robust predictions. The algorithm mitigates the weaknesses of individual decision trees, such as overfitting and high variance, by introducing randomness in the tree-building process. Random subsets of the training data are created by sampling with replacement (bootstrap sampling) and each tree in the forest is trained on a different subset of the data. At each split within a tree, a random subset of features is selected, and the best split is determined only from this subset. This introduces diversity among the trees and reduces the correlation between them. For classification, the final output is determined by majority voting among the predictions of all trees, while for regression the final output is the average of the predictions of all trees. By averaging the predictions of multiple trees, Random Forests are less prone to overfitting than individual decision trees, and less sensitive to noise and outliers compared to single trees (Breiman, 2001; Schonlau & Zou, 2020).

Extra Trees (Extremely Randomized Trees) (Kocev et al., 2020) is also an ensemble learning method similar to Random Forests, but with difference in how the decision trees are built. It also uses multiple decision trees to create an ensemble model, but it introduces more randomness during the tree construction process to increase diversity and improve generalization. Extra Trees also use bootstrap sampling, at each split in the tree not only select a random subset of features (like Random Forests) but also randomly select the threshold for the split, rather than choosing the best possible split for each feature. Extra Trees can be faster to train than Random Forests because of the randomization in the tree-building process, which avoids the need to search for the best split. This can significantly reduce computation time (Kocev et al., 2020).

Neural Networks

Neural Networks (Bishop, 2016; Mehlig, 2021) is a set of machine learning algorithms inspired by the structure and functionality of the human brain. Depending on their architecture, they are used to solve complex problems such as classification, regression, image analysis, and natural language processing. The basic unit of a neural network is the neuron. Each neuron receives inputs, combines them using weights, adds a constant (bias), and applies an activation function:

$$y=\varphi(\sum wi\bullet xi+bni=1)$$

where w_i is the weight connecting the neuron of the previous layer to the current neuron, x_i is the input from the previous neuron, n is the number of features and b is the bias term. The activation functions used include the sigmoid function and the softmax function, which are applied in logistic regression, as well as the RELU function RELU(x)=max (0, x) (Rectified Linear Unit), which takes the value 0 for $x<0$ and the value x for $x>0$, and the hyperbolic tangent function (Bishop, 2016; Mehlig, 2021):

$$\tanh(x)=\frac{e^x - e^{-x}}{e^x + e^{-x}},$$

Figure 3. Representation of the mathematical processes of an artificial neural network.

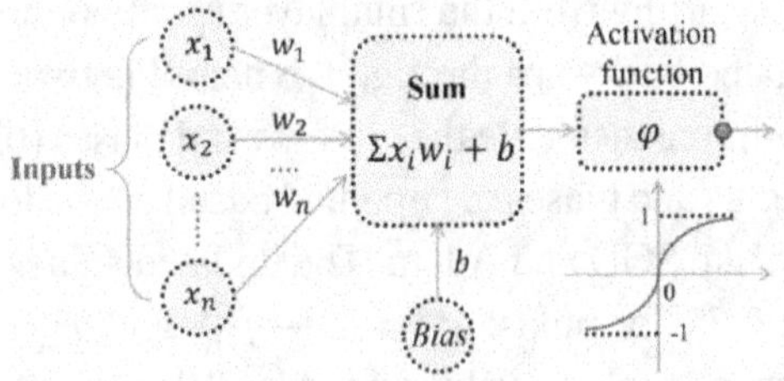

Available at: https://www.linkedin.com/pulse/neural-networks-from-biological-artificial-muhammed-nizam-pcxwf/

A fundamental building block of a neural network (Bishop, 2016) is the layers. There are three main types of layers: the Input Layer, where data is introduced; the Hidden Layers, where most of the computational tasks are performed and which are not visible at the input or output; and the Output Layer, which produces the network's final output. Defining the network's architecture involves determining the number of layers, the neurons in each layer (Dense), and the activation functions. Data is propagated through the network (Forward Propagation) from the input layer to the output layer via neurons and their connections. The error between the predicted output and the actual value is calculated using the loss function (such as the Mean Squared Error or Cross-Entropy), and this error is used to adjust the connection weights to improve the network's accuracy in future predictions through the process of Backpropagation. Backpropagation (backward error propagation) (Bishop, 2016) is an algorithm used to train neural networks, enabling the update of the network's weights to minimize prediction error. The algorithm relies on calculating the derivatives of the cost function with respect to the network weights using the chain rule. Let $z^{(l)}=w^{(l)}a^{(l-1)}+b$ represent the input vector to a layer, where $w^{(l)}$ is the weight vector and $a^{(l-1)}$ is the activation vector of the previous layer. Using the chain rule, the weights that minimize the cost function can be determined:

$$\frac{\partial L}{\partial w_{ij}^{(l)}} = \frac{\partial L}{\partial z_j^{(l)}} \text{?} \frac{\partial z_j^{(l)}}{\partial w_{ij}^{(l)}}$$

where i corresponds to the neuron in the previous layer l-1 from which the connection (weight) originates, and j corresponds to the neuron in the current layer l, w_{ij} is the weight connecting the i-th neuron in layer $l-1$ to the j-th neuron in layer l, and L is the loss function, mean square error, $L = MSE = \frac{1}{n}\sum_{i=1}^{n}(y_i - \hat{y}_i)^2$, where y_i and $\hat{y}_i$ represent the observed and predicted values, respectively for regression or Cross-Entropy function for classification (see above, Logistic Regression).

From the definition of $z^{(l)}$ it follows that: $\frac{\partial z_j^{(l)}}{\partial w_{ij}^{(l)}} = a^{(l-1)}$, where $a^{(l-1)}$ is the activation of the i-th neuron in the previous layer, and the term $\frac{\partial L}{\partial z_j^{(l)}}$ represents the error $\delta_j^{(l)}$ of the j-th neuron in layer l. Starting from the final layer (prediction/output), the error is calculated and propagated backward through the neural network. This backward propagation uses the chain rule to update the weights and biases (bias) in each layer. During this process, weights and biases are updated based on the loss function L and optimization algorithms, like SGD and Adam. The updated values of weights w and bias b are calculated using the formulas: $w_{new} = w_{old} - \eta\frac{\partial L}{\partial w}$ and $b_{new} = b_{old} - \eta\frac{\partial L}{\partial b}$, where η is the learning rate, typically taking values between 10^{-1} and 10^{-6} (Bishop, 2016).

Figure 4. Learning process of neural network.

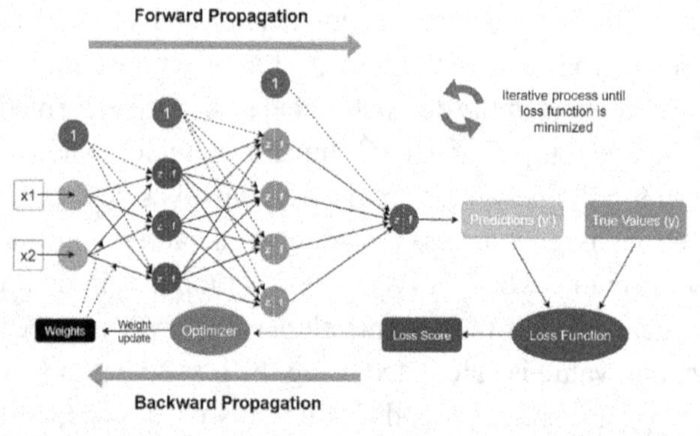

Available at: https://medium.com/data-science-365/overview-of-a-neural-networks-learning-process-61690a502fa

The training of a neural network is conducted over successive epochs. Each epoch involves processing the entire training dataset by the model, and training is repeated for multiple epochs to achieve better model performance. During an epoch, the network updates its weights based on the error calculated. Weight updates are performed on subsets of data called batches. To avoid overfitting, a technique called Dropout is employed during each epoch, where a percentage of neurons in the network is temporarily "dropped out" or ignored. This prevents the model from becoming overly reliant on specific neurons, enabling it to learn more generalized representations. After sufficient epochs, the loss function reaches a minimum, and the changes in weights approach zero. At this point, the training process converges to a solution (Bishop, 2016).

One training method for neural networks is Batch Gradient Descent (BGD), where the total error is calculated for all data samples (batches) in the training set, and weights are updated based on the average of the samples once per epoch. A variation of this method is Mini-batch Gradient Descent, which, instead of using the entire dataset (as in BGD) or a single sample (as in Stochastic Gradient Descent, SGD), uses small subsets (mini-batches) of the data to compute derivatives and update weights. This strikes a balance between computational efficiency and the ability to escape local minima or saddle points during optimization (Bishop, 2016).

Stochastic Gradient Descent (SGD)

Stochastic Gradient Descent (SGD) (Sharma, 2018; Wang & Joshi, 2021) is an optimization algorithm used in machine learning. Its purpose is to find the minimum of a loss function by adjusting the model's parameters θ through iterative steps determined by the gradient of the function. The process is described by the equation:

$$\theta_{t+1}=\theta_t-\eta\cdot\nabla_\theta L(\theta_t)$$

where $\nabla_\theta L$ represents the gradient of the loss function, η is the learning rate, and t denotes the iteration step. SGD randomly selects a single sample from the training dataset and updates the parameters based on the above equation until convergence is achieved or a predetermined number of iterations is completed. SGD is a fundamental algorithm for training artificial neural networks, as well as for linear classification models such as Logistic Regression or Support Vector Machines (SVM) (Sharma, 2018; Wang & Joshi, 2021).

Challenges and Proposed Solutions in the Application of Machine Learning

Machine learning involves several challenges (L'heureux et al., 2017) that can affect the performance and accuracy of models. One of the fundamental problems is overfitting (Lopez et al., 2022), where the model learns the training data too well, including noise and irrelevant details, but resulting in poor performance on unseen data. This happens because the model memorizes the training data instead of learning general patterns, making it unable to generalize to new inputs. On the other hand, underfitting occurs when the model is too simple to capture the underlying patterns in the data, leading to low accuracy both on training and test sets. Additionally, issues with data quality such as missing values, imbalanced datasets, or irrelevant features can significantly affect the algorithm's ability to learn effectively. Computational inefficiencies, including high training times and memory usage, often arise with complex models or large datasets, making it challenging to scale. Finally, the choice of hyperparameters can also impacts performance, where incorrect configurations may prevent the model from reaching its optimal potential. Identifying and addressing these problems is crucial for achieving robust and efficient machine learning models (Yang & Shami, 2020).

To address performance challenges in machine learning algorithms, several solutions can be employed. Regularization techniques (Ng, 2004), such as L1(Lasso) or L2 (Ridge) regularization, can help reduce overfitting by penalizing overly complex models. Additionally, techniques like cross-validation can ensure that the model generalizes well to unseen data. For underfitting, using more complex models or incorporating additional features can enhance the model's ability to capture data patterns (Pedregosa et al., 2011). Addressing data quality issues is critical, and this can involve preprocessing steps such as imputation for missing values, feature selection to remove irrelevant attributes, class-weighted loss functions to balance datasets. To moderate computational inefficiencies, dimensionality reduction techniques like PCA (Principal Component Analysis) are used, and algorithms designed to handle high-dimensional data, such as tree-based models are chosen. Hyperparametes (Yang & Shami, 2020) can be optimized through GridSearch method or Bayesian optimization. By combining these strategies, machine learning algorithms can achieve improved accuracy, efficiency, and generalization.

Performance Evaluation of Supervised Machine Learning Algorithms

The performance evaluation of supervised machine learning algorithms (supervised learning) is based on various metrics, which differ depending on the type of problem (classification or regression). For regression problems, measures like Mean Squared Error (MSE), $MSE = \frac{1}{n}\sum_{i=1}^{n}(y_i - \hat{y}_i)^2$ and Mean Absolute Error (MAE), $MAE = \frac{1}{n}\sum_{i=1}^{n}|y_i - \hat{y}_i|$, where $y_i, \hat{y}_i$ are the observed and predicted values respectively and n is the number of samples, assess the accuracy of predictions against the actual values (Carvalho et al., 2020). In classification problems, to evaluate the performance of a classifier, we use vectors for which the class membership is known. The decisions made by the classifier for these vectors are organized into a confusion matrix.

Table 1. Confusion matrix for binary classification

	Predicted Positive	Predicted Negative
Actual Positive	TP	FN
Actual Negative	FP	TN

TP: True Positive, FN: False Negative, FP: False Positive, TN: True Negative

The evaluation of a classifier's performance (or a classification model) is based on the number of records in the test set that were correctly or incorrectly predicted by the classifier. To compare the performance of different models, various performance metrics are used (Carvalho et al., 2020; Grandini et al., 2020; Lopez et al., 2022), such as:

Accuracy $=\frac{\text{TP+TN}}{TP+TN+FP+FN}$, which refers to the percentage of correct predictions out of the total predictions, and conversely, the error rate evaluation,

Error rate $=\frac{\text{FP+FN}}{TP+TN+FP+FN}$, which refers to the percentage of incorrect predictions out of the total predictions,

Precision $=\frac{\text{TP}}{TP+FP}$, which refers to the percentage of actual positives out of those predicted as positive,

Recall $=\frac{\text{TP}}{TP+FN}$, which refers to the percentage of those predicted as positive out of the actual positives,

F-score ή F1 $=2\cdot\frac{\text{precision}\times\text{recall}}{precision+recall}$, which is the harmonic mean of precision per class and recall,

G-score $=\sqrt{\text{precision}\times\text{recall}}$, which is the geometric mean of precision per class and recall,

Matthews Correlation Coefficient MCC

$=\frac{\text{TP}\times\text{TN} - \text{FP}\times\text{FN}}{\sqrt{(TP+FP)\times(TP+FN)\times(TN+FP)\times(TN+FN)}}$, which ranges from -1 to 1, where 1 corresponds to perfect prediction, 0 to random prediction, and -1 to inverse prediction (correct instances are predicted as incorrect,

Balanced accuracy BA $=\frac{\frac{TP}{P}+\frac{TN}{N}}{2}$, which is the mean of recall per class,

Cohen's Kappa K $=\frac{P_o-P_e}{1-P_e}$, where P_o=model's accuracy and $P_e=P_{Positive}+P_{Negative}$, $P_{Positive}=\frac{\text{TP+FN}}{N}\times\frac{\text{TP+FN}}{N}$ and $P_{Negative}=\frac{\text{TN+FP}}{N}\times\frac{\text{TN+FN}}{N}$.

In case of binary classification, two additional metrics are used: ***sensitivity***, which is essentially the recall of the first class or the true positive rate (TPR), calculated as: ***sensitivity*** $=\frac{TP}{TP+FN}$, and ***specificity***, which is the recall of the second class or the true negative rate (TNR), calculated as: ***specificity*** $=\frac{TN}{TN+FP}$. For the visual comparison of classification models, the ROC (Receiver Operating Characteristic) curve can also be employed, showing the relationship between the true positive rate (sensitivity) and the false positive rate, calculated as 1-*specificity*. The area under the ROC curve (AUC) is a measure of the model's accuracy. The closer the ROC curve is to the diagonal, with an AUC of 0.5, the less accurate the model is. A higher AUC indicates better model performance. AUC provides a single scalar value to summarize the trade-off between sensitivity and specificity across all thresholds. When AUC=1 means perfect classification, that is the model correctly distinguishes between all positive and negative cases without error. AUC=0.5 means random guessing, that is the model has no discriminative ability. When AUC<0.5 means worse than random guessing, that is the model is systematically predicting the wrong classes, which suggests issues in the model or data. When comparing models, a higher AUC generally reflects better overall performance in classification tasks (Carvalho et al., 2020; Grandini et al., 2020).

Figure 5. Comparison of algorithm's accuracy based on the ROC curve.

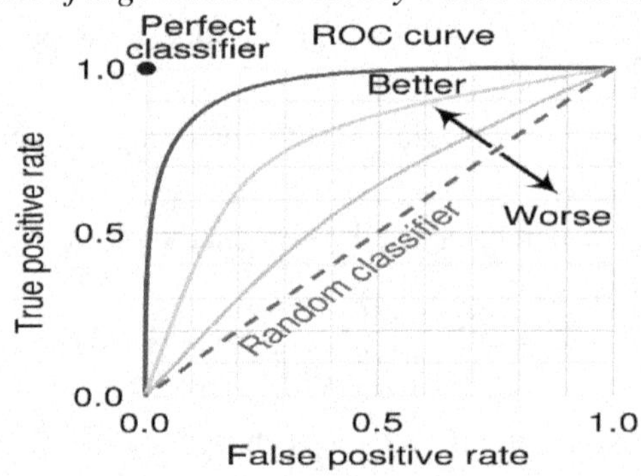

Available at: https://commons.wikimedia.org/w/index.php?curid=109730045

In multiclass classification problems, accuracy is affected when dealing with imbalanced classes. In such cases, we use metrics like Balanced Accuracy, which is the average of the recalls for each category, and Balanced Accuracy Weighted, which is the weighted average of the recalls, with the weights being the relative frequencies of each class (Grandini et al., 2020):

Balanced Accuracy $= \frac{\sum_{i=1}^{k} recall_i}{k}$, where k is the number of classes,

Weighted Balanced Accuracy $= \sum_{i=1}^{k} w_{i} recal$, where $w_i = n_i / N$, n_i the frequency of class i and N the total number of samples.

Also, in case of multiclass classification, the metrics Macro/Micro average precision, recall, and F1 score are used, and they are defined as follows:

Macro Average Precision $= \frac{\sum_{k=1}^{K} precision_k}{K}$, where k is the number of classes,

$$\textbf{Macro Average Recall} = \frac{\sum_{k=1}^{K} recall_k}{K}$$

$$\textbf{Macro F1 score} = \frac{\sum_{k=1}^{K} F1_k}{K}$$

$$\textbf{Micro Average Precision} = \frac{\sum_{k=1}^{K} TP_k}{\sum_{k=1}^{K} (TP_k + FP_k)}$$

$$\textbf{Micro Average Recall} = \frac{\sum_{k=1}^{K} TP_k}{\sum_{k=1}^{K} (TP_k + FN_k)}$$

$$\textit{Micro F1 score} = 2 \cdot \frac{\text{(Micro Average Precision)x (Micro Average Recall)}}{\text{Micro Average Precision+ Micro Average Recall}}$$

The difference between Macro/Micro performance metrics (Lee et al., 2024) lies in how they handle the classes during their calculation, particularly in the case of imbalanced class distributions. Macro metrics compute the average of the metrics for each class, giving equal weight to all classes regardless of their size. On the other hand, Micro metrics take all classes into account during the calculation, weighting the classes based on their frequency. Similarly, in correspondence with binary classification, the following metrics are also used:

$$\textbf{\textit{MCC}} = \frac{c \times s - \sum_{k=1}^{K} p_k \times t_k}{\sqrt{\left(s^2 - \sum_{k=1}^{K} p_k^2\right)(s^2 - \sum_{k=1}^{K} t_k^2)}}$$, where $c = \sum_{k=1}^{K} C_{kk}$ the sum of the elements along the main diagonal of the confusion matrix C, s the total number of samples, p_k the total of column k, and t_k the total of row k,

$$\textbf{Cohen's Kappa K} = \frac{c \times s - \sum_{k=1}^{K} p_k \times t_k}{s^2 - \sum_{k=1}^{K} p_k \times t_k}.$$

MATERIALS AND METHODS

The Machine Learning algorithms applied in this study included the Support Vector Machine (SVM) algorithm, specifically the Support Vector Classifier (SVC), the Stochastic Gradient Descent (SGDClassifier) algorithm, the Gaussian Naïve Bayes (GaussianNB) algorithm, the K-Nearest Neighbor (K-NN) algorithm (Mahesh, 2020), the Decision Tree Classifier algorithm (Liu et al., 2020), the Logistic Regression Multinomial algorithm (Bisong, 2019), as well as Neural Networks. Additionally, the ensemble methods Random Forest Classifier and ExtraTrees (Yadav & Pal, 2020), along with the Adaboosting method, were utilized. Also, the optimal parameters of the SVC, Decision Tree and K-NN algorithms were searched through the command GridSearch (Python 3.10), as well as the features with the greatest contribution through the command feature importances (Eidelman, 2020). Furthermore, the characteristics were standardized through the StandardScaler command, so that they have a mean value of zero and a standard deviation of one, while cross-validation (cross-validation with k-folds=5) was applied to each algorithm to strengthen the reliability in performance of the model and reducing the possibility of overfitting.

The dataset to which the above algorithms were applied was the Forest Cover type from the UCI Machine Learning Repository (Blackard, 1998). This dataset includes information on tree type, shadow coverage, distance to nearby landmarks, soil type, and local topography. It consists of thirteen (13) variables, ten (10) quantitative and three (3) qualitative, and contains 581,012 rows. The quantitative variables included Elevation measured in meters, Horizontal_ Distance_to_Hydrology (H_D_To_H) and Vertical_Distance_to_Hydrology (V_D_To_H) which are measure in meters the distance to the nearest surface water features, Horizontal_Distance_to_Roadways (H_D_To_R), and Horizontal_Distance_to_Fire_Points (H_D_To_FP), which are measure in meters the distance to nearest roadway and to nearest wildfire ignition points respectively, Aspect, which is measured in degrees' azimuth, Slope, which is measured in degrees, and the degree of Hillshade during the summer solstice at 9 a.m., noon, and 3 p.m. (Hillshade 9am, HS_9am; Hillshade Noon, HS_Noon; Hillshade 3pm, HS_3pm), which are measured with an index from 0 to 255, where 0 indicates complete shadow with no sunlight reaches the terrain and 255 represents full illumination, where the terrain receives the maximum light. The qualitative variables included Wilderness Area (W_A) with four (4) categories (Rawah, Neota, Comanche Peak, Cache la Poudre), Soil Type (S_T) with forty (40) categories, and Cover Type (Cover_Type) with seven (7) categories (Spruce-Fir, Lodgepole Pine, Ponderosa Pine, Cottonwood/Willow, Aspen, Douglas Fir, Krummholz).

Figure 6. The seven cover types of forest trees

Available at: https://rpubs.com/aravindkr/160297

The dataset was split into two parts: the training subset (train) and the test subset (test, size=25%), which was used to assess prediction accuracy. The accuracy of the algorithms was evaluated on both the original and standardized data, and the results were compared. Additionally, applying the algorithms produced visualized results, such as confusion matrices, decision trees, and charts indicating feature importance for each predictive algorithm. Predictions were also made for a random vector on both the original and standardized data.

For the implementation of the algorithms, Python 3.10 was used through the Anaconda platform and Jupyter Notebook 6.4.5 (Eidelman, 2020).

RESULTS

At the first stage, the algorithms were applied to the original data. The highest accuracy was achieved by the K-NN algorithm (97%), with the precision of each class ranging from 0.92 to 0.98, the recall ranging from 0.81 to 0.98, and the F1 score ranging from 0.86 to 0.97, while the Macro average measures were found to be 0.95, 0.93 and 0.94 for the precision, recall and F1 measures respectively. The Neural Networks achieved the lowest accuracy (53%) with the precision of each class ranging from 0.00 to 0.63, the recall ranging from 0.00 to 0.83, and the F1 score ranging from 0.00 to 0.63, while the Macro average measures were found to be 0.23, 0.28 and 0.25 for precision, recall and F1 respectively.

High accuracy was also achieved by the Decision Tree algorithm (94%), with class precision ranging from 0.82 to 0.95, recall ranging from 0.79 to 0.95, and F1 score ranging from 0.81 to 0.95. The Macro average metrics were 0.90, 0.89, and 0.89 for precision, recall, and F1, respectively. Similarly, the Random Forest (RF) and Extra Tree (ET) algorithms showed high accuracy (96%), with class precision ranging from 0.92 to 0.98 for RF and 0.93 to 0.97 for ET, recall ranging from 0.82 to 0.97 for RF and 0.84 to 0.98 for ET, and F1 score ranging from 0.88 to 0.97 for RF and 0.89 to 0.97 for ET. The Macro average metrics were 0.95 for RF and 0.96 for ET, 0.92 for RF and 0.93 for ET, and 0.94, 0.94 for precision, recall, and F1, respectively.

Low accuracy was observed for the SGD (63%) and Naïve Bayes (NB) (62%) algorithms, with class precision ranging from 0.05 to 0.81 for SGD and 0.18 to 0.75 for NB, recall ranging from 0.01 to 0.91for SGD and 0.24 to 0.79 for NB, and F1 score from 0.02 to 0.70 for SGD and 0.21 to 0.65 for NB. The Macro average metrics for these algorithms were 0.39 for SGD and 0.43 for NB, 0.38 for SGD and 0.55 for NB, and 0.34 for SGD and 0.46 for NB, for precision, recall, and F1, respectively.

Moderate accuracy was recorded for the Logistic Regression (LR) multinomial (71%) and SVC (72%) algorithms, with class precision ranging from 0.00 to 0.74 for LR and 0.00 to 0.91 for SVC, recall ranging from 0.00 to 0.79 for LR and 0.00 to 0.82 for SVC, and F1 score ranging from 0.00 to 0.76 for LR and 0.00 to 0.78 for SVC. The Macro average metrics for these algorithms were 0.56 for LR and

0.42 for SVC, 0.46 for LR and 0.37 for SVC, and 0.49 for LR and 0.37 for SVC, for precision, recall, and F1, respectively (Table 2).

Table 2. Accuracy and ranging values of Precision, Recall, and F1 score for model's performance to original data

Model	**Accuracy**	**Precision's class range**	**Recall's class range**	**F1 score's class range**
Logistic Regression	0.71	0.00-0.74	0.00-0.79	0.00-0.76
SVC	0.72	0.00-0.91	0.00-0.82	0.00-0.78
K-NN	0.97	0.92-0.98	0.81-0.98	0.86-0.97
Decision Tree	0.94	0.82-0.95	0.79-0.95	0.81-0.95
Extra Tree	0.96	0.92-0.98	0.82-0.97	0.88-0.97
Random Forest	0.96	0.93-0.97	0.84-0.98	0.89-0.97
SGD	0.63	0.05-0.81	0.01-0.91	0.02-0.70
Naive Bayes	0.62	0.18-0.75	0.24-0.79	0.21-0.65
Neural Networks	0.53	0.00-0.63	0.00-0.83	0.00-0.63

Subsequently, the algorithms were applied to the standardized data (*z*-scores) using the StandardScaler command, while optimal parameters were searched through GridSearch command. Standard scaling is part of the preprocessing, necessary in case to prevent features with larger numeric ranges from dominating the objective function, ensuring equal contribution of all features. Algorithms like SVM, LR, K-NN, NN are sensitive to feature scales, in contrast with tree-based algorithms where has not any influence (Sharma, 2022). Cross-validation with five *k*-folds was also performed and mean accuracy with standard deviation for each algorithm were recorded. For the SVC algorithm, the parameter search determined the best parameters to be C=0.1, gamma=1, and kernel function = '*rbf*' (Radial Basis Function). For the Decision Tree algorithm, the parameter search determined the best parameters to be criterion= 'entropy' and max_depth=None, while for the K-NN algorithm, the parameter search identified the best value for K as one (K=1), where this value resulted in the highest accuracy. For Neural Networks, all three activation functions (sigmoid, ReLU, and hyperbolic tangent) were applied, and it was observed that the hyperbolic tangent resulted in the highest accuracy. Also, for Neural Network algorithm, were used an input layer with Dense=64, two (2) hidden layers with Dense 64 and 128 respectively, and with Dropout=0.5, an output layer with Dense=7 (as the categories were predicted) and with activation function 'softmax', optimizer= 'adam' and loss function= 'sparse categorical cross entropy', epochs=50, and batch_size=32.

Figure 7. A diagram of K values (1 to 30) versus Accuracy Score for the K-NN algorithm

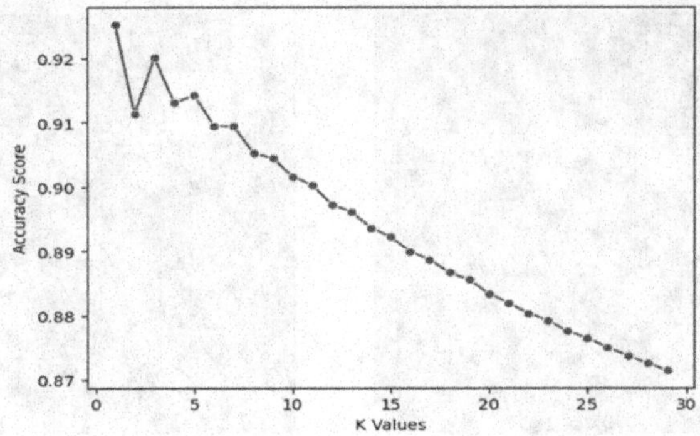

Figure 8. Decision tree plot with criterion entropy and max_depth=2

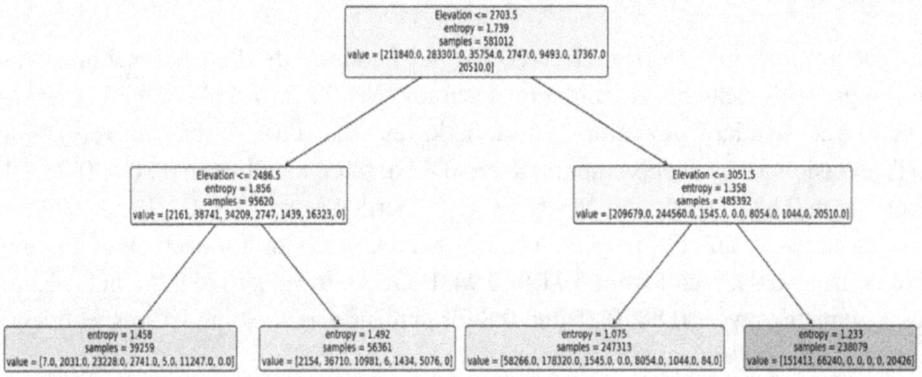

The highest accuracy on the standardized data was achieved by the Random Forest and Extra Tree algorithms, both of them (96%), with mean cross-validation accuracy (M=95.6% and 95.7%, respectively) and standard deviation (SD=0.0009 and 0.0010, respectively) (Table 3). Also, for RF, precision for each class ranging from 0.92 to 0.97, recall from 0.81 to 0.98, F1 score from 0.86 to 0.97, and Macro average metrics were 0.95, 0.93 and 0.92 for precision, recall, and F1, respectively. For ET, precision for each class ranging from 0.93 to 0.97, recall from 0.84 to 0.97, F1 score from 0.89 to 0.97, and Macro average metrics were 0.95, 0.92 and 0.94 for precision, recall, and F1, respectively (Table 3).

Figure 9. Confusion matrices of Random Forest and Extra Tree algorithms, when they were applied to the standardized data

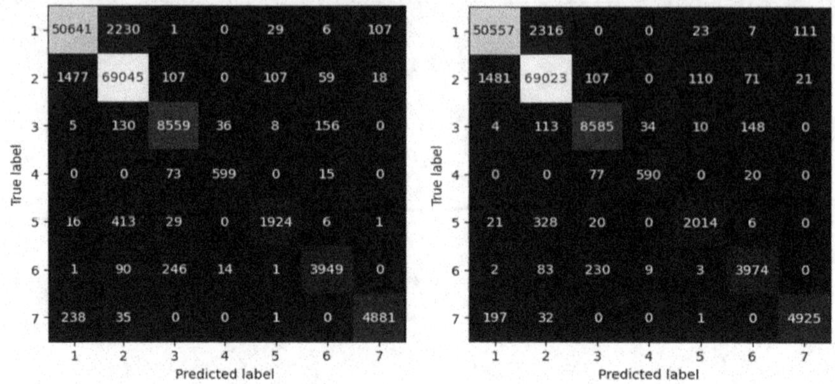

Lighter Shades Representing Higher Frequencies

The accuracy of Decision Tree (DT) (94%) and K-NN (93%) algorithms was also high, with mean cross-validation accuracy (M=93.7% and M=89.4%, respectively) and standard deviation (SD=0.0006 and SD=0.0152, respectively). For DT, precision for each class ranging from 0.82 to 0.95, recall from 0.80 to 0.95, F1 score from 0.82 to 0.95, and Macro average metrics were 0.89, 0.89 and 0.89 for precision, recall, and F1, respectively. For K-NN, precision for each class ranging from 0.81 to 0.94, recall from 0.71 to 0.94, F1 score from 0.76 to 0.94, and Macro average metrics were 0.88, 0.85 and 0.8 for precision, recall, and F1, respectively.

Figure 10. Confusion matrices of Decision Tree and K-NN algorithms, when they were applied to the standardized data

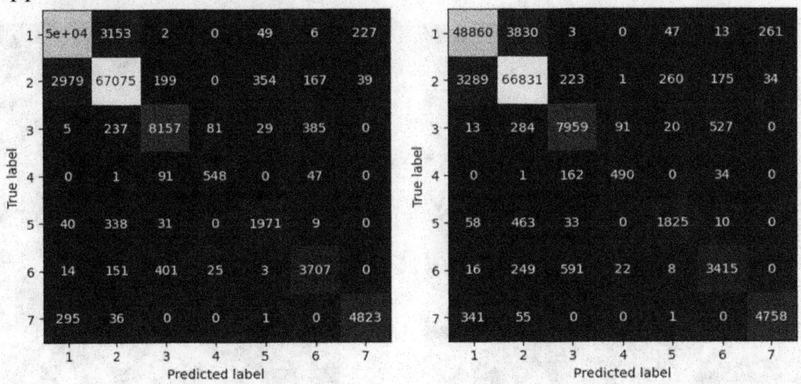

Lighter Shades Representing Higher Frequencies

Similarly, the accuracy of the SVC algorithm was quite high (86%), with mean cross-validation accuracy (M=84,8%) and standard deviation (SD=0.0012), while precision for each class ranging from 0.78 to 0.95, recall from 0.31 to 0.92, F1 score from 0.46 to 0.88, and Macro average metrics were 0.86, 0.67 and 0.72 for precision, recall, and F1, respectively. The accuracy of Neural Network algorithm was also quite high (77%), with mean cross-validation accuracy (M=62,3%) and standard deviation (SD=0.0155), while precision for each class ranging from 0.56 to 0.82, recall from 0.12 to 0.90, F1 score from 0.20 to 0.81, and Macro average metrics were 0.72, 0.53 and 0.55 for precision, recall, and F1, respectively.

Figure 11. Confusion matrices of SVC and Neural Network algorithms, when they were applied to the standardized data

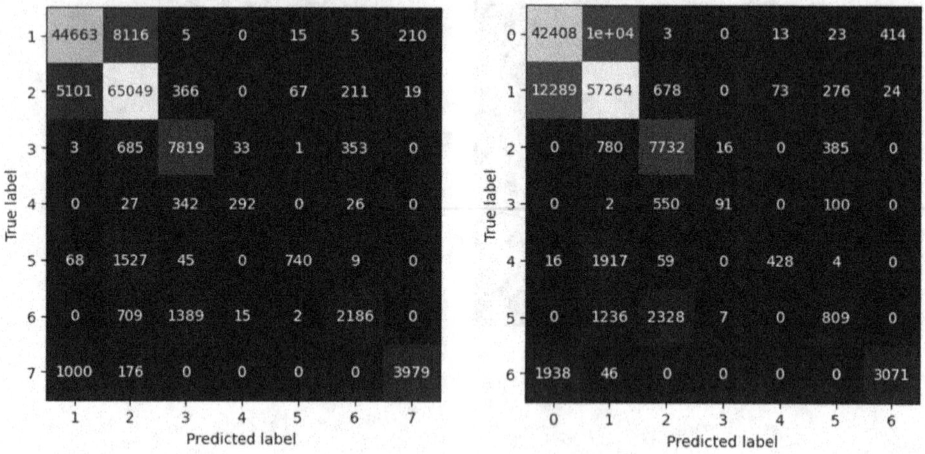

Lighter Shades Representing Higher Frequencies

The accuracy of the algorithm SGD remained low (67%) with mean cross-validation accuracy (M=66,5%) and standard deviation (SD=0.034), while precision for each class ranging from 0.14 to 0.89, recall from 0.02 to 0.84, F1 score from 0.04 to 0.72, and Macro average metrics were 0.52, 0.33 and 0.34 for precision, recall, and F1, respectively. The accuracy of the algorithm Gaussian NB also remained low (60%), with mean cross-validation accuracy (M=59,7%) and standard deviation (SD=0.0101), while precision for each class ranging from 0.09 to 0.75, recall from 0.25 to 1.00, F1 score from 0.16 to 0.65, and Macro average metrics were 0.40, 0.53 and 0.40 for precision, recall, and F1, respectively. The accuracy of the LR algorithm remained unchanged (71%), with mean cross-validation accuracy (M=69,1%) and standard deviation (SD=0.0246), while precision for each class ranging from 0.00 to 0.74, recall from 0.00 to 0.79, F1 score from 0.00 to 0.76, and Macro average metrics were 0.55, 0.48 and 0.50 for precision, recall, and F1, respectively (Table 3).

Figure 12. Confusion matrices of SGD, Gaussian Naïve Bayes, and Logistic Regression algorithms, when they were applied to the standardized data

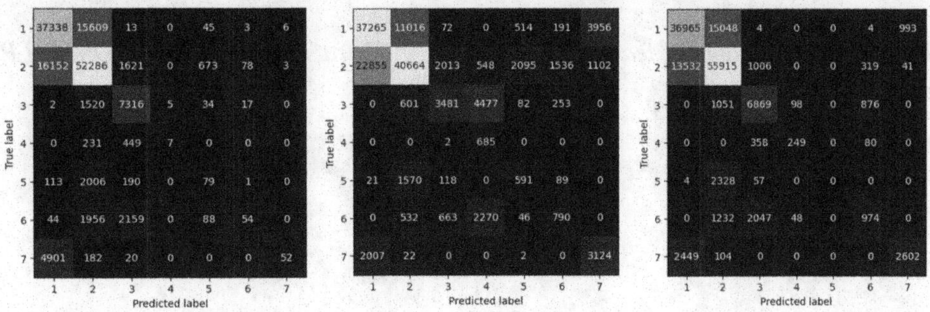

Lighter Shades Representing Higher Frequencies

Table 3. Accuracy and ranging values of Precision, Recall, and F1 score for model's performance to standardized data

Model	Accuracy	Precision's class range	Recall's class range	F1 score's class range
Logistic Regression	0.71	0.00-0.74	0.00-0.79	0.00-0.76
SVC	0.86	0.78-0.95	0.31-0.92	0.46-0.88
K-NN	0.93	0.81-0.94	0.71-0.94	0.76-0.94
Decision Tree	0.94	0.82-0.95	0.80-0.95	0.82-0.95
Extra Tree	0.96	0.93-0.97	0.84-0.97	0.89-0.97
Random Forest	0.96	0.92-0.97	0.81-0.98	0.86-0.97
SGD	0.67	0.14-0.89	0.02-0.84	0.04-0.72
Naive Bayes	0.60	0.09-0.75	0.25-1.00	0.16-0.65
Neural Networks	0.77	0.56-0.82	0.12-0.90	0.20-0.81

By searching for the most significant features using the 'feature_importance' command in the algorithms where this was applicable, it was found that Elevation had the greatest impact on classification in the Extra Tree, Decision Tree, Gaussian NB, and Logistic Regression algorithms. Additionally, the S_T (Soil Type) and W_A (Wildness Area) were also important, with W_A (Wildness Area) being the most significant feature in the SGD algorithm. In the Decision Tree, Gaussian NB, and SGD algorithms, features related to horizontal distance from hydrological points (such as rivers and lakes) and roads had a notable influence. In contrast, in the Logistic Regression algorithm, the Hillshade features had also a significant impact (Figure 13).

Figure 13. Diagrams of features importance in Extra Tree, DT, NB, SGD and LR algorithms

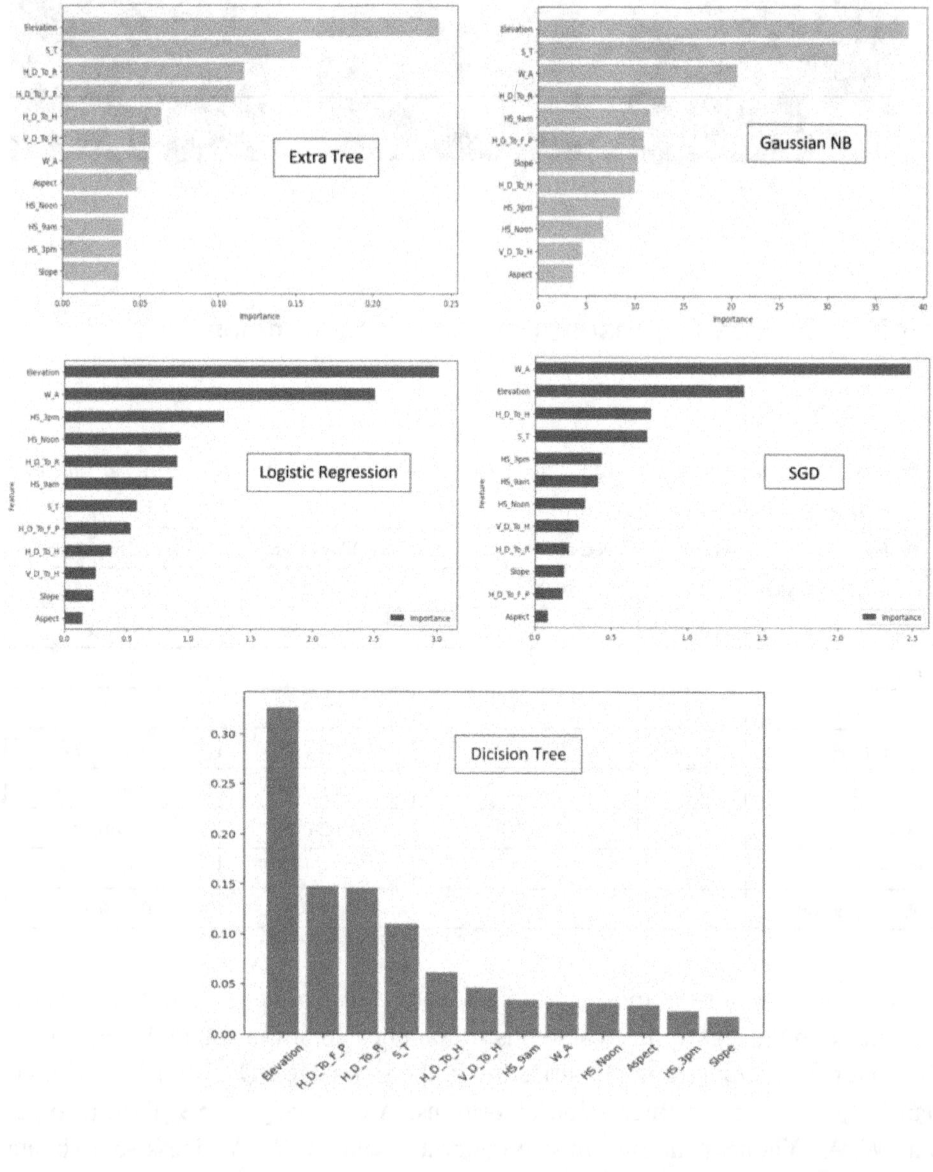

Finally, predictions were made on a random feature vector for each algorithm, both for the original and standardized data, which yielded the final column of the above table (Table 4). Based on the data in the table, apart from the K-NN and Extra

Tree algorithms, no differences in the predicted category were observed between the original and standardized data for the remaining algorithms, where based on the input vector $\boldsymbol{x}$ = (2, 20, 2959.4, 2.7, 0.2, 269.4, 46.4, 2350.1, 1980.3, 212.1, 223.3, 142.5), which related to the mean value of the attributes, the prediction was class 2 (Lodgepole Pine).

Table 4. Comparison of algorithm's accuracy to original and standardized data, mean accuracy of cross-validation, and prediction of a random vector

Model	Accuracy	Mean accuracy of cross-validation	Standard deviation	Prediction of a random vector
Logistic Regression	0.71-**0.71**	0.691	0.0246	2-2
SVC	0.72-**0.86**	0.848	0.0012	2-2
K-NN	0.97-**0.93**	0.894	0.0152	1-2
Decision Tree	0.94-**0.94**	0.937	0.0006	2-2
Extra Tree	0.96-**0.96**	0.957	0.0009	1-2
Random Forest	0.96-**0.96**	0.956	0.0010	2-2
SGD	0.63-**0.67**	0.665	0.0340	2-2
Naive Bayes	0.62-**0.60**	0.597	0.0101	2-2
Neural Networks	0.53-**0.77**	0.623	0.0155	2-2

Accuracy and prediction for standardized data are marked in bold

DISCUSSION

The comprehensive evaluation of multiple machine learning algorithms on both original and standardized datasets yielded significant insights into their performance for classification, as measured by accuracy, precision, recall, and F1 scores.

On the original data, the K-Nearest Neighbors (K-NN) algorithm demonstrated the highest accuracy (97%) with excellent precision, recall, and F1 scores across all classes, as evidenced by macro averages of 0.95, 0.93, and 0.94, respectively. Similarly, Random Forest (RF) and Extra Tree (ET) algorithms also performed exceptionally well, achieving 96% accuracy. These results underline their robustness in handling complex datasets. In contrast, Neural Networks showed the weakest performance (53%), with all evaluation metrics falling below acceptable thresholds. Naïve Bayes and Stochastic Gradient Descent (SGD) algorithms also yielded suboptimal results, with accuracies of 62% and 63%, respectively. Moderate performance was observed with Logistic Regression (71%) and Support Vector Classifier (72%), reflecting limitations in capturing non-linear relationships within the data. However, the class-wise ranges of precision, recall, and F1 scores for these algorithms revealed significant variability, particularly for SVC, which had a precision range of 0.00 to 0.91 and recall range of 0.00 to 0.82.

Standardization using z-scores, combined with parameter optimization and cross-validation, significantly influenced the performance of most algorithms. The Neural Networks' accuracy improved markedly from 53% to 77%, while SVC's accuracy improved from 72% to 86%, highlighting the critical role of data preprocessing in enhancing model efficacy. Tree based algorithms (DT, RF, ET) maintained their high accuracy across both original and standardized datasets, reflecting their invariance to feature scaling. The Random Forest and Extra Tree algorithms continued to outperform other models on standardized data, achieving 96% accuracy with minimal standard deviations 0.0009 and 0.0010, respectively. The accuracy of Logistic Regression algorithm also was maintained (71%), while SGD and Gaussian Naïve Bayes struggled to achieve comparable improvements.

The analysis of feature importance across algorithms revealed Elevation as the most influential feature for classification, particularly for the Extra Tree, Decision Tree, Gaussian NB, and Logistic Regression models. Additionally, features related to Soil Type and Wilderness Area were significant, with Wilderness Area being the primary determinant for the SGD algorithm, while for the Logistic Regression algorithm, the Hillshade features also played a crucial role. This analysis underscores the relevance of these attributes in predicting class labels, providing potential approaches for feature engineering in future studies.

Predictions on a random feature vector confirmed consistency in classification outcomes between original and standardized data for most algorithms, with exceptions noted for K-NN and Extra Tree, indicating sensitivity to data standardization.

CONCLUSIONS

This study evaluated the performance of various machine learning algorithms on a classification task, analyzing their effectiveness on both original and standardized datasets. The results highlighted the superiority of tree-based models, particularly Random Forest and Extra Tree, which consistently achieved high accuracy, precision, recall, and F1 scores. These models demonstrated robustness and adaptability, performing well across diverse conditions and with minimal variation in cross-validation metrics. In contrast, some algorithms, including SGD and Gaussian Naïve Bayes, struggled to achieve competitive performance, indicating their limitations in handling complex datasets. These results underscore the necessity of selecting models based on dataset characteristics and performance requirements.

The findings also emphasized the critical role of data preprocessing and parameter tuning in enhancing algorithm performance. Standardization significantly improved the accuracy of algorithms like Neural Networks and SVC, illustrating the potential of such techniques in addressing variability and scaling challenges in datasets. Furthermore, feature importance analysis identified key attributes such as Elevation, Wilderness Area, and Soil Type as major contributors to accurate classification, providing valuable insights for future feature engineering efforts.

The study highlights several limitations and implications for further research. Algorithms like K-NN, SVC, and Neural Networks showed significant sensitivity to data preprocessing, emphasizing the importance of scaling, while tree-based models were unaffected, showing their robustness. Performance disparities across classes, particularly for Neural Networks and SGD, suggest the need for techniques to address class imbalance, while high accuracy in models like K-NN and Extra Tree raises potential concerns about overfitting, warranting further validation on diverse datasets. Feature importance analysis revealed a strong dependence on specific features, such as Elevation, which may impacts generalizability. Additionally, Neural Networks required extensive hyperparameter tuning, making them with limited interpretability compared to tree-based models. Future research should explore alternative neural network architectures, ensemble methods, and advanced feature engineering to enhance model performance. Finally, optimizing hyperparameters for underperforming algorithms and investigating alternate scaling methods could further improve predictive performance.

ACKNOWLEDGEMENT

Competing Interests:

The authors of this publication declare there are no competing interests.

Funding:

This research received no specific grant from any funding agency in the public, commercial, or not-for-profit sectors. Funding for this research was covered by the author(s) of the article.

REFERENCES

Aziz, G., Minallah, N., Saeed, A., Frnda, J., & Khan, W. (2024). Remote sensing based forest cover classification using machine learning. *Scientific Reports*, *14*(1), 69. DOI: 10.1038/s41598-023-50863-1 PMID: 38167902

Bishop, C. M. (2016). *Pattern recognition and machine learning*. Springer.

Bisong, E. (2019). Logistic regression. *Building machine learning and deep learning models on google cloud platform: A comprehensive guide for beginners*, 243-250.

Blackard, J. (1998). Covertype [Dataset]. UCI Machine Learning Repository. DOI: 10.24432/C50K5N

Blackard, J. A., & Dean, D. J. (1999). Comparative accuracies of artificial neural networks and discriminant analysis in predicting forest cover types from cartographic variables. *Computers and Electronics in Agriculture*, *24*(3), 131–151. DOI: 10.1016/S0168-1699(99)00046-0

Breiman, L. (2001). Random forests. *Machine Learning*, *45*(1), 5–32. DOI: 10.1023/A:1010933404324

Carvalho, D. V., Pereira, E. M., & Cardoso, J. S. (2019). Machine Learning Interpretability: A survey on methods and metrics. *Electronics (Basel)*, *8*(8), 832. DOI: 10.3390/electronics8080832

Charbuty, B., & Abdulazeez, A. (2021). Classification based on decision tree algorithm for machine learning. *Journal of Applied Science and Technology Trends*, *2*(01), 20–28. DOI: 10.38094/jastt20165

Dong, S. (2021). Multi class SVM algorithm with active learning for network traffic classification. *Expert Systems with Applications*, *176*, 114885. DOI: 10.1016/j.eswa.2021.114885

Eidelman, A. (2020). Python Data Science Handbook by Jake VANDERPLAS (2016). *Statistique Et Société, 8*(2), 45–47. https://statistique-et-societe.fr/article/view/792

García-Gonzalo, E., Fernández-Muñiz, Z., Nieto, P. J. G., Sánchez, A. B., & Fernández, M. M. (2016). Hard-Rock Stability Analysis for Span Design in Entry-Type Excavations with Learning Classifiers. *Materials (Basel), 9*(7), 531. DOI: 10.3390/ma9070531 PMID: 28773653

Grandini, M., Bagli, E., & Visani, G. (2020). *Metrics for multi-class classification: an overview.* arXiv preprint arXiv:2008.05756.

Hassan, J., & Alsamhi, S. (Eds.). (2024). *Applications of Machine Learning in UAV Networks.* IGI Global., DOI: 10.4018/979-8-3693-0578-2

Jahromi, A. H., & Taheri, M. (2017, October). A non-parametric mixture of Gaussian naive Bayes classifiers based on local independent features. In *2017 Artificial intelligence and signal processing conference (AISP)* (pp. 209-212). IEEE._DOI: 10.1109/AISP.2017.8324083

Jain, A. (2024). *SVM kernels and its type.* [Infographic]_https://medium.com/@abhishekjainindore24/svm-kernels-and-its-type-dfc3d5f2dcd8

Jiang, M., Liang, Y., Feng, X., Fan, X., Pei, Z., Xue, Y., & Guan, R. (2016). Text classification based on deep belief network and softmax regression. *Neural Computing & Applications, 29*(1), 61–70. DOI: 10.1007/s00521-016-2401-x

Johri, P., Saxena, V. S., & Kumar, A. (2021). Rummage of Machine Learning Algorithms in Cancer Diagnosis. [IJEHMC]. *International Journal of E-Health and Medical Communications, 12*(1), 1–15. DOI: 10.4018/IJEHMC.2021010101

Joshi, V., Ayushi, A., & Agarwal, N. (2023). Forest Cover Type Prediction using Automatic Machine Learning. *2022 13th International Conference on Computing Communication and Networking Technologies (ICCCNT).* DOI: 10.1109/ICCCNT56998.2023.10307426

Kamel, H., Abdulah, D., & Al-Tuwaijari, J. M. (2019, June). Cancer classification using gaussian naive bayes algorithm. In *2019 international engineering conference (IEC)* (pp. 165-170). IEEE. DOI: 10.1109/IEC47844.2019.8950650

Kamilaris, A., & Prenafeta-Boldú, F. X. (2018). Deep learning in agriculture: A survey. *Computers and Electronics in Agriculture, 147*, 70–90. DOI: 10.1016/j.compag.2018.02.016

Kishore, R. R., Narayan, S. S., Lal, S., & Rashid, M. A. (2016, December). Comparative accuracy of different classification algorithms for forest cover type prediction. In *2016 3rd Asia-Pacific World Congress on Computer Science and Engineering (APWC on CSE)* (pp. 116-123). IEEE. DOI: 10.1109/APWC-on-CSE.2016.029

Kocev, D., Ceci, M., & Stepišnik, T. (2020). Ensembles of extremely randomized predictive clustering trees for predicting structured outputs. *Machine Learning*, *109*(11), 2213–2241. DOI: 10.1007/s10994-020-05894-4

Kolasa, T., & Raja, A. (2016). *Forest Cover Type Machine Learning Study.* [Infographic] https://rpubs.com/aravindkr/160297

Kumar, A., & Sinha, N. (2020). Classification of forest cover type using random forests algorithm. In *Advances in Data and Information Sciences: Proceedings of ICDIS 2019* (pp. 395-402). Springer Singapore. DOI: 10.1007/978-981-15-0694-9_37

Kumar, R., Nandy, S., Agarwal, R., & Kushwaha, S. P. S. (2014). Forest cover dynamics analysis and prediction modeling using logistic regression model. *Ecological Indicators*, *45*, 444–455. DOI: 10.1016/j.ecolind.2014.05.003

L'heureux, A., Grolinger, K., Elyamany, H. F., & Capretz, M. A. (2017). Machine learning with big data: Challenges and approaches. *IEEE Access : Practical Innovations, Open Solutions*, *5*, 7776–7797. DOI: 10.1109/ACCESS.2017.2696365

Lee, M. C. H., Braet, J., & Springael, J. (2024). Performance Metrics for multilabel emotion Classification: Comparing micro, macro, and weighted F1-Scores. *Applied Sciences (Basel, Switzerland)*, *14*(21), 9863. DOI: 10.3390/app14219863

Liu, Y., Liu, Y., & Zhao, Y. (2020). Research on the Application of Decision Tree Algorithm in Credit Risk Evaluation. In *2020 International Conference on Intelligent Transportation, Big Data & Smart City (ICITBS)* (pp. 1-5). IEEE.

López, O. M., López, A. M., & Crossa, J. (2022). Overfitting, model tuning, and evaluation of prediction performance. In *Springer eBooks* (pp. 109–139). DOI: 10.1007/978-3-030-89010-0_4

Mahesh, B. (2020). Machine learning algorithms-a review. [IJSR]. *International Journal of Scientific Research*, *9*, 381–386.

Mehlig, B. (2021). *Machine learning with neural networks: an introduction for scientists and engineers*. Cambridge University Press. DOI: 10.1017/9781108860604

Mitchell, T. M. (1997). New York: McGraw-hill. *Machine Learning*, *1*(9)

Murphy, K. P. (2012). *Machine learning: a probabilistic perspective*. MIT press.

Ng, A. Y. (2004, July). Feature selection, L 1 vs. L 2 regularization, and rotational invariance. In *Proceedings of the twenty-first international conference on Machine learning* (p. 78). DOI: 10.1145/1015330.1015435

Nizam, M. (2024). *Neural Networks: From Biological to Artificial.* [Infographic] https://www.linkedin.com/pulse/neural-networks-from-biological-artificial-muhammed-nizam-pcxwf/

Parmar, A., Katariya, R., & Patel, V. (2019). A review on random forest: An ensemble classifier). In *International Conference on Intelligent Data Communication Technologies and Internet of Things (ICICI)* (pp. 758-763). Springer International Publishing. DOI: 10.1007/978-3-030-03146-6_86

Pedregosa, F., Varoquaux, G., Gramfort, A., Michel, V., Thirion, B., Grisel, O., & Duchesnay, É. (2011). Scikit-learn: Machine learning in Python. *the Journal of machine Learning research, 12*, 2825-2830. https://dl.acm.org/citation.cfm?id=1953048.2078195

Pramoditha, R. (2022). *Overview of a Neural Network's Learning Process.* [Infographic]. https://medium.com/data-science-365/overview-of-a-neural-networks-learning-process-61690a502fa

Punia, S. K., Kumar, M., Stephan, T., Deverajan, G. G., & Patan, R. (2021). Performance analysis of machine learning algorithms for big data classification: Ml and ai-based algorithms for big data analysis. [IJEHMC]. *International Journal of E-Health and Medical Communications, 12*(4), 60–75. DOI: 10.4018/IJEHMC.20210701.oa4

Sakr, C., Patil, A., Zhang, S., Kim, Y., & Shanbhag, N. (2017, March). Minimum precision requirements for the SVM-SGD learning algorithm. In *2017 IEEE International Conference on Acoustics, Speech and Signal Processing (ICASSP)* (pp. 1138-1142). IEEE. DOI: 10.1109/ICASSP.2017.7952334

Sarker, I. H. (2021). Machine learning: Algorithms, real-world applications and research directions. *SN Computer Science, 2*(3), 160. DOI: 10.1007/s42979-021-00592-x PMID: 33778771

Schonlau, M., & Zou, R. Y. (2020). The random forest algorithm for statistical learning. *The Stata Journal, 20*(1), 3–29. DOI: 10.1177/1536867X20909688

Sen, P. C., Hajra, M., & Ghosh, M. (2020). Supervised classification algorithms in machine learning: A survey and review. In *Emerging Technology in Modelling and Graphics: Proceedings of IEM Graph 2018* (pp. 99-111). Springer Singapore. DOI: 10.1007/978-981-13-7403-6_11

Sharma, A. (2018). Guided Stochastic Gradient Descent Algorithm for inconsistent datasets. *Applied Soft Computing*, *73*, 1068–1080. DOI: 10.1016/j.asoc.2018.09.038

Sharma, V. (2022). A study on data scaling methods for machine learning. *International Journal for Global Academic & Scientific Research*, *1*(1), 31–42. DOI: 10.55938/ijgasr.v1i1.4

Sjöqvist, H., Längkvist, M., & Javed, F. (2020). An analysis of fast learning methods for classifying forest cover types. *Applied Artificial Intelligence*, *34*(10), 691–709. DOI: 10.1080/08839514.2020.1771523

Suyal, M., & Goyal, P. (2022). A review on analysis of k-nearest neighbor classification machine learning algorithms based on supervised learning. *International Journal of Engineering Trends and Technology*, *70*(7), 43–48. DOI: 10.14445/22315381/IJETT-V70I7P205

Tangirala, S. (2020). Evaluating the Impact of GINI Index and Information Gain on Classification using Decision Tree Classifier Algorithm*. *International Journal of Advanced Computer Science and Applications*, *11*(2). DOI: 10.14569/IJACSA.2020.0110277

Thoma, M. (2016). Receiver operating characteristic. [Infographic] https://commons.wikimedia.org/w/index.php?curid=109730045

Wang, J., & Joshi, G. (2021). Cooperative SGD: A unified framework for the design and analysis of Local-Update SGD algorithms. *Journal of Machine Learning Research*, *22*(213), 1–50. https://jmlr.org/papers/volume22/20-147/20-147.pdf

Wang, W., & Sun, D. (2021). The improved AdaBoost algorithms for imbalanced data classification. *Information Sciences*, *563*, 358–374. DOI: 10.1016/j.ins.2021.03.042

Yadav, D. C., & Pal, S. (2020). Prediction of thyroid disease using decision tree ensemble method. *Human-Intelligent Systems Integration*, *2*(1-4), 89–95. DOI: 10.1007/s42454-020-00006-y

Yang, L., & Shami, A. (2020). On hyperparameter optimization of machine learning algorithms: Theory and practice. *Neurocomputing*, *415*, 295–316. DOI: 10.1016/j.neucom.2020.07.061

Zhang, Y. (2012). Support vector machine classification algorithm and its application. In *Communications in computer and information science* (pp. 179–186). DOI: 10.1007/978-3-642-34041-3_27

Zhang, Z. (2016). Introduction to machine learning: K-nearest neighbors. *Annals of Translational Medicine*, *4*(11), 218. DOI: 10.21037/atm.2016.03.37 PMID: 27386492

Zhou, L., Pan, S., Wang, J., & Vasilakos, A. V. (2017). Machine learning on big data: Opportunities and challenges. *Neurocomputing*, *237*, 350–361. DOI: 10.1016/j.neucom.2017.01.026

KEY TERMS AND DEFINITIONS

SVC: Finds the hyperplane that best separates data points of different classes in a high dimensional space.

SGD: Optimizes a loss function by iteratively updating model parameters using random data subsets.

Naïve Bayes: A probabilistic classifier using Bayes' Theorem, assuming conditional independence of features.

K-NN: Assigns class or value based on the majority or average of 'k' nearest neighbors.

Decision Tree: Splits data based on feature values to make decisions, with each leaf representing a predicted class or value.

Logistic Regression: Estimates outcome probabilities using a sigmoid function on a linear combination of features.

Neural Networks: Layers of interconnected nodes (neurons) transform inputs to outputs for complex pattern recognition.

Chapter 6
Analyzing European Elections:
A Multivariate Perspective (2019–2024)

Georgia Panagiotidou
https://orcid.org/0000-0002-7264-1187
Aristotle University of Thessaloniki, Greece

Theodore Chadjipadelis
Aristotle University of Thessaloniki, Greece

ABSTRACT

European elections provide an opportunity to explore the political dynamics of the European Union (EU), voter behavior, party competition, and regional alignments. This chapter introduces a multivariate framework, combining Hierarchical Cluster Analysis (HCA) and Factorial Correspondence Analysis (FCA), to analyze the 2024 European elections as compared to 2019. By uncovering key axes of polarizations and grouping constituencies based on voting patterns, the analysis reveals significant shifts in party alignments and voter preferences. The findings highlight increasing polarization between progressive and nationalist forces and the consolidation of nationalist and Eurosceptic movements such as Patriots for Europe (PfE) and the European Conservatives and Reformists (ECR). Regional variability is also evident, with a rising alignment of constituencies around clear ideological blocs. This study underscores the multidimensional nature of electoral competition in Europe and offers a robust methodological framework for understanding shifts in voter behavior and political competition.

DOI: 10.4018/979-8-3693-9400-7.ch006

INTRODUCTION

European elections are a critical mechanism for understanding the political dynamics of the European Union (EU), offering insights into voter behavior, party competition, and the broader trajectory of European integration. Held every five years, these elections provide EU citizens with the opportunity to shape the composition of the European Parliament, thereby influencing legislative priorities, governance, and the political direction of the Union. Beyond their immediate electoral significance, European elections reflect the connection between domestic political concerns and transnational issues, highlighting the complexities of multi-level governance in the EU.

The analysis of electoral results has traditionally relied on descriptive statistics and geographical mappings to provide insights into voting patterns and party performance. While these approaches offer valuable perspectives, they often fail to capture the multidimensional nature of electoral competition, particularly in a union as diverse as the EU. Recent advances in multivariate statistical techniques, such as Hierarchical Cluster Analysis (HCA) and Factorial Correspondence Analysis (FCA), present an opportunity to deepen our understanding of electoral dynamics. These methods allow for the exploration of complex relationships between voters, parties, and political contexts, offering a more comprehensive framework for analyzing and reporting election results.

This chapter introduces a original application of multivariate analysis to the study of European elections, focusing on the 2024 elections in comparison to those of 2019. First, it examines the most recent European election results, focusing on the unique political and social dynamics of the post-COVID-19 era. Second, it contrasts these findings with the 2019 elections, identifying changes and continuities in voter behavior, party strategies, and political alignments. The proposed approach is designed to uncover patterns and dimensions of political competition that are often overlooked in conventional analyses. By employing HCA and FCA in a two-step process, this study aims to:

1. Identify key axes of political competition in the European Parliament elections.
2. Group countries based on electoral outcomes to reveal regional and ideological affinities across the EU.
3. Map the political landscape of Europe, illustrating how constituencies and parties align along ideological and regional lines.

To address these challenges, this chapter introduces a original methodological approach that combines Hierarchical Cluster Analysis (HCA) and Factorial Correspondence Analysis (FCA) to analyze and report the results of European elections. This two-step process not only deepens our understanding of electoral dynamics but also provides a comparative framework for examining changes over time and across member states. By applying these methods to the 2024 European elections and comparing them with the results of 2019, this study seeks to uncover the evolving dimensions of political competition and voter preferences in a rapidly changing political landscape.

The scope of this chapter is threefold. First, it examines the most recent European election results, focusing on the unique political and social dynamics of the post-COVID-19 era. Second, it contrasts these findings with the 2019 elections, identifying changes and continuities in voter behavior, party strategies, and political alignments. Finally, it investigates the connection between European and national electoral outcomes, offering insights into how voter preferences shift across different electoral contexts. The analysis highlights the multidimensional nature of political competition in Europe, revealing not only the dominance of traditional cleavages, such as the liberal-conservative divide, but also the emergence of new dynamics, such as the growing polarization between progressive and nationalist forces. The findings from the 2019 and 2024 elections underscore significant transformations in voter behavior and party alignments, including the rising influence of progressive platforms like the Greens and Renew Europe, the consolidation of nationalist movements such as Identity and Democracy (PfE) and the European Conservatives and Reformists (ECR), and the decline of traditional party dominance.

The significance of this study lies in its ability to provide a detailed understanding of European elections, moving beyond surface-level analyses to uncover deeper patterns and trends. The comparative perspective enabled by HCA and FCA enhances our ability to map the trajectories of political competition across the EU, offering valuable insights for scholars, policymakers, and political practitioners. Moreover, the originality of this approach lies in its capacity to integrate multivariate techniques into the study of elections, bridging the gap between methodological innovation and applied political analysis.

By introducing and applying this analytical framework, this chapter contributes to the broader discourse on electoral studies, emphasizing the importance of robust, data-driven methods for understanding the complexities of voter behavior and political competition. It also identifies critical regional and ideological shifts that have implications for European governance, such as the growing salience of environmental concerns, the increasing polarization between progressive and nationalist agendas, and the impact of geopolitical tensions on voter alignments. The findings are expected to not only enrich academic research but also inform the development of more effective policies and strategies within the European political arena, particularly as the EU navigates an increasingly polarized and fragmented political environment.

LITERATURE REVIEW

The Role and Function of European Elections

European elections have long been a subject of scholarly interest, serving as a window into the political dynamics of the European Union (EU). Initially envisioned as a means of strengthening the democratic legitimacy of EU institutions, these elections have evolved into complex political events that reflect both domestic and European-level issues. Scholars such as Reif and Schmitt (1980) introduced the concept of "second-order elections," arguing that European elections are often overshadowed by national concerns, with lower stakes leading to reduced voter turnout and protest voting. This framework remains influential, although recent studies, such as those by Hobolt and De Vries (2016), highlight how increasing European integration and transnational challenges are shifting the nature of these elections.

In the past two decades, European elections have become a battleground for competing narratives about the future of the EU. The rise of Eurosceptic parties, polarization between pro-European and nationalist forces, and the increasing salience of issues such as migration, economic inequality, and climate change have transformed these elections into a critical reflection of broader societal trends (Wallashek, 2020). Scholars such as Hix and Marsh (2007) have emphasized the growing importance of party competition at the European level, while other research (Panagiotidou & Chadjipadelis 2021) have analyzed the role of voter mobilization and issue salience in shaping electoral outcomes.

Despite their evolving significance, European elections continue to grapple with challenges such as declining voter turnout and the persistence of domestic concerns dominating the electoral agenda (Schmitt & Toygür, 2016). Recent research underscores the need for innovative analytical frameworks to capture the multidimensional nature of these elections, moving beyond the dichotomy of national versus European-level politics (Fischer, Fieldhouse et al., 2018).

The 2019 European Elections

The 2019 European elections marked a pivotal moment in EU politics, with high voter turnout and significant shifts in the political landscape. Held from May 23 to May 26, the elections saw a 50.6% turnout, the highest since 1994, reflecting growing public interest in European-level politics (European Parliament, 2019). Key drivers of this increased engagement included rising concerns over climate change, Brexit, and the emergence of populist and nationalist parties (Hutter & Kriesi, 2020).

The elections featured a diverse array of political parties, representing both traditional and emerging political forces. The two largest political groups, the European

People's Party (EPP) and the Progressive Alliance of Socialists and Democrats (S&D), saw their dominance eroded, losing their combined majority for the first time in decades. This decline highlighted the fragmentation of the political landscape and the growing influence of smaller groups such as the Greens/European Free Alliance (Greens/EFA) and Renew Europe (RE), formerly the Alliance of Liberals and Democrats for Europe (ALDE) (Brack & Marié, 2024).Eurosceptic parties, such as Italy's League and France's National Rally, made significant gains, underscoring the polarization between pro-European and nationalist forces. At the same time, the Greens/EFA group capitalized on the rising salience of environmental issues, particularly among younger voters, achieving their best-ever results.

Turnout varied widely across member states, from over 88% in Belgium, where voting is mandatory, to less than 25% in Slovakia. Countries like Germany and Spain saw moderate increases in voter participation, while others, such as the United Kingdom (participating for the last time due to Brexit), experienced relatively stable turnout rates (European Parliament, 2019). The election results painted a complex picture of the EU's political dynamics. The EPP remained the largest group in the European Parliament, securing 182 seats, followed by the S&D with 154 seats. Renew Europe emerged as a key player with 108 seats, while the Greens/EFA secured 74 seats. Eurosceptic and far-right groups also increased their presence, with Identity and Democracy (ID) winning 73 seats (Brack & Marié, 2024).

The 2019 elections were characterized by several noteworthy phenomena:

- The "Green Wave": The surge in support for Green parties highlighted the growing importance of climate change as a political issue, particularly among younger and urban voters (Hutter & Kriesi, 2020).
- Brexit's Shadow: The United Kingdom's participation added complexity, with the Brexit Party achieving a strong showing, reflecting frustration over the prolonged Brexit process (De Vries et al., 2021).
- Fragmentation of Traditional Power: The decline of the EPP and S&D marked a shift away from the traditional dominance of centrist parties, paving the way for more diverse and coalition-driven governance (European Parliament, 2019).

The 2024 European Elections

The 2024 European elections, held from June 6 to June 9, were conducted in a markedly different political environment shaped by the aftermath of the COVID-19 pandemic, the ongoing war in Ukraine, and economic challenges such as inflation and energy crises (Dennison & Kriesi, 2023). These elections were pivotal in reflecting the EU's resilience and adaptability in the face of unprecedented challenges. The major political groups—the EPP, S&D, Renew Europe, and the Greens/EFA—continued to play dominant roles, although the dynamics of competition were influenced by the emergence of new political forces and shifting alliances. The far-right Identity and Democracy group and the European Conservatives and Reformists (ECR) capitalized on public discontent over economic instability and migration, while the Greens/EFA faced challenges in maintaining their momentum from 2019 (European Parliament, 2024).

Turnout in 2024 remained relatively stable, averaging around 51%, with notable variations across member states. Mandatory voting countries like Belgium and Luxembourg saw the highest participation rates, while Eastern European states such as Slovakia and the Czech Republic continued to report lower turnout (European Parliament, 2024). The results reinforced the trends of political fragmentation and polarization. The EPP regained some ground, securing 190 seats, while the S&D stabilized with 157 seats. Renew Europe and the Greens/EFA faced modest declines, while the ECR and ID groups expanded their influence.

Several significant phenomena shaped the 2024 elections:

- Post-COVID Recovery: Public sentiment around economic recovery and healthcare reforms played a central role, influencing the electoral success of parties with strong policy proposals in these areas (European Parliament, 2024).
- Geopolitical Concerns: The war in Ukraine and energy security issues heightened the salience of foreign policy, benefiting parties advocating for strong EU cohesion and defense policies.
- Digital and Green Transformation: The elections saw a growing emphasis on digitalization and sustainability, although these issues were unevenly prioritized across member states (Hutter & Kriesi, 2020).

Methods for Analyzing and Mapping Electoral Results

The analysis of electoral results has undergone significant advancements in recent decades, moving beyond traditional descriptive statistics to incorporate innovative quantitative and spatial methods. These advancements have enhanced researchers' ability to uncover complex relationships in voter behavior, political competition, and electoral trends. European elections, with their transnational scope and multidimensional nature, have particularly benefited from the integration of these advanced methods.

Multivariate statistical techniques have emerged as essential tools for analyzing electoral data. Among them, Correspondence Analysis (CA) and its extension, Factorial Correspondence Analysis (FCA), have proven invaluable for exploring the underlying dimensions of electoral competition. By transforming categorical data into geometric spaces, these methods reveal latent structures, such as ideological divides or issue-based cleavages, within electoral datasets. CA has been frequently used to analyze voter-party relationships, while FCA offers the added capacity to incorporate multiple categorical variables, making it particularly suited to mapping complex electoral behaviors. Studies such as those by Le Roux and Rouanet (2010) have demonstrated how FCA can be used to uncover ideological alignments across diverse political systems, while Chadjipadelis and Teperoglou (2021) have applied CA to highlight the axes of political competition in European elections, such as the polarization between Eurosceptic and pro-European forces.

Hierarchical Cluster Analysis (HCA) complements these methods by classifying regions, political parties, or voter groups based on shared characteristics. This approach allows for the grouping of countries or constituencies into clusters that reflect regional or ideological affinities. For example, HCA has been used to create typologies of EU member states based on their electoral behavior, offering insights into cross-country variations and shared trends. The integration of HCA with FCA enhances the analytical depth, as the clusters derived from HCA can be visualized alongside the dimensions identified through FCA, offering a comprehensive perspective on electoral competition.

Spatial and geographic methods also play a pivotal role in analyzing and mapping electoral results. Geographic Information Systems (GIS) enable researchers to visualize electoral outcomes spatially, shedding light on regional variations in voter preferences and party performance. By employing techniques such as electoral cartography and spatial regression, researchers can explore how socio-economic factors, regional identities, and urbanization influence electoral outcomes. Johnston and Pattie (2006) have demonstrated how electoral maps can reveal spatial disparities in party support, while other studies have incorporated spatial regression to account for geographic dependencies, such as the "spillover effects" of voter behavior between neighboring regions.

In addition to statistical and spatial methods, machine learning and network analysis have gained traction in electoral studies. Predictive modeling using machine learning algorithms, such as Random Forests and Neural Networks, allows researchers to identify key predictors of electoral outcomes and model voter behavior with remarkable accuracy. Similarly, network analysis has been used to study the relationships between political actors, issues, and voter groups, revealing the structural dynamics of political competition. These approaches have proven particularly valuable in the context of European elections, where the connection between national and European-level politics creates complicate networks of alliances and conflicts.

The integration of these methods provides a robust framework for analyzing electoral results, yet challenges remain in capturing the full complexity of elections, particularly in transnational contexts like those of the European Union. This chapter proposes a original approach that combines FCA and HCA to analyze and map the results of the 2019 and 2024 European elections. By employing FCA, this study identifies the key dimensions of political competition, such as ideological polarization and the tension between European integration and national sovereignty. These dimensions are then complemented by HCA, which groups countries into clusters based on their electoral outcomes, providing a comparative perspective that highlights regional affinities and differences.

The rationale for this approach lies in its ability to go beyond traditional analyses by integrating multivariate methods with spatial visualization. FCA and HCA not only uncover latent structures and patterns in the data but also enable the creation of detailed electoral maps that visualize the dimensions of competition and the clustering of political systems. These maps offer a powerful tool for understanding the spatial and ideological dynamics of European elections, bridging the gap between advanced statistical analysis and practical political insights. By adopting this integrated framework, the chapter aims to contribute to the growing body of research on electoral studies, offering a methodologically innovative and empirically grounded approach to analyzing and reporting election results. The use of FCA and HCA in conjunction with mapping techniques provides a more detailed understanding of voter behavior, party competition, and regional variations, addressing the need for deeper and more comprehensive analyses in the study of European elections.

METHODOLOGY

This study employs a two-step methodology, combining Factorial Correspondence Analysis (FCA) and Hierarchical Cluster Analysis (HCA), to analyze and interpret the results of the 2019 and 2024 European elections. These methods are applied to a dataset where the rows represent constituencies (classified at the NUTS2 level) and the columns represent political groups, including eight official European political groups and independent parties divided into NI-Left, NI-Right, and NI. The purpose of this approach is to reveal underlying structures in electoral competition, identify clusters of constituencies and political groups, and provide a comprehensive graphical representation of the results. Data analysis was conducted using the software M.A.D. (Méthodes d'Analyse des Données) which has been developed by Professor Dimitris Karapistolis (2010).

Factorial Correspondence Analysis (also referred to as Multiple Correspondence Analysis, or MCA, when applied to categorical variables) serves as the initial phase of the analysis. This multivariate statistical technique is ideal for examining relationships within categorical data Greenacre (2017), such as voter preferences across political groups, by reducing the data into a smaller number of dimensions that explain the most variability.

FCA analyzes contingency tables where rows represent constituencies (NUTS2 regions) and columns represent political groups, including independent parties. Key components include row and column masses, which reflect the distribution of data, and inertia, which measures the variability explained by the dimensions. The method begins with a decomposition of the contingency table using singular value decomposition (SVD), a technique that extracts principal components. The primary goal is to identify the polarizations and antagonisms within the data, with a focus on the inertia explained by the extracted dimensions.

To ensure the relevance and reliability of the extracted dimensions, the empirical criterion of Benzecri is applied (1973). Specifically, two conditions are verified:

- A COR (contribution) value exceeding 200.
- A CTR (contribution to inertia) value greater than 10001+Total Number of Categories\frac{1000}{1 + \text{Total Number of Categories}}1+Total Number of Categories1000.

FCA generates graphical representations that map constituencies and political groups onto a factorial plane, providing insights into their relationships and highlighting key dimensions of electoral competition. This visualization helps uncover latent patterns, such as the polarization between pro-European and Eurosceptic parties or the alignment of independent groups.

The second step of the methodology involves applying Hierarchical Cluster Analysis (HCA) to the coordinates derived from the FCA. This step clusters constituencies and political group categories based on their positions in the factorial space, providing a complementary layer of analysis.

1. Distance Matrix: The first step in HCA involves computing a distance matrix using the Euclidean distance between the principal coordinates obtained from FCA. The distance D_{ij} between two points *i* and *j* is calculated as:

$$D_{ij} = \sqrt{\sum_{k=1}^{p} \left(F_{ik} - F_{jk}\right)^2}$$

where *p* represents the number of dimensions retained in the analysis.

2. Clustering Algorithm: HCA uses Ward's method, a clustering algorithm that minimizes the total within-cluster variance (inertia) at each step by merging

clusters. The objective is to minimize the increase in inertia (ΔE) as clusters are combined:

$$\Delta E = \sum_{i \in C_1 \cup C_2} d_i - \sum_{i \in C_1} d_i - \sum_{i \in C_2} d_i$$

where d_i is the squared distance of observation *i* from the centroid of its cluster.

3. Ward's Criterion: Clusters C_i and C_j are merged to minimize the following criterion:

$$\Delta E\left(C_i, C_j\right) = \frac{r_i r_j}{r_i + r_j} \|\mathbf{a}_i - \mathbf{a}_j\|^2$$

where r_i and r_j are the weights (masses) of clusters C_i and C_j , and a_i and a_j are the centroids of these clusters.

4. Output: HCA produces a dendrogram that visualizes the hierarchical structure of clusters. Each merge represents a step in the clustering process, with the height of the merge indicating the distance (inertia) at which clusters were combined. The final clustering groups constituencies and political group categories, offering insights into their associations.

The synergy between FCA and HCA is central to the analytical framework of this study. While FCA identifies the key dimensions of political competition, HCA clusters constituencies and political group categories based on these dimensions, providing a layered understanding of electoral dynamics. The combined approach allows for:

- The identification of latent structures and patterns in electoral results.
- The clustering of constituencies and political groups into meaningful categories.
- The creation of a biplot on the first factorial plane, visualizing the relationships between clusters and dimensions of competition.

This integrated methodology provides a framework for analyzing and mapping electoral results. By combining FCA and HCA, the study moves beyond descriptive statistics to uncover deeper insights into voter behavior, regional variations, and ideological alignments. The visualization of results on the factorial plane enhances the interpretability of complex data, making this approach particularly suited to the study of European elections, where the connection of national and European-level politics creates unique analytical challenges. The proposed framework enables the exploration of key questions, such as the polarization between political groups, the regional clustering of constituencies, and the evolution of electoral dynamics from 2019 to 2024. It offers a replicable model for future studies of electoral competition in transnational contexts.

RESULTS

Elections 2019

The Factorial Correspondence Analysis (FCA) of the 2019 European elections revealed four significant dimensions (factors) of political competition, collectively explaining the relationships and polarizations among European political groups (Table 1). These dimensions account for distinct portions of the total inertia and provide insight into the underlying axes of competition that shaped the electoral dynamics during this period.

Table 1. Axes of political competition in 2019 European Parliamentary elections

IND	#G1	COR	CTR	#G2	COR	CTR	#G3	COR	CTR	#G4	COR	CTR
EPP	-121	49	11	331	374	104	185	117	75	212	154	112
S&D	-44	6	2	61	12	4	414	596	319	5	0	1
RE	339	160	44	346	166	59	-170	40	32	-576	459	418
ECR	-1653	912	682	-48	0	1	-426	60	132	-177	10	26
LEFT	311	71	20	462	157	55	84	5	5	55	2	3
PfE	336	92	46	-984	789	495	-157	20	29	181	26	44
GREEN	503	247	85	397	154	68	-622	377	381	376	138	161
NI-LEFT	-591	141	24	-1177	559	126	335	45	23	325	42	25
OTHER	380	172	45	-459	251	85	25	0	1	-443	233	208
NI-RIGHT	-969	293	41	-198	12	3	-145	6	3	127	5	2
INERTIA		32,2%			25%			11%			9.60%	

The first factor explains 32.2% of the total inertia and represents the dominant axis of political competition in 2019. This dimension represents the Liberal-Conservative polarization between conservative nationalism and progressive, environmentally focused values. The European Conservatives and Reformists (ECR) are positioned at the extreme negative end of this axis, reflecting their strong conservative and Eurosceptic stance. In contrast, the Greens/European Free Alliance (GREEN) occupy the opposite, progressive end, emphasizing environmentalism and pro-European values. The more centrist groups, such as the European People's Party (EPP) and the Progressive Alliance of Socialists and Democrats (S&D), are situated closer to the center, illustrating their moderate roles within this competition. This factor encapsulates the liberal-conservative divide that remains a defining feature of European politics.

The second factor, which accounts for 25% of the total inertia, captures the tension between European integration and national isolation. On this axis, the EPP and Renew Europe (RE) are strongly aligned with the pro-European pole, advocating for deeper integration and multilateral governance. On the opposing side, nationalist and populist forces, such as the Identity and Democracy Party (PfE) and independent leftist parties (NI-LEFT), dominate the Eurosceptic end. This dimension reflects a critical ideological conflict within the European Union, revolving around the future balance between national sovereignty and shared governance.

The third factor, explaining 11% of the total inertia, highlights an internal competition within the left-wing spectrum. The S&D group is positioned at the extreme positive end, representing the traditional center-left agenda focused on social and economic justice through moderate reforms. On the other hand, the GREEN group occupies the most negative position, reflecting a progressive and radical approach that prioritizes environmental sustainability and systemic change. This axis underscores the growing influence of Green politics and the challenge it poses to traditional social democratic platforms.

The fourth factor, which explains 9.6% of the total inertia, reveals a competition between European liberalism and nationalist populism. Renew Europe (RE) is positioned at the extreme negative end of this axis, representing a pro-European, centrist liberal perspective that emphasizes economic openness and EU unity. In contrast, NI-RIGHT, which represents nationalist populist forces, is positioned at the positive end, reflecting a focus on identity politics and anti-European Union rhetoric. This dimension reflects the broader tension between globalization and nationalism in European politics.

Overall, the FCA reveals a fragmented and multidimensional landscape of political competition in the 2019 European elections (Diagram 1). The first two factors represent broad ideological and policy divides, while the third and fourth factors reveal more detailed conflicts within specific political families. The positioning of political groups along these dimensions provides a comprehensive understanding of their roles in the broader electoral dynamics. The analysis highlights the growing prominence of environmentalism, Euroscepticism, and nationalist populism, alongside the challenges faced by traditional political actors in maintaining their dominance in a rapidly evolving political environment.

Figure 1. Principal plane for 2019 European elections

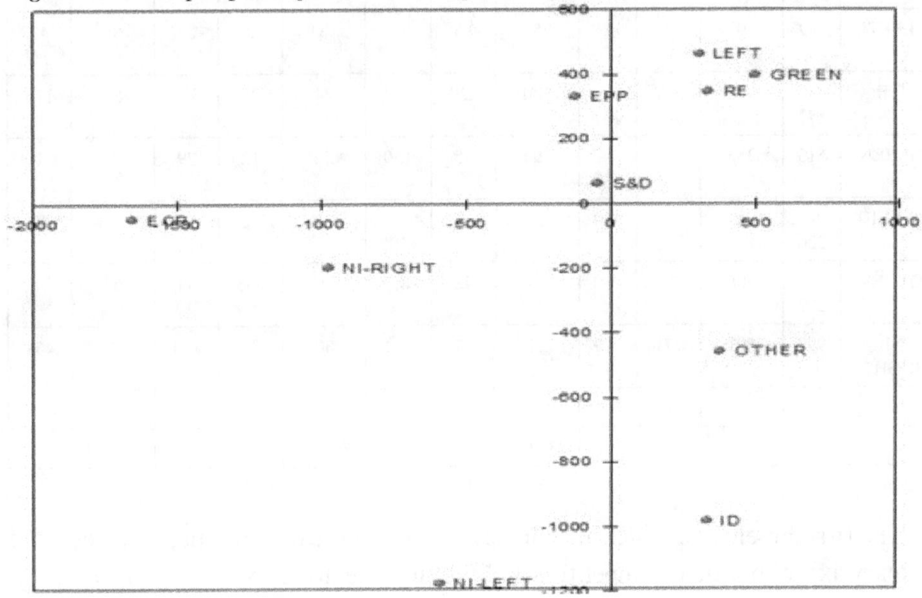

Elections 2024

Following the methodology, the FCA of the 2024 European elections reveals four significant dimensions of political competition, each representing distinct axes of polarization among European political groups (Tabe 2). These dimensions collectively explain key electoral dynamics, offering insights into the ideological and strategic divides shaping voter preferences in a politically charged post-COVID-19 and geopolitically tense environment.

Table 2. Axes of political competition in 2024 European Parliamentary elections

IND	#G1	COR	CTR	#G2	COR	CTR	#G3	COR	CTR	#G4	COR	CTR
EPP	317	341	106	159	86	29	-13	0	1	-257	223	161
S&D	-19	1	1	-57	12	3	242	227	80	-161	99	49
RE	18	0	1	-542	340	124	-185	39	23	-96	10	9
ECR	-645	260	186	960	577	443	382	91	109	29	0	1
LEFT	-99	9	3	-496	253	65	-60	3	2	-293	88	49
PfE	-877	492	317	-707	320	220	194	24	27	361	83	125
GREEN	513	355	93	-380	194	55	150	30	13	29	1	1
NI-LEFT	-236	3	2	450	13	5	-1593	174	82	-1227	103	69
OTHER	-202	40	20	280	78	42	-873	756	636	260	67	78
NI-RIGHT	999	501	271	219	24	14	244	30	27	853	365	458
		24.70%			23.10%			14.80%			10.60%	

The first dimension, which accounts for 24.7% of the total inertia, represents the primary axis of political competition and captures the divide between pro-European collaboration and nationalist isolation. On this axis, the European People's Party (EPP) takes a prominent position, aligned with pro-European policies and advocating for deeper integration. Similarly, the Greens/European Free Alliance (GREEN) and NI-RIGHT, a nationalist group surprisingly supportive of European unity, also align on the pro-European pole. By contrast, the European Conservatives and Reformists (ECR) and the Identity and Democracy Party (PfE) dominate the opposing side, representing nationalist agendas that prioritize sovereignty over EU collaboration. This dimension underscores the enduring tension between those advocating for a cohesive European framework and those championing a return to national sovereignty.

The second dimension, explaining 23.1% of inertia, highlights a growing ideological polarization between centrist pro-European parties and right-wing conservative forces. Renew Europe (RE) is strongly positioned on the positive side of this axis, reflecting its advocacy for liberal, centrist policies that emphasize European integration. The EPP, though similarly aligned with centrist pro-European ideals, exhibits a slightly more moderate position. Conversely, ECR and PfE consolidate their positions on the negative side, emphasizing traditional conservative values, nationalism, and skepticism toward further EU integration. This dimension reflects the increasing influence of right-wing coalitions in regions affected by economic instability and migration challenges, contributing to the rightward shift in parts of the EU's political landscape.

The third dimension, which explains 14.8% of inertia, highlights a significant divide between progressive and populist political movements. The Progressive Alliance of Socialists and Democrats (S&D) occupies a leading position on the positive side, championing policies focused on social justice, equality, and progressive reform. In contrast, populist parties, particularly NI-LEFT, are situated at the extreme negative end, reflecting their alignment with anti-establishment rhetoric and opposition to progressive agendas. This dimension highlights the challenges faced by traditional progressive parties in countering populist narratives, which continue to resonate strongly with voters experiencing economic and social discontent.

The fourth and final dimension, accounting for 10.6% of inertia, captures the economic and identity-based divide between liberal centrists and nationalist populists. Renew Europe is positioned at the extreme negative end of this axis, symbolizing its commitment to liberal economic policies, globalization, and free trade. On the opposing side, nationalist populist groups such as NI-RIGHT and PfE emphasize protectionist policies and identity-based narratives, reflecting their focus on preserving national interests and resisting globalization. This dimension reveals the tension between economic liberalization and nationalist protectionism, which has been further amplified by the EU's economic recovery efforts and ongoing energy security challenges.

Analysis of the 2024 European elections portrays again a complicated and differentiated political landscape, characterized by multidimensional axes of competition. The first two dimensions highlight broad ideological and policy divides, particularly around the degree of European integration and the rising influence of right-wing conservatism. Meanwhile, the third and fourth dimensions expose more subtle conflicts, such as those between progressive and populist movements, as well as the divide between economic liberalization and nationalist protectionism. These findings underscore the growing polarization within European politics, driven by a combination of geopolitical tensions, economic recovery efforts, and the continued resonance of nationalist and populist rhetoric. The analysis provides a detailed perspective on voter preferences and party alignments, reflecting the complicate dynamics shaping the EU's political framework in 2024.

Figure 2. Principal plane for 2024 European elections

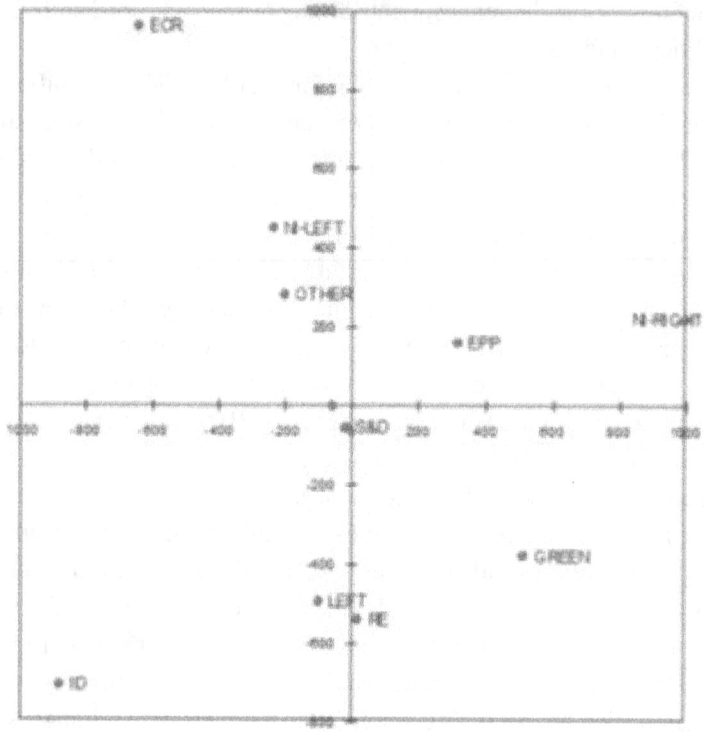

Transition Matrix for Voter Movement (2019-2024)

The transition matrix (Andreadis & Chadjipadelis, 2009) provides a detailed view of voter movements between political groups from the 2019 to the 2024 European elections. Each row represents the distribution of voters who supported a specific political group in 2019 across the political groups in 2024, expressed as percentages. This analysis reveals patterns of loyalty, defection, and shifts among voters, highlighting the dynamic nature of European electoral behavior.

The transition matrix for the 2019 and 2024 European elections reveals significant insights into voter behavior, loyalty, and shifts between political groups over the five-year period. Each row in the matrix represents the distribution of voters who supported a particular political group in 2019 and shows the proportion of these voters who transitioned to other parties in 2024. This analysis highlights both the stability of certain parties and the fluidity of voter preferences across the political spectrum:

- The European People's Party (EPP) demonstrates a high degree of voter retention, with 73.3% of its 2019 voters remaining loyal in 2024. Despite this stability, the party experiences notable losses, particularly to unaffiliated or "Other" groups, which attract 11.1% of former EPP voters, as well as to Renew Europe (3.7%). These shifts suggest that while the EPP remains a key anchor of European politics, some of its voters are exploring centrist or alternative political options, potentially reflecting dissatisfaction with its policies or a search for more dynamic platforms.
- The Progressive Alliance of Socialists and Democrats (S&D) also retains a solid majority of its base, with 63.7% of its voters from 2019 continuing their support in 2024. However, the party faces challenges in retaining all of its voters, with significant defections to the EPP (15.8%) and to smaller, unaffiliated groups (7.5%). These transitions indicate that S&D continues to struggle with voter attrition, as some of its base gravitates toward the stability of the EPP or explores non-traditional political alternatives.
- Renew Europe (RE) exhibits a more fragmented voter base. Only 34.6% of its 2019 supporters remain loyal, while the party attracts voters from diverse sources, including the EPP (3.7%), S&D (4.2%), and the Greens (17.3%). This dynamic reflects Renew Europe's appeal as a centrist, pro-European alternative, capable of attracting voters across ideological boundaries. However, the party also loses significant support to Identity and Democracy (PfE), which captures 19.6% of its former voters, and to unaffiliated groups (14.1%). These losses reveal the vulnerability of centrist platforms to both progressive and Eurosceptic competitors.
- The European Conservatives and Reformists (ECR) shows moderate voter retention, with 45.3% of its 2019 base remaining intact in 2024. It gains voters from NI-LEFT (58.4%) and the EPP (22.4%), suggesting an ability to attract those disillusioned with mainstream conservative or independent leftist platforms. However, the party experiences leakage to smaller groups (12.1%) and to the Greens (7.8%), signaling challenges in broadening its appeal beyond its core constituency.
- The Left (LEFT) retains 54.6% of its 2019 voters, maintaining a relatively stable base. The party attracts voters from S&D (8.9%) and unaffiliated groups (6.3%), reflecting its position as a progressive alternative for disenchanted center-left voters. However, its losses to Identity and Democracy (5.7%) and unaffiliated groups suggest difficulties in consolidating support among its progressive base.
- The Identity and Democracy Party (PfE) retains 28.5% of its 2019 voters, reflecting a highly fragmented base. Nevertheless, it successfully attracts support from Renew Europe (19.6%) and NI-RIGHT (22.2%), highlighting

its growing appeal among centrist defectors and right-wing independents. Despite these gains, its low retention rate suggests that its base remains fluid and lacks long-term loyalty.

- The Greens/European Free Alliance (GREEN) demonstrates moderate voter retention, with 53.1% of its 2019 voters continuing their support in 2024. It gains significant support from Renew Europe (17.3%), S&D (3.2%), and the Left (8.7%), showcasing its position as a key progressive force. However, the Greens lose some voters to PfE (3.5%) and NI-RIGHT (10.7%), highlighting the challenges of maintaining cohesion among a diverse and often ideologically varied base.
- Independent groups, particularly NI-LEFT and NI-RIGHT, show considerable instability. NI-LEFT retains only 2.1% of its 2019 voters, with most shifting to ECR (58.4%) and unaffiliated groups (28.2%). This collapse reflects the difficulties independent left-wing parties face in maintaining their relevance. NI-RIGHT retains just 9.8% of its 2019 base but attracts voters from PfE (22.2%) and unaffiliated groups (39.4%). Its voter gains suggest a consolidation of right-wing independent movements, even as its overall voter base remains fluid.
- The "Other" category, which encompasses smaller and unaffiliated parties, exhibits high voter turnover, with its 2019 voters dispersing widely. S&D captures 28.6% of these voters, followed by PfE (15.9%) and NI-RIGHT (10.0%). This reflects the diverse and fragmented nature of smaller political groups, which often act as transitional spaces for voters exploring new alignments.

Hierarchical Cluster Analysis (HCA) for Constituencies and Parties (2019)

The Hierarchical Cluster Analysis (HCA) of the 2019 European elections provides a detailed examination of how constituencies (NUTS regions) and political parties are grouped based on voting patterns. The analysis identified eight clusters, each comprising specific countries and associated political parties. These clusters reflect regional voting behaviors and ideological alignments, revealing the complex connection between geography and political competition in Europe.

Figure 3. HCA dendrogram for countries and parties 2019

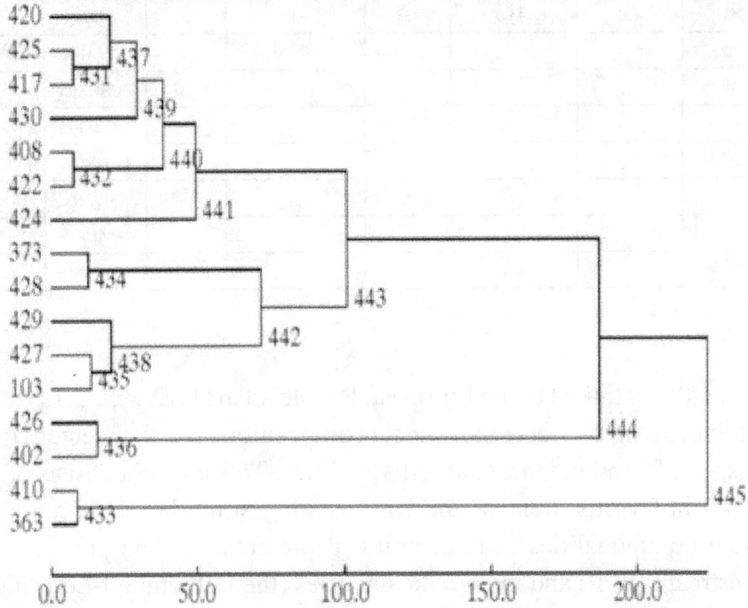

Table 3. 2019-HCA for countries and parties

INT	424	430	432	433	434	436	437	438
EPP	3362.96	995.5808		1745.73	2216.6166		5035.390	
S&D		5901.233				1345.75	1988.679	
RE		708.0613	3265.6759				2275.080	4649.0934
ECR			5500.8574	20958.7				
LEFT	13778.7	3343.976			185.1874			701.2552
PfE						9328.375		1921.5093
GREEN			874.6146		5960.5791			2288.1281
NI-LEFT	1887.19			2935.91		6128.907		
OTHER						3535.505	928.1886	4033.0146
NI-RIGHT				4874.72		231.5439	2304.392	
	CYPRUS	PORTUGAL	SWEDEN	POLAND	FINLAND	ITALY	ESTONIA	FINLAND
	IRELAND	SPAIN	CHECH		DEUTSCHLAND		FINLAND	DENMARK
	GREECE		BELGIUM		AUSTRIA		LATVIA	SPAIN
			NETHERLAND				LIETUVA	BELGIUM
							BULGARIA	LUXEMBOURG

continued on following page

Table 3. Continued

INT	424	430	432	433	434	436	437	438
							ROMANIA	FRANCE
							HUNGARY	
							SLOVAKIA	
							CROATIA	
							SPAIN	
							MALTA	
							SLOVENIA	

- Cluster 424 is linked to the European People's Party (EPP), the Left, and NI-Left, indicating a diverse ideological composition spanning center-right, progressive left, and independent leftist politics. This cluster consists of constituencies in Cyprus, Ireland, and Greece, suggesting that these regions share certain commonalities in their political preferences. The presence of both mainstream (EPP) and more radical parties (the Left and NI-Left) highlights a degree of polarization within these constituencies.
- Cluster 430 is associated with the Socialists and Democrats (S&D), Renew Europe (RE), and NI-Right, indicating a mix of moderate center-left, centrist liberal, and independent right-wing alignments. This cluster encompasses constituencies in Portugal, Malta, and Austria, reflecting the broad appeal of both progressive and centrist pro-European platforms in these regions, as well as some representation of right-wing independence. The inclusion of NI-Right suggests an undercurrent of nationalist or anti-establishment sentiment in certain areas.
- Cluster 432 is strongly linked with Renew Europe (RE), the Greens/European Free Alliance (GREEN), and EPP, capturing regions with a strong centrist, environmentally focused, and center-right leaning. This cluster includes constituencies in Luxembourg, Denmark, and the Netherlands, highlighting the ideological diversity within these areas. The overlap between RE and GREEN suggests that centrist and environmentalist platforms resonate strongly with voters in these constituencies, while the EPP presence indicates continued support for traditional pro-European values.
- Cluster 433 is associated with the European Conservatives and Reformists (ECR), NI-Left, and NI-Right. This cluster comprises constituencies in the Czech Republic, Poland, and Hungary, reflecting a strong Eurosceptic and nationalist presence. The combination of ECR with independent left and right groups underscores the fragmented nature of political competition in

these regions, where nationalist sentiments coexist with more localized political movements.

- Cluster 434 is linked to the Left, the Greens, and NI-Left, representing progressive and environmentally focused politics. This cluster includes constituencies in Germany, Sweden, and Finland. The presence of the Greens alongside the Left highlights a strong progressive and reform-oriented voter base, particularly in environmentally conscious regions such as Sweden and parts of Finland. However, the inclusion of NI-Left suggests a more fragmented leftist alignment, with some regions leaning toward less structured, independent movements.
- Cluster 436 is associated with Identity and Democracy (PfE) and NI-Right, emphasizing nationalist and populist politics. This cluster includes constituencies in Italy, France, and Bulgaria, highlighting regions where anti-European and identity-based platforms hold significant sway. The presence of both PfE and NI-Right underscores the strength of populist and nationalist rhetoric in these areas, often rooted in dissatisfaction with the EU's policies on issues such as migration and economic governance.
- Cluster 437 is linked with the Greens, NI-Right, and NI-Left. This cluster is unique in its inclusion of constituencies from Finland, demonstrating the regional variability within the country. While parts of Finland align with environmentally focused policies (Greens), other regions show support for independent political movements on both the left and right, reflecting a fragmented political identity within the nation.
- Cluster 438 is associated with NI-Right, S&D, and Renew Europe, comprising constituencies in Spain and Belgium. This cluster highlights the ideological diversity within these two countries, with constituencies supporting a mix of moderate center-left (S&D), centrist liberal (RE), and nationalist right (NI-Right) platforms. The inclusion of NI-Right suggests the presence of strong nationalist or regionalist movements within certain parts of these countries, contrasting with the more pro-European orientation of S&D and RE.

The HCA results reveal significant regional and ideological diversity within the European Union, with constituencies grouped based on shared political affinities and voting behaviors. The clusters demonstrate varying degrees of cohesion, with some reflecting broad ideological alignments and others highlighting the fragmentation and coexistence of contrasting political movements (Figure 1).

Figure 4. Map of clusters of political alignments in the 2024 European elections

- 424: EPP, Left, NI-Left – Pro-European, fragmented with leftist influence.
- 430: S&D, Renew Europe, NI-Right – Center-left and centrist-liberal alliances.
- 432: Renew Europe, Greens, EPP – Centrist liberal and environmentally focused regions.
- 433: ECR, NI-Left, NI-Right – Conservative and nationalist Eurosceptic bloc.
- 434: Greens, EPP, NI-Left – Environmental and fragmented progressive areas.
- 436: Identity and Democracy (PfE), ECR, NI-Right – Populist-nationalist strongholds.
- 437: Greens, NI-Right, NI-Left – Progressive and independent-leaning regions.

For example, Cluster 424, which includes Cyprus, Ireland, and Greece, reflects the coexistence of mainstream center-right politics (EPP) with progressive and independent leftist movements. This suggests that while traditional parties maintain a strong presence, there is also room for alternative voices, particularly those advocating systemic change. Similarly, Cluster 434, which spans Germany, Sweden, and Finland, underscores the growing importance of progressive and environmentalist politics, although regional differences within Finland complicate the overall picture.

In contrast, clusters like 433 and 436 reveal the strength of Eurosceptic and nationalist politics, particularly in Eastern and Southern Europe. The alignment of constituencies in Poland, Hungary, and Italy with parties like ECR and PfE highlights the challenges faced by pro-European forces in these regions, where dissatisfaction with the EU often translates into support for nationalist and populist platforms. The division of countries like Finland and Belgium into multiple clusters highlights significant intra-country variability in voter behavior. In Finland, some regions align with progressive and Green politics, while others lean toward nationalist or independent movements. Similarly, Belgium's inclusion in Cluster 438 reflects the fragmentation of its political landscape, with constituencies supporting a mix of pro-European, centrist, and nationalist platforms.

Hierarchical Cluster Analysis (HCA) for Constituencies and Parties (2024)

The 2024 HCA results (Table 4) provide a comprehensive view of the evolving European political landscape. The seven clusters reveal a complex connection between traditional pro-European parties, progressive movements, and nationalist forces. While mainstream parties like EPP and S&D retain their influence, the rise of populist and Eurosceptic rhetoric underscores the challenges faced by pro-European platforms in maintaining unity and appeal. The analysis highlights the importance of regional variability, as countries like Belgium and Finland appear across multiple clusters, reflecting diverse and localized political dynamics.

Figure 5. HCA dendrogram for countries and parties 2024

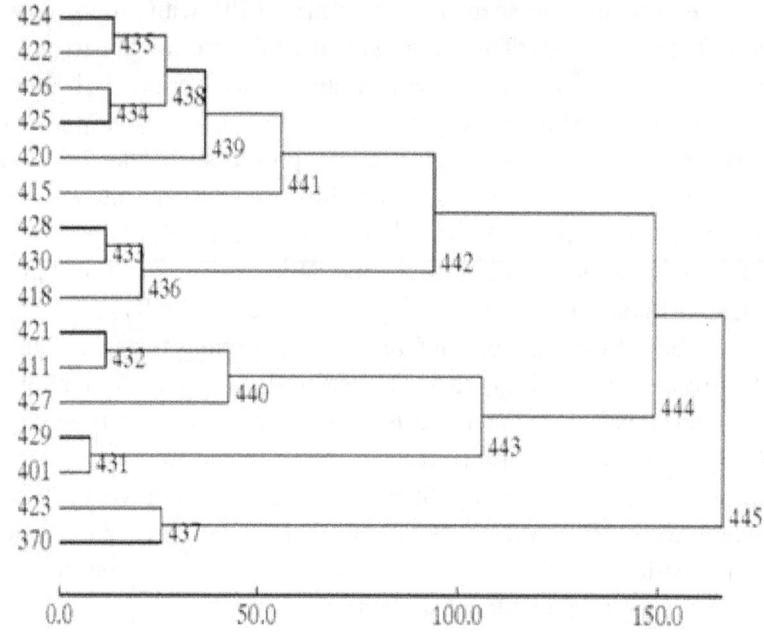

Table 4. 2024-HCA for countries and parties

INT	415	420	427	431	432	434	435	436	437
EPP	1534.2	4582.063	5753.5				1102.006		2655.6591
S&D		4752.638			2027.1	1117.47	2231.1946		
RE						4780.36	3175.1273	2346.7691	
ECR	1684.3		11458		8404.0		1724.2479		
LEFT	10426	248.271				4464.58		1422.0636	
PfE		615.4029			288.70			8111.1208	
GREEN						2549.53		403.5926	3896.306
NI-LEFT	17841			4225.241			1210.7467		
OTHER			1069.28	17551.03	67.1842				
NI-RIGHT					2242.05				6856.1336
	GREECE	MALTA	POLAND	CHECH	ESTONIA	FINLAND	ESTONIA	AUSTRIA	DEUTSCHLAND
		PORTUGAL		HUNGARY	LATVIA	SWEDEN	LIETUVA	BELGIUM	SLOVENIA
		SPAIN		SLOVAKIA	ITALY	DENMARK	BULGARIA	NETHERLAND	

continued on following page

Table 4. Continued

INT	415	420	427	431	432	434	435	436	437
						CYPRUS	ROMANIA	FRANCE	
						IRELAND	CROATIA		
						BELGIUM	LUXEMBOURG		

- Cluster 415 is associated with the European People's Party (EPP), the Socialists and Democrats (S&D), and NI-Right, reflecting a mix of center-right, center-left, and nationalist tendencies. This cluster includes constituencies from Cyprus, Malta, and Ireland. These regions display relatively stable electoral preferences, where mainstream pro-European parties like EPP and S&D dominate, while nationalist groups like NI-Right maintain a modest presence. The coexistence of mainstream and nationalist parties suggests a degree of ideological diversity within these constituencies.
- Cluster 420 is linked to Renew Europe (RE), the Greens/European Free Alliance (GREEN), and the Left, highlighting a progressive and centrist alignment. Constituencies from the Netherlands, Denmark, and Luxembourg form this cluster. The presence of both RE and GREEN emphasizes the continued salience of liberal and environmentally focused politics in these regions. The inclusion of the Left reflects the presence of progressive voters advocating for systemic reforms and social justice, alongside the liberal and environmentalist platforms.
- Cluster 427 is associated with Identity and Democracy (PfE) and NI-Left, signaling the influence of nationalist and independent left-wing movements. It includes constituencies from Poland, Hungary, and Italy. The alignment with PfE highlights the continued appeal of nationalist and populist rhetoric in these regions, while the inclusion of NI-Left suggests a fragmented political landscape, where left-leaning voters may align with independent or localized movements.
- Cluster 431 is linked to the ECR and NI-Right, representing conservative and Eurosceptic forces. Constituencies from the Czech Republic, Bulgaria, and Austria are included in this cluster. The dominance of ECR reflects the persistence of traditional conservative values and skepticism toward deeper European integration, while the presence of NI-Right highlights the role of nationalist and regionalist sentiments within these areas.
- Cluster 432 is associated with the Left, S&D, and Renew Europe, reflecting a mix of progressive and centrist pro-European tendencies. Germany, Finland, and Spain are represented in this cluster. The diversity of associated political groups indicates the coexistence of progressive (Left, S&D) and centrist lib-

eral (RE) platforms, suggesting that these constituencies encompass a broad spectrum of pro-European voter preferences. The presence of both S&D and RE highlights a balance between moderate progressivism and centrist pragmatism.

- Cluster 434 is dominated by the Greens, EPP, and NI-Left, capturing constituencies from Sweden and parts of Belgium. The presence of the Greens emphasizes the influence of environmentally focused politics, while the EPP's inclusion reflects continued support for traditional pro-European values. NI-Left's presence indicates a more fragmented political environment in certain regions, where independent leftist movements retain a foothold alongside mainstream and progressive parties.
- Cluster 436 is associated with Identity and Democracy (PfE), ECR, and NI-Right, highlighting the strength of nationalist and Eurosceptic forces. Constituencies from Italy and France dominate this cluster, reflecting the growing appeal of identity-based and conservative politics in these regions. The inclusion of ECR alongside PfE and NI-Right underscores the consolidation of Eurosceptic and nationalist rhetoric, particularly in areas where dissatisfaction with European policies on migration and economic recovery is pronounced.

The results for 2024 (Figure 2) reflect that traditional pro-European forces like EPP, S&D, and RE maintain a strong presence in clusters such as 415 and 420, particularly in regions like Cyprus, Malta, and the Netherlands.

Figure 6. Map of clusters of political alignments in the 2024 European elections

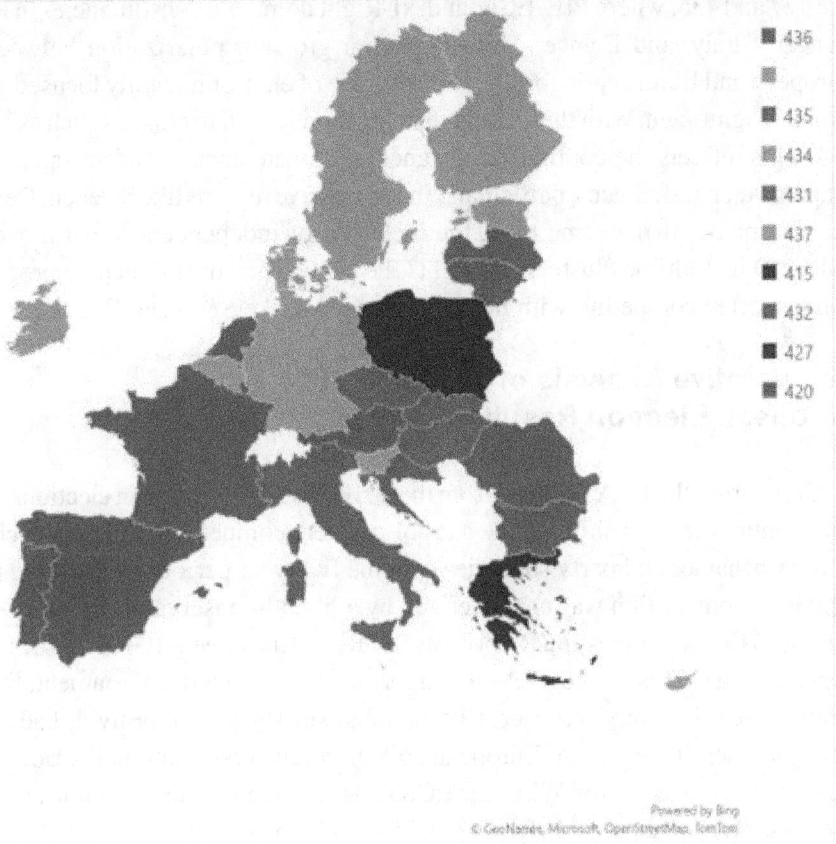

- 415: EPP, S&D, NI-Right – Pro-European regions with a mix of center-right, center-left, and nationalist elements.
- 420: Renew Europe, Greens, S&D – Progressive and centrist liberal regions characterized by environmentally focused and pro-European policies.
- 427: Identity and Democracy (PfE), NI-Left – Nationalist and left-leaning independent regions.
- 431: ECR, NI-Right – Conservative and nationalist Eurosceptic constituencies.
- 432: Renew Europe, Greens, S&D – Centrist liberal and environmentally focused regions with progressive tendencies.
- 434: Greens, EPP, NI-Left – Environmentalist and fragmented progressive regions.
- 436: Identity and Democracy (PfE), ECR, NI-Right – Nationalist-populist regions.

However, the rise of nationalist and populist movements is evident in clusters like 427 and 436, where PfE, ECR, and NI-Right dominate constituencies in Poland, Hungary, Italy, and France. This suggests a growing polarization between pro-European and Eurosceptic forces. The presence of environmentally focused politics remains significant, with the Greens strongly represented in clusters such as 420 and 434. This reflects the continued relevance of climate change and sustainability in shaping voter preferences, particularly in progressive regions like Sweden, Denmark, and Germany. At the same time, the inclusion of independent leftist movements (NI-Left) in multiple clusters highlights the persistence of fragmented progressive politics, often competing with mainstream parties like S&D and RE.

Comparative Analysis of 2019 and 2024 European Election Results

Comparing the FCA findings from the 2019 and 2024 European elections reveals both continuities and shifts in the axes of political competition, reflecting changes in voter behavior and party strategies over the five-year period. In 2019, the primary axis of competition was characterized by a liberal-conservative divide, with the European Conservatives and Reformists (ECR) and the Greens (GREEN) occupying opposite poles. This ideological split largely revolved around environmentalism and progressive values versus conservative nationalism. By 2024, this divide had evolved into a broader dimension of European collaboration versus nationalist isolation, as captured by the first factor. While the ECR and PfE continued to champion nationalist agendas, groups such as NI-RIGHT, which had a marginal role in 2019, emerged as unexpected supporters of European collaboration, albeit with nationalist undertones. This evolution indicates a reframing of political competition in response to challenges such as the war in Ukraine and the EU's energy security policies, which have reshaped the discourse on national versus transnational priorities.

The second axis in 2019 highlighted the polarization between pro-European integration groups like EPP and RE, and nationalist forces such as PfE and NI-LEFT. In 2024, this dimension persisted but became increasingly dominated by right-wing conservative forces like ECR and PfE, reflecting their consolidated influence across several EU member states. This shift suggests that the right-wing coalition has gained ground, particularly in regions facing economic pressures and migration challenges, amplifying their appeal to voters disillusioned with the status quo.

The third axis in 2019, which focused on competition within the left between S&D and GREEN, evolved in 2024 into a more explicit divide between progressive and populist forces. While the S&D retained its progressive agenda, the rise of NI-LEFT as a significant populist challenger highlights the growing appeal of anti-establishment rhetoric. This shift underscores a deepening struggle for the left to maintain its influence in an increasingly fragmented political spectrum.

The fourth axis in both elections highlights a tension between economic liberalism and nationalist populism. However, in 2024, this dimension took on greater significance as Renew Europe (RE) strongly represented the liberal pole, advocating for globalization and economic openness, while PfE and NI-RIGHT solidified their positions as nationalist populist forces. This reflects the continued salience of economic recovery policies and the polarization around globalization versus protectionism, particularly in the context of post-pandemic recovery and geopolitical uncertainties.

The most notable difference between the two elections is the emergence of a more fragmented and complex political landscape in 2024. While the 2019 elections were marked by relatively clear-cut ideological divides, the 2024 elections reveal a more multidimensional competition, shaped by external shocks such as the pandemic, geopolitical tensions, and economic instability. Additionally, the increased influence of populist and nationalist rhetoric across several dimensions indicates a shift in voter priorities, driven by disenchantment with traditional political structures and a growing focus on national interests. These changes underline the evolving nature of European politics, as parties adapt their strategies to respond to shifting voter concerns and broader contextual challenges.

The transition matrix highlights key patterns of voter movement between the 2019 and 2024 European elections. Traditional parties like the EPP and S&D exhibit relatively strong voter retention, maintaining their roles as stabilizing forces in European politics. However, both parties face significant competition, with notable losses to centrist groups like Renew Europe and unaffiliated categories, signaling the increasing fluidity of voter preferences. Centrist platforms such as Renew Europe attract voters from across the political spectrum, underscoring their role as a bridge between traditional parties and emerging alternatives. However, Renew Europe's vulnerability to defections, particularly to populist parties like PfE, illustrates the challenges centrist parties face in navigating a polarized political environment.

The rising influence of populist and nationalist groups, such as PfE and NI-RIGHT, is a key trend reflected in the transition matrix. These parties successfully attract disenchanted voters from a range of political affiliations, reflecting their appeal to those dissatisfied with mainstream politics. However, their low voter retention rates indicate a lack of deep-rooted loyalty, suggesting that their support may remain volatile. Finally, the data underscore the fragmentation and instability of independent and smaller groups, particularly NI-LEFT, which loses a significant portion of its base to more structured parties like ECR. These shifts highlight the ongoing consolidation of political forces, as voters gravitate toward parties with clearer ideological or policy platforms.

The Hierarchical Cluster Analysis (HCA) for the 2019 and 2024 European elections provides a fascinating comparative lens to explore the evolution of electoral dynamics across constituencies and political parties. Both analyses reveal a fragmented European political landscape, yet they highlight significant differences in the creation of clusters, party alignments, and regional political dynamics over the five-year period. These differences reflect changing voter preferences, the rise of new political issues, and shifting party strategies.

In 2019, eight distinct clusters were identified, while in 2024, the analysis yielded seven clusters. The reduction in the number of clusters in 2024 suggests a degree of consolidation in political alignments, as constituencies with similar voting behaviors and party affiliations became more cohesive. However, this does not indicate a simplification of the political landscape. Instead, it reflects an increasing polarization, where constituencies and parties align more clearly along ideological or issue-based lines. For instance, in 2019, Finland and Belgium were divided across multiple clusters, reflecting significant intra-country variability. While this fragmentation persists in 2024, it appears less pronounced as constituencies in these countries show more alignment with specific party groupings, such as the Greens or Renew Europe (RE). This suggests that voter preferences in these countries may have become more consolidated around certain issues, such as climate change or pro-European liberalism.

One of the most notable differences between 2019 and 2024 is the shifting prominence of political parties within the clusters. In 2019, traditional pro-European parties such as the European People's Party (EPP) and the Progressive Alliance of Socialists and Democrats (S&D) featured prominently in multiple clusters, reflecting their strong, stable presence across Europe. However, by 2024, their dominance has diminished, with parties like Renew Europe (RE) and the Greens playing increasingly central roles in shaping cluster dynamics.

Renew Europe, which struggled with voter retention and cluster representation in 2019, emerges in 2024 as a key player, appearing in clusters such as 420, 432, and 415. This growth highlights the appeal of centrist, pro-European liberalism in a period marked by economic recovery and geopolitical challenges. Similarly, the Greens, while influential in 2019, expand their footprint in 2024, appearing in clusters linked to progressive constituencies in Sweden, Denmark, and Germany. This reflects the rising salience of environmental issues, particularly in the wake of climate crises and sustainability debates.

On the other hand, Eurosceptic and nationalist parties such as Identity and Democracy (PfE) and the European Conservatives and Reformists (ECR) show stronger and more consolidated cluster representation in 2024. In 2019, their influence was fragmented, with PfE and ECR appearing in separate clusters. By 2024, their alignment in clusters like 427 and 436 underscores a growing coalition of nationalist and populist forces, particularly in regions like Poland, Hungary, Italy, and France. This trend highlights the continued resonance of Eurosceptic rhetoric and identity-based politics in certain parts of Europe.

In 2019, intra-country variability was particularly pronounced, with countries like Finland, Belgium, and Spain appearing in multiple clusters. This reflected the diversity of voter preferences within these nations, driven by regional identities and localized political issues. While this variability persists in 2024, it appears less fragmented. For example, Finland in 2024 is predominantly associated with progressive and environmentalist clusters (e.g., Greens, Left), while its earlier alignment with nationalist and independent movements has diminished. Similarly, Belgium shows a stronger association with centrist and pro-European parties in 2024, reflecting a potential shift toward greater regional coherence in its voting behavior. However, regions like Italy and France demonstrate increasing alignment with nationalist and Eurosceptic clusters in 2024, as opposed to their more fragmented representation in 2019. This suggests that political polarization has intensified in these countries, with voters gravitating toward more defined ideological blocs.

The polarization of European politics becomes more pronounced in 2024. In 2019, many clusters represented a mix of ideological groups, such as the coexistence of EPP, S&D, and NI-Left within Cluster 424. By 2024, the clusters show greater ideological coherence, with constituencies aligning more distinctly with either progressive, centrist, or nationalist platforms. For example, Cluster 436 in 2024 is dominated by PfE, ECR, and NI-Right, reflecting the consolidation of Eurosceptic forces, whereas similar groups were more dispersed in 2019. This ideological polarization is also evident in the increasing influence of environmental and progressive politics, as seen in the growth of the Greens and Renew Europe, alongside the simultaneous strengthening of nationalist and populist parties. This bifurcation highlights a Europe divided not only by traditional left-right dynamics but also by emerging cleavages around issues like climate change, globalization, and migration.

The political "maps" derived from the 2019 and 2024 HCA analyses differ significantly in their representation of Europe's political landscape. The 2019 clusters capture a relatively fragmented and multipolar environment, where traditional parties like EPP and S&D maintained a central role despite growing competition from smaller or independent movements. By 2024, the clusters illustrate a more polarized and consolidated political map, with centrist, progressive, and nationalist forces competing more distinctly. This shift reflects broader societal and political changes, including the aftermath of the COVID-19 pandemic, heightened geopolitical tensions, and increasing voter engagement with issues like sustainability and identity. The strengthening of both progressive and nationalist platforms, alongside the declining dominance of traditional parties, underscores the evolving nature of European political competition and the need for parties to adapt to shifting voter priorities.

DISCUSSION

The analysis of the European elections from 2019 to 2024 reveals a dynamic and evolving political landscape characterized by increasing polarization, ideological consolidation, and the growing influence of both progressive and nationalist forces. Through the application of Factorial Correspondence Analysis (FCA) and Hierarchical Cluster Analysis (HCA), this study highlights the multidimensional nature of political competition and the complex connection between regional contexts and party alignments. The findings not only capture shifts in voter behavior and party dynamics but also shed light on broader societal, economic, and geopolitical developments shaping European politics during this period.

The results demonstrate clear transformations in the European political landscape between 2019 and 2024. In 2019, the primary axis of competition revolved around the liberal-conservative divide, with parties such as the Greens (GREEN) and European Conservatives and Reformists (ECR) representing opposite poles. By 2024, this divide had evolved into a broader polarization between pro-European collaboration and nationalist isolation, reflecting the EU's challenges in addressing pressing issues like the COVID-19 pandemic, economic recovery, and geopolitical tensions.

The rise of progressive and environmentalist platforms, particularly the Greens and Renew Europe (RE), underscores the increasing importance of climate change and sustainability in voter preferences. At the same time, nationalist and Eurosceptic forces such as Identity and Democracy (PfE) and ECR have consolidated their influence, particularly in regions facing economic or migratory pressures. This polarization is further reflected in the clustering patterns, with constituencies increasingly aligning along distinct ideological lines, whether progressive, centrist, or nationalist. Notably, the study highlights a reduction in the number of clusters from eight in 2019 to seven in 2024, suggesting a degree of consolidation in voter alignments. However, this does not indicate simplification; rather, it points to greater ideological coherence within clusters, with regions more clearly defined by their political and regional identities. The persistence of intra-country variability, particularly in countries like Finland, Belgium, and Spain, underscores the role of localized factors in shaping electoral outcomes.

The findings of this study are closely tied to broader political and social developments in Europe. The polarization observed between 2019 and 2024 reflects the increasing salience of issues such as environmental policy, migration, and national sovereignty. For example, the rising prominence of the Greens and RE can be attributed to growing public awareness of climate crises and the EU's role in addressing sustainability challenges. Conversely, the consolidation of nationalist and populist forces reflects dissatisfaction with EU policies on migration, economic inequality, and perceived overreach into national affairs. The geopolitical landscape, particularly the aftermath of Brexit and the war in Ukraine, has further shaped voter preferences. Pro-European parties like RE and EPP have benefited in regions prioritizing unity and collaboration, while nationalist forces like PfE and ECR have capitalized on fears of economic instability and cultural erosion. These dynamics are particularly evident in countries like Italy, France, and Hungary, where Eurosceptic rhetoric has gained traction among disillusioned voters.

Intra-country variability in countries like Finland and Belgium also reflects the complex connection of national and regional identities. For instance, Finland's alignment with both progressive and nationalist clusters highlights the tension between urban, environmentally conscious constituencies and rural areas prioritizing national sovereignty. Similarly, Belgium's fragmentation mirrors its deeply rooted linguistic and regional divisions, which continue to influence voting behavior. The key findings of the analysis, reflect the shifting contours of European politics:

- Increased Polarization: The divide between pro-European and Eurosceptic forces has deepened, with fewer constituencies falling into mixed ideological clusters. This polarization is driven by diverging views on the EU's role in addressing economic, social, and geopolitical challenges.
- Rising Importance of Environmental Issues: The growing influence of the Greens and their presence across multiple clusters reflect the prioritization of sustainability and climate policies among European voters.
- Consolidation of Nationalist Movements: Nationalist and populist parties like PfE and ECR have strengthened their foothold, particularly in regions facing economic inequality or migratory pressures.
- Decline of Traditional Party Dominance: While the EPP and S&D remain significant, their dominance has been eroded by the rise of centrist, progressive, and nationalist alternatives, reflecting a more fragmented political landscape.
- Regional Variability: Intra-country differences persist, with countries like Finland and Belgium continuing to exhibit significant variability in voter alignments across regions.

These findings provide a comprehensive understanding of the evolving nature of European politics, highlighting the connection between ideological, regional, and contextual factors in shaping electoral outcomes. By identifying key dimensions of competition and clustering patterns, the study contributes to a deeper understanding of how voter preferences and party strategies have adapted to societal changes. This research also underscores the importance of addressing polarization and regional variability in fostering a more cohesive European Union. The insights gained from this study are valuable for policymakers, political analysts, and party strategists seeking to navigate the complexities of European electoral dynamics. Understanding the drivers of voter behavior and the clustering of constituencies can inform more targeted policies and campaigns, particularly in addressing the concerns of disillusioned or polarized voters.

While the study provides valuable insights, several limitations must be acknowledged. The reliance on FCA and HCA, while robust, may oversimplify the complexity of voter behavior and party alignments. Additionally, the analysis is based on aggregate data, which may mask individual-level variations in voter preferences. Intra-country variability, while highlighted, requires further exploration to account for local factors that influence voting behavior. Finally, the study focuses on two electoral cycles, limiting its ability to capture longer-term trends and shifts.

The findings of this study open several avenues for future research. A more detailed analysis of intra-country variability, particularly in fragmented nations like Belgium and Finland, could provide deeper insights into the drivers of regional differences. Longitudinal studies examining multiple electoral cycles would allow for a more comprehensive understanding of trends and their persistence over time. Additionally, integrating qualitative methods, such as voter surveys or interviews, could enrich the analysis by capturing the motivations and perceptions underlying voter behavior. Future research could also explore the impact of external factors, such as geopolitical tensions, economic crises, or technological advancements, on voter alignments. Finally, examining the role of digital communication and social media in shaping party strategies and voter preferences would provide a more detailed understanding of modern electoral dynamics.

Overall, the study manages to highlight the dynamic and fragmented nature of European politics between 2019 and 2024, emphasizing the connection between polarization, regional variability, and emerging political trends. The findings underscore the importance of addressing voter disillusionment and ideological divides in fostering a more cohesive and inclusive European Union. While challenges persist, the study provides a robust framework for understanding and navigating the evolving political landscape, offering valuable insights for researchers, policymakers, and political actors alike.

REFERENCES

Andreadis, I., & Chadjipadelis, T. (2009). A Method for the Estimation of Voter Transition Rates. *Journal of Elections, Public Opinion, and Parties*, *19*(2), 203–218. DOI: 10.1080/17457280902799089

Benzécri, J. P. (1973). *L'Analyse des données*. Tome 2: *L'Analyse des correspondances*. Dunod.

Brack, N., & Marié, A. (2024). From fringe to front? Assessing the voting influence of the radical right in the European Parliament. *European Union Politics*, *25*(4), 748–771. DOI: 10.1177/14651165241268127

Chadjipadelis, T., & Teperoglou, E. (2021). What Was Really the Case? Party Competition in Europe at the Occasion of the 2019 European Parliament Elections. In *Data Analysis and Rationality in a Complex World 16* (pp. 9-16). Springer International Publishing.

De Vries, C. E., Hobolt, S. B., Proksch, S. O., & Slapin, J. B. (2021). *Foundations of European politics: a comparative approach*. Oxford university press. DOI: 10.1093/hepl/9780198831303.001.0001

Dennison, J., & Kriesi, H. (2023). Explaining Europe's transformed electoral landscape: Structure, salience, and agendas. *European Political Science Review*, *15*(4), 483–501. DOI: 10.1017/S1755773923000085

European Parliament. (2019). *Results of the 2019 European Elections*. Retrieved from https://results.elections.europa.eu

European Parliament. (2024). *Provisional Results of the 2024 European Elections*. Retrieved from https://results.elections.europa.eu

Fisher, J., Fieldhouse, E., Franklin, M. N., Gibson, R. K., Cantijoch, M., & Wlezien, C. (Eds.). (2018). *The Routledge handbook of elections, voting behavior and public opinion* (pp. 280–292). Routledge.

Greenacre, M. (2017). *Correspondence analysis in practice*. chapman and hall/crc.

Hix, S., & Marsh, M. (2007). Punishment or protest? Understanding European parliament elections. *The Journal of Politics*, *69*(2), 495–510. DOI: 10.1111/j.1468-2508.2007.00546.x

Hobolt, S. B., & De Vries, C. (2016). Turning against the Union? The impact of the crisis on the Eurosceptic vote in the 2014 European Parliament elections. *Electoral Studies*, *44*, 504–514. DOI: 10.1016/j.electstud.2016.05.006

Hutter, S., & Kriesi, H. (2020). Politicizing Europe in times of crisis. In *The European Union Beyond the Polycrisis?* (pp. 34–55). Routledge. DOI: 10.1201/9781003002215-3

Karapistolis, D. (2010). *Software Method of Data Analysis MAD* [Internet]. Retrieved from https://www.pylimad.gr/

Le Roux, B., & Rouanet, H. (2010). *Multiple correspondence analysis* (Vol. 163). Sage. DOI: 10.4135/9781412993906

Panagiotidou, G., & Chadjipadelis, T. (2021). First-time voters in Greece: views and attitudes of youth on Europe and democracy. In *Data Analysis and Rationality in a Complex World 16* (pp. 225–232). Springer International Publishing. DOI: 10.1007/978-3-030-60104-1_25

Reif, K., & Schmitt, H. (1980). Nine second-order national elections – A conceptual framework for the analysis of European election results. *European Journal of Political Research*, *8*(1), 3–44. DOI: 10.1111/j.1475-6765.1980.tb00737.x

Schmitt, H., & Toygür, I. (2016). European Parliament elections of May 2014: Driven by national politics or EU policy making? *Politics and Governance*, *4*(1), 167–181. DOI: 10.17645/pag.v4i1.464

Wallaschek, S. (2020). *Analyzing the European parliamentary elections in 2019: Actor visibility and issue-framing in transnational media*. Springer Fachmedien Wiesbaden.

Chapter 7
Political Marketing Strategies in the 2024 European Elections

Vasiliki Bouranta
https://orcid.org/0009-0003-1717-7048
Aristotle University of Thessaloniki, Greece

Georgia Panagiotidou
https://orcid.org/0000-0002-7264-1187
Aristotle University of Thessaloniki, Greece

Theodore Chadjipadelis
Aristotle University of Thessaloniki, Greece

ABSTRACT

The 2024 European Elections marked a pivotal post-Brexit moment, influenced by socio-economic challenges like climate change, migration, and the energy crisis. This study examines the evolution of political marketing as a strategic electoral tool, focusing on its role in voter engagement and party positioning at national and supranational levels. By analyzing the dual-level dynamics of political competition, the research explores how European parties employed marketing strategies to shape key political axes and address diverse voter concerns. Using a comparative multivariate analysis, it investigates variations in campaign approaches, alignment with European political groups, and thematic priorities during the 2024 elections.

DOI: 10.4018/979-8-3693-9400-7.ch007

INTRODUCTION: POLITICAL MARKETING STRATEGIES IN THE 2024 EUROPEAN ELECTIONS

The 2024 European Elections marked a pivotal moment in the post-Brexit European Union, shaping the political landscape amid socio-economic challenges such as climate change, the energy crisis, and migration pressures. Political marketing has evolved into a crucial element of electoral strategy, playing a significant role in influencing voter behavior and party positioning.

Political marketing represents an evolving and interdisciplinary field that combines political goals with marketing approaches. In the era of rapid digital transformation and the reshaping of public discourse through online platforms, political marketing has become essential for political parties, candidates, and governing bodies. Its primary purpose is to anticipate, comprehend, and respond to the electorate's needs, enhancing political engagement and communicating ideas and policy platforms effectively. Scholars such as Wright (1999) have highlighted its role in increasing political participation, while some critics, like Crompton & Lamb (1986), remain skeptical of its comprehensive strategic nature. This comprehensive approach emphasizes the integration of long-term plans that extend to organizational structure, leadership, and day-to-day operations.

Rather than being limited to advertisements or public relations efforts, modern political marketing involves strategic communication supported by data-driven insights. It encompasses voter behavior studies and employs advanced marketing tools to create resonant messages, campaigns, and initiatives that align with societal concerns. With technological progress enabling targeted outreach through big data and digital campaigns, political entities can tailor their messaging to reach specific groups more effectively. Political marketing is built upon reciprocal communication flows, fostering connections between political organizations and the electorate. It serves as a democratic mechanism, encouraging dialogue and collaboration.

A key aspect of political marketing is its application during electoral campaigns. It includes various techniques designed to understand voter preferences, craft compelling narratives, and manage resources efficiently. Prior to being recognized as an academic discipline, political marketing was primarily associated with campaign organization and execution (Lees-Marshment, 2009). Core activities include orchestrating pre-election strategies, developing political personas, utilizing media platforms, and allocating campaign resources at both national and local levels. Activities such as organizing pre-election campaigns, shaping messages and profiles, using media outlets, and strategically allocating resources at local and national levels remain key stages in planning and implementing electoral campaigns.

For more than four decades, voters across Europe have participated in elections to select representatives for supranational bodies, particularly the European Parliament. This raises the question of whether political marketing strategies, tools, and methods are employed at this level of elections with the same rigor as in national contests. European elections are often perceived as “second-order” elections (Reif & Schmitt, 1980), characterized by lower voter turnout and campaigns that are less vigorous compared to national elections. These elections often center around local issues and serve as opportunities for citizens to express dissatisfaction with government policies without threatening national stability. As a result, pre-election periods are typically marked by fewer confrontations between political parties, leading to more subdued campaign dynamics.

Simultaneously, political and electoral competition also occurs among European parliamentary political groups, or Euro parties, shaping legislative priorities and influencing the broader political agenda. Each national party that sends representatives to the European Parliament aligns itself with a Euro party or remains independent, coordinating its platform on European policy matters with the broader ideological framework of its Euro party. Consequently, political competition unfolds at two levels: within national contexts and at the supranational European level. Given the dual-level dynamics of political competition, understanding the strategies employed at both national and European levels is essential.

Therefore, this study investigates the extent to which European parties adopted strategic political marketing to position themselves on key political axes and engage with diverse audiences. By employing a comparative multivariate analysis, this analysis aims to answer the following questions:

How do European parties differ in their marketing strategies?

To what extent do national parties align with their European political groups?

What thematic priorities dominated the issue agendas during the 2024 elections?

LITERATURE REVIEW: THE EVOLUTION AND IMPACT OF POLITICAL MARKETING

Defining Political Marketing

Political marketing is an interdisciplinary field that integrates political science, marketing principles, communication studies, and behavioral insights to develop strategies aimed at understanding and influencing voter behavior. Scholars such as Newman (1999) and Lees-Marshment (2001) have characterized political marketing as encompassing not only campaign promotion but also market research, voter segmentation, and policy branding. It includes strategic positioning and long-term branding efforts by political actors. According to Big (2017), political marketing frameworks facilitate understanding voter behavior and party strategies, building on foundational theories from Blumenthal (1980) and Lees-Marshment (2001).

Historical Development

Political marketing has undergone substantial transformations over time, shaped by technological advancements, media evolution, and changing political landscapes. Historically, political communication was limited to in-person events, print, and broadcast media. The initial stages of political marketing focused on building candidate visibility through newspapers, radio, and later, television. For instance, the 1952 U.S. presidential campaign of Dwight Eisenhower was a milestone in political advertising, using television ads to reach a broader audience at a fraction of the cost compared to traditional rallies (Spenkuch & Toniati, 2018). As political communication research grew, scholars such as Lippmann (1922) emphasized the persuasive role of mass media in shaping public opinion.

However, modern political marketing incorporates data-driven strategies, social media engagement, and targeted online advertisements. In the 2010s, the introduction of advanced data analytics and micro-targeting led to a paradigm shift in political campaigning. Political advertisements became increasingly tailored to specific voter segments, using online platforms to disseminate emotionally charged messages. The Cambridge Analytica scandal exemplified the ethical concerns surrounding the use of voter data to influence elections (Abid et al., 2022; Lopez-Lopez et al., 2023). Data-driven strategies have shown mixed outcomes (Spenkuch & Toniati, 2018). While some research indicates that digital political ads can reinforce voter biases and affect decision-making, other studies suggest that their long-term influence may be limited without broader engagement strategies (Abid et al., 2022).

Another communication tool that has emerged by the expansion of Web 2.0 was online networks and communities. Social media has become a dominant channel for political marketing, enabling real-time interaction between politicians and voters. Platforms like Facebook, Twitter, and YouTube have allowed political brands to co-create content with their supporters, fostering a sense of community and loyalty. However, social media has also facilitated the spread of disinformation and political polarization, as highlighted by studies on political discourse during events like the Brexit campaign and U.S. elections.

THEORETICAL FRAMEWORKS

The field of political marketing relies on several foundational theoretical models that collectively explain its structure and operational dynamics, offering insights into how political actors convey their messages, construct narratives, and adjust their strategies to sustain voter engagement and participation.

The Marketing Mix Model (4Ps), originally developed for commercial marketing, has been adapted to the political sphere where policy corresponds to the product, promotion represents campaign communication, place refers to the distribution channels for messaging, and people include both political representatives and voters as key actors in this strategic dynamic. Another theoretical model that is used to analyze party activities and stances is the Market-Oriented Party Model, as outlined by Lees-Marshment. This model presents political parties as market-driven entities that adjust their strategies, policies, and communications based on continuous feedback loops and voter preferences. This model emphasizes strategic planning and adaptation to the evolving political market. The agenda-setting and framing theories highlight how political actors and media outlets shape public perception by emphasizing certain issues and presenting them in ways that influence public discourse and voter priorities. These frameworks illustrate the power of political communication to direct attention and create context around key topics during campaigns. Another model focuses on the party leaders or opinion leaders' profiles. This two-step flow of communication model emphasizes the role of opinion leaders or influential figures who interpret political messages and relay them to wider audiences, demonstrating how campaigns can have an indirect yet considerable influence on voter perceptions and behavior. Finally, there is the push-pull strategy in campaigning model of analysis which refers to the blend of outbound communication, such as campaign advertisements, and inbound engagement techniques that encourage voter interaction, such as interactive content on digital platforms. This dual approach enables campaigns to proactively share information while simultaneously fostering organic connections with the electorate. Each of these frameworks contributes to understanding how political entities craft and refine their strategies to resonate with voters in diverse socio-political contexts.

Political Marketing in the European Context

Political marketing in Europe faces distinct challenges due to the diverse cultural, linguistic, and political landscapes across the European Union's member states. Unlike national elections, European Parliament elections are often perceived as 'second-order' contests, with lower voter turnout and less intense media coverage (Reif & Schmitt, 1980). This perception influences political marketing strategies, requiring parties to balance local, national, and supranational priorities.

Theories such as the marketing mix (Kotler & Levy, 1969) highlight that 'place' in the political context refers to both physical spaces—such as public debates, constituency events, and rallies—and digital spaces, including social media platforms and news outlets (Lees-Marshment, 2001). Agenda-setting and framing theories (McCombs & Shaw, 1972) emphasize that how and where messages are delivered impacts voter engagement and issue salience. The increasing prominence of digital platforms has redefined 'place,' making virtual engagement spaces essential to modern campaigns (Abid et al., 2023).

Euro parties and national parties often collaborate to align campaign messaging, aiming to present cohesive platforms while addressing specific national concerns. This dual-level strategy involves using digital campaigns for mass reach and localized events for direct interaction, reflecting the push-pull dynamic of political communication (Lopez-Lopez et al., 2023). As technology continues to evolve, future political marketing in the European context will require even greater integration of data analytics, micro-targeting, and real-time voter feedback to address the nuanced expectations of a diverse electorate, particularly in response to pressing issues such as climate change, economic recovery, and migration policy.

Political Competition in the EU

European elections are characterized by a unique interplay of national and supranational dynamics. Unlike national elections, these contests involve both domestic political debates and broader European issues, leading to a complex electoral environment. The axes of political competition - such as "Left-Right," "Liberalism-Authoritarianism," "European Cooperation-National Isolation," and "Modernization-Conservatism" - frame the ideological landscape within which parties maneuver.

Recent European elections, including the 2019 and 2024 contests, illustrate the diversity of political interests across the EU's 27 member states. In 2019, voter turnout surged to 50.6%, the highest in two decades, signaling increased public interest. Key themes included climate change, migration, economic inequality, and the rise of populist movements. The election resulted in a fragmented European Parliament, with traditional center-right and center-left parties losing their dominance, while Green, liberal, and far-right parties gained significant ground.

In the 2024 elections, political competition was further shaped by the aftermath of the COVID-19 pandemic, economic recovery policies, and the war in Ukraine. Euro-skeptic parties continued to push narratives centered on national sovereignty, while pro-European coalitions focused on unity and collective security. These elections also saw intensified use of digital campaign strategies, reflecting broader trends in political marketing. These dynamics underscore the multi-layered nature of political competition in the EU, where electoral outcomes are influenced not only by domestic concerns but also by pan-European challenges that require cross-border cooperation.

RESEARCH METHODOLOGY

For this study, data was collected and analyzed on the election campaigns of major party formations in each of the EU's 27 member states during the pre-election period of the European elections held from June 6 to 9, 2024 - the 10th EU elections and the first after Brexit. Observational data was gathered from online platforms and political discourse, complemented by a 35-question survey designed to capture insights on pre-election issues, marketing tools, and communication strategies used by parties. A total of 182 surveys were analyzed, focusing on the criteria that may have contributed to the prominence of certain candidates over others.

The primary objective of the study is to explore the relationship between marketing strategies, issue prioritization, and electoral outcomes across European nations within the framework of the 2019 and 2024 European parliamentary elections. By examining public priorities, campaign approaches, and changes in parliamentary seat distribution, the research aims to shed light on the forces shaping voter behavior and the performance of political groups within an evolving European political landscape.

- Campaign tools: TV, radio, social media, and printed materials.
- Issue salience across sixteen thematic categories (e.g., climate change, economic policy, migration).
- Party positioning on four political axes ("Left-Right", "Liberalism-Authoritarianism", "European Cooperation-National Isolation", and "Modernization-Conservatism")

The study employed a multi-step quantitative framework to investigate the relationship between marketing strategies, voter priorities, and electoral outcomes for European political parties. The methodology integrated data from Eurobarometer surveys, election results, and marketing strategies, analyzed through a combination of statistical and exploratory techniques. These methods were designed to capture both individual aspects of interest and their interconnections across European nations and political groups.

The primary data sources included:

- **Eurobarometer survey data (April/May 2024)** capturing citizens' issue priorities across EU countries.
- **European parliamentary election results (2024)**.
- **Secondary sources** outlining the marketing strategies adopted by major political groups.

All data were harmonized and standardized to ensure cross-variable and cross-country comparability. Survey responses were grouped into thematic categories (e.g., "Health," "Climate/Environment") based on the most frequently mentioned issues.

To analyze the data, we employed Correspondence Analysis (CA) followed by Hierarchical Cluster Analysis (HCA) on the factors identified in the initial step. This methodological combination offers a robust approach for uncovering patterns and relationships within complex datasets.

Specifically, the integration of CA and HCA allows us to:

1. **Identify Clusters of Political Entities:** These clusters represent groups of political parties with similar behavior profiles and positions on key political issues. By detecting these clusters, we can discern commonalities and distinctions in party stances and strategies.
2. **Link Behavior Profiles with Variable Categories:** Correspondence analysis enables us to associate these behavioral profiles with the broader categories of variables, revealing the dynamics underlying the studied phenomenon. For instance, it illustrates how issue salience corresponds to electoral strategies and marketing tools employed by parties.
3. **Provide a Comprehensive Visualization of Dynamics:** The combination of these methods produces a detailed, interpretable visual representation of the identified dynamics. Biplots generated through CA offer an insightful low-dimensional mapping of categorical data, while the HCA dendrograms illustrate the hierarchical relationships between clusters.

This approach enhances our ability to interpret the alignment (or divergence) of party agendas with broader political group priorities and voter concerns. By visualizing both the clustering of political actors and the relationships between key variables, this analysis provides a nuanced understanding of how parties position within the European political landscape and adapt their strategies to resonate with voter priorities.

The role of marketing strategies was examined through correspondence analysis of campaign approaches and political groups. Campaign strategies were classified as "traditional," "digital," or "mixed" based on qualitative assessments of campaign styles and linked to electoral outcomes to evaluate the influence of strategic choices on voter behavior. Visual representations such as biplots, factor loading matrices, and cluster maps were generated to aid interpretation. Patterns of alignment between countries, issues, political groups, and marketing strategies were analyzed to draw insights into voter behavior and campaign effectiveness.

The statistical analyses were conducted using Python and IBM SPSS.

Marketing Strategies

The correspondence analysis visualization in Fig. 1, illustrates distinct categories of political marketing strategies and the varying degrees to which parties implemented pre-electoral activities. The data points reflect different tools and strategic choices, revealing clear clusters.

At one end, we observe a group (44) that reflects the robust implementation of marketing tools such as professional consultants, research-based messaging, and the engagement of volunteers. This cluster represents organized and data-driven political campaigns that demonstrate significant resource allocation and strategic planning. These campaigns tend to leverage both digital and traditional marketing channels, indicating a comprehensive approach to voter outreach.

Conversely, another prominent cluster (Group B and C) demonstrates minimal to no use of organized marketing strategies. These campaigns lack professional consultation, volunteer engagement, and targeted outreach tools such as email campaigns. This suggests a limited or outdated approach to political communication, relying on ad-hoc activities rather than a cohesive plan.

Within the intermediate spectrum, there are three notable subcategories. The first subcategory (Group B) excludes the use of professionals, volunteers, and email as primary outreach methods, indicating a reliance on more general communication without personalized voter engagement. The second subcategory shows the avoidance of modern technological tools such as websites and social media, as well as a lack of open events aimed at broad public engagement. This signals resistance to adopting digital trends or engaging mass audiences interactively. Finally, the third subcategory omits traditional media tools such as TV and radio advertisements, SMS campaigns, and printed materials, reflecting a departure from legacy methods in favor of other forms of outreach—or potentially signaling a lack of overall resource availability.

Overall, this analysis highlights a spectrum of marketing approaches, from highly structured and research-informed campaigns to minimal and fragmented efforts. The categories provide insights into how different political entities balance modern, traditional, and hybrid communication strategies within their pre-electoral activities. Additionally, the clustering suggests that the integration of professional expertise and technological tools significantly differentiates the sophistication of campaign strategies across political parties.

Figure 1. Correspondence analysis of political marketing strategies and tools

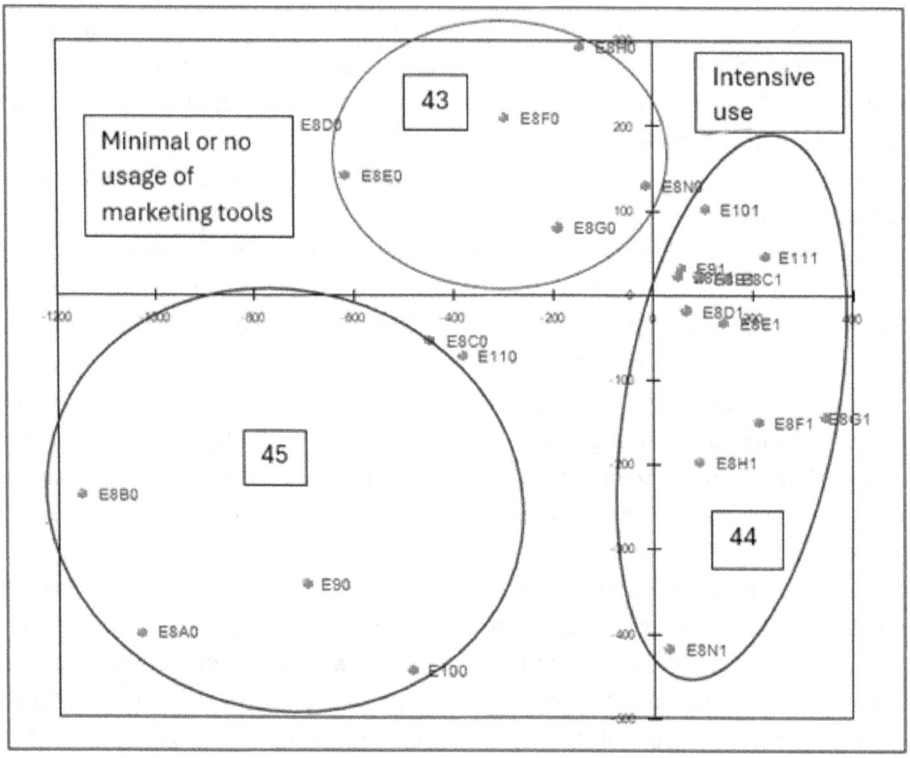

Table 1 presents the correspondence analysis loadings for various political marketing tools, showing their association with specific factors and strategic implementation levels. The values, both negative and positive, reflect the extent to which each tool contributes to the axes of political communication strategies.

Negative loadings, such as those for websites (-1030), social media (-1150), and SMS campaigns (-687), indicate minimal integration in certain campaign strategies. This suggests a preference for traditional approaches, where digital tools are not prioritized, potentially due to limited technological adaptation or a focus on older voter segments that are more responsive to conventional media. In contrast, positive loadings for tools like TV and radio spots (+207), research-based messaging (+229), and professional consultancy (+213) highlight organized, research-driven approaches. These values imply that parties using these tools adopt structured strategies that emphasize broad audience outreach and data-informed decision-making.

A closer examination of the data reveals that some communication tools, such as events (+291) and printed materials (-445), show mixed associations across different contexts. While events have a strong positive impact in fostering public engagement, traditional print materials may not always yield the same level of responsiveness in the digital age. This contrast reflects the evolving preferences of the electorate, where personalized and interactive experiences are gaining prominence over static forms of communication.

The differentiation between push and pull strategies is also evident in the table. Tools associated with push marketing, such as SMS and print advertisements, show high negative values, while pull-oriented strategies, such as volunteer engagement (+128) and research-based outreach, display positive values. This pattern indicates a shift in some campaigns toward participatory marketing, focusing on fostering interaction rather than merely broadcasting messages. The higher positive scores for professional consultancy and research-based messaging further underscore the growing reliance on expert support to refine campaign narratives and enhance voter targeting.

This analysis highlights distinct campaign approaches, ranging from minimal, traditional methods to highly organized, research-backed strategies. The negative associations for digital outreach tools in some instances suggest a potential underinvestment in modern communication channels. Conversely, the positive loadings for volunteer engagement, professional consultancy, and research-based tools indicate recognition of the importance of personalization and precision in voter engagement. These findings suggest that campaigns with greater investment in interactive and data-driven approaches are more likely to resonate with broader electorates, demonstrating the increasing significance of adaptive and research-focused political marketing in contemporary elections.

Table 1. Correspondence analysis loadings table of political marketing tools and their strategic implementation

	IND	#F1	COR	CTR	#F2	COR	CTR	#F3	COR	CTR	#F4	COR	CTR
Website	**E8A0**	-1030	473	75	-401	71	34	-669	199	181	82	3	4
Website	**E8A1**	53	473	4	20	71	2	34	199	11	-5	3	1
Social Media	**E8B0**	-1150	749	145	-238	32	19	-376	79	89	-224	28	36
Social Media	**E8B1**	95	749	13	19	32	2	31	79	8	18	28	4
Emails	**E8C0**	-445	554	72	-56	8	4	222	139	102	-175	85	72
Emails	**E8C1**	149	554	24	18	8	2	-75	139	34	58	85	24
Events	**E8D0**	-733	449	67	200	33	15	-71	4	4	-687	394	394
Events	**E8D1**	70	449	7	-20	33	2	6	4	1	65	394	37
Spot online	**E8E0**	-614	665	103	140	34	16	-245	105	93	160	45	47
Spot online	**E8E1**	145	665	24	-34	34	4	57	105	22	-38	45	11
TV radio spots	**E8F0**	-295	505	52	207	249	77	-20	2	2	-2	0	1
TV radio spots	**E8F1**	214	505	38	-151	249	55	13	2	1	1	0	1
SMS	**E8G0**	-188	547	32	78	95	16	26	11	4	31	15	7
SMS	**E8G1**	348	547	61	-146	95	31	-50	11	8	-59	15	11
Print material	**E8H0**	-143	130	11	291	545	145	83	45	23	-32	6	4
Print material	**E8H1**	97	130	9	-198	545	99	-57	45	15	21	6	3
Other	**E8N0**	-11	3	1	128	546	53	41	55	10	47	75	16
Other	**E8N1**	34	3	1	-420	546	175	-134	55	34	-156	75	54
Volunteers	**E90**	-692	386	56	-343	95	40	579	271	227	81	5	6
Volunteers	**E91**	61	386	6	30	95	4	-52	271	20	-8	5	1

Figure 2. Dendrogram of the groups using different marketing strategies.

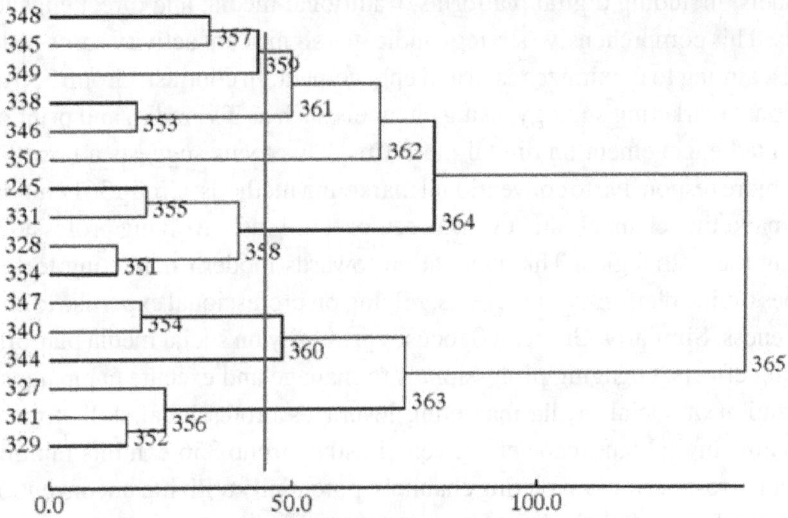

Table 2. Groups of different marketing strategies utilizing various promotion tools.

	358	350	361		360		356
	358	350	359	353	344	354	356
Website	0,7052	0,7052	0,7052	0,7052	0,7052	0,7052	
Social Media	1,115	1,115	1,115	1,115	1,115		
Emails	4,0522			4,0522			
Events	1,2821	1,2821	1,2821	1,2821			
Spot online	2,9728	0,1317		2,9728	1,2345		
TV radio spots	5,56	1,0342					
SMS	5,5595						
Print material	5,7962	0,7258	1,0063				
Other	4,9105				2,7459	1,8327	5,8913
Volunteers	1,1983		1,1983	1,1983	1,1983		
Professionals	2,3756			2,4501	1,1458		
Research	4,6021			1,8772	4,6655		

HCA has identified five distinct groups—358, 350, 361, 360, and 356 - each representing a unique marketing strategy characterized by the utilization of various channels such as websites, social media, emails, and events.

Group 358 employs a mixed strong marketing approach, utilizing a diverse range of channels, including digital platforms, traditional media, and direct engagement methods. This comprehensive strategy indicates significant activity across multiple channels, aiming to maximize reach and engagement. In contrast, Group 350 adopts a traditional marketing strategy, using channels such as TV, radio, and print media, with limited engagement on digital platforms. This focus suggests a target demographic more responsive to conventional marketing methods. Group 361 emphasizes digital marketing channels and event-based promotions, involving professionals in executing their strategies. This inclination towards modern marketing techniques combines online platforms with events, relying on professional expertise to enhance effectiveness. Similarly, Group 360 focuses primarily on social media platforms for marketing efforts, engaging professionals to manage and execute campaigns. This prioritization of social media marketing leverages professional skills to navigate the dynamic digital landscape effectively. Lastly, Group 356 exhibits minimal engagement across various marketing channels, potentially utilizing unconventional or niche methods not extensively covered in the data. This suggests limited marketing activities or unique strategies diverging from mainstream channels with more focus on political programs and ideological identities.

The analysis of political marketing strategies across various clusters highlights a complex and dynamic landscape where parties adopt differing approaches based on their goals, available resources, and target demographics. The data reveal both patterns of effective outreach and significant gaps in strategic communication.

The groups identified in the analysis (Groups 358, 350, 361, 360, and 356) underscore the diversity of pre-electoral campaign approaches. Group 358 represents a mixed, robust strategy that combines digital platforms, traditional media, and direct voter engagement methods such as events and volunteer networks. This group's campaigns appear well-resourced, aiming for broad and personalized voter outreach. By contrast, Group 356 shows minimal engagement across traditional and digital channels, likely indicating a focus on ideological messaging or resource constraints. Campaigns in this category may rely more on grassroots initiatives or direct communication through local political networks rather than large-scale media investments.

Group 350 primarily utilizes traditional media, including TV, radio, and print materials, with minimal digital engagement. This approach suggests a focus on older voter demographics who are more responsive to conventional communication methods. Conversely, Groups 360 and 361 lean toward digital-first strategies, emphasizing social media platforms, professional consultants, and event-based promotions. These groups prioritize interactivity and data-driven insights to attract tech-savvy voters, particularly younger demographics who engage with campaigns online.

Groups like 361 and 358 highlight the importance of hybrid approaches that integrate both online and offline tools. These groups make strategic use of professionals to manage complex digital campaigns and ensure that events complement online outreach. The emphasis on research-based messaging and consultancy in these clusters indicates the growing reliance on expertise to craft effective narratives and refine voter targeting. This contrasts with other clusters, such as 350, where the absence of consultants suggests a more traditional or resource-constrained approach.

The analysis reveals a clear differentiation between "push" and "pull" communication strategies. Push strategies, such as SMS and print advertisements, rely on unidirectional broadcasting of messages. These tools, while effective in past decades, may appear impersonal and fail to foster meaningful voter engagement. Pull strategies, such as volunteer engagement and events, reflect participatory marketing that invites voters into direct interactions with the campaign. This trend points to a broader shift toward building community-oriented engagement, fostering loyalty, and increasing voter mobilization.

The analysis shows that some groups, particularly 358 and 361, invest in multiple communication tools and professional support to ensure a broad and data-informed campaign. In contrast, the low engagement in digital and traditional marketing tools by Group 356 suggests a niche or limited-budget approach, potentially focusing more on policy-based messaging than comprehensive voter outreach. This divergence highlights how resource allocation significantly shapes a party's ability to run a versatile and high-impact campaign.

The reliance on specific tools correlates with varying levels of voter interaction. For example, campaigns that avoid digital tools, such as websites and social media, may struggle to engage younger, urban voters. Conversely, parties that invest heavily in social media outreach and events but neglect traditional media could risk alienating older voters who are less active online. A balanced approach that integrates targeted digital messaging with broader conventional outreach appears to be the most effective strategy for reaching diverse voter segments.

The analysis of the 2024 European elections demonstrates how political marketing has evolved in response to changing voter expectations and geopolitical dynamics. The findings suggest that Euroskeptic and far-right parties, often positioned within minimalist marketing clusters, may prioritize direct voter appeals and ideological narratives over complex communication strategies. Conversely, pro-European coalitions, especially those in Groups 358 and 361, appear to adopt broader, research-informed approaches designed to emphasize collaboration and socio-economic solutions.

The findings highlight a multi-tiered approach to political marketing across Europe, where parties navigate between digital innovation, traditional outreach, and limited-resource campaigns. The most successful campaigns, as demonstrated by Groups 358 and 361, leverage a hybrid strategy that integrates data-driven insights, professional expertise, and participatory engagement. In contrast, minimalist approaches, as seen in Group 356, may be constrained by resources or strategic choices, focusing more on symbolic messaging rather than broad outreach. These distinctions underscore the importance of adaptability and strategic planning in shaping voter perceptions and influencing electoral outcomes.

Future campaigns should consider the importance of balancing traditional and digital outreach, investing in both professional consultation and grassroots participation. The evolving preferences of voters, especially in the post-Brexit EU, highlight the need for parties to adopt flexible, research-backed marketing strategies that resonate across diverse socio-political landscapes.

National vs. European Group Alignment

As the research focuses on the 2024 European Parliament elections, our analysis of the convergence between national parties and their respective European Parliament (EP) groups gains renewed significance. The elections resulted in a notable shift within the EP's composition, with the European People's Party (EPP) securing 188 seats (26.11%), the Progressive Alliance of Socialists and Democrats (S&D) obtaining 136 seats (18.89%), and the newly formed Patriots for Europe (PfE) group, comprising far-right parties, capturing 84 seats (11.67%)

This evolving political landscape underscores the importance of understanding the alignment between national parties and their EP groups.

Figure 3. Correspondence analysis biplot of policy priorities in political campaign agendas

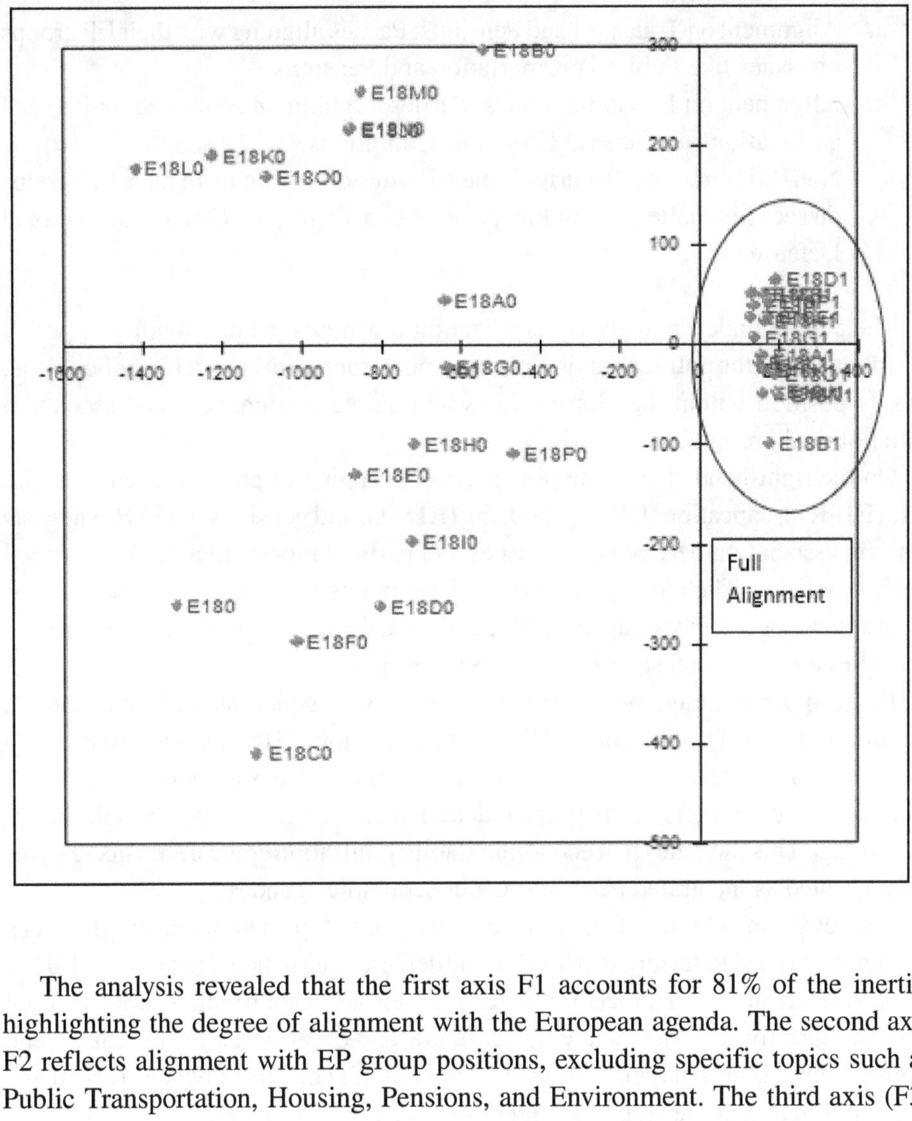

The analysis revealed that the first axis F1 accounts for 81% of the inertia, highlighting the degree of alignment with the European agenda. The second axis F2 reflects alignment with EP group positions, excluding specific topics such as Public Transportation, Housing, Pensions, and Environment. The third axis (F3) and fourth axis (F4) further delineate non-alignment in areas related to security, economy, and social policies.

The correspondence analysis to assess party behavior concerning thematic agendas, identified distinct categories:

1. Full Alignment: Parties that align with their EP group's positions across all policy areas.

2. No Alignment: Parties that do not align with their EP group's agenda in any policy area.
3. Selective Alignment:
    a. Alignment on Transport and Pensions: Parties aligning with their EP groups on issues like Public Transportation and Pensions.
    b. Alignment on Economic Issues: Parties aligning on economic topics such as Taxation and General Economic Conditions.
    c. Non-Alignment on Security Issues: Parties diverging from their EP groups on security matters, including positions on Terrorism, Crime, and National Defense.

The correspondence analysis visualization illustrates the distribution of policy priorities across the political landscape. Each point represents a specific policy issue, and its position within the plot indicates its relative prominence and association with political groups.

On the right-hand side of the plot, a dense grouping of points related to health care (E18K1), education (E18L1), housing (E18J1), and pensions (E18M1) suggests a strong association between these issues and party platforms that prioritize social welfare policies. The close proximity of these points indicates that these policies often appear together in the agendas of specific parties or groups, reflecting cohesive strategies around social security and public services.

In the upper quadrant, policies related to economic concerns, such as economic conditions (E18C1), inflation (E18D1), and taxation (E18E1), are prominently positioned. Their relative closeness implies that these issues are interconnected in party narratives and often form the foundation of economic policy proposals during campaigns. This indicates that economic stability, inflation control, and tax reforms are presented as integrated parts of broader economic agendas.

The lower left section of the plot contains points representing crime (E18A1), immigration (E18I1), terrorism (E18G1), and defense/international relations (E18H1). These issues tend to cluster together, indicating their association with security-focused political narratives. Parties that emphasize law and order or national security tend to group these issues in their platforms, often linking concerns about terrorism and immigration to defense strategies and crime prevention policies.

The environment (E18N1) and energy (E18O1) occupy distinct positions on the left side of the plot, separate from the economic and security clusters. This separation suggests that environmental policies and energy concerns form distinct thematic priorities, often championed by parties focused on sustainability and climate change initiatives.

continued on following page

Table 3. Continued

The overall spread of points across the plot highlights the multi-dimensional nature of policy prioritization in political campaigns. Social welfare, economic management, and security concerns each form cohesive thematic groups, while issues related to the environment and state debt occupy more isolated positions, reflecting their distinct and sometimes less integrated roles in political agendas. This distribution underscores the importance of strategic issue framing, as parties seek to align their messaging with voter priorities and create clear, issue-based platforms to differentiate themselves from their competitors.

The 2024 election outcomes, particularly the gains by far-right parties and the formation of the PfE group, may influence these alignment patterns. For instance, national parties joining the PfE might exhibit distinct thematic agendas compared to their previous affiliations, potentially affecting their alignment on factors like F3 (security-related issues).

Understanding these dynamics is crucial for comprehending the interplay between national and supranational political agendas, as well as the factors that facilitate or hinder policy alignment within the EP. As the EP's composition evolves, continuous analysis is essential to monitor how these shifts impact legislative processes and policy outcomes.

The next step was the cluster analysis to examine the alignment between national parties and their respective European Parliament (EP) groups across various policy issues. The analysis focused on five clusters, labeled 346, 355, 359, 360, and 361, with each cluster representing a group of parties sharing similar positions on specific policy areas.

Table 3. Policy alignment scores across clusters for key European policy areas

		346	355	359	360	361
Overall	E181				1,2262	1,3502
Crime	E18A1				2,7348	
Public Transportation	E18B1		4,0131		2,1374	
Economic Conditions	E18C1			1,6033	0,9775	1,6033
Inflation/Price Increase	E18D1				1,7476	
Taxation	E18E1			0,0365	1,563	
Unemployment	E18F1				1,2316	0,4163
Terrorism	E18G1				2,8484	
Defense/International Relations	E18H1				2,8107	
Housing	E18J1				2,326	
Immigration	E18I1				0,8304	1,6042

		346	355	359	360	361
Health/Care	E18K1				1,5228	0,4905
Education	E18L1				1,1852	1,4341
Pensions	E18M1		2,4812		1,0883	2,0427
Environment	E18N1				2,2867	
Energy	E18O1				1,9166	0,1478
National Debt	E18P1				1,9563	

Table 4. Policy non-alignment scores across clusters for key European policy areas

		346	355	359	360	361
Overall	E180	29,2315	29,2315	10,6949		
Crime	E18A0	12,7634	1,444	16,5708		9,419
Public Transportation	E18B0	20,3396		6,7533		7,123
Economic Conditions	E18C0	31,3848	36,5492			
Inflation/Price Increase	E18D0	20,7714	24,6307	1,6314		0,9534
Taxation	E18E0	27,4238	19,1689			0,9745
Unemployment	E18F0	30,9052	17,365	0,8838		
Terrorism	E18G0	17,7953	5,9392	2,023		13,9327
Defense/International Relations	E18H0	17,3721	5,6847	12,1895		10,0212
Housing	E18J0	25,6814	5,9392	20,424		2,2376
Immigration	E18I0	23,8325	11,2761	5,6828		
Health/Care	E18K0	34,4316	9,9366	14,8622		
Education	E18L0	38,9865	6,5554	17,3658		
Pensions	E18M0	28,056		22,4627		
Environment	E18N0	25,146	13,4587	17,3721		1,1367
Energy	E18O0	29,411	12,0477	17,8355		
National Debt	E18P0	15,3763	8,9195	7,8653		2,3747

The cluster analysis reveals significant insights into the alignment between national parties and their corresponding European Parliament (EP) groups across various policy domains. The five identified clusters (346, 355, 359, 360, and 361) represent distinct sets of parties that exhibit similar stances on specific policy areas. This analysis sheds light on the degree of policy coherence and divergence, highlighting the complexities of European political collaboration and the strategic implications for policy negotiations and coalition-building.

Cluster 346 demonstrates relatively strong alignment on social welfare policies, particularly pensions, with a score of 2.4812. This indicates that parties within this cluster prioritize pension reforms and social security measures in line with their EP groups. However, there is a notable divergence on public transportation, where the high score of 4.0131 suggests differences in funding priorities, infrastructure projects, or regional policy preferences. This divergence points to potential challenges in forming cohesive transportation policies at the European level.

Cluster 355 shows a similar pattern of alignment on pensions (2.4812) and divergence on public transportation (4.0131). The fact that both Clusters 346 and 355 share similar patterns in these areas suggests that pension policies may be a unifying issue within broader political coalitions, while public transportation remains contentious. This could reflect regional disparities in transportation infrastructure needs or differences in national transportation agendas versus European-wide initiatives.

Cluster 359 exhibits minimal alignment across most policy areas in the provided data, suggesting either a lack of strong policy cohesion or an intentional focus on divergent national interests. The noticeable divergence in public transportation (4.0131) further supports the idea that transportation policies may be a significant point of contention. This lack of alignment could hinder efforts to form stable coalitions and may point to challenges in achieving policy consensus within this group.

Cluster 360 demonstrates alignment in economic conditions (1.6033) and taxation (0.0365), indicating that parties within this cluster share common priorities related to economic stability and tax reforms. However, the cluster diverges in areas such as public transportation (2.1374) and pensions (1.0883). This partial alignment in economic matters suggests that, while fiscal policies may serve as a common ground for collaboration, social welfare policies and public infrastructure remain areas of potential friction.

Cluster 361 shows significant alignment in economic conditions (1.6033) and pensions (2.0427), indicating a shared focus on economic security and social welfare. However, the cluster diverges on public transportation (2.1374) and taxation (1.563), reflecting differing views on the role of public spending and fiscal policy. The focus on economic stability across both Clusters 360 and 361 may point to a broader trend among certain EP groups to prioritize economic reforms and social security measures, despite disagreements on infrastructure investments and tax policies.

The accompanying data on policy alignment and non-alignment (Figures 7 and 8) provide further context for these findings. For example, in non-alignment scores, Cluster 346 records high values for crime (12.7634) and public transportation (20.3396), indicating significant divergence from the EP group positions in these areas. In contrast, Clusters 360 and 361 show lower scores for crime and defense-related policies, suggesting a closer alignment with the broader European security

continued on following page

Table 5. Continued

agenda. Similarly, the high non-alignment scores for health care (34.4316 in Cluster 346) and education (38.9865) suggest that social service policies may still vary significantly across national and supranational platforms.

The data also highlight differences in environmental and energy policies. For example, Cluster 361 shows alignment in environmental policies (2.2867) and energy issues (1.9166), whereas Cluster 346 records much higher non-alignment scores (25.146 for the environment and 29.411 for energy). This divergence suggests that while some EP groups are cohesive in their sustainability goals, others face internal disagreements over environmental regulations and energy transition strategies.

Overall, this analysis underscores the varying degrees of alignment and divergence across key policy areas, such as economic conditions, social welfare, public infrastructure, and security. The findings suggest that some clusters, particularly Clusters 360 and 361, are more cohesive on fiscal and economic matters but may face internal tensions over public spending priorities and social policies. Clusters 346 and 355 demonstrate unity in social welfare priorities like pensions but diverge on infrastructure-related issues. In contrast, Cluster 359 appears more fragmented, reflecting a weaker policy alignment that could complicate coalition-building and policy negotiations.

Understanding these patterns of policy alignment and divergence is essential for analyzing the coherence of political agendas within the European Parliament. These insights can inform strategies for forming alliances, negotiating policy compromises, and addressing areas of contention within and across political groups. By identifying key areas of agreement and divergence, stakeholders can better anticipate challenges in legislative decision-making and develop more targeted approaches to fostering policy cohesion at both the national and European levels.

Analysis of the Parties of the Twenty-Seven Member States

The last step of the analysis was to connect the data with the parties of the twenty-seven member states and their respective parties. Again, distinct clusters were identified based on the marketing strategies employed by political parties and their alignment with European Parliament (EP) groups across various policy areas.

Table 5. Mapping of national party alignment and political marketing strategies in European elections

72	85		84		86	
72	74	77	80	48	82	83
BG	ES	CZ	AT	EE	CY	DK
DE	Alignment on transportation and pension policies	SE	ID	ALDE	Non-alignment on security policy	FR
intensive marketing mix	HU	EPP	BE		GR	RENEW
PT	PL	Not elected	Non alignment		ECR	elected
FI		Alignment on all policies	IT		Alignment on economy policy	IE
		digital / events and use of professionals	traditional marketing		LV	LEFT
		MT			social media and use of professionals	LT
		RO			NL	mild marketing mix / other
		SD				SI
		LU				
		GREENS_EFA				

The biplot in Fig. 10 illustrates the clustering of national parties and their alignment on policy issues and political marketing strategies within the European electoral landscape. The plot visualizes how parties group together based on shared characteristics, policy stances, and communication approaches, with ellipses representing clusters of similar positions.

In the top-left quadrant, we observe a grouping of parties such as **Germany (DE)**, **Portugal (PT)**, and **Bulgaria (BG)**, which share a strategic emphasis on economic policies and employ comprehensive marketing approaches. These parties appear to cluster around robust communication strategies involving professional consultancy and research-based messaging, as indicated by the label “strong marketing mix.” Their proximity to the axis suggests a balanced approach to policy priorities, particularly on economic issues and defense.

On the top-right side, the grouping around **Cyprus (CY)**, **Netherlands (NL)**, and **Denmark (DK)** illustrates parties that diverge on security and defense policies but emphasize digital outreach tools such as social media. This cluster suggests the adoption of more modern communication methods, with a "mild marketing mix" that leverages digital strategies but may lack the intensive resource allocation of more comprehensive campaigns. The proximity of this group to terms such as "non-alignment on security policy" reflects differences in their defense narratives.

The central and lower regions of the biplot capture the dynamics of parties from **Finland (FI)**, **Czechia (CZ)**, and **Italy (IT)**, indicating balanced alignment across multiple policy areas, such as education, pensions, and public services. These parties' positioning close to the axis indicates policy consistency and the use of hybrid marketing approaches that combine digital engagement with traditional outreach tools. Their inclusion within the "alignment on all policies" label suggests they pursue cohesive narratives that resonate across economic, social, and infrastructure priorities.

The bottom-left grouping, including **Hungary (HU)**, **Romania (RO)**, and **Austria (AT)**, shows parties that exhibit divergent views on public transportation and pensions. These parties are associated with traditional marketing approaches, relying more on conventional media such as print and broadcast, rather than modern digital outreach methods. The ellipse enclosing these parties suggests limited alignment with broader European Parliament narratives, particularly on infrastructure and welfare policies.

The outliers, such as **France (FR)** and **Poland (PL)**, reflect unique policy stances or communication strategies that set them apart from the main clusters. For example, Poland's position near the lower right indicates possible alignment on infrastructure policies but divergence in terms of digital marketing adoption. France's placement suggests an emphasis on specific domestic issues rather than alignment on broader European policies.

This biplot highlights the heterogeneity of political approaches within the European Parliament, where some parties adopt cohesive, data-driven strategies aligned with European group narratives, while others maintain distinct national priorities. The analysis also underscores the role of communication strategies in shaping these alignments, with digital-first campaigns clustering separately from traditional marketing approaches. The clustering patterns further reveal how certain policy areas, such as pensions, public transportation, and economic stability, act as focal points for coalition-building, while security policies and defense narratives remain more fragmented.

Overall, the visualization provides valuable insights into how parties navigate the dual pressures of national and supranational politics, using policy alignment and marketing strategies to position themselves within the European political landscape.

By identifying key clusters and their corresponding policy narratives, stakeholders can better understand the challenges and opportunities for consensus-building and political cohesion within the European Parliament.

Figure 4. National parties' policy alignment and marketing strategies in the European political space

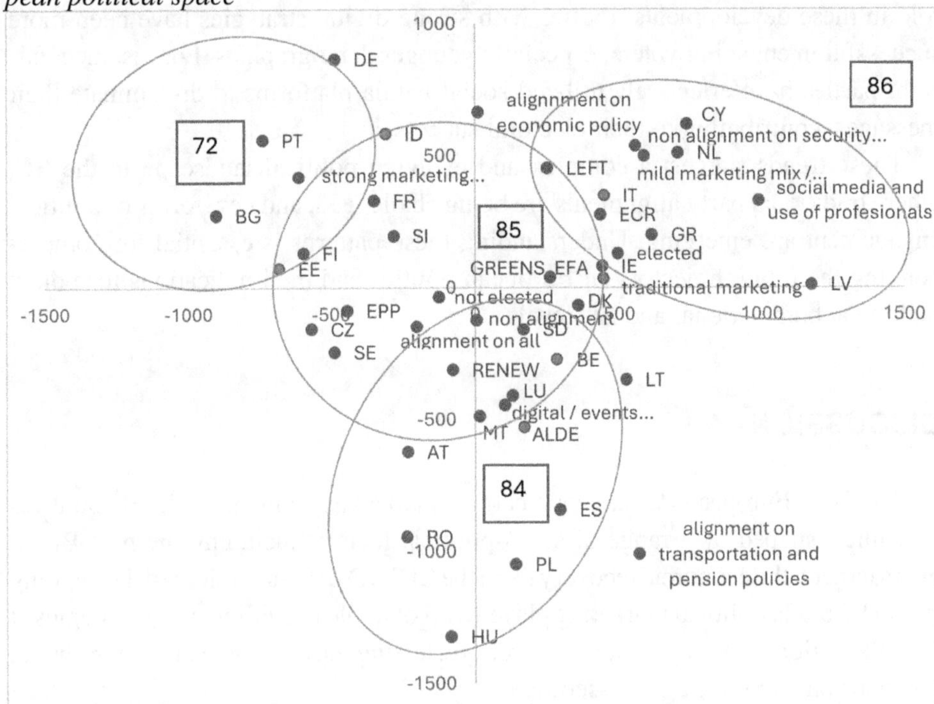

The biplot provides a comprehensive overview of how national parties align with their EP groups across key policy areas and the corresponding communication strategies they employ. The analysis highlights both areas of policy consensus, such as social welfare, and points of divergence, particularly in security and environmental policies. These findings highlight the diverse marketing strategies and policy alignments among political parties across the EU, reflecting varying degrees of engagement with European integration and differing approaches to national versus European priorities. The varying levels of marketing intensity and digital engagement further illustrate how communication approaches shape policy narratives and influence political positioning. These insights are essential for understanding the complexities of European political dynamics and for identifying opportunities to strengthen policy coherence and build effective coalitions within the European Parliament.

In the context of the 2024 European Parliament elections, these dynamics have become even more pronounced. Far-right parties have made significant gains, particularly in countries like France and Germany, where they have outperformed traditional centrist parties. This shift indicates a growing emphasis on national sovereignty and skepticism towards deeper European integration among certain electorates.

Moreover, the increasing use of digital political marketing has played a crucial role in these developments. Parties with strong digital strategies have been more successful in engaging voters, especially younger demographics. For instance, far-right parties have effectively utilized social media platforms to disseminate their messages, contributing to their electoral successes.

These trends suggest a complex and evolving political landscape in the EU, where traditional party alignments are being challenged, and new forms of political engagement are emerging. Understanding these patterns is essential for comprehending the future trajectory of European politics and the implications for policy making at both national and EU levels.

DISCUSSION

The 2024 European Parliament elections marked a turning point in EU political dynamics, shaped by a range of socio-political factors, including the post-Brexit environment, the economic recovery from the COVID-19 pandemic, and the ongoing war in Ukraine. Political marketing played a pivotal role in guiding party strategies at both the national and supranational levels, reflecting the evolving nature of political competition and issue agenda-setting.

The findings from this analysis highlight the significant evolution of political marketing strategies. Political parties moved beyond traditional outreach methods such as television, radio, and print to adopt more hybrid and digital-first approaches. These shifts were driven by the need to engage younger voters and capitalize on the opportunities provided by digital platforms. Data-driven strategies allowed parties to micro-target their messaging and tailor their communication to specific audience segments. Within this context, three distinct clusters of political marketing strategies emerged: digital-first innovators, traditional broadcasters, and hybrid strategists. The digital-first innovators prioritized social media engagement, real-time interactions, and the use of advanced data analytics. In contrast, traditional broadcasters continued to rely on conventional media outlets, which proved effective for reaching older, more traditional voter bases. Hybrid strategists, on the other hand, balanced traditional and digital media to expand their reach while maintaining credibility with diverse demographic groups.

The elections also underscored the importance of issue agenda-setting, with key themes such as climate change, economic inequality, and national sovereignty dominating campaign narratives. Nationalist and far-right parties emphasized migration control, economic protectionism, and national defense, often diverging from the more cohesive messaging of pro-European coalitions that focused on climate policy, economic recovery, and collective security. The alignment between national parties and their respective Euro parties varied significantly. While pro-European parties maintained a consistent thematic focus across member states, nationalist parties exhibited greater heterogeneity in their platforms, often tailoring their messaging to domestic concerns.

Digital campaigning played a decisive role in shaping electoral outcomes, particularly among younger voters who responded to interactive, real-time content disseminated on social media. In several cases, far-right parties achieved substantial electoral gains by leveraging digital platforms to amplify their narratives and foster a sense of community among supporters. Conversely, parties that adhered to traditional outreach methods faced difficulties in adapting to the demands of a more dynamic communication landscape and struggled to connect with younger voters. The findings suggest that while traditional methods remain relevant for certain voter groups, the increasing prominence of digital tools has transformed the dynamics of political competition.

The analysis reveals that hybrid campaigns - combining traditional and digital strategies -tended to be the most effective, as they allowed parties to appeal to a broad range of voters. Parties that successfully integrated data-driven insights into their outreach efforts were better able to align their thematic priorities with voter concerns. In contrast, parties that relied solely on outdated communication strategies often failed to resonate with the electorate. The prominence of research-based messaging and professional consultation highlighted the growing importance of precision in political communication.

The 2024 elections also illustrated the complexities of policy alignment between national parties and their European Parliament groups. Some parties demonstrated full alignment with their Euro party's policy positions, while others displayed selective or no alignment across key issues. The findings indicated that parties aligned more closely on topics such as public transportation, pensions, and economic recovery, but diverged on security-related issues, such as defense policy and counter-terrorism measures. The formation of the far-right "Patriots for Europe" (PfE) group further demonstrated the fragmentation within the European Parliament, as these parties prioritized national agendas over broader European objectives.

The role of the media in agenda-setting also remains critical. Social media platforms have become central to political discourse, enabling real-time interactions and amplifying campaign messages. However, these platforms have also contributed to

REFERENCES

Abid, A., Roy, S. K., & Lees-Marshment, J.. (2023). *Political social media marketing: a systematic literature review and agenda for future research*. Electron Commer Res., DOI: 10.1007/s10660-022-09636-7

Blumenthal, S. (1980). *The Permanent Campaign*. Simon and Schuster.

Jörg, L. (2018, November). Spenkuch, David Toniatti, Political Advertising and Election Results. *The Quarterly Journal of Economics*, *133*(4), 1981–2036. DOI: 10.1093/qje/qjy010

Kotler, P., & Levy, S. J. (1969). Broadening the Concept of Marketing. *Journal of Marketing*, *33*(1), 10–15. DOI: 10.1177/002224296903300103 PMID: 12309673

Lees-Marshment, J. (2001). The Marriage of Politics and Marketing. *Political Studies*, *49*(4), 692–713. DOI: 10.1111/1467-9248.00337

López-López, P. C., Barredo-Ibáñez, D., & Jaráiz-Gulías, E. (2023). Research on Digital Political Communication: Electoral Campaigns, Disinformation, and Artificial Intelligence. *Societies (Basel, Switzerland)*, *13*(5), 126. DOI: 10.3390/soc13050126

McCombs, M. E., & Shaw, D. L. (1972). The Agenda-Setting Function of Mass Media. *Public Opinion Quarterly*, *36*(2), 176–187. DOI: 10.1086/267990

Nimmo, D. (1999). *Political Communication and Public Opinion in America*. Goodyear Publishing Company.

Reif, K., & Schmitt, H. (1980). Nine second-order national elections: A conceptual framework for the analysis of European election results. *European Journal of Political Research*, *8*(1), 3–44. DOI: 10.1111/j.1475-6765.1980.tb00737.x

Sparrow, N., & Turner, J. (2001). The Permanent Campaign: Marketing as a Governing Tool. *European Journal of Marketing*, *35*(9/10), 984–1002. DOI: 10.1108/03090560110400605

Chapter 8
Political Participation and Voting Behavior Among Greek University Students

Georgia Panagiotidou
https://orcid.org/0000-0002-7264-1187
Aristotle University of Thessaloniki, Greece

ABSTRACT

This chapter investigates the dynamics of political participation and electoral behavior among university students in Thessaloniki, Greece, offering a detailed analysis of the factors shaping engagement within this critical demographic. Using advanced multivariate methodologies, the study identifies distinct respondent clusters based on their voting patterns, political knowledge, interest, mobilization strategies, and issue prioritization. The findings reveal two primary axes of polarization: engagement versus disengagement and progressive versus conservative ideologies. The study also highlights the transformative role of social media as a mobilizing tool for progressive voters, contrasting with the reliance on traditional media among conservative respondents. Issue-based voting patterns further delineate the ideological divides, with pragmatic voters prioritizing safety and economic stability, and progressive voters focusing on societal and cultural reforms. These insights contribute to a deeper understanding of voter segmentation and the factors influencing political behavior.

INTRODUCTION

Political participation and voting behavior are cornerstones of democratic engagement. It shapes political outcomes and reflects citizens' commitment to societal well-being (Verba, Nie, & Kim, 1978). The decision to vote or abstain is influenced

DOI: 10.4018/979-8-3693-9400-7.ch008

by, and in turn influences, trust in political institutions and a sense of political efficacy (Rosenstone & Hansen, 1993). However, in recent years, many democracies have observed declining voter turnout and growing political apathy, especially among younger generations. Understanding the drivers of political participation has thus become more urgent than ever. Understanding the factors that drive or hinder political participation is vital, especially among youth, a demographic that is in the process of forming its political identity. This chapter examines the interaction of political knowledge, interest, media consumption, and issue-based voting in shaping political participation and voting intentions among university students.

Youth represent a distinct demographic in political behavior studies. As individuals in the process of forming their political identities, young voters are less influenced by long-standing socio-economic factors, making them an ideal group for studying the effects of knowledge, interest, and ideology. Their engagement patterns often forecast broader societal shifts, as their political mobilization reflects emerging trends in media consumption and issue prioritization.

While prior research has extensively examined political participation, few studies focus on the intersection of youth behavior, issue voting, and the role of media in shaping electoral choices. This study fills that gap by focusing on university students in Thessaloniki. By exploring how political interest, issue prioritization, and mobilization strategies influence voting intention, it contributes to a richer understanding of modern democratic behavior. In Greece, the political landscape has experienced significant upheaval over the past decade, shaped by economic crises, political polarization, and shifting voter allegiances. Thessaloniki, as Greece's second-largest city and a hub of youth activity, offers a unique setting to investigate these trends. Youth in particular represent a crucial yet under-researched demographic in political science. Unlike older generations, younger voters are still forming their political identities, making them an ideal group to study how factors like political knowledge, interest, and ideology shape participation.

This research, based on a face-to-face survey of 1,572 university students in Thessaloniki, Greece, conducted between May 15 and May 24, 2024, explores these dynamics. Respondents were asked to evaluate their political interests, mobilization strategies, voting intentions, issue-based voting criteria, and political knowledge. Additionally, they positioned themselves on the left-right ideological spectrum, enabling a complex analysis of ideological influence on voting behavior. The study was designed with the following objectives:

1. To investigate the factors influencing the political participation of university students in Thessaloniki, including voting intention, political interest, and knowledge.

2. To classify political behaviors into distinct typologies based on variables such as gender, voting intention, ideology, and issue criteria.
3. To explore the connection between political participation and electoral choices, and how factors like media consumption and mobilization methods shape these behaviors.

Through these objectives, the study aims to uncover the complexities of youth political engagement and contribute to the broader understanding of voter behavior in modern democracies. Using a combination of Hierarchical Cluster Analysis (HCA) and Factorial Correspondence Analysis (FCA), the study identifies clusters of youth voters, highlighting distinct behavioral patterns. For instance, some groups prioritize economic and security issues, aligning with conservative or far-right ideologies, while others emphasize education and justice, aligning with progressive values. This approach allows for a deeper exploration of youth political behavior. This approach allows for the identification of behavioral profiles and their connections to political and electoral behaviors. By synthesizing these analyses, this chapter constructs a comprehensive framework for understanding youth political behavior and its implications for political engagement.

The methodology employed in this study, combining Factorial Correspondence Analysis (FCA) and Hierarchical Cluster Analysis (HCA), is particularly well-suited for comparative analysis due to its ability to uncover multidimensional relationships and categorize complex data. FCA enables the identification of underlying patterns and associations between categorical variables, making it ideal for examining the interaction between political behavior, ideological alignment, and voting criteria across different respondent groups. By mapping variables and respondents onto a spatial framework, FCA allows for a clear visualization of how distinct clusters relate to key factors. HCA complements FCA by grouping respondents into clusters based on shared characteristics, enabling the comparison of behavioral and ideological typologies across different segments. This dual approach provides a robust framework for comparative analysis, as it captures both the diversity within respondent groups and the broader structural patterns across the dataset. The flexibility of these methods also allows for adaptation to other contexts, making them valuable tools for analyzing political participation and electoral behavior in diverse populations and settings. Together, FCA and HCA offer a powerful means of identifying and comparing the multifaceted dynamics that define voter behavior, issue prioritization, and political engagement.

Preliminary analysis reveals that youth voters in Thessaloniki fall into distinct clusters based on their political behaviors and priorities. For instance, one group demonstrates high political interest and aligns with progressive parties, while another exhibits low engagement and abstains or casts invalid votes. These findings underscore the diversity within youth political behavior and challenge assumptions about political apathy in this demographic. This research comes at a critical juncture when youth engagement is both a challenge and an opportunity for democracies worldwide. As younger generations increasingly turn to social media for political mobilization, traditional models of understanding voter behavior are evolving. By focusing on the various correlations between issue criteria, ideological alignment, and media consumption, this study provides actionable insights into how political participation can be fostered among youth.

The findings of this study validate and extend existing research by providing an empirical examination of political participation among university students. The segmentation of respondents into clusters based on their political knowledge, interest, mobilization, and voting preferences reveals a complex interaction of ideological, informational, and behavioral factors. These insights underscore the importance of addressing the polarization in political participation, both in terms of ideological alignment and engagement levels, and highlight the transformative role of digital platforms in fostering political mobilization among younger voters. The study thus contributes to a deeper understanding of the mechanisms driving political participation and electoral behavior in contemporary democratic systems. The following sections develop the relevant literature, methodology, findings, and implications of this study, offering a comprehensive view of youth political participation in Thessaloniki and its broader relevance to modern democracies.

LITERATURE REVIEW

Political participation is widely recognized as a cornerstone of democracy. It encompasses the ways in which individuals engage with the political system, including voting, campaigning, protesting, and engaging in political discourse. As Verba, Nie, and Kim (1978) articulate, political participation reflects citizens' commitment to societal well-being and shapes political outcomes, ensuring that diverse voices are heard in the democratic process. The act of participation is not merely symbolic; it plays a vital role in legitimizing political institutions and maintaining social cohesion. High levels of political participation are often seen as indicators of a healthy democracy, as they reflect an engaged and informed citizenry. Conversely, declining participation rates, particularly in elections, can undermine trust in institutions, weaken political efficacy, and threaten the stability of democratic governance (Rosenstone & Hansen, 1993). Political participation also serves as a mechanism for accountability. Through voting, citizens express their approval or disapproval of government policies and leadership. In broader forms, such as activism and protests, participation ensures that marginalized voices are represented, fostering inclusivity and pluralism. Thus, understanding the factors that influence participation is critical for policymakers, scholars, and practitioners aiming to strengthen democratic systems.

Research on political participation has identified several key factors that influence the extent and nature of individuals' engagement. These factors range from personal attributes, such as political knowledge and interest, to external influences, such as media consumption and socio-economic conditions. Political knowledge and interest are consistently highlighted as central drivers of participation (Chadjiadelis & Panagiotidou, 2021). Informed citizens are more likely to engage in political activities, as knowledge equips them with the tools to navigate complex political systems and make informed decisions. Galston (2001) argues that higher levels of political knowledge correlate with increased participation, fostering a sense of competence and agency among voters. Political interest acts as a motivational force, driving individuals to seek information, engage in discussions, and ultimately participate in elections or other political activities (Verba et al., 1978). Moreover, knowledge and interest interact synergistically: those who are interested in politics tend to acquire more knowledge, which in turn enhances their likelihood of participating. Delli Carpini and Keeter (1996) describe political knowledge as a "gatekeeper" for meaningful engagement, enabling individuals to prioritize relevant issues and align their actions with their values.

The role of media in shaping political participation has evolved dramatically in recent years. Social media, in particular, has emerged as a powerful tool for political mobilization, especially among younger populations. Platforms like Twitter and Facebook provide accessible spaces for political discourse, activism, and engagement, often serving as entry points for individuals who might otherwise remain disengaged (Bode, 2016). Traditional media, such as television and newspapers, while still influential, have been shown to correlate with lower levels of engagement compared to digital platforms (Prior, 2007). Social media not only enhances exposure to political content but also encourages interaction, making it a critical driver of participation in both online and offline contexts. However, the impact of media consumption is not universally positive; as Kahne and Bowyer (2018) note, exposure to partisan content on social media can contribute to political polarization, potentially diminishing the quality of democratic engagement.

Another key factor influencing participation is the salience of specific issues to voters. Miller et al. (1996) argue that voters are mobilized by issues they perceive as personally or socially significant, such as the economy, education, and justice. This concept, known as issue-based voting, highlights the importance of aligning political campaigns with the priorities of the electorate. Ideological alignment also plays a crucial role in issue prioritization. Left-leaning voters, for example, are more likely to prioritize education and social justice, while right-leaning voters focus on issues like national security and unemployment (Abramowitz, 2018). Understanding these preferences is essential for interpreting voting behavior and designing effective political strategies.

Gender differences in political participation have been a focus of scholarly attention for decades (Ferrin et al., 2020). Research indicates that women are more likely to engage in political activism and support progressive ideologies, often utilizing social media as a primary tool for mobilization (Inglehart & Norris, 2003). Social media offers women a platform to amplify their voices, challenge traditional power dynamics, and engage in discussions on issues they prioritize. Conversely, men tend to rely on traditional media and exhibit lower levels of political engagement in certain contexts. Their preferences often align with conservative or far-right ideologies, reflecting a narrower focus on issues like security and economic stability. These gendered patterns underscore the need for a complex understanding of how demographic factors shape participation.

Youth are often considered a politically distinct group due to their unique position in the lifecycle. Unlike older generations, younger individuals are in the process of forming their political identities, making them more susceptible to the influence of political knowledge, interest, and media exposure. Sapiro (2013) argues that the relative absence of long-term socio-economic pressures among youth allows researchers to isolate critical factors driving their participation. The potential of youth to shape the future of democracy is immense, as they bring fresh perspectives and energy to political discourse (Marangudakis et al, 2021). However, their participation levels often lag behind older age groups, emphasizing the need for targeted interventions to enhance their engagement.

In recent years, political participation has undergone significant changes, characterized by both decline and transformation. Traditional forms of participation, such as voting in elections, have shown a marked decrease in many democracies. Voter turnout has been on a steady decline globally, with abstention rates rising even in established democracies. This trend is often attributed to growing political apathy, disillusionment with political institutions, and a perceived lack of efficacy among voters, especially younger generations (Dalton, 2020). Scholars suggest that declining trust in political systems, coupled with the increasing complexity of global challenges, has alienated citizens from traditional forms of engagement (Rosenstone & Hansen, 1993). However, this decline in conventional participation does not necessarily indicate a complete disengagement from politics. Instead, it reflects a transformation in the ways individuals choose to engage. Contemporary research highlights the rise of alternative forms of political participation, particularly through social media and digital platforms. Social media has emerged as a critical tool for mobilization, enabling users to participate in online campaigns, sign petitions, join digital protests, and amplify their voices on political issues (Bode, 2016). These platforms have become especially significant for youth, who often bypass traditional institutions in favor of direct, issue-based activism.

This transformation is partly driven by the affordances of digital technologies, which allow for decentralized, immediate, and often global forms of political engagement. Unlike traditional participation, which often requires institutional structures, digital participation is accessible, cost-effective, and tailored to individual preferences (Kahne & Bowyer, 2018). However, while these new forms of mobilization have expanded participation in certain ways, they also come with challenges, such as the potential for misinformation, political polarization, and performative activism. The decline in traditional political participation and the concurrent rise of digital engagement suggest a shift in how citizens interact with the political sphere. This shift reflects broader societal changes, including the erosion of traditional political structures, the increasing influence of technology, and the evolving expectations of citizens in a globalized world. Understanding this transformation requires recognizing both the opportunities and challenges posed by new forms of participation, as well as addressing the underlying causes of disillusionment with traditional political systems (Fiorina, 2017).

The current research builds on a critical gap in the existing literature by focusing on youth, a demographic often perceived as politically disengaged but, in reality, one that increasingly engages through alternative channels like social media. Unlike older generations, younger individuals are less constrained by long-term socio-economic pressures, allowing researchers to isolate key factors driving their behavior (Panagiotidou & Chadjipadelis, 2021). Furthermore, the study addresses the need to link behavioral patterns with broader ideological and electoral trends, offering insights into how emerging forms of participation interact with traditional systems. Against the backdrop of declining voter turnout and the rise of new forms of digital mobilization, it aims to uncover how political knowledge, interest, media consumption, issue salience, and ideological alignment shape participation, adopting a multivariate comparative analysis.

METHODOLOGY

This study employs a quantitative research design to investigate the political participation and mobilization of university students in Thessaloniki. The primary aim is to identify the factors influencing voting intention, classify political behaviors into distinct typologies, and examine the relationship between political participation and electoral choices. The research focuses on a demographic that plays a critical role in shaping future democratic engagement while also representing a population that is still forming its political identity.

Data were collected through a face-to-face survey conducted between May 15 and May 24, 2024. The survey targeted university students in Thessaloniki, with a sample size of 1,572 respondents. This demographic was selected for its relevance in understanding youth political behavior and its potential impact on broader democratic processes. The questionnaire included multiple sections designed to capture a range of political and demographic variables, including:

- **Political Interest**: Respondents were asked to evaluate their level of interest in politics.
- **Political Knowledge**: Questions assessed respondents' understanding of political systems and current affairs.
- **Voting Intention**: Respondents indicated whether they intended to vote, abstain, or cast invalid votes in the upcoming elections and identified their preferred political party.
- **Issue Criteria**: Respondents specified the key issues influencing their voting decisions, such as the economy, education, or justice.
- **Ideological Positioning**: Participants positioned themselves on a left-right ideological scale ranging from 0 (far-left) to 10 (far-right).
- **Media Consumption**: The survey explored respondents' primary sources of political information, including traditional media, social media, or informal networks.
- **Political Mobilization**: Participants described their methods of mobilization, such as participating in protests, addressing political authorities, or engaging through social media.

To analyze the collected data, the study employs a two-step multivariate methodological approach that combines Hierarchical Cluster Analysis (HCA) and Factorial Correspondence Analysis (FCA) (Benzécri, 1973). Data analysis was conducted using the software M.A.D. (Méthodes d'Analyse des Données) which has been developed by Professor Dimitris Karapistolis (2010). In the first step, HCA was utilized to group respondents and variables into distinct clusters based on their similarities. This approach identified clusters of respondents sharing common characteristics, such as their levels of political interest, knowledge, and voting intentions. It also produced clusters of variables, such as issue criteria and media consumption patterns, to explore their interrelationships.

In the second step, FCA was applied to analyze the Burt table, as described by Greenacre (2017). This table incorporated all variables, including the cluster membership variables derived from HCA. FCA enabled the reorganization of categorical data into a multidimensional space, uncovering the underlying dimensions that describe the relationships among variables. These dimensions highlighted the broader patterns that influence political participation, while grouping categories with similar profiles into clusters. The results of this combined analysis revealed distinct behavioral profiles and ideological tendencies, linking respondents' political engagement to their voting behavior.

The final output of the analysis was synthesized into a semantic map that visualizes the connections between variables, clusters, and respondents. This map provides a comprehensive overview of how political participation is shaped by a combination of individual and contextual factors. The methodology is particularly suited for this research, as it allows for an in-depth examination of the interaction between voting intentions, political interest, knowledge, and mobilization methods.

Ethical approval for the study was obtained before data collection commenced. All participants provided informed consent, and their responses were anonymized to maintain confidentiality. The survey design carefully avoided leading questions and minimized biases to ensure the validity and reliability of the findings. This rigorous methodological approach ensures that the study provides robust and complex insights into youth political behavior in Thessaloniki.

RESULTS

Sources of Information on Politics

The FCA analysis extracted four key factors (Table 1), explaining the variance in how respondents use different sources for political information. These factors were derived from variables such as traditional media (TV, newspapers), social media, informal networks (family and friends), and formal or organizational sources (e.g., political parties).

Table 1. Factor loadings of variables on information sources for politics

IND		#F1/24%	#F2/18%	#F3/15%	#F4/14,5%
E1511	TV-Radio	221	194	206	318
E1521	Newpapers	809	-47	258	-237
E1531	Family	-250	389	66	-91
E1541	Friends	-310	136	95	-322
E1551	Social media	-206	-100	74	74

continued on following page

Table 1. Continued

IND		#F1/24%	#F2/18%	#F3/15%	#F4/14,5%
E1561	Party	582	19	212	-704
E1571	Internet	14	-332	-42	-26
E1591	Not interested	284	514	-1745	109

- First Factor (F1 - 24% of Total Variance): Represents the contrast between traditional media (TV, newspapers) and informal networks such as family and friends. This axis highlights a divergence in reliance on institutionalized media versus personal or relational sources.
- Second Factor (F2 - 18% of Total Variance): Captures the role of social media and the internet in opposition to traditional and formal sources like newspapers or political party information.
- Third Factor (F3 - 15% of Total Variance): Differentiates between respondents who prioritize informal sources (family, friends) versus those relying on the internet or showing disinterest in political information.
- Fourth Factor (F4 - 14.5% of Total Variance): Indicates the degree of political disengagement, represented by respondents who identify as "not interested" versus those engaging with traditional media or social media.

Clustering the variables based on their factor loadings yielded five distinct clusters of information sources (Table 2, Diagram 1). These clusters represent groups of sources with similar usage patterns, such as social media, family, and traditional media. This clustering provides insight into how different sources are grouped and utilized by respondents.

Figure 1. Clustering dendrogram of information sources

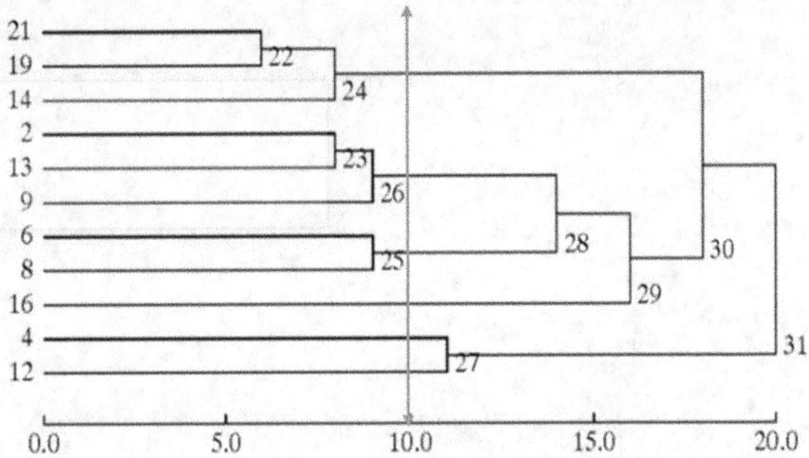

Table 2. Clusters of information sources

24	25	26	27	16
E1551	E1531	E1511	E1521	E1591
E1571	E1541		E1561	

The FCA biplot (Diagram 2) further visualizes the relationships among these variables. The horizontal axis (F1) emphasizes the opposition between institutional sources (e.g., newspapers) and informal or relational networks (family, friends), while the vertical axis (F2) underscores the distinction between disinterested respondents and those relying on digital platforms like the internet.

Figure 2. Biplot of information sources by factorial correspondence analysis

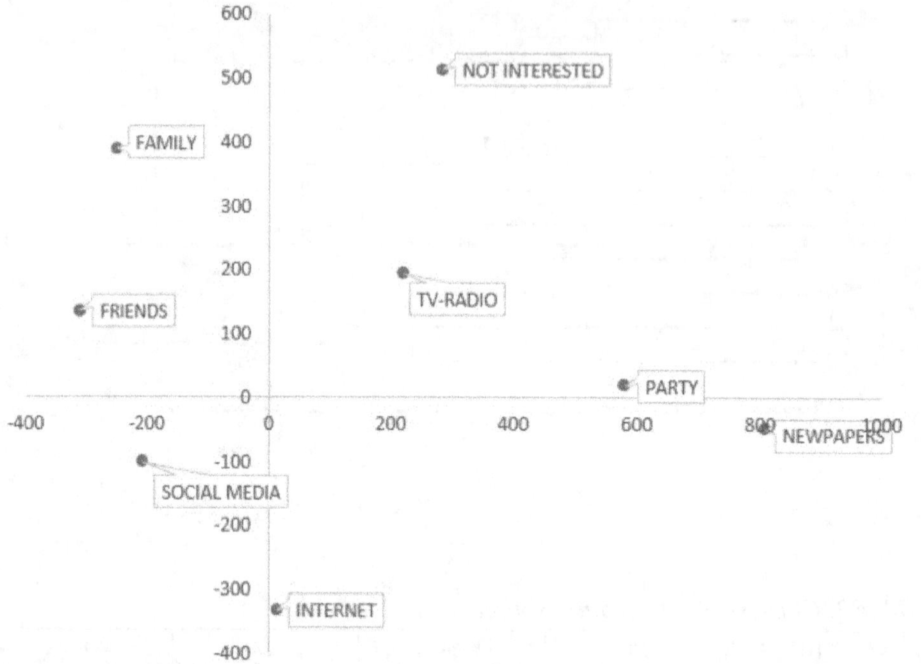

HCA was then applied to group respondents into seven clusters based on their use of political information sources (Table 3, Diagram 3).

Figure 3. Hierarchical clustering of respondents by information source

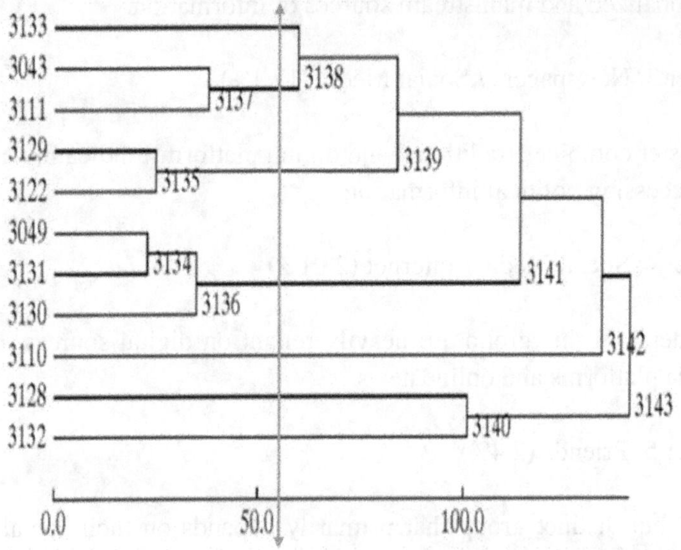

Table 3. Clusters of respondents by information sources

3110	CL1	Not informed	16.3%
3128	CL2	TV-Radio / Newspapers	23.1%
3132	CL3	Newspapers / Social Media	17.1%
3133	CL4	Social Media / Internet	23.1%
3135	CL5	Friends	2.4%
3136	CL6	Family / Friends	14.4%
3137	CL7	TV-Radio / Social Media	3.6%

These clusters reveal diverse behavioral patterns in how respondents access and prioritize political information:

- Cluster 1: Not Informed (16.3%)

Respondents in this group show minimal engagement with political information sources, reflecting a general disinterest in politics.

- Cluster 2: TV-Radio / Newspapers (23.1%)

These individuals rely primarily on traditional media, suggesting a preference for institutionalized and mainstream sources of information.

- Cluster 3: Newspapers / Social Media (17.1%)

This cluster combines traditional and digital platforms, indicating a hybrid approach to accessing political information.

- Cluster 4: Social Media / Internet (23.1%)

Respondents in this group are heavily reliant on digital sources, particularly social media platforms and online news.

- Cluster 5: Friends (2.4%)

A small but distinct group that primarily depends on their social circles for political information.

- Cluster 6: Family / Friends (14.4%)

This cluster reflects a reliance on informal networks, with respondents seeking information from both family members and friends.

- Cluster 7: TV-Radio / Social Media (3.6%)

This group combines elements of traditional and digital media consumption, blending older and newer approaches to accessing information.

While traditional media remains significant for some, the growing reliance on digital platforms, particularly social media, underscores a shift in how younger demographics access political content. Informal networks, such as family and friends, play a crucial role for others, particularly those with lower political interest or knowledge. The clusters reveal a spectrum of engagement, from the "not informed" group with minimal political activity to the "social media/internet" cluster characterized by high digital engagement. These patterns reflect broader societal trends, where traditional sources coexist with emerging digital channels, each catering to different segments of the population.

Issue-Based Voting Criteria

FCA extracted four key factors (Table 4), explaining 87.5% of the total variance in the dataset. Issues like unemployment, justice, and the economy dominate the first factor, indicating their centrality in shaping voting decisions. Global issues like the European Union and Ukraine war feature prominently in the second factor, reflecting their salience for certain respondent groups. Defense and security concerns emerge in the third factor, highlighting a subset of voters prioritizing national stability. These factors illustrate the relationships between various issue criteria, providing a multidimensional understanding of how respondents prioritize issues.

- First Factor (F1 - 74% of Total Variance): The first and dominant factor captures the divide between respondents who deem issues unimportant (ending in "0") and those who find them significant (ending in "1"). Respondents prioritizing justice, unemployment, education, and the economy (high positive scores) are contrasted with those dismissing these concerns (high negative scores).
- Second Factor (F2 - 6% of Total Variance): This factor emphasizes global versus localized concerns. Respondents assigning importance to the European Union, Ukraine war, and pandemic management (positive scores) differ from those who deprioritize these issues (negative scores).
- Third Factor (F3 - 4.5% of Total Variance): The third factor highlights differences between defense and security-related priorities (positive scores) versus respondents who neglect these issues (negative scores).
- Fourth Factor (F4 - 3% of Total Variance): This factor captures a complex divergence between respondents prioritizing daily life and justice (positive scores) and those emphasizing education and culture (negative scores).

Table 4. Factor loadings for issue-based voting criteria

IND		#F1/74%	#F2/6%	#F3/4,5%	#F4/3%
E1210	Economy	-1378	-241	112	-16
E1211		117	20	-10	1
E1220	Pandemic Management	-282	213	2	-76
E1221		195	-149	-2	52
E1230	Crime	-808	-125	-164	-154
E1231		177	27	35	33

continued on following page

Table 4. Continued

IND		#F1/74%	#F2/6%	#F3/4,5%	#F4/3%
E1240	Management of Daily Life	-375	-7	201	-277
E1241		166	2	-90	122
E1250	Universities	-1268	-283	113	81
E1251		130	29	-12	-9
E1260	Relations with Turkey	-470	98	-255	-58
E1261		213	-45	116	26
E1270	Security	-807	-142	-299	-31
E1271		177	31	65	6
E1280	War in Ukraine	-319	218	59	26
E1281		221	-153	-42	-19
E1290	European Union	-541	216	-4	124
E1291		195	-79	1	-46
E12100	Culture	-565	109	76	112
E12101		173	-34	-24	-35
E12110	Unemployment	-1322	-202	178	159
E12111		138	21	-19	-17
E12120	Justice	-1374	-235	180	188
E12121		133	22	-18	-19

Table 5 presents the clusters of variables obtained by applying Hierarchical Cluster Analysis (HCA) to the factor scores derived from the FCA. The clustering process, as shown in the dendrogram of Diagram 4, resulted in four primary clusters (35, 34, 45, and 44), which are further grouped into two overarching clusters, labeled 41 and 46. Each cluster represents a set of issues that share similar salience among respondents, reflecting coherent patterns in voting criteria.

Figure 4. Dendrogram of HCA for variables

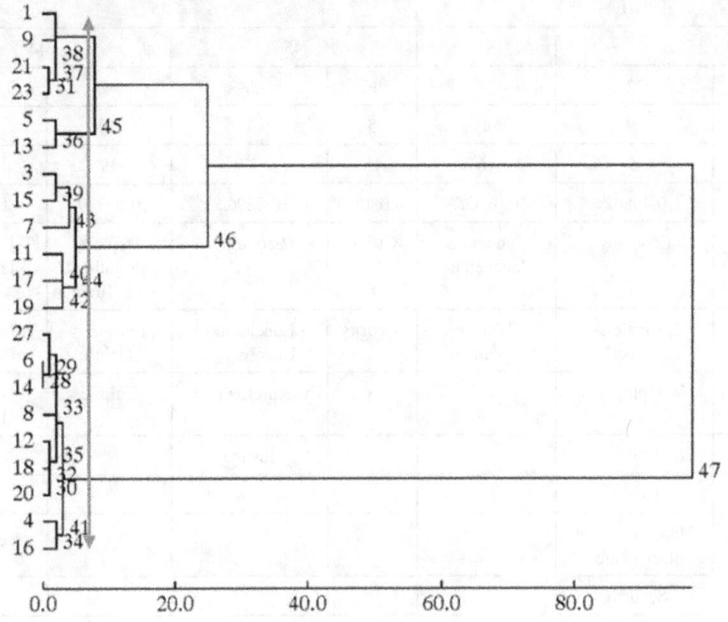

The first major cluster, Cluster 41, combines Clusters 35 and 34. Cluster 35 is associated with issues like defense, relations with Turkey, and security, which reflect concerns related to national stability and external threats. This cluster is indicative of respondents prioritizing conservative, security-focused values. Cluster 34, on the other hand, emphasizes practical concerns such as unemployment and daily life management. When combined, Cluster 41 highlights a pragmatic and governance-focused perspective, where voters are concerned with both security and economic stability.

The second major cluster, Cluster 46, integrates Clusters 45 and 44. Cluster 45 includes justice, education, and culture, aligning with progressive and societal priorities. Cluster 44 adds global and external issues such as the European Union and the Ukraine war, reflecting an international outlook. Together, Cluster 46 captures a broader and more ideologically progressive perspective, encompassing both societal reforms and global concerns.

The clustering results illustrate a clear division in issue prioritization among respondents. Cluster 41 represents practical and conservative values, focusing on economic and security-related matters. In contrast, Cluster 46 reflects a progressive and global perspective, emphasizing justice, education, and international issues.

Table 5. Clusters of issues derived from hierarchical cluster analysis

41			**46**			
35		34	45		44	
32	33	34	36	38	42	43
12	29	4	5	1	40	39
30	8	16	13	37	19	7
0.18219	0.49802	0.09828	0.03005	0.02962	0.06779	0.09398
Defense - Relations with Turkey	Economy	Pandemic Management	Crime	Economy	Defense - Relations with Turkey	Pandemic Management
European Union	Education - Universities	War in Ukraine	Security	Education - Universities	European Union	Management of Daily Life
Culture	Unemployment			Unemployment	Culture	War in Ukraine
	Justice			Justice		
	Crime					
	Management of Daily Life					
	Security					

Table 6 summarizes the fundamental oppositions identified in the biplot of Diagram 5, with the factors highlighting the diverse priorities among respondents regarding issue-based voting criteria. The first factor distinguishes between disengaged voters who find no issues important and those who assign significant importance to specific issues such as justice, unemployment, and education. This reflects a divide between low and high engagement in issue-based voting. The second factor contrasts respondents who prioritize global concerns, such as the pandemic and the Ukraine war, with those who focus on localized, practical concerns like daily life management. This factor highlights the tension between international and domestic priorities. The third factor captures the opposition between respondents emphasizing security, defense, and relations with Turkey and those who deprioritize these issues in favor of justice and societal concerns. This divide reflects ideological differences, with one side leaning towards conservative values and the other towards progressive ideals. The fourth factor underscores the contrast between voters focused on practical, daily life governance and those who prioritize broader ideological concerns like justice and education.

Figure 5. Biplot of issue-based voting criteria by factorial correspondence analysis

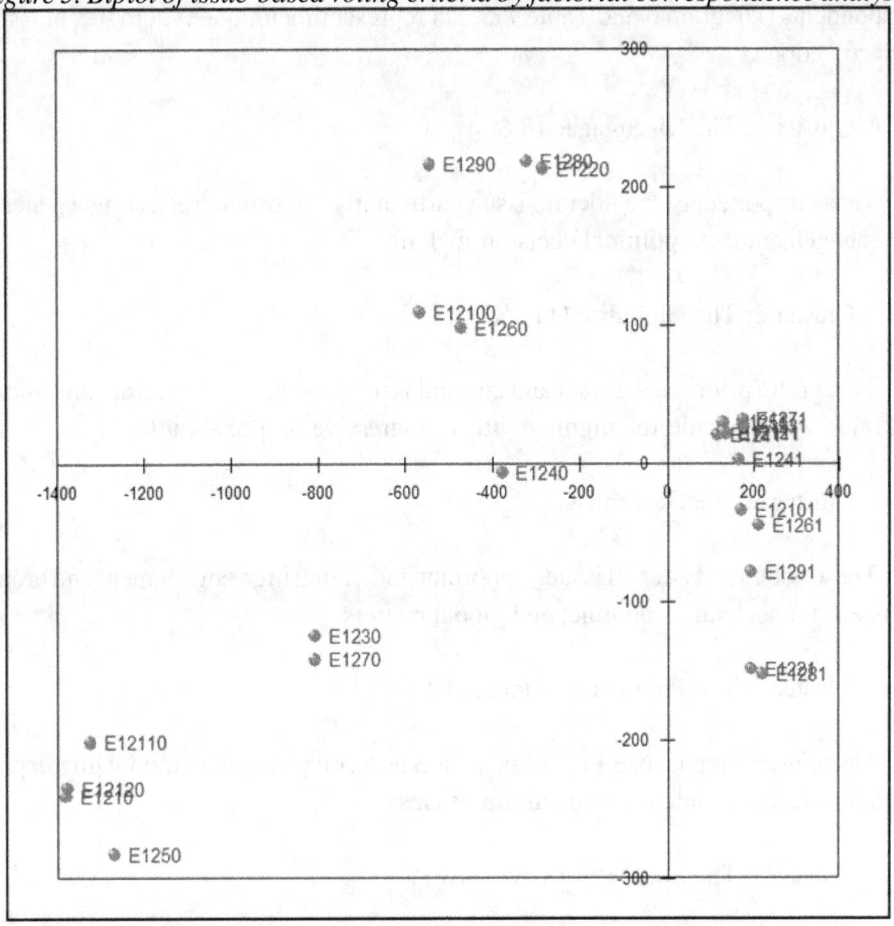

Table 6. Polarity of responses by factor

	(-)	(+)
F1	none important	
F2	pandemic and war in Ukraine are important, not daily life	not important
F3	not important	crime, security, defense, and relations with Turkey are important
F4	not important	daily life is important

The second application of HCA for the subjects produced six (6) clusters of respondents (Diagram 6 and Table 7), each representing unique typologies of issue prioritization:

- Cluster 1: The Disengaged (8.8%)

These respondents consider no issue particularly important, reflecting a general disengagement from political decision-making.

- Cluster 2: The Globalists(11.1%)

This group prioritizes global and cultural issues, including the European Union, Ukraine war, and culture, aligning with a progressive or global outlook.

- Cluster 3: The Generalists (37.7%)

These voters consider all issues important, indicating high engagement and broad concern for societal, economic, and global matters.

- Cluster 4: The Pragmatic Voters (14.1%)

This cluster emphasizes economic concerns, crime, and education while deprioritizing security, defense, and cultural issues.

- Cluster 5: The Local Prioritizers (18.8%)

These respondents focus on economic stability, education, and justice while neglecting global issues like the pandemic and the Ukraine war.

- Cluster 6: The Justice-Oriented Voters (9.6%)

This group prioritizes justice above all other issues, showing strong alignment with progressive values.

Figure 6. Dendrogram of HCA for respondents

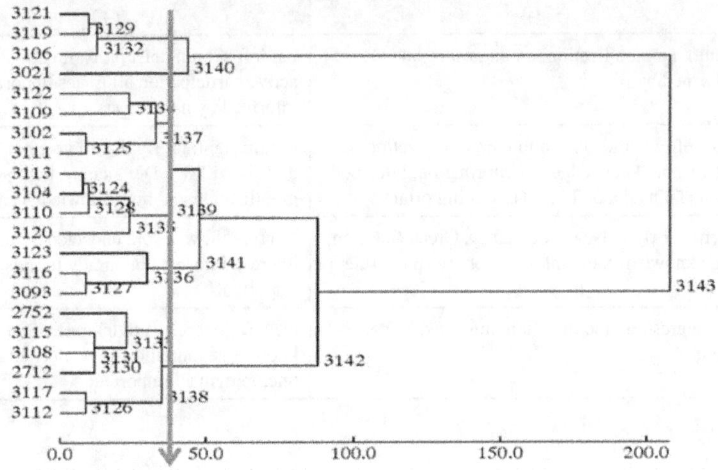

Table 7. Clusters of respondents based on issue-based voting criteria

3140	**none important**	**8.8%**
3126	special importance to EU and culture	11.1%
3133	all important	37.7%
3135	economy, crime, education, justice / not crime, security, defense, culture	14.1%
3136	economy, crime, education, security, unemployment, justice / not pandemic, Ukraine, EU	18.8%
3137	justice / not all the rest	9.6%

Factors of Political Participation and Electoral Behavior

FCA revealed four factors that explain the relationships between political mobilization, voting intention, knowledge, interest, and information sources. Each factor represents a distinct axis of variation, with the positive and negative sides highlighting contrasting respondent profiles and their associated variables (Table 8). Factor 1 highlights the contrast between active far-left mobilization and passive centrists, while Factor 2 emphasizes the divide between disengaged far-left voters and ideologically motivated center-right and far-right respondents. Factors 3 and 4 further reveal distinctions based on gender, knowledge, and issue salience, illustrating the complexity of political participation across different groups.

Table 8. Descirption of the four factors of FCA for all variables

	(-)	(+)
F1 /13,2%	Centre no mobilization, no interest or knowledge informed by friends	Far left, Communist party, high mobilization, active participation all issues important informed by newspapers
F2 /6,2%	Far Left, Invalid, abstention, no mobilization, no interest, no knowledge, no information, informed by social media or TV, no issues important	Centre right, far right, will vote but undecided, New Democracy, high personal mobilization and high knowledge and interest
F3 /5%	Men, Far right, New Democracy, Greek Solution, high knowledge and interest, not informed, rule of law	Centre left, will vote undecided, little interest, informed by family friends, no issue important
F4 /4,5%	No interest, no mobilization, informed by social media	New Democracy, not interested, no knowledge, not informed, safety and unemployment important

- First Factor (F1 - 13.2% of Total Variance): The first factor captures the divide between respondents with low and high political engagement. On one side are centrists who exhibit no political interest or knowledge, rely on friends as their primary source of information, and show minimal engagement with political issues. On the other side are far-left respondents, particularly those aligned with the Communist Party, characterized by high mobilization, active participation, and reliance on newspapers for information. These individuals consider all issues important in their political decision-making, reflecting a highly engaged and ideologically aligned profile.
- Second Factor (F2 - 6.2% of Total Variance): The second factor highlights the contrast between far-left disengagement and center-right and far-right mobilization. One side includes respondents who abstain from voting, cast invalid votes, and show no mobilization, knowledge, or interest. These individuals rely on social media or TV for political information and do not prioritize any specific issues. The opposite side comprises center-right and far-right voters, including New Democracy supporters, who exhibit strong personal mobilization, high levels of knowledge and interest, and rely on traditional or conservative information sources. This factor emphasizes the ideological and engagement-based divide between inactive far-left voters and active right-leaning participants.
- Third Factor (F3 - 5% of Total Variance): The third factor reflects differences in gender, ideology, and information sources. One side features men aligned with far-right ideologies, such as New Democracy and Greek Solution supporters, who demonstrate high levels of political knowledge and interest but are not actively informed and prioritize issues like the rule of law. The other side includes centre-left voters who are undecided about their voting inten-

tions, exhibit low political interest, and rely on family and friends for information. These respondents tend to dismiss most political issues as unimportant, indicating a low level of political engagement.

- Fourth Factor (F4 - 4.5% of Total Variance): The fourth factor focuses on the role of interest and mobilization in shaping issue prioritization. One side represents respondents with no political interest or mobilization who are informed predominantly through social media and do not prioritize political issues. The other side includes New Democracy supporters who, despite limited political interest and knowledge, prioritize practical concerns such as safety and unemployment. This factor underscores the link between specific issue salience and moderate levels of engagement.

Figure 7. 3D view of the positionings of all items on the first 3 factors.

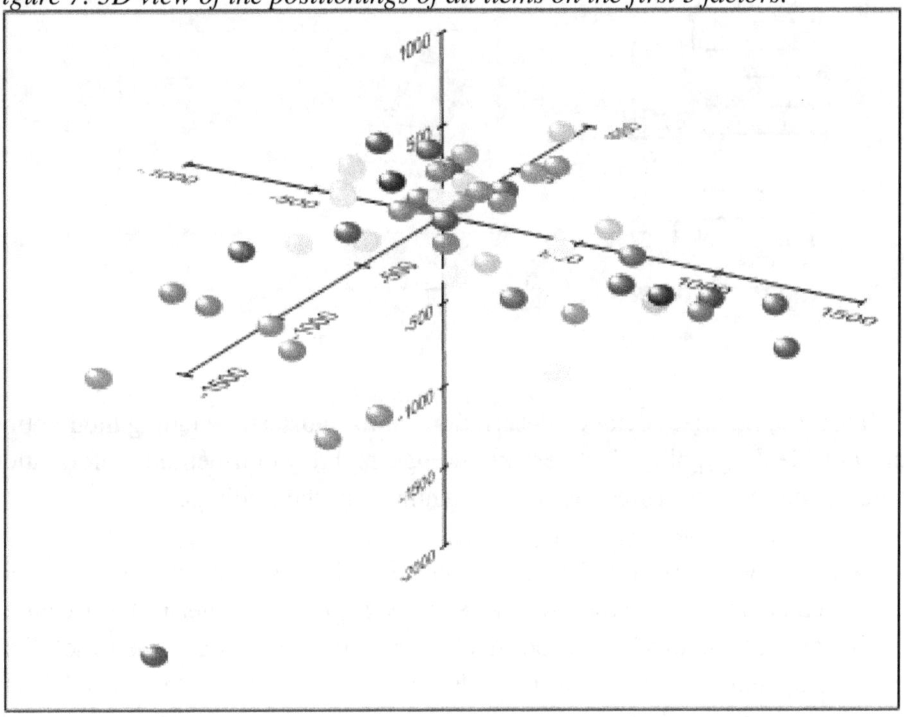

Hierarchical Cluster Analysis (HCA) applied to the factor loadings identified six primary clusters of respondents (104, 102, 95, 94, 101, and 57) as shown in the dendrogram in Diagram 8, which were further grouped into larger clusters: 105 and 107.

Figure 8. Clustering the four factors of FCA

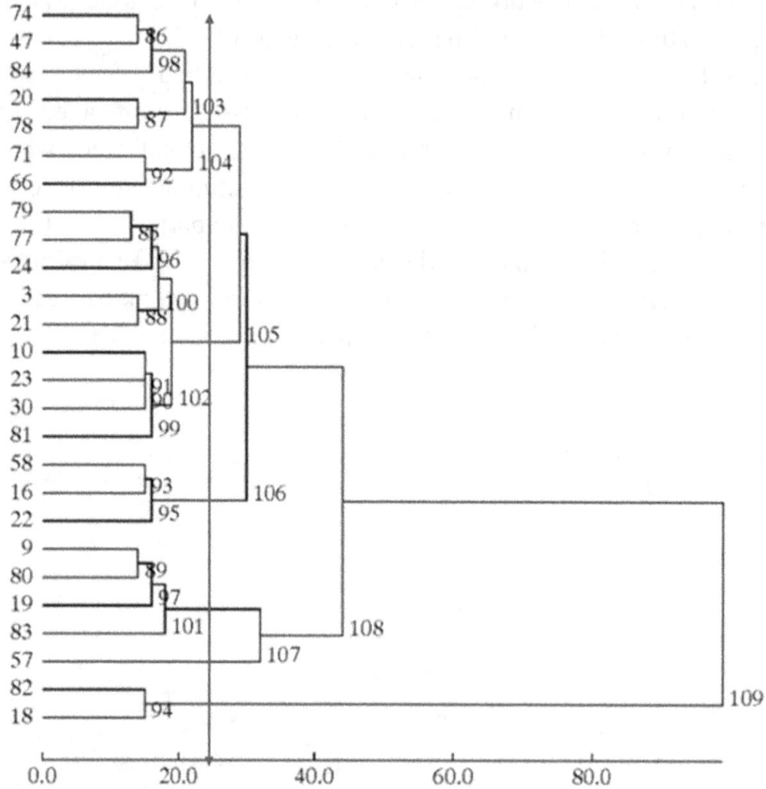

Table 9 summarizes detailed descriptions of the clusters, including their voting behavior, ideology, political interest, knowledge, mobilization methods, information sources, and issue priorities. Below is a synthesis of the findings:

- Cluster 104 (Part of 105): This cluster includes respondents who are undecided about whether they will vote. They align with center-right or centrist ideologies and exhibit low political interest and knowledge. Their mobilization approach is passive, as they "let the people in charge do their job," and they rely on family and friends for information. Their issue priorities include safety and unemployment but exclude global concerns like the pandemic, the Ukraine war, and the European Union.
- Cluster 102: This group represents decided voters leaning toward Syriza. They exhibit moderate to high political knowledge and interest, identifying with center-left ideologies. Their mobilization methods are more active, with respondents addressing authorities directly. They rely on TV-radio and social

media for political information and prioritize a broad range of issues, particularly those related to European and cultural concerns.

- Cluster 95: This cluster includes abstainers and PASOK-KINAL voters, primarily composed of men with low to moderate political knowledge. They prioritize the rule of law while disregarding broader issues like the pandemic or the EU. Their mobilization methods are traditional, involving engagement with newspapers and television.
- Cluster 94: Respondents in this cluster are aligned with far-left parties, including the Communist Party and New Left. They are highly politically interested and knowledgeable, mobilizing actively through protests and social media. They emphasize justice and education while deprioritizing issues like crime, safety, and defense.
- Cluster 101 (Part of 107): This cluster represents disengaged respondents, characterized by invalid or blank votes and low political interest or mobilization. They do not align with any specific ideology and consider no issues as important. Their information sources are minimal or nonexistent, reflecting their detachment from political activity.
- Cluster 57 (Part of 107): Similar to Cluster 101, this group exhibits political disengagement and minimal mobilization. They are unlikely to participate actively in political processes or align with any ideological group.

Regarding the combined clusters n the next level of the analysis, Cluster 105 (104 and 95), reflects voters who are primarily undecided, abstaining, or aligned with centrist ideologies. Their political engagement is low to moderate, and they prioritize practical concerns like safety and unemployment while avoiding global issues. Cluster 107 (101 and 57) encompasses respondents who are politically disengaged, with no interest, mobilization, or alignment to any party. They are characterized by invalid or blank voting behavior and a lack of prioritization for any political issues.

Table 9. Description of the the six (6) clusters of political behavior.

	105				**95**	**107**		**94**
	104		102		95	101	57	94
	92	103	99	100	95	101	57	94
VOTING BEHAVIOUR	I haven't decided yet if I will vote	Abstention / Pasok-Kinal	I will vote (decided) / Syriza	I will vote (not decided yet) / Foni Logikis / Plefsi	New Democracy / Greek Solution	Invalid/Blank / Niki		Communist Party / New Left Other
SEX		Men		Women / Other				

continued on following page

Table 9. Continued

	105				95	107		94
IDEOLOGY	Centre-Right / Centre			Centre Left	Far Right	N/A		Far Left
POLITICAL INTEREST	a little			quite		not at all	N/A	very much
POLITICAL KNOWLEDGE		low to moderate	high	quite		none		
POLITICAL MOBILIZATION	I let the people in charge do their job	I am addressing a television channel, a newspaper	I am personally addressing the authorities	I am active through social networks (FB, Instagram, etc.)		Nothing, I don't care	N/A	I take part with others in protests
INFORMATION SOURCES	Family / Friends	Friends and TV-Radio / Newspapers	TV-Radio / Social Media	Social Media / Internet		Not informed		Newspapers / Social Media Other
ISSUE CRITERIA		[yes to safety, unemployment but no to pandemic, Ukraine war or EU] / [Rule of law]	EU and culture	Every issue		No issue important		yes to education and justice but not crime, safety, defense

The biplot in diagram 9, illustrates the spatial distribution of clusters (104, 102, 95, 94, 101, and 57) along the factorial axes derived from Factorial Correspondence Analysis (FCA). These axes reveal critical dimensions of polarization in political behavior, including engagement, ideology, and voting preferences, alongside variations in political interest, knowledge, and information sources. The horizontal axis (F1) captures the division between low political engagement and high political engagement. The antithesis along F1 reflects the stark contrast between highly engaged voters (94, 102) and disengaged or passive respondents (101, 57, 104). This divide highlights the role of political interest, knowledge, and mobilization in shaping electoral behavior.

More specifically, Clusters 94 and 102 occupy the positive side of F1, reflecting high political engagement. Cluster 94 represents far-left voters, including supporters of the Communist Party and other progressive ideologies. These respondents exhibit high political interest and knowledge, actively participating through protests and other forms of mobilization. They rely on newspapers and social media as their primary information sources and consider justice, education, and every issue important. Cluster 102, meanwhile, consists of center-left voters (primarily Syriza supporters) with moderate to high engagement. These individuals are politically active, using direct mobilization methods and relying on TV-radio and social media for information. They prioritize a wide range of issues, including European integration and cultural concerns.

Clusters 101, 57, and 104 are situated on the negative side of F1, representing low political engagement. Cluster 101 includes voters who cast invalid or blank votes, display minimal political interest or knowledge, and lack mobilization altogether. They do not align with any specific ideology and consider no issues important. Similarly, Cluster 57 is characterized by complete disengagement, with minimal political activity or alignment. Cluster 104, although slightly more engaged, comprises undecided voters who demonstrate low interest and knowledge. These respondents rely on family and friends for political information and focus on practical concerns like safety and unemployment.

The vertical axis (F2) represents the ideological divide between progressive and conservative voters. The antithesis along F2 lies in the ideological clash between progressive voters (94, 102) and conservative voters (95, 104). This divide is reflected in their electoral behavior, issue prioritization, and reliance on different information sources.

Clusters 94 and 95 are positioned on opposite ends of F2, showcasing the ideological polarization between far-left and far-right voters. Cluster 94 aligns with far-left ideologies, including voters who prioritize justice, education, and societal reforms. These individuals are highly knowledgeable, politically active, and rely on newspapers and social media for information. In contrast, Cluster 95 represents far-right voters, including supporters of New Democracy and Greek Solution. These respondents emphasize safety, unemployment, and national security as critical issues. They rely on traditional media like TV-radio and newspapers and show moderate to high political knowledge and interest.

Clusters 102 and 104 represent the center-left and center-right, respectively. Cluster 102 is located closer to the lower side of F2, reflecting its alignment with progressive ideologies (Syriza supporters). These respondents demonstrate moderate political engagement and prioritize cultural and European issues. Cluster 104, positioned on the upper side of F2, aligns with center-right ideologies, including undecided voters leaning toward PASOK-KINAL. These individuals focus on safety and unemployment but exhibit passive mobilization and low political knowledge.

Further study of the data highlights the following key polarizations that configure the voting behavior of the respondents:

- Clusters 94 vs. 95 (Far Left vs. Far Right): These clusters represent the stark ideological opposition between progressive (Communist Party supporters) and conservative (New Democracy and Greek Solution voters) ideologies. The far-left emphasizes justice and education, while the far-right focuses on safety and national security.

- Clusters 102 vs. 104 (Center Left vs. Center Right): These clusters illustrate the competition between moderate center-left (Syriza) and center-right (PASOK-KINAL) voters. The center-left prioritizes broad societal concerns, while the center-right focuses on practical governance issues.
- Clusters 94 and 102 vs. 101 and 57 (Engaged vs. Disengaged): Engaged clusters (94, 102) demonstrate clear voting preferences and issue prioritization, while disengaged clusters (101, 57) show minimal political activity and no alignment with specific parties or issues.
- Clusters 95 and 104 vs. 94 and 102 (Pragmatic vs. Ideological): Clusters 95 and 104 focus on pragmatic concerns like safety and unemployment, contrasting with the ideological emphasis of clusters 94 and 102, which prioritize justice, education, and cultural issues.

Figure 9. The biplot brings together all the different categories (characteristics) on the first factorial plane, depicting how profiles are positioned on the two first factors of FCA. The visualized output is used as a behavioral (semantic) map, which can assist In the deeper comprehension of electoral behavior, the shaping factors of political participation and the dynamics among various political behavioral profiles

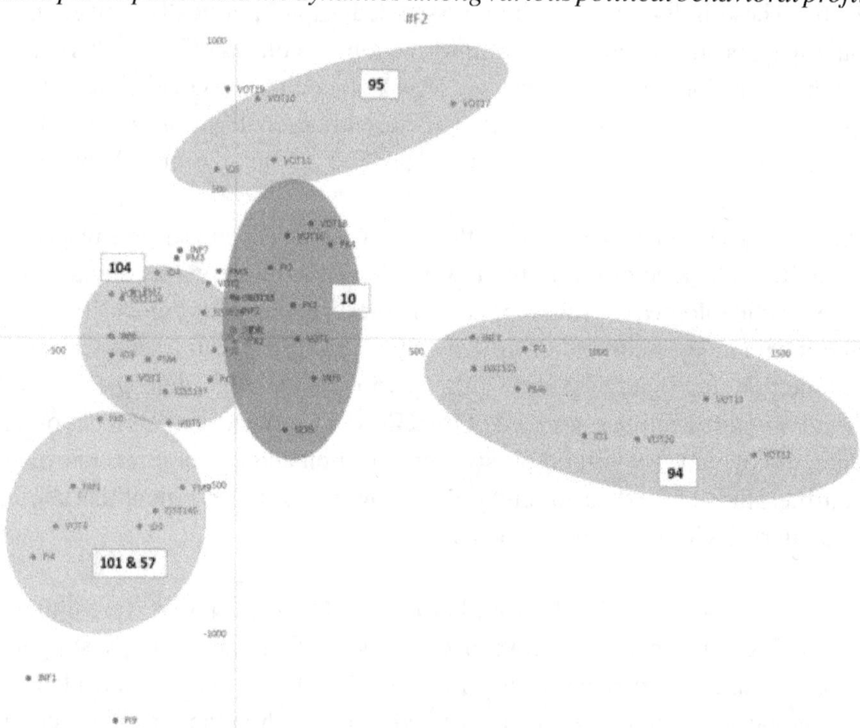

DISCUSSION

The analysis of political participation and electoral behavior among university students in Thessaloniki has provided a comprehensive understanding of the multidimensional factors influencing their engagement and decision-making. Using advanced multivariate techniques, such as Factorial Correspondence Analysis (FCA) and Hierarchical Cluster Analysis (HCA), the study revealed critical insights into voting behavior, political mobilization, issue prioritization, and media consumption. This section synthesizes the most important findings, reflecting the polarizations and interactions that define political participation and its broader implications.

The findings highlight a clear divide between highly engaged and disengaged voters, captured along the first factorial axis (F1). At one extreme, highly engaged respondents, represented by Clusters 94 and 102, demonstrate significant political interest and knowledge, active mobilization, and clear ideological alignment. Cluster 94, aligned with far-left ideologies such as the Communist Party, is characterized by high participation through protests and social media, emphasizing justice, education, and societal reforms. Cluster 102, associated with center-left ideologies (primarily Syriza), reflects a more moderate yet active group, prioritizing cultural and European issues while utilizing both social media and traditional information sources. At the opposite extreme, disengaged voters in Clusters 101, 57, and 104 exhibit minimal political mobilization, interest, or knowledge. Cluster 101, in particular, includes individuals casting invalid or blank votes, showing no ideological alignment or prioritization of issues. Cluster 57 mirrors this disengagement, while Cluster 104, although slightly more active, represents undecided voters with low engagement, relying on family and friends for information and focusing on pragmatic concerns like safety and unemployment. This polarization highlights the broader societal challenge of engaging politically passive groups, whose disconnection limits their influence on democratic processes. The second factorial axis (F2) captures the ideological polarization between progressive and conservative voters, reflecting significant differences in issue prioritization and voting behavior. Cluster 94 epitomizes the far-left segment, emphasizing justice, education, and comprehensive issue prioritization. These respondents align strongly with far-left political parties and are driven by ideological commitment. In contrast, Cluster 95 represents the far-right, including supporters of New Democracy and Greek Solution, who prioritize safety, unemployment, and national security. These individuals demonstrate moderate to high political knowledge and interest, relying on traditional media for information. Clusters 102 and 104 occupy intermediate positions, reflecting the competition between center-left and center-right ideologies. Cluster 102, leaning toward Syriza, emphasizes European integration and cultural concerns, demonstrating active mobilization and moderate engagement. Cluster 104, associated with undecided voters leaning toward PASOK-KINAL, prioritizes practical governance issues such as safety and unemployment while remaining passive in terms of mobilization and information consumption.

The spatial distribution of these clusters underscores the ideological competition shaping electoral behavior, with progressive and conservative voters clashing over issue salience and mobilization strategies. The far-left (Cluster 94) emphasizes social reforms and equality, while the far-right (Cluster 95) focuses on national stability and security, reflecting a broader polarization in democratic systems. The analysis of issue-based voting criteria further highlights the diversity in respondent priorities:

- Broad Prioritizers: Clusters 94 and 102 prioritize a wide range of issues, including justice, education, and cultural concerns, reflecting progressive values and active political engagement.
- Pragmatic Prioritizers: Clusters 95 and 104 focus on practical concerns like safety, unemployment, and economic stability, aligning with conservative ideologies and governance-oriented perspectives.
- Disengaged Respondents: Clusters 101 and 57 exhibit no clear issue prioritization, emphasizing their disconnection from political processes.

The alignment of these issue priorities with ideological stances further reinforces the polarization in political competition, with far-left and far-right voters emphasizing contrasting sets of priorities. While progressive voters champion broader societal and cultural reforms, conservative voters focus on immediate and tangible governance outcomes.

The study also reveals significant differences in media consumption and mobilization methods across clusters. Social media emerges as a critical tool for mobilizing highly engaged voters, particularly in Clusters 94 and 102. These respondents use digital platforms for activism and information, demonstrating a strong connection between digital engagement and political participation. In contrast, traditional media (TV-radio and newspapers) is more prominent among moderate and conservative voters in Clusters 95 and 104, correlating with lower levels of engagement and a focus on pragmatic concerns. Clusters 101 and 57, characterized by disengagement, exhibit minimal reliance on any media sources, further reflecting their detachment from political processes. This divide in media consumption patterns highlights the generational and technological shifts influencing political participation, with digital platforms enabling new forms of activism while traditional media maintains its role among more passive groups.

The clusters also reflect distinct patterns in voting behavior and mobilization strategies, highlighting the diversity in political engagement across groups. Active voters, represented by Clusters 94 and 102, demonstrate clear voting preferences and strong ideological alignment. Cluster 94, for instance, consists of far-left supporters, including those aligned with the Communist Party, who engage actively through protests, direct interaction with authorities, and a significant reliance on social media as their primary tool for mobilization. Similarly, Cluster 102, which leans toward Syriza, is characterized by a more moderate yet engaged voter base, utilizing a combination of traditional and digital platforms to participate politically. In contrast, undecided and passive voters, grouped in Clusters 104, 101, and 57, exhibit low political engagement and minimal mobilization. These respondents include undecided voters and individuals casting invalid or blank votes. Their reliance on informal information sources, such as family and friends, or the absence of any significant media use reflects their lack of political efficacy and detachment from active participation. Finally, Cluster 95 comprises conservative voters who support far-right parties, such as New Democracy and Greek Solution. This group prioritizes issues like safety and economic stability and relies heavily on traditional media, including TV and newspapers, as their primary source of political information. Together, these clusters reveal the varied approaches to political participation, shaped by ideological leanings, information access, and mobilization strategies.

The findings of this study align with and expand upon existing research on political participation and electoral behavior. The multidimensional analysis of university students in Thessaloniki demonstrates patterns consistent with established theories while highlighting nuances in engagement, ideological alignment, and issue prioritization. The observed polarization in political engagement, with distinct differences between active and disengaged voters, mirrors previous studies emphasizing the central role of political knowledge, interest, and mobilization in shaping participation (Verba, Nie, & Kim, 1978). Highly engaged clusters, such as Clusters 94 and 102, exemplify the theoretical connection between high political efficacy and active participation, as these respondents demonstrate clear voting preferences, strong ideological alignment, and diverse mobilization methods, including protests and social media activism. Their reliance on multiple information sources, such as newspapers and social media, corroborates findings that media diversity enhances political interest and participation (Delli Carpini & Keeter, 1996; Bode, 2016). In contrast, the disengaged clusters, such as Clusters 101 and 57, reflect the challenges described in the literature regarding low efficacy and minimal mobilization. These respondents exhibit the characteristics of politically apathetic individuals, as identified by Rosenstone and Hansen (1993), who argue that a lack of trust in institutions and perceived inefficacy contribute to disengagement. The absence of significant issue prioritization among these clusters further underscores their detachment, a phenomenon also linked to insufficient political knowledge (Galston, 2001). The findings demonstrate that disengaged respondents rely minimally on media or other information sources, further limiting their ability to navigate political systems and make informed decisions.

The ideological polarization observed along the second axis of the biplot provides empirical support for Downs' (1957) theory of ideological alignment as a motivator for political participation. Far-left respondents in Cluster 94 align with justice- and equality-driven agendas, while far-right respondents in Cluster 95 emphasize national stability and security. This ideological divide echoes Abramowitz's (2018) findings on the increasing polarization of political participation, where ideological extremity correlates with heightened voter activity. Moreover, the prioritization of issues such as justice, education, and cultural concerns among far-left voters contrasts sharply with the pragmatic focus on safety, unemployment, and governance among far-right respondents, revealing a clear segmentation in voter priorities and mobilization strategies.

The study also underscores the role of media consumption in shaping political behavior, adding depth to Prior's (2007) assertion that traditional media consumption is associated with lower engagement levels. Social media, as highlighted in this study, emerges as a critical tool for active political participation, particularly among highly engaged clusters like 94 and 102. This supports Bode's (2016) findings on the mobilizing power of digital platforms, particularly for younger voters. Conversely, the reliance on traditional media by conservative and center-right voters (Clusters 95 and 104) reflects a more passive engagement style, consistent with earlier studies emphasizing the contrasting effects of traditional versus digital information sources on political mobilization (Kahne & Bowyer, 2018).

Issue-based voting patterns identified in this study further enrich the understanding of voter behavior. Clusters prioritizing broad issues, such as justice and education (94 and 102), align with Miller et al's (1996) observation that issue salience drives turnout among ideologically motivated voters. On the other hand, the narrow issue prioritization seen in conservative clusters, such as 95 and 104, aligns with Dalton's (2020) argument that pragmatic concerns dominate the preferences of more traditionally engaged voters. The disengagement of Clusters 101 and 57, characterized by a lack of issue prioritization, reinforces the notion that political apathy undermines meaningful participation, as suggested by Verba et al. (1978).

This study provides a deeper understanding of the factors shaping political participation and electoral behavior among university students, a demographic often underrepresented in political analyses. The segmentation of respondents into distinct clusters based on their political engagement, ideological alignment, and issue prioritization has significant implications for understanding democratic participation and the challenges of mobilizing younger voters. Firstly, the study reinforces the critical role of political knowledge and interest in fostering active participation. Clusters with high levels of knowledge and interest, such as Clusters 94 and 102, demonstrate the transformative potential of political education and engagement. These findings suggest that increasing access to political knowledge and fostering interest through targeted educational interventions could reduce disengagement among younger populations, aligning with democratic goals of inclusivity and representation.

Secondly, the polarization observed in this study highlights the growing divide between progressive and conservative ideologies in political behavior. The ideological alignment of far-left and far-right voters, coupled with their distinct issue prioritization and mobilization strategies, underscores the importance of addressing polarization in democratic systems. Policies and interventions aimed at fostering dialogue and understanding between ideological groups could help mitigate the risks of increasing polarization and its effects on governance and political stability. Additionally, the findings emphasize the transformative role of social media as a mobilizing tool, particularly for engaged and progressive voters. This highlights the need for policymakers and political actors to recognize and leverage the potential of digital platforms to reach and mobilize younger voters. However, the reliance on traditional media by conservative and disengaged groups suggests that a dual approach to communication strategies is essential to bridge generational and ideological divides in political engagement. Finally, the study draws attention to the disengagement of certain clusters, particularly those characterized by low political knowledge, interest, and mobilization. This highlights the need for targeted outreach efforts to engage these voters, addressing barriers to participation such as perceived inefficacy, lack of trust in political institutions, and limited access to relevant information. Engaging disengaged voters is not only a democratic imperative but also a critical step in ensuring the representativeness and legitimacy of electoral outcomes.

The findings of this study open several avenues for practical applications and future research in political science and electoral studies. One key application is the design of targeted civic education programs aimed at increasing political knowledge and interest among young voters. Such programs could focus on equipping students with the tools to critically analyze political information, fostering a sense of political efficacy and engagement. Additionally, educational campaigns utilizing digital platforms could capitalize on the mobilizing potential of social media, particularly among progressive and ideologically aligned voters. The study also has implications for political communication strategies. Political actors and parties could use the insights from this research to develop tailored messaging that resonates with specific voter clusters. For example, messages emphasizing justice and societal reforms could appeal to progressive voters, while those focusing on safety and economic stability might better engage conservative or center-right audiences. Incorporating digital media campaigns alongside traditional media outreach could help bridge the engagement gap between younger, tech-savvy voters and more traditionally inclined demographics.

The methodology employed in this study, combining Factorial Correspondence Analysis (FCA) and Hierarchical Cluster Analysis (HCA), provides a powerful framework for comparative analysis of political participation and electoral behavior. FCA uncovers underlying patterns and relationships between categorical variables, enabling a complex examination of how voting behavior, ideological alignment, and issue prioritization interact within a multidimensional space. By visualizing these associations, FCA offers insights into the structural dynamics of political behavior. HCA complements this approach by clustering respondents based on shared characteristics, facilitating the comparison of behavioral and ideological typologies across distinct groups. Together, these methods provide a robust and flexible analytical framework, capable of capturing the diversity within respondent groups while identifying overarching patterns, making them ideal tools for understanding the complexities of political engagement in different contexts.

Future research could explore the long-term trends in political participation among younger populations, investigating whether the patterns identified in this study persist over time or evolve with changing political and social dynamics. Comparative studies across different countries or regions could provide valuable insights into the generalizability of these findings and the influence of cultural and institutional contexts on political behavior. Another promising direction for research is the exploration of the role of misinformation and political polarization in shaping voter behavior. Given the reliance on social media by certain clusters, future studies could investigate how exposure to biased or misleading information influences political interest, knowledge, and mobilization. Similarly, examining the effectiveness of counter-polarization strategies, such as cross-ideological dialogue initiatives, could provide actionable insights for reducing polarization and fostering inclusivity in democratic systems. Finally, future studies could focus on the psychological and emotional dimensions of political participation, investigating how factors such as trust, perceived efficacy, and emotional resonance with political messaging influence voter engagement. Understanding these underlying mechanisms could provide a more holistic view of the factors driving political behavior and inform the development of more effective strategies for enhancing participation and reducing disengagement.

In conclusion, the findings of this study not only contribute to the academic understanding of political participation but also offer practical insights for improving democratic engagement. By addressing the gaps identified in this research and exploring new avenues for inquiry, future studies can build on these findings to foster more inclusive, representative, and vibrant democratic systems.

REFERENCES

Abramowitz, A. I. (2018). *The great alignment: Race, party transformation, and the rise of Donald Trump*. Yale University Press. DOI: 10.2307/j.ctvhrczh3

Benzécri, J. P. (1973). *L'Analyse des données*. Tome 2: *L'Analyse des correspondances*. Dunod.

Bode, L. (2016). Political news in the news feed: Learning politics from social media. *Mass Communication & Society, 19*(1), 24–48. DOI: 10.1080/15205436.2015.1045149

Carpini, M. X. D., & Keeter, S. (1996). *What Americans know about politics and why it matters*. Yale University Press.

Chadjipadelis, T., & Panagiotidou, G. (2021). Students Teaching Democracy to Other Students: The Effects on Political Engagement. *GLOBALLY*, 185.

Ferrín, M., Fraile, M., García-Albacete, G. M., & Gómez, R. (2020). The gender gap in political interest revisited. *International Political Science Review*, *41*(4), 473–489. DOI: 10.1177/0192512119860260

Fiorina, M. P. (2017). *Unstable majorities: Polarization, party sorting, and political stalemate*. Hoover press.

Galston, W. A. (2001). Political knowledge, political engagement, and civic education. *Annual Review of Political Science*, *4*(1), 217–234. DOI: 10.1146/annurev.polisci.4.1.217

Greenacre, M. (2017). *Correspondence analysis in practice*. chapman and hall/crc.

Inglehart, R., & Norris, P. (2003). *Rising tide: Gender equality and cultural change around the world*. Cambridge University Press. DOI: 10.1017/CBO9780511550362

Kahne, J., & Bowyer, B. (2018). The political significance of social media activity and social networks. *Political Communication*, *35*(3), 470–493. DOI: 10.1080/10584609.2018.1426662

Karapistolis, D. (2010). *Software Method of Data Analysis MAD* [Internet]. Available from: https://www.pylimad.gr/ [Accessed: October 25, 2024]

Marangudakis, M., Chadjipadelis, T., & Panagiotidou, G. (2021). The Political Culture of Young Adults in Greece. *The Journal of Mediterranean and European Linguistic Anthropology, 3*(IKEEART-2022-1217), 24-51.

Miller, W. E., Shanks, J. M., & Shapiro, R. Y. (1996). *The new American voter*. Harvard University Press.

Panagiotidou, G., & Chadjipadelis, T. (2021). First-time voters in Greece: views and attitudes of youth on Europe and democracy. In *Data Analysis and Rationality in a Complex World 16* (pp. 225–232). Springer International Publishing. DOI: 10.1007/978-3-030-60104-1_25

Prior, M. (2007). *Post-broadcast democracy: How media choice increases inequality in political involvement and polarizes elections*. Cambridge University Press. DOI: 10.1017/CBO9781139878425

Rosenstone, S. J., & Hansen, J. M. (1993). *Mobilization, participation, and democracy in America*. Macmillan.

Sapiro, V. (2013). Gender, Social Capital, and Politics 1. In *Gender and social capital* (pp. 151-183). Routledge.

Verba, S., Nie, N. H., & Kim, J.-O. (1978). *Participation and political equality: A seven-nation comparison*. University Chicago Press.

Chapter 9
Maritime Security Risk Assessment Using the Bow–Tie Method

Panayiotis Papageorgiou
University of Peloponnese, Greece

Zacharias Dermatis
https://orcid.org/0000-0001-5605-3169
University of Peloponnese, Greece

Panagiotis Liargovas
University of Peloponnese, Greece

ABSTRACT

This study proposes a risk assessment method for the maritime industry to address international threats such as terrorism, piracy, robbery, cyber-attacks, smuggling, and drug trafficking. Due to limited literature on maritime risk acceptance criteria, the method employs the Bow-Tie diagram for Risk Identification, Analysis, Evaluation, and Cost-Benefit Analysis. After identifying the Hazard and Top Event, the risk is calculated based on the defined probability and contribution of each threat, and the effectiveness of each prevention barrier. Similarly, risk for each consequence is computed using the defined risk value and the effectiveness of each mitigation barrier. The total cost of risk treatment is calculated from the cost of all prevention and mitigation measures and compared to the cost in case of the Top Event and consequences. Results are visualized in the Bow-Tie diagram using a color-coded scheme, offering a clear overview of risks and associated costs.

DOI: 10.4018/979-8-3693-9400-7.ch009

1. THE CONCEPT OF MARITIME SECURITY

When the term security is used, it includes all intentional acts. In contrast, when the term safety is used, it refers to unintentional events. The traditional approach to safety mainly deals with accidental and unintended failures, unintentional errors and malfunctions, as well as all possible damages or losses arising from them. On the contrary, the analysis of intentional and planned actions aimed at harmful results in specific targets, is the subject of security. With this approach, any act of piracy, vandalism, terrorism, theft, espionage etc., are considered as a security issue. In addition, the concept of security includes organized crime, such as smuggling, human trafficking, tax evasion, blackmail, etc., which benefit criminals while harming people, property and the state (Nyman et al., 2010).

Based on who uses the term or in what context it is used, there are many definitions for the terms general "security" and especially "maritime security" (Klein, 2011). In Natalie Klein's book *Maritime Security and the Law of the Sea* defines security in many ways from an academic point of view. Thus, for example the Copenhagen School defines security as: *"Security is a socially constructed concept and that discourse is a key element in the construction and identification of security issues. Based on the discourse which surrounds it, a public policy issue can be classified as non-politicized, politicized or securitized. [...] A securitized issue is identified as a potential threat to the continued existence of the state. Once securitized, issues are perceived to be of such immediate importance that they are elevated above the ordinary norms of the political debate and the state acquires special rights to adopt extraordinary measures in order to protect itself"*. (Bradford, 2004).

Moreover, Genserik Reniers has defined security as: *"taking all preventive measures in order to avoid harmful incidents caused by unauthorized (internal or external) persons who intend to seriously damage the company, as well as controlling such incidents and their adverse effects"*. (Reniers, 2011)

Also, the concept of security is defined by The Finnish Ministry of Defence as follows: *"The comprehensive concept of security comprises security issues which, if exacerbated, may turn into threats that can jeopardize or seriously harm Finland, Finns or the functions vital to Finnish society. Wide-ranging threats include premeditated action such as the use of military force, terrorism or interference with information networks. They can also occur spontaneously, such as widespread failures of the electric grid or extreme forces of nature"*. (Ministry of Defence of Finland, 2011).

Natalie Klein finds that those working in different fields use the term "maritime safety" differently. Thus, the army treats the concept of maritime security differently than the companies in the shipping industry. For example, the US Naval Operations Concept uses the following phraseology when referring to shipping security business issues: *"ensuring the freedom of navigation, the flow of commerce and the protec-*

tion of ocean resources, as well as securing the maritime domain from nation-state threats, terrorism, drug trafficking and other forms of transnational crime, piracy, environmental destruction and illegal seaborne immigration". (Bradford, 2004; US Navy, 2007).

On the other hand, ship-owners believe that the existence of maritime security is primarily necessary to ensure the transport of cargo without any problems related to criminal activity (Raymond & Morriën, 2008). Regarding this view of ship-owners, Steven Jones (Jones, 2006), in his book *Maritime Security* considers that the importance of security for them is *"the state of a shipping company/vessel/crew/port, being of feeling secure"*, or *"the safety of a shipping company/vessel/crew/port against such threats as terrorism, piracy, and other criminal activities".* Moreover, the UN Secretary-General concludes that there is no commonly accepted definition of maritime security, and seeks to identify actions that are commonly considered to deliberate threats to maritime security, rather than clearly defining the term "security" (United Nations, 2008).

2. MARITIME SECURITY REGULATIONS

In response to the events of September 11, 2001 in the United States, as well as a host of ongoing problems in maritime transport of people and cargo, due to rising terrorism, crime, piracy, etc., several regulations have been drawn up that are mandatory or voluntary, aiming at security of the maritime industry (Chalk, 2008; Bichou, 2008).

The main regulatory measures that have been adopted and implemented are included in the IMO security package, including a new Chapter XI-2 on the Convention on the Safety of Life at Sea (SOLAS) Convention (1974/1988), which specifically mentions measures to achieve security in maritime transport. This is the of International Ships and Port Facilities Security (ISPS) Code, (DNV-GL, 2016) which describes the minimum security measures for vessels, port facilities and public organizations, to identify security threats and take the necessary preventive measures to deal with incidents that affect vessels or ports. The Code is a two-part document. The first part includes mandatory requirements, while the second part provides instructions for implementation. The Code was adopted by 108 members of the SOLAS Convention at a meeting in London on 12 December 2002. The Code entered into force on 1 July 2004, with application for ships for international travel, such as passenger ships, cargo ships with a total capacity of 500 GT or more, mobile offshore drilling, as well as in port facilities for such ships.

The ISPS code aims to: a) Cooperation between governments, public services, local governments and shipping industry companies to identify security threats and implement security measures. b) Determine the relevant roles and responsibilities of collaborating governments, public services, local government and shipping companies, nationally and internationally to achieve maritime security. c) The timely and effective collection and exchange of information related to security. d) The preparation of a procedure for security evaluation, by creating plans and procedures for the various forms of security. e) Create the necessary confidence for taking appropriate and specialized security measures at sea.

It is clear that this code aims to create appropriate collaborations for a timely and effective collection and exchange of security information, through a security assessment process, with appropriate plans and procedures for the various forms of security (IMO, 2003). The Automatic Identification System (AIS) and the Long Range Identification and Tracking (LRIT) can be used in the context of IMO, which improve navigation security levels and the marine industry security in general. These systems are designed to provide useful information (such as identity, type, location, course, speed, navigation status) of the ship to other ships and to port authorities (IMO, 2004; IMO, 2009).

However, the previously mentioned procedures for the implementation of security measures have not been introduced in a scientific way and scientific justification, while the issue of cost and benefit assessment has not been estimated systematically and in advance. In addition, the contribution of security measures to reduce risk should be made clearer. Given that both the proposed security measures and their implementation are based on procedures and criteria for assessing the acceptable level of risk, it is clear that a risk-based approach is needed to assess the effectiveness of security measures.

3. SECURITY VULNERABILITIES EXISTING IN THE MARITIME INDUSTRY

Terrorism, piracy, robbery, cyber-threats on ships and port facilities, smuggling and drug trafficking through cargo and containers, etc., are some of the international security problems due to which the national maritime security services must provide solutions. **Cyber-attack is considered particularly important due to the possibility to be combined with all the other security vulnerabilities in the shipping industry**.

It is clear that there is a range of vulnerabilities in the maritime supply chain, which makes it particularly vulnerable to malicious acts resulting from criminal and terrorist elements. It is therefore necessary for the national maritime security services to take appropriate security measures in order to reduce the threats posed by these malicious elements. Figure 1 shows the vulnerabilities in maritime transport and port facilities.

Figure 1. Security vulnerabilities of maritime transport and ports.

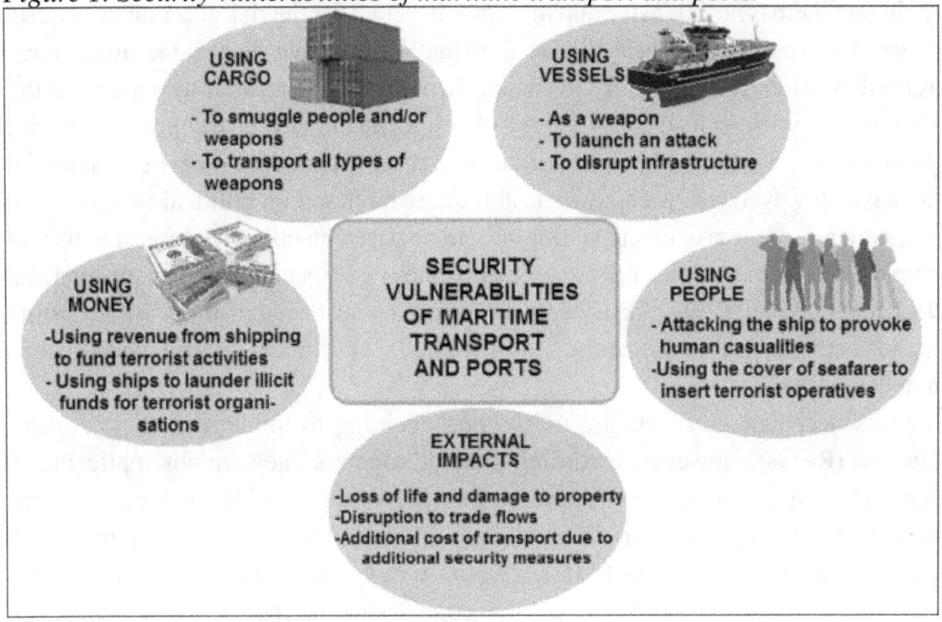

Source: Created by the author, based on OECD (2003).

4. MARITIME SECURITY ASSESSMENT

The historical past of maritime safety in terms of regulations, methods and guidelines goes back to IMO research and methodology specifically for the Formal Safety Assessment (FSA) procedure. This is a systematic methodology structured to use risk assessment and economic analysis in order to improve maritime safety, which includes the protection of human life and health, as well as the protection of property and the marine environment. However, it seems necessary to develop a similar security risk management process to be combined with that of the security risk management. Bichou (2008) and Brooks & Pelot (2008) consider that the safety risk assessment and management procedures could also be used as security risk assessment and management procedures. Also, Parnell et al. (2007) argue that it is possible to assess, evaluate and manage together security and safety risks, which can be implemented in a common framework that generally covers the social, environmental and economic dimensions. In addition, Lambert (2007) examines the possibility of combined actions and the allocation of financial resources to address both security and safety in the maritime industry, given their interdependence. It is therefore logical to use similar approaches and practices to address both security and safety issues in the shipping industry.

In fact there is no extensive and in-depth literature for the risk acceptance criteria concerning maritime security. Basic information is drawn from other areas, such as civil aviation. It is necessary to standardize these criteria, as is the case with the FSA, where practitioners know how to gather information, to make comparisons with previous experience and to make decisions that are often based on experience from the past. However, risk acceptance criteria may be based on political or subjective assessments, while risk identification should be based on objective assessments and be performed prior to the FSA. Their usefulness is especially important for final decision making. Both risk acceptance criteria and decision-making determine the measurable risk assessment, as well as the level of necessary financial cost to reduce the risk.

Risk acceptance criteria are used when deciding to implement Risk Control Options (RCOs). However, according to some scholars, they are not applicable in cases of saving human life or reducing marine pollution, while they can be used only when making decisions about choosing, for example, rescue equipment and not the rescue operation itself. This is because in a rescue operation, all resources for human lives must be available without considering the cost (Skjong et al., 2007).

Cost-benefit Analysis (CBA) is usually added to these processes for the best possible decision making. It is essentially a basic complement to the process for the optimization of the result. The FSA process introduced the CBA through its official guidelines approved by the IMO in 2001 and incorporated into many well-known maritime security assessment systems (OECD, 2003; US Maritime Transportation Security Act, 2002; UK Regulatory Impact Assessments, 2005)

Therefore, regarding maritime security, it is necessary to set specific and specially defined risk criteria, as the security risk assessment is not comparable to the traditional risks of maritime safety. Thus, as with the FSA case, it is proposed by some researchers, a **five-step risk assessment process** for the maritime security case: i) Hazards identification (HAZID), ii) Risk assessment (RA), iii) Risk management with alternative risk control options (RCOs), iv) Cost-benefit analysis (CBA) and iv) Recommendations for decision making. Also, Yang et al. (2016) proposed a six steps Quantitative Maritime Security Assessment (QMSA).

Regarding the **existing tools for Risk Assessment**, usually the Bayesian Network (BN) Influence Diagrams, the Fault Tree analysis (FTA), the Event Tree analysis (ETA) and the Risk Contribution Tree (RCT) are used.

A far as for the existing **methods for Risk Assessment** concerns, a 4╳7, 5╳5 or 3╳3 or any other type of Risk Index Matrix (or Severity/Frequency Matrix) for initial ranking of different scenarios, is constructed (see Figure 2).

Figure 2. Example of calculated Risk Index Matrix of type 5X5.

FREQUENCY \ SEVERITY	Negligible	Minor	Moderate	Significant	Severe
Very Likely	Low Med	Medium	Med Hi	High	High
Likely	Low	Low Med	Medium	Med Hi	High
Possible	Low	Low Med	Medium	Med Hi	Med Hi
Unlikely	Low	Low Med	Low Med	Medium	Med Hi
Very Unlikely	Low	Low	Low Med	Medium	Medium

Source: Created by the author.

Also, to meet the **requirements for risk criteria**, different types of risk expression are commonly used. However, the "As Low As Reasonably Practicable" (ALARP) principle is considered to be the commonly accepted principle, which accepts a maximum level of risk as a limit above which the risk is considered "intolerable", cannot be justified and should be limited, regardless of economic cost. There is also a lower limit where, according to the same principle, the level of risk is considered as "broadly accepted", so the risk is considered negligible and no reduction is required. Between these two extreme limits of this principle, and based on the economically permissible possibility, the risk must be reduced to reasonable levels (see Figure 3) (Kristiansen, 2004). It is noted that the proposed risk reduction measures should have a technical basis for their implementation, while, as mentioned above, there should be a balance between the benefits of risk reduction and the financial costs required. This is usually examined during the CBA.

Figure 3. The ALARP principle.

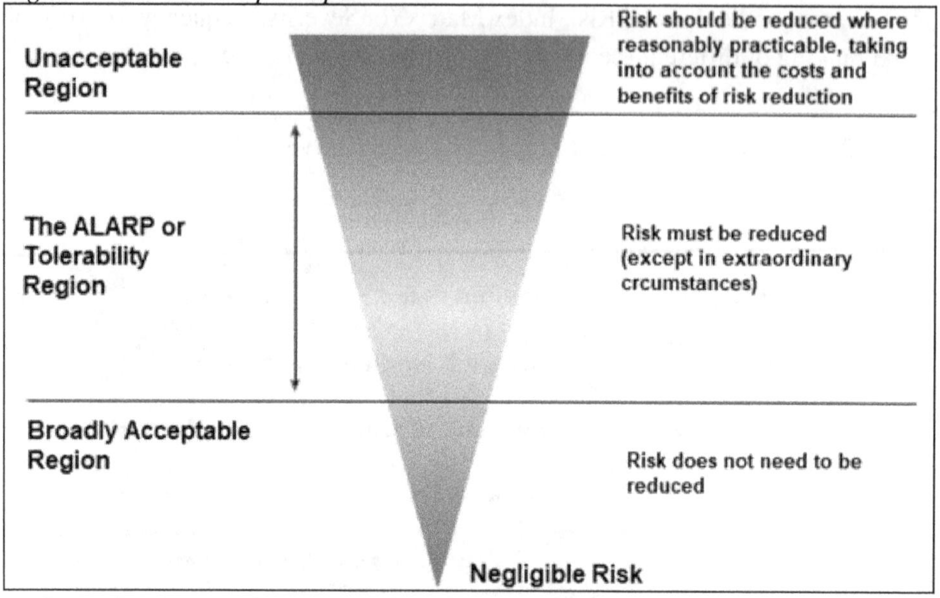

Source: Common carrot diagram, modified graphically and descriptively by the author.

Risk management regulations were implemented in the United Kingdom in the 1970s and in the European Union and Australia in the 1990s, and have been incorporated into ISO 31000 standard. Many high-risk industries use this risk management standard and through this, they develop, implement and improve their risk management framework, following the provided guidance (ISO, 2018a). Also, ISO 20858 for Maritime port facility security assessments and security plan development was developed (ISO, 2007).

Additionally, the ISO 27000 family of standards provides a complete system for **information security risk management**. It uses the existing experience from other existing quality system standards, but in the same time it introduces new approaches for the management of complex cases of cyber and information security (Risk Review, 2018).

ISO 27005 standard was prepared by the ISO / IEC JTC1 Joint Technical Committee, Information Technology, SC 27 subcommittee, IT Security Techniques. ISO 27005 and includes guidance for security Risk Management (ISO, 2011). This standard includes both qualitative and quantitative risk analysis options, but does not provide detail about methods and their application (Agrawal, 2017).

Although ISO 31000 and ISO 27005 risk management processes does not specify any specific risk management method, they imply a continual information risk management process based on six key components:

1. Context establishment
2. Risk assessment
3. Risk treatment
4. Risk acceptance
5. Risk communication and consultation
6. Risk monitoring and review

Figure 4 provides an overview of ISO 31000 and ISO 27005 (ISO, 2018; ISO, 2011) and Figure 5 provides a risk management programme according to ISO 31000 and ISO 27005.

Figure 4. Overview of ISO 31000 and ISO 27005.

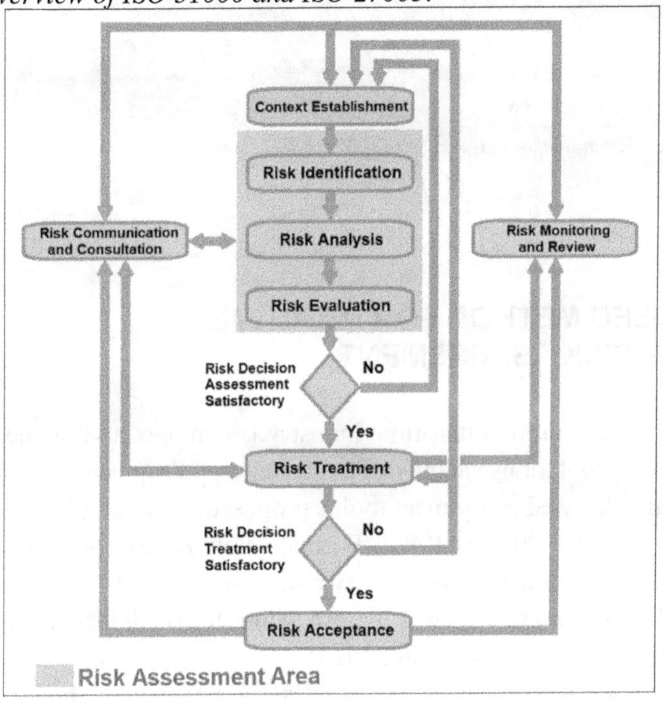

Source: Created by the author, based on ISO 31000 and ISO 27005.

Figure 5. Risk management programme according to ISO 31000 and ISO 27005.

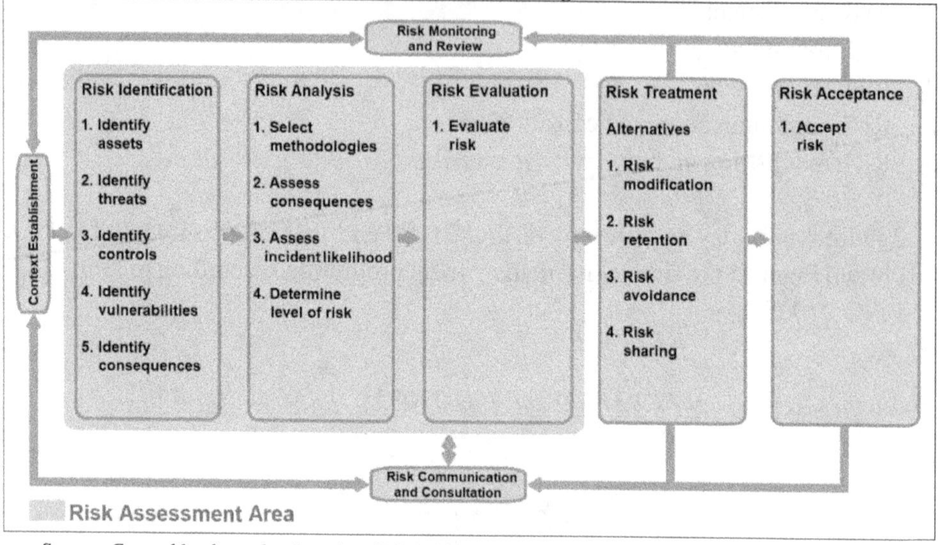

Source: Created by the author based on ISO 31000 and ISO 27005.

5. PROPOSED METHOD FOR MARITIME SECURITY RISK ASSESSMENT

Taking into account that maritime industry is complex and the necessity for a decision-making tool for use at the different stages of design and operation is obvious, a special risk-based assessment tool is proposed, following Yang et al. (2016), ISO 31000 and ISO 27005 and IMO's Formal Safety Assessment guidelines based on studies focusing on maritime security risk quantification. This framework will include **Risk Assessment** which is the procedure for risk identification, analysis, and evaluation (Makrodimitris et al., 2014).

For our research work the Bow-Tie visual risk assessment method will be applied using the **Bow-Tie diagram** technique by which both incident prevention barriers and consequence reduction barriers are identified (DNV-GL, 2016). Specifically: "*A Bow-Tie diagram visualizes the risk and creates a clear differentiation between the proactive and reactive side of risk management. Specifically, the bowtie diagram provides an overview of multiple plausible incident scenarios and shows the barriers controlling these scenarios. Bow-Tie software will get us started to identify and manage our risks in 6 simple steps: 1) Add a Hazard, 2) Add a Top Event, 3) Add Threats, 4) Add Consequences, 5) Add Prevention Barriers, 6) Add Mitigation Barriers.*".

A **specially developed software (MarSec)** was used, that enables the creation of bowties diagrams and the necessary calculations to assess risk in Visual Basic 6 programming language. As it was mentioned previously, the proposed method provides a **technical approach** to the required prevention and mitigation measures. **Security Cost-Benefit Analysis** (as C-B) is also required since the required measures can increase costs and required shipping time and additionally, possible negative economic effect for supporting rational security policymaking needs to be investigated.

5.1. Framework's General Description

The **special risk-based assessment tool is proposed for development,** which performs **security risk assessment, through risk identification, risk analysis and risk evaluation,** integrating studies related with the quantification of maritime security risk, is shown descriptively and schematically in Figure 6. As it is shown:

Risk Identification consists of:

1. Hazard identification (Hazard addition in the Bow-Tie diagram).
2. Top Event identification (To Event addition in the Bow-Tie diagram).
3. Threats identification (Threats addition in the Bow-Tie diagram).
4. Consequences identification (Consequences addition in the Bow-Tie diagram).
5. Prevention measures identification (Prevention barriers addition in the Bow-Tie diagram).
6. Mitigation measures identification (Mitigation barriers addition in the Bow-Tie diagram).

Barriers are a means of prevention or mitigation for any negative outcome and can reduce the occurrence likelihood of the latter. Prevention barriers aim to prevent the top event and ultimately the release of the hazard. Mitigation barriers aim to reduce the likelihood or minimise the severity of the consequences.

Risk analysis consists of:

1. Top Event risk estimation separately for People, Environment, Assets and Reputation (using a 5×5 risk matrix in the Bow-Tie diagram).
2. Consequences risk estimation separately for People, Environment, Assets and Reputation (using a 5×5 risk matrix in the Bow-Tie diagram).

Risk evaluation consists of:

1. Check prevention barriers effectiveness (Prevention barriers effectiveness calculations in the Bow-Tie diagram).
2. Check mitigation barriers effectiveness (Mitigation barriers effectiveness calculations in the Bow-Tie diagram).

From the previous description it is clear that prevention and mitigation measures (barriers) identification is very important in order to reduce the estimated Top Event risk and Consequences risk to a desired level. Thus, the focus in the Bow-Tie analysis is mainly on the barriers.

Figure 7 represents graphically the compatibility of the proposed Risk Assessment procedure using the Bow-Tie method, with ISO 31000 and ISO 27005. This Figure could be compared with Figure 6, illustrating graphically a Risk management programme according to ISO 31000 and ISO 27005.

Figure 6. Proposed flow diagram for risk assessment using the Bow-Tie method.

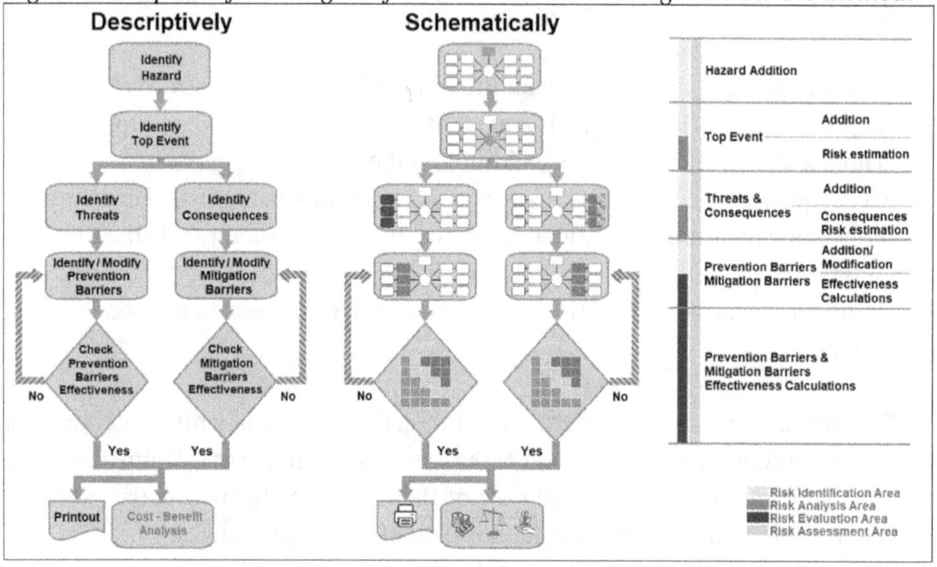

Source: Created by the author.

Figure 7. Compatibility of proposed risk assessment procedure with ISO 31000 and ISO 27005, using the Bow-Tie method.

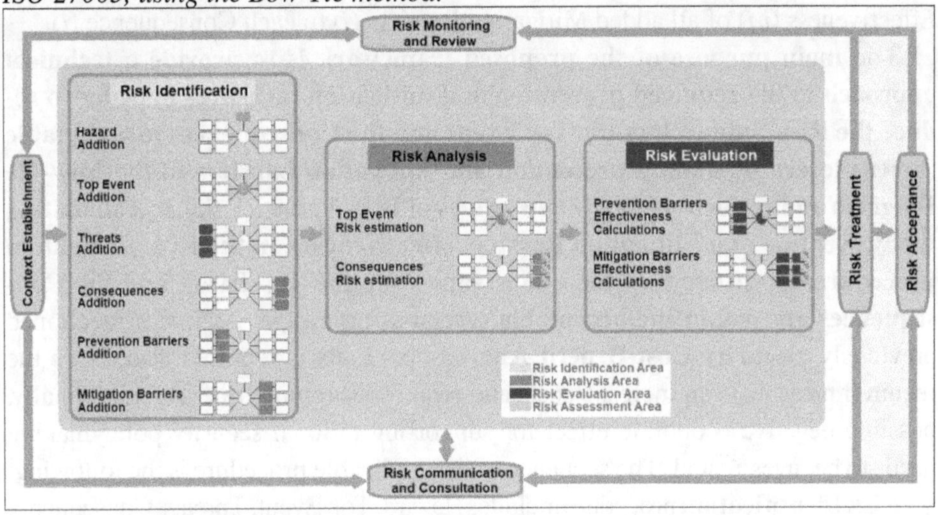

Source: Created by the author.

5.2. Required Calculations and Used Data

For the purpose of this study, we define the following: The **Risk Value (*RV*)**, which includes information about the Frequency and the Severity of the main unwanted event (Top Event) and all resulting Consequences. For this scope we use a 5╳5 risk matrix (see Figure 8). Thus, the range of arithmetic values of Risk Value is [1 – 25].

The **Threat Index (TI_i)** for each Threat (*i*), which express the partial effect of each Threat (*i*) on the Top Event, resulting from the Probability and Contribution of each Threat (*i*) (see Figure 9). Arithmetic values of probabilities given by the user are integer numbers within the range [1 – 6], while arithmetic values of contributions given also by the user are integer numbers within the range [1 – 3]. The Threat's Probability values vs the described verbally probability is almost logarithmic (see diagram in Figure 9). The sum of all Threat Indices expressing the total effect of all Threats on the Top Event is one (1). It is clear that there are no similar indices for the Consequences.

The **Risk Value Reduction coefficient (R_{redi})** for each Threat (*i*), which represents the degree of the reduction of the Risk Value, resulting from the Effectiveness (*i,j*) of all added Prevention Barriers (*j*) to each Threat (*i*) (see Figure 10). Arithmetic values of Effectiveness given by the user are integer numbers within the range [1 – 6]. The Barrier's Effectiveness values vs the described verbally Effectiveness is assumed as linear (see diagram in Figure 10). Similarly, there are separate **Risk Value Reduction coefficients (R_{red})** for each Consequence (*i*), resulting from the Effectiveness (*i,j*) of all added Mitigation Barriers (*j*) to each Consequence (*i*).

The main purpose of the proposed framework is to provide a technical approach to the required prevention and mitigation measures in order to reduce the Risk Value *RV* of the Top Event and the Consequences to acceptable (green) levels, by adding prevention and mitigation barriers in the Bow-Tie Diagram and recalculating the new values of Risk Value *RV* (i.e. by calculating the Prevention and Mitigation Barriers effectiveness), in the Risk Evaluation procedure, to ensure that these new values for the Top Event and the Consequences are within the acceptable (green area) range. As it was mentioned previously, **Security Cost-Benefit Analysis** (as C-B) is required also, since the required measures can increase costs and required shipping time and additionally, possible negative economic effect for supporting rational security policymaking needs to be investigated. The sequence of the applicable procedure is the following:

Risk Identification procedure includes Hazard, Top Event, Threats, Consequences, Prevention Barriers and Mitigation Barriers additions in the Bow-Tie Diagram. **Risk Analysis** procedure – which is carried out in the same time with Top Event and Consequences additions during Risk Identification procedure – provides the Top Event risk estimation and the Consequences risk estimation, separately for People, Environment, Assets and Reputation, by the **Risk Value *RV*** calculation, combining **Frequency** and **Severity**, as follows:

$$RV = Frequency \times Severity \qquad (1)$$

The selection of the proper Frequency and the proper Severity depends on the user's experience. As shown in Figure 8, the range of Risk Values is [1 – 25]. The range of the acceptable green area is [1 – 6], the range of the tolerable yellow area is (6 – 15) and the range of the unacceptable red area is [15 – 25].

Figure 8. Risk analysis for top event and consequences, using 5x5 risk matrix.

	Frequency	Severity
People	A: Very Unlikely B: Unlikely C: Possible D: Likely E: Very Likely	1: Slight Injury 2: Minor Injury 3: Major Injury 4: Single Death 5: Multiple Deaths
Environment	A: Very Unlikely B: Unlikely C: Possible D: Likely E: Very Likely	1: Slight Effect 2: Minor Effect 3: Moderate Effect 4: Major Effect 5: Massive Effect
Assets	A: Very Unlikely B: Unlikely C: Possible D: Likely E: Very Likely	1: Slight Damage 2: Minor Damage 3: Moderate Damage 4: Major Damage 5: Extensive Damage
Reputation	A: Very Unlikely B: Unlikely C: Possible D: Likely E: Very Likely	1: Slight Impact 2: Minor Impact 3: Moderate Impact 4: National Impact 5: Global Impact

5×5 Risk Matrix

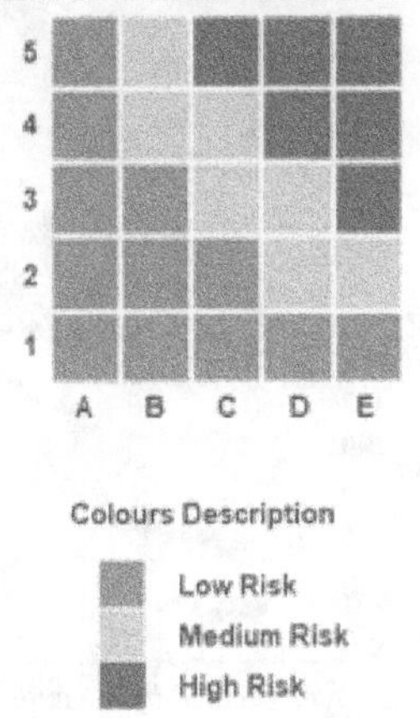

Arithmetic values for Frequencies	A: 1, B: 2, C: 3, D: 4, E: 5
Arithmetic values for Severities	1: 1, 2: 2, 3: 3, 4: 4, 5: 5
Risk Value (RV) = Frequency × Severity	Range: [1 – 25]
Green Area: [1 – 6], Yellow Area: (6 – 15),	Red Area: [15 – 25]

Source: Created by the author.

Figure 9. Threat's probability and contribution.

	Probability P	Value	Contribution C	Value
Scaling of Threats' Probabilities and Contributions	Once per year or more	1		
	Once per six months	2	Low	1
	Once per three months	3	Medium	2
	Once per month	4	High	3
	Once per week	5		
	Once per day	6		
	Undefined	0	Undefined	0

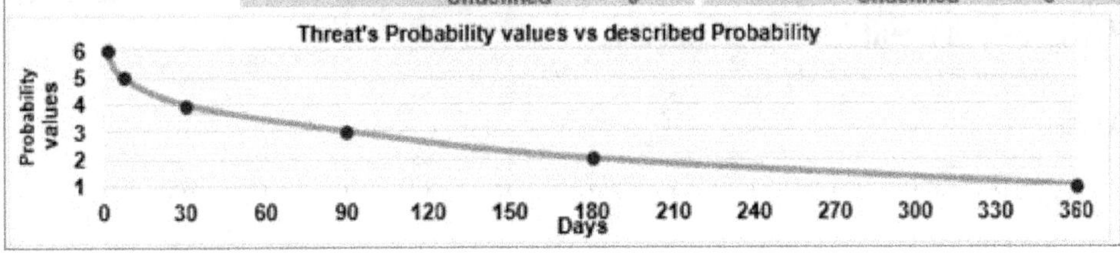

Source: Created by the author.

Figure 10. All types of barriers' effectiveness.

	Effectiveness E	Value	Value
Barrier's Effectiveness Scaling	Very Poor	1/6	= 0.17
	Poor	2/6	= 0.34
	Moderate	3/6	= 0.50
	Good	4/6	= 0.67
	Very Good	5/6	= 0.84
	Excellent	6/6	= 1.00
	Undefined	0	

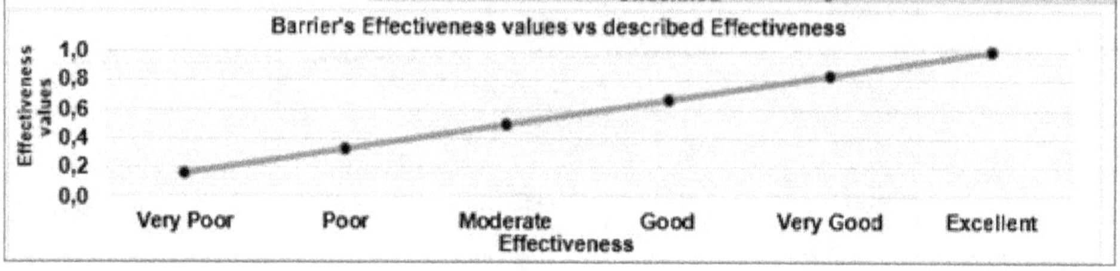

Source: Created by the author.

During the Threats addition, user must define by experience, the **Probability** $\mathbf{P_i}$ and the **Contribution** $\mathbf{C_i}$ of each **Threat (i)**, using the probability and contribution scaling presented in Figure 9. Also, during the Prevention Barriers addition, user must define by experience, the **Effectiveness** value $\mathbf{E_{ij}}$, of each **Prevention Barrier (j)** for the same defined **Threat (i)**, using the effectiveness scaling presented in Figure 12. It is clear that risk evaluation is based mainly on the accuracy of the definition of the **Probability** $\mathbf{P_i}$ and the **Contribution** $\mathbf{C_i}$ of each **Threat (i)** as well as on the accuracy of the definition of the **Effectiveness** value $\mathbf{E_{ij}}$, of each **Prevention Barrier (j)** for the same defined **Threat (i)**, set by the user and based on experience.

Risk Evaluation for the Top Event is the next step, by calculating the Prevention Barriers Effectiveness. For the Top Event Risk Evaluation we define **Threat Index** $\boldsymbol{TI_i}$ as the partial effect of each **Threat** $(\boldsymbol{i})$ on the Top Event. Thus, the **Threat Index** $\boldsymbol{TI_i}$ for each **Threat** $(\boldsymbol{i})$, is expressed as:

$$TI_i = \frac{P_i \times C_i}{\sum_{i=1}^{n} [P_i \times C_i]} \quad (2)$$

i.e.

$$TI = TI_1 + TI_2 + TI_3 + \cdots\cdots + TI_n = 1 \quad (3)$$

To estimate the Top Event **resulting Risk Value** $\boldsymbol{RV_{final,}}$ for People, Environment, Assets and Reputation, after the introduction of prevention barriers, we use the **Prevention Barrier (*j*) Effectiveness value E_{ij}**, for **Threat (*i*)** to calculate the **Risk Value Reduction coefficient R_{redi}** of **Threat (*i*)** as follows:

$$R_{redi} = TI_i \times \left[\sum_{j=1}^{k_i} E_{ij}\right] / k_i \tag{4}$$

where (k_i) is the total **Number of Prevention Barriers** for **Threat (*i*)**.

Then the total Top Event **Risk Value Reduction coefficient R_{red}** is calculated from:

$$R_{red} = \sum_{i=1}^{n} R_{redi} \tag{5}$$

And the resulting **Final Top Event Risk Value RV_{final}**, for People, Environment, Assets and Reputation, after the introduction of all prevention barriers is:

$$RV_{final} = (1 - R_{red})\, RV \tag{6}$$

During the Consequences addition, user must define by experience, the **Concern N_i** of each **Consequence (*i*)**, using the concern scaling presented in Figure 11. This value for each ***Consequence (i)*** will be used for Cost-Benefit Analysis later. Also, during the Mitigation Barriers addition, user must define by experience, the **Effectiveness** value E_{ij}, of each **Mitigation Barrier (*j*)** for the same defined **Consequence (*i*)**, using the effectiveness scaling presented in Figure 10. It is clear that risk evaluation is based mainly on the accuracy of **Effectiveness** value E_{ij}, of each **Mitigation Barrier (*j*)** for the same defined **Consequence (*i*)**, set by the user and based on experience.

Figure 11. Consequence's concern.

	Concern	Value
Graduations of Consequences' Concerns	Minor	1
	Medium	2
	Major	3
	Undefined	0

Source: Created by the author.

Risk Evaluation for each one of the Consequences is the next step by calculating the Mitigation Barriers Effectiveness. To estimate the **resulting Consequence Risk Value $RV_{finali,}$** for People, Environment, Assets and Reputation, for **Consequence (*i*)**, after the introduction of mitigation barriers, we use the **Mitigation Barrier (*j*) Effectiveness value E_{ij}**, for **Consequence (*i*)** to calculate the **Risk Value Reduction coefficient R_{redi}** of **Consequence (*i*)** as follows:

$$R_{redi}\left[\sum_{j=1}^{k} E_{ij}\right] / k_i \quad (7)$$

where (k_i) is the total **Number of Mitigation Barriers** for **Consequence** (***i***).

And the resulting **Final Consequence Risk Value** RV_{finali}, for People, Environment, Assets and Reputation, for **Consequence** (***i***), after the introduction of all mitigation barriers is:

$$RV_{finali} = (1 - R_{redi})\, RV_i \quad (8)$$

In the case of undefined probabilities and contributions of threats, undefined levels of effectiveness of prevention and mitigation barriers and undefined risk levels of the Top Event and Consequences, risk evaluation could not be carried out and Bow-Tie Diagram remains as an information diagram only.

Since the use of primary data would not provide the ability to check the accuracy of the results of the framework used, secondary data will be used, mainly from similar frameworks, such as the information and details provided by DNV-GL for similar cases [60], **so that the accuracy of the obtained results can be checked and compared with results obtained for similar cases. DNV-GL provides hints for the way of identification of threats and consequences and lists with details for what prevention and mitigation barriers may include, specifically for cyber-security resilience management for ships and mobile offshore units in operation** [60].

5.3. Cost-Benefit Analysis

The Total Cost of Risk Treatment is the cost of all Prevention Barriers (***CP***) and Mitigation Barriers (***CM***), i.e., the sum of the Cost of each Prevention Barrier (CP_{ij}) and the Cost of each Mitigation Barrier (CM_{ij}) as follows:

Total Cost of Risk Treatment = CP + CM =

$=$

$\sum_{i=1}^{nt}$

$\sum_{j=1}^{kp_i} CP_{ij}$

$+$

$$\sum_{i=1}^{nc}$$

$$\sum_{j=1}^{km_i} CM_{ij} \tag{9}$$

where ($\boldsymbol{kp_i}$) is the total number of Prevention Barriers for Threat ($\boldsymbol{i}$) and ($\boldsymbol{km_i}$) is the total number of Mitigation Barriers for Consequence ($\boldsymbol{i}$), (***nt***) is the total number of Threats and (***nc***) is the total number of Consequences. For similar barriers their cost is taken once.

The Total Cost in the case of occurrence of the Top Event and all the Consequences is given by:

CTE + CCO =

$$= CTE_p + CTE_e + CTE_a + CTE_r +$$

$$\sum_{i=1}^{nc} \left[CCOpi + CCOei + CCOai + CCOri\right] \tag{10}$$

where (***CTE***) is the cost of the Top Event, (***CCO***) is the cost of all Consequences (from $\boldsymbol{i}$ = 1 to ***nc***) and indicators (***p***), (***e***), (***a***) and (***r***) denote people, environment, assets and reputation.

Then for a Factor ($\boldsymbol{f < 1}$) set by the maritime company, according to its economic policy, and ethics, it must be:

$$\textit{Total Cost of Risk Treatment} < f \cdot [CTE + CCO] \tag{11}$$

6. CONTRIBUTION TO KNOWLEDGE AND NOVELTY

The new scenario-based risk model, and the corresponding mitigation measures will be recorded as a 'live' database [40]. Specifically, the proposed 'scenario-based' framework: i) provides the user with quick results and information, while requiring a limited amount of data, ii) provides the possibility of immediate changes in the data provided if the user so wishes, iii) provides the ability to create and maintain a history of changes that the user can refer to, iv) the way of calculating the effectiveness of the prevention and mitigation barriers is direct and understandable enabling the user to make improvements whenever required, v) compared with the existing frameworks (mainly for safety risk assessment), it is easier and flexible. Additionally, the novelty of using the Bow-Tie technique for risk identification [60] "*lies in the fact that:* a) *it is developed with the end-user in mind, making it one of the most visually helpful risk assessment tools, b) it reduces complexity to a manageable size without losing context and focus on the critical elements, c) filters allow users to decide which information they want to display on their diagram, without deleting any information, d) allows users to link their management systems to their bowtie diagram by creating activities and document links, e) enables users to maintain the diagram so it will always represent the current status of safety barriers, f)* it is *possible to run reports based on users' customized template.*"

7. A CYBER SECURITY EXAMPLE: MALWARE RELATED INCIDENT

This application is related to malware attacks in a medium-sized maritime company and thus, malware related incidents are the Top Event of our example. Threats are related to the way that the malware can get into the system. Normally, there are mainly three ways: i) through the network (worms), ii) through removable storage media and iii) through ways based on user's behaviour. In this example, the barriers to prevent a worm are network segregation and antivirus software. Blocking USB ports and the use of antivirus software are barriers to prevent malware through removable storage media, but also awareness training is considered as a prevention barrier too. Lastly, e-mail washing and proper antivirus software are the proper barriers to avoid malware.

The consequence of these malware related incidents is the system unavailability. To avoid this consequence, a system disconnection procedure is required, as well as a backup and restore system and an antivirus alarm as mitigation barriers.

All the required data and details for the Risk Identification and Cost Benefit Analysis, are summarized in the following Tables.

Table 1. Hazard and top event data.

Hazard and Top Event	Frequency F	Severity S	Risk Value	Cost CTE
Hazard: Electronic system vulnerable to cyber security incidents				
Top Event: Malware related incident				
Risk Value for People	**1**	**1**	**1 → A1**	**0 €**

continued on following page

Table 1. Continued

Hazard and Top Event	Frequency F	Severity S	Risk Value	Cost CTE
Risk Value for Environment	1	1	1 → A1	0 €
Risk Value for Assets	4	4	16 → D4	2,000,000 €
Risk Value for Reputation	4	3	12 → D3	500,000 €

Table 2. Threats' data.

Threats	Probability P	Contribution C
Malware spread through network	Once per Year	High
Antivirus Software	Once per Year	High
Awareness Training	Once per 6 months	High

Table 3. Consequences data.

Consequences	Frequency F	Severity S	Risk Value	Cost CCO
System unavailability → Concern: Major				
Risk Value for People	1	1	1 → A1	0 €
Risk Value for Environment	1	1	1 → A1	0 €
Risk Value for Assets	4	4	16 → D4	1,800,000 €
Risk Value for Reputation	4	4	16 → D4	1,200,000 €

Table 4. Required barriers' data.

Required Barriers	Effectiveness E	Capital Cost	5 years maintenance	Total CP + CM
Blocking USB ports	Good	10,000 €	15,000 €	~25,000 €
Antivirus Software	Very Good	25,000 €	50,000 €	~75,000 €
Awareness Training	Good	5,000 €	40,000 €	~45,000 €
Network Segregation (VLANs, firewalls)	Very Good	40,000 €	30,000 €	~70,000 €

continued on following page

Table 4. Continued

Required Barriers	Effectiveness E	Capital Cost	5 years maintenance	Total CP + CM
e-mail Washing (secure e-mail gateway)	Very Good	15,000 €	40,000 €	**~55,000 €**
System Disconnection (air-gapping)	Very Good	20,000 €	10,000 €	**~30,000 €**
Backup-Restore System (on-site/cloud hybrid)	Good	30,000 €	50,000 €	**~80,000 €**
Antivirus Software Alert (SIEM System)	Moderate	10,000 €	20,000 €	**~30,000 €**

Hazard and Top Event are defined and their cost is estimated, using the forms of Figures 12 and 13, while the complete Bow-Tie diagram for this application is shown in the following Figure 14, as generated by the previously mentioned software MarSec.

Figure 12. Risk Matrices for the selection of frequency and severity for people, environment, assets & reputation.

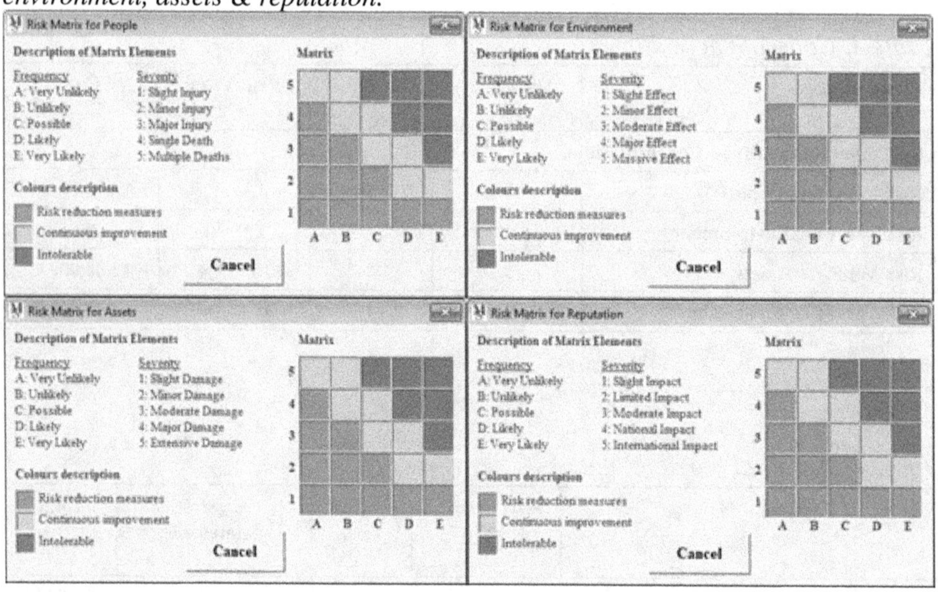

Figure 13. Hazard identification and top event form before the introduction of threat barriers.

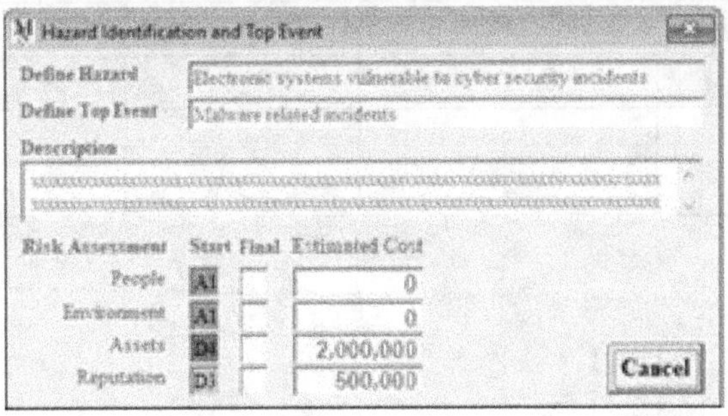

Figure 14. Bow-Tie diagram for electronic systems vulnerable to unauthorized external user.

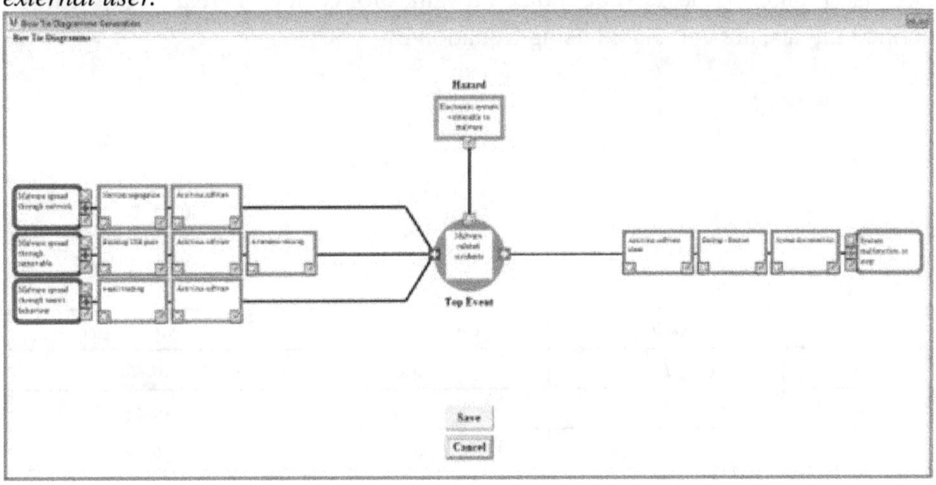

For the Top Event, the **Frequency** and **Severity** of each one of People, Environment, Assets and Reputation is selected from the corresponding Risk Matrix form as shown in Figure 12, by clicking the proper square in the Risk Matrix (see also Figure 8).

The proper colour appears in the lower part of the Hazard identification and Top Event Form of Figure 13. The **Risk Value** for each one of People, Environment, Assets and Reputation is calculated using Equation 1:

$RV = Frequency \times Severity$

Table 5. Frequency and severity

	Frequency	Severity	Risk Value RV
People	1	1	1
Environment	1	1	1
Assets	4	4	16
Reputation	4	3	12

i.e.:

$RV_P = A1 = 1 \times 1 = 1$

$RV_E = A1 = 1 \times 1 = 1$

$RV_A = D5 = 4 \times 4 = 16$

$RV_R = D4 = 4 \times 3 = 12$

Three threats are then defined, using the forms of Figures 15, 16 and 17. From the described three threats (**n = 3**) and their **Probabilities P_i** and **Contributions C_i** (see also Figure 9), the **Threat Index TI_i** values for each of Threat (i) of the three defined threats, are calculated using Equation 2:

$$TI_i = \frac{P_i \times C_i}{\sum_{i=1}^{n} [P_i \times C_i]}$$

Table 6. Threat index

Threat i	Probability P_i	Contribution C_i	Threat Index TI_i
1	1	3	0.25
2	1	3	0.25
3	2	3	0.50

i.e.:

$TI_1 = (1 \times 3)/12 = 0.25$

$TI_2 = (1 \times 3)/12 = 0.25$

$TI_3 = (2 \times 3)/12 = 0.50$

and:

$TI = TI_1 + TI_2 + TI_3 = 0.25 + 0.25 + 0.50 = 1$

Figure 15. 1st Threat identification.

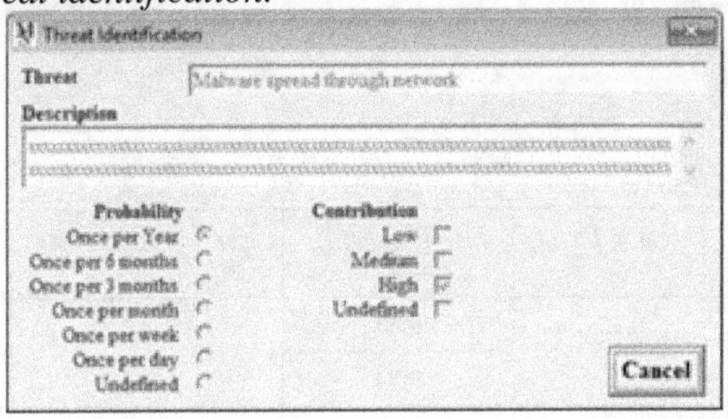

Figure 16. 2nd Threat identification.

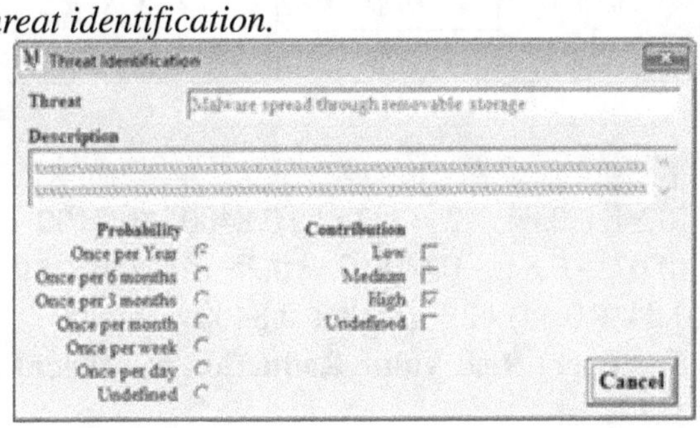

Figure 17. 3rd Threat identification.

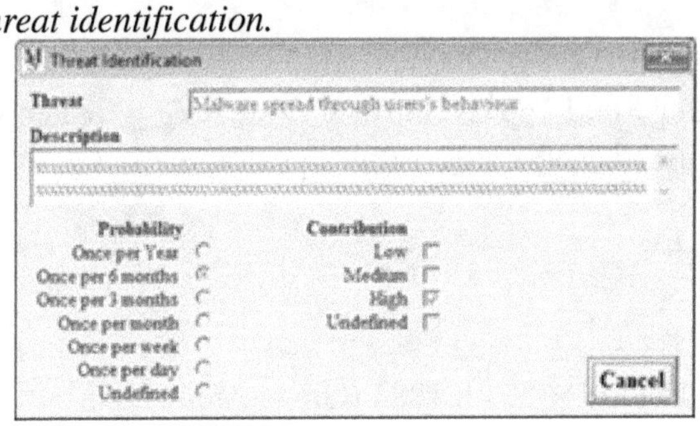

The prevention barriers for each threat are then defined and their cost is estimated, using the forms of Figures 18, 19, 20, 21, 22, 23 and 24. From the values of the **Effectiveness E_{ij}** of each Barrier (j) for each Threat (i) (see also Figure 10), the **Risk Value Reduction coefficient R_{redi}** of each Threat (i) are calculated using Equation 4:

$$R_{redi} = TI_i \times \left[\sum_{j=1}^{k_i} E_{ij}\right] / k_i$$

Table 7. Barrier effectiveness

Threat i	*Barrier j*	*Threat Index TI_i*	*Effectiveness E_{ij}*	*Number of Barriers k_i*	*Reduction Coeff. R_{redi}*
1	**1**	**0.25**	**0.84**	**2**	**0.2100**
	2		**0.84**		
2	**1**	**0.25**	**0.67**	**3**	**0.1817**
	2		**0.84**		
	3		**0.67**		
3	**1**	**0.50**	**0.84**	**2**	**0.4200**
			0.84		

i.e.:

$R_{red1} = 0.25 \times (0.84 + 0.84) / 2 = 0.25 \times 0.8400 = 0.2100$

$R_{red2} = 0.25 \times (0.67 + 0.84 + 0.67) / 3 = 0.25 \times 0.7267 = 0.1817$

$R_{red3} = 0.50$ x $(0.84 + 0.84) / 2 = 0.50 \times 0.8400 = 0.4200$

Then, the total Top Event **Risk Value Reduction coefficient R_{red}** is calculated using Equation 5:

$$R_{red} = \sum_{i=1}^{n} R_{redi}$$

i.e.:

$R_{red} = 0.2100 + 0.1817 + 0.4200 = 0.8117$

Then the resulting **Final Top Event Risk Value RV_{final}**, for People, Environment, Assets and Reputation is calculated using Equation 6:

$$RV_{final} = (1 - R_{red})\ RV$$

i.e.:

$RV_{finalP} = (1 - 0.8117) \times 1 = 0.1883$ (Green Area)

$RV_{finalE} = (1 - 0.8117) \times 1 = 0.1883$ (Green Area)

$RV_{finalA} = (1 - 0.8117) \times 16 = 3.0128$ (Green Area)

$RV_{finalR} = (1 - 0.8117) \times 12 = 2.2596$ (Green Area)

Figure 18. 1st Barrier for 1st threat.

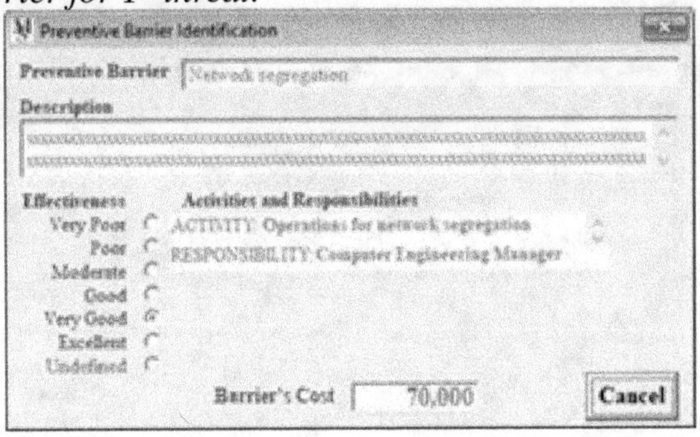

Figure 19. 2nd Barrier for 1st threat.

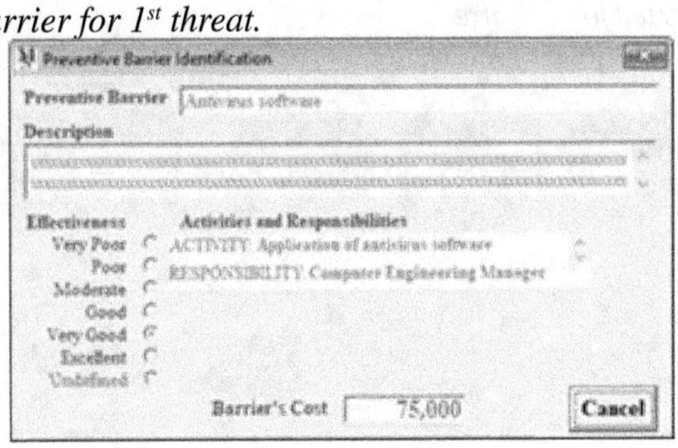

Figure 20. 1st Barrier for 2nd threat.

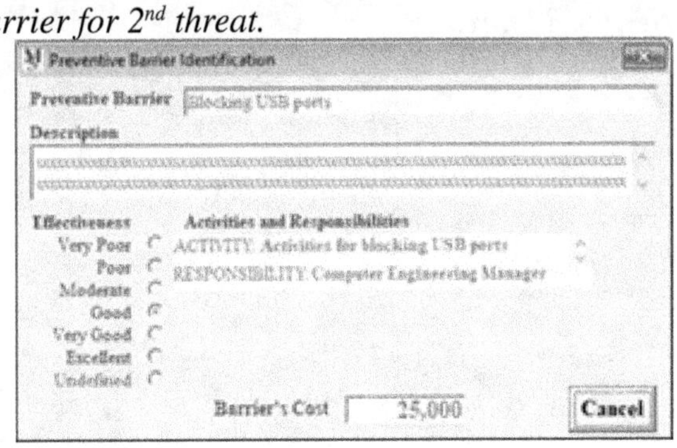

Figure 21. 2nd Barrier for 2nd threat.

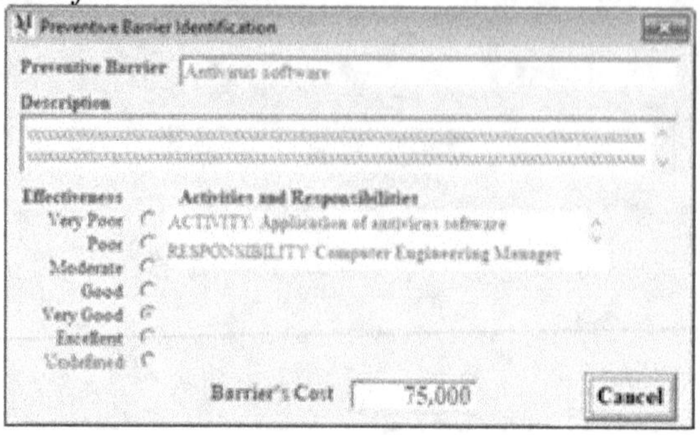

Figure 22. 3rd Barrier for 2nd threat.

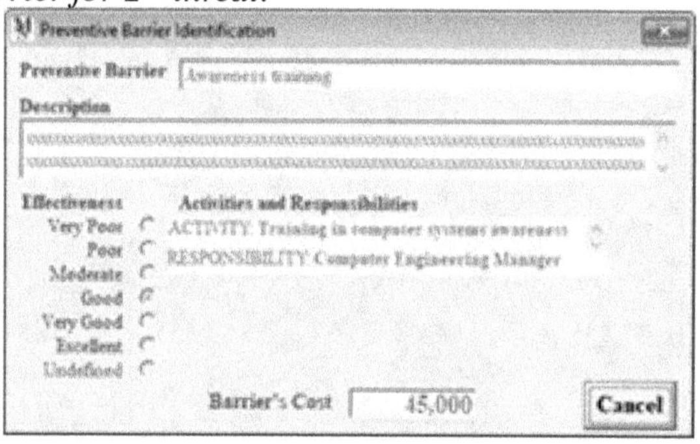

Figure 23. 1st Barrier for 3rd threat.

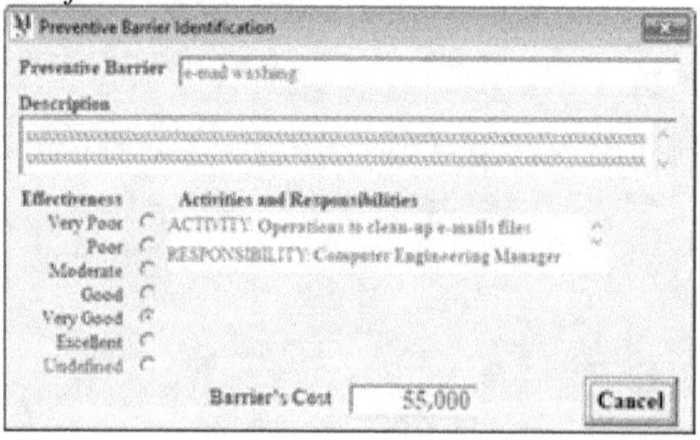

Figure 24. 2nd Barrier for 3rd threat.

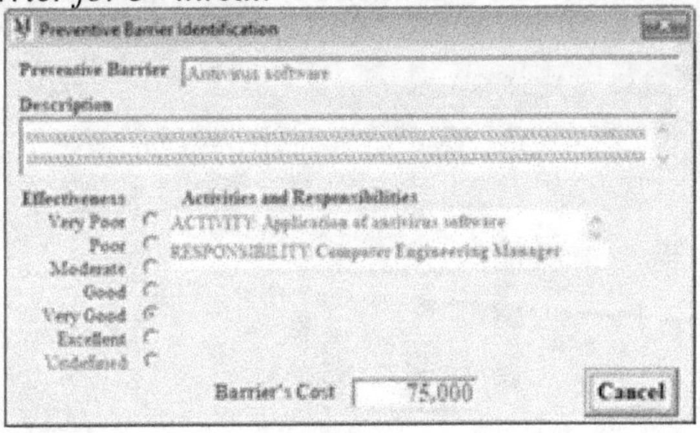

The resulting Hazard Identification and Top Event form, after the introduction of the Threats Barriers, takes the form of Figure 25, with all Final Top Event Risk Values colours in green.

Figure 25. Hazard identification and top event after the introduction of all threat barriers.

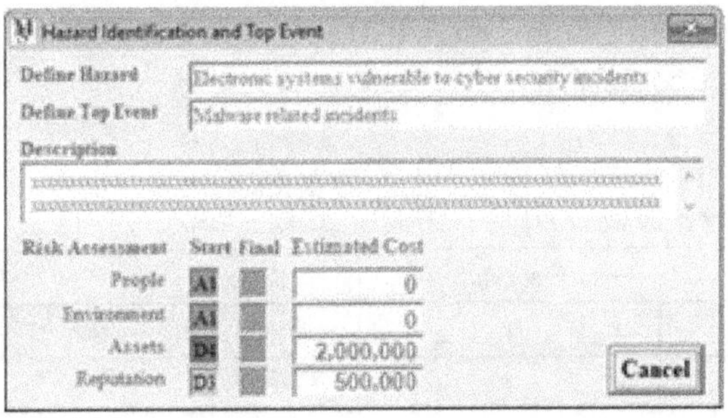

Similarly, the 1st (and single) consequence is defined and its cost id estimated, using the form of Figure 26.

Figure 26. 1st Consequence identification before the introduction of consequence barriers.

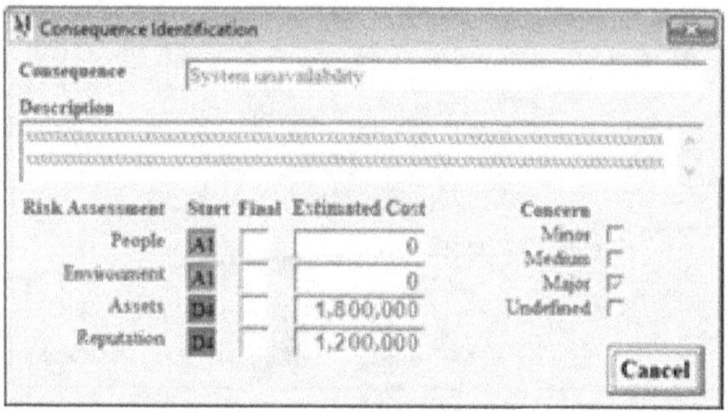

For the 1st Consequence, the **Frequency** and **Severity** of each one of People Environment, Assets and Reputation is selected from the corresponding Risk Matrix form, as shown in Figure 12, by clicking the proper square in the Risk Matrix (see also Figure 8). Also a Concern level is selected (see Figure 11) for the 1st Consequence.

The proper colour appears in the lower part of the 1st Consequence identification of Figure 26. The **Risk Value** for each one of People, Environment, Assets and Reputation is calculated using Equation 1:

$RV = Frequency \times Severity$

Table 8. Frequency and severity

	Frequency	*Severity*	*Risk Value RV*
People	*1*	*1*	*1*
Environment	*1*	*1*	*1*
Assets	*4*	*4*	*16*
Reputation	*4*	*4*	*16*

i.e.:

$RV_P = A1 = 1 \times 1 = 1$

$RV_E = A1 = 1 \times 1 = 1$

$RV_A = D4 = 4 \times 4 = 16$

$RV_R = D4 = 4 \times 4 = 16$

The mitigation barriers for the 1st consequence are then defined, using the forms of Figures 27, 28 and 29. From the values of the **Effectiveness E_{ij}** of each Barrier (j) for the single Consequence (see also Figure 10), the **Risk Value Reduction coefficient R_{redi}** of the 1st Consequence are calculated using Equation 16:

$$R_{redi}\left[\sum_{j=1}^{k} E_{ij}\right] / k_i$$

Table 9. Effectiveness of barriers

Conseq. i	*Barrier j*	*Effectiveness E_{ij}*	*Number of Barriers k_i*	*Reduction Coeff. R_{redi}*
1	**1**	**0.84**	**3**	**0.6700**
	2	**0.67**		
	3	**0.50**		

i.e.:

$R_{red1} = (0.84 + 0.67 + 0.50) / 3 = 0.6700$

Then the resulting **1st Consequence Risk Value $\mathbf{RV_{final}}$**, for people, environment, assets and reputation, is calculated using Equation 17:

$RV_{finali} = (1 - R_{redi})\ RV_i$

i.e.:
$RV_{finalP} = (1 - 0.6700) \times 1 = 0.3400$ (Green Area)
$RV_{finalE} = (1 - 0.6700) \times 1 = 0.3400$ (Green Area)
$RV_{finalA} = (1 - 0.6700) \times 16 = 5.4400$ (Green Area)
$RV_{finalR} = (1 - 0.6700) \times 16 = 5.4400$ (Green Area)

Figure 27. 1st Barrier for 1st consequence.

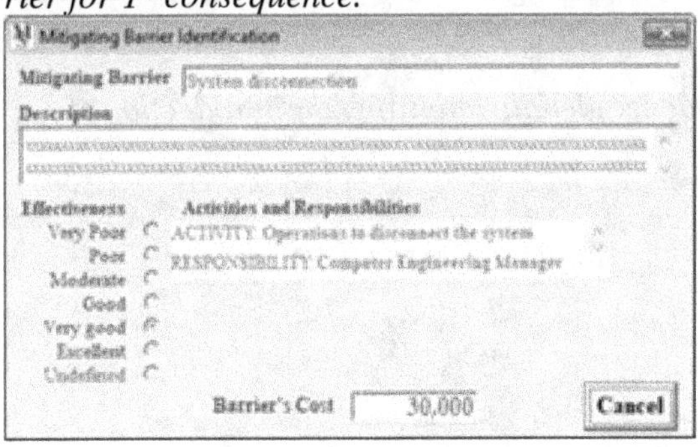

Figure 28. 2nd Barrier for 1st consequence.

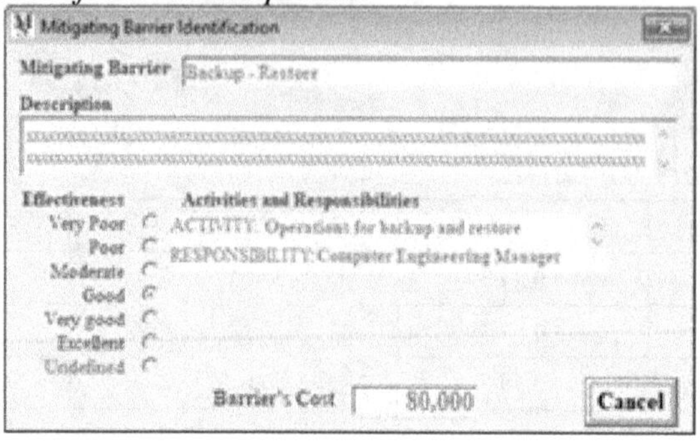

Figure 29. 3rd Barrier for 1st consequence.

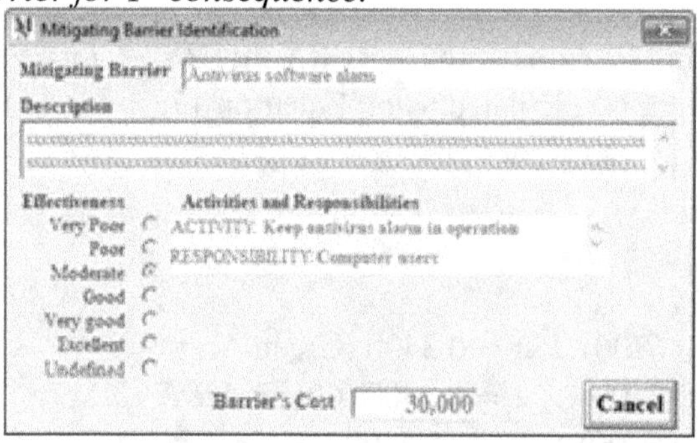

The resulting 1st Consequence form, after the introduction of the Consequence Barriers, takes the form of Figure 30, with all Risk Values colours in green.

Figure 30. Consequence identification after the introduction of consequence barriers.

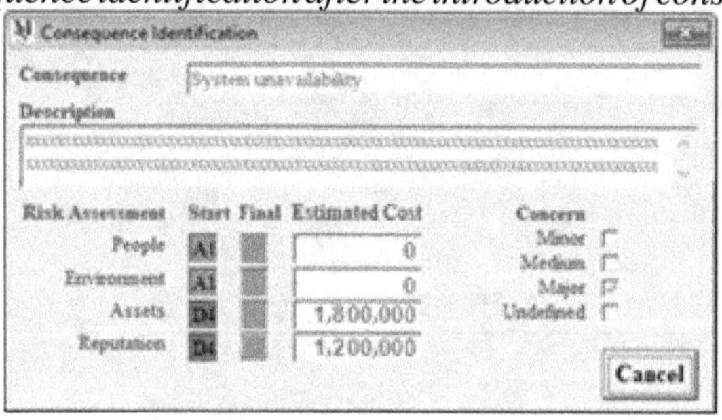

Finally, a Cost Benefit Analysis is carried out automatically, using Equations 9, 10 and 11, as shown in Figure 31.

Figure 31. Cost benefit analysis final result summary.

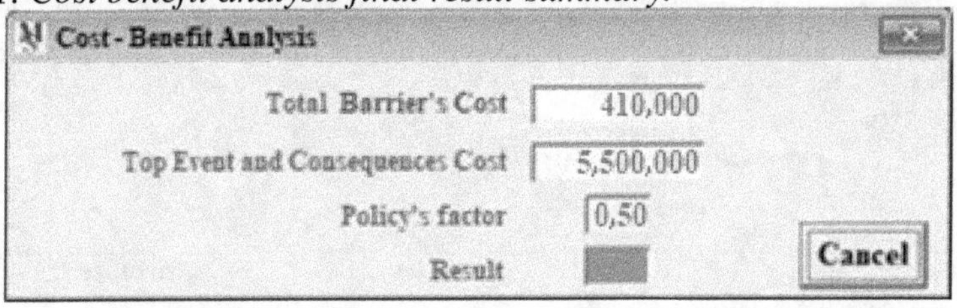

From the defined Cost of Risk Treatment (see Figures 18, 19, 20, 21 22, 23, 24 and Figures 27, 28 29) the Total Cost of Risk Treatment is (for similar barriers their cost is taken once):

Total Cost of Risk Treatment = CP + CM = 410,000 €

From the defined Cost of Top Event and Cost of Consequence (see Figures 13 and 26) the total event cost is:

CTE + CCO = 5,500,000 €

For a factor f=0.50, it is:

Total Cost of Risk Treatment $<$ f · [CTE + CCO]

Finally, the resulting printouts are presented in the following Figure 31 (Main Printout about Threats and Consequences) and Figure 32 (Printout of Activities and Responsibilities resulting from Prevention & Mitigation Barriers).

Figure 32. Main printout about threats and consequences.

Main Printout about Threats and Consequences

Hazard: Electronic systems vulnerable to cyber security incidents

Top Event: **Malware related incidents**

Risk Values				**Cost of Event**
People	**Environment**	**Assets**	**Reputation**	**€ 2,500,000**

Threats

Description	**Frequency**	**Contribution**
1. Malware spread through network	Once per year or less	High
2. Malware spread through removable storage	Once per year or less	High
3. Malware spread through users' s behaviour	Once per 6 months	High

Consequences

Description	**Risk Value for**	**Concern**	**Cost €**
1. System malfunction or stop	P E A R	Major	**3,000,000**

Figure 33. Printout of activities and responsibilities resulting from prevention & mitigation barriers.

Printout of Activities and Responsibilities for Prevention and Mitigation Barriers

Hazard: Electronic systems vulnerable to cyber security incidents

Top Event: Malware related incidents

1st Threat Prevention Barriers				
Description	**Effectiveness**	**Activities**	**Responsibilities**	**Cost €**
1. Network segregation	Very good	Operations for network segregation	Computer Engineering Manager	**70,000**
2. Antivirus software	Very good	Application of antivirus software	Computer Engineering Manager	**75,000**
2nd Threat Prevention Barriers				
Description	**Effectiveness**	**Activities**	**Responsibilities**	**Cost €**
1. Blocking USB ports	Good	Activities for blocking USB ports	Computer Engineering Manager	**25,000**
2. Antivirus software	Very good	Application of antivirus software	Computer Engineering Manager	75,000
3. Awareness training	Good	Training in computer awareness	Computer Engineering Manager	**45,000**
3rd Threat Prevention Barriers				
Description	**Effectiveness**	**Activities**	**Responsibilities**	**Cost €**
1. e-mail washing	Very good	Operations to clean-up e-mail files	Computer Engineering Manager	**55,000**
2. Antivirus software	Very good	Application of antivirus software	Computer Engineering Manager	75,000
1st Consequence Mitigation Barriers				
Description	**Effectiveness**	**Activities**	**Responsibilities**	**Cost €**
1. System disconnection	Very good	Operations to disconnect the system	Computer Engineering Manager	**30,000**
2. Backup - Restore	Good	Operations for backup and restore	Computer Engineering Manager	**80,000**
3. Antivirus software alarm	Moderate	Keep antivirus alarm in operation	Computer users	**30,000**

8. CONCLUSIONS AND RECOMMENDATIONS

As a general conclusion, **the aim of the proposed study, to address the issue of maritime security through the development of a method applicable to the maritime industry that evaluates and manages security related risks, was achieved**. The proposed framework provides the user optically with all necessary information concerning threats and consequences, all necessary prevention and mitigation barriers, all related actions to reduce risk and all responsible persons and finally, all necessary related printouts.

Risk computation is carried out easily in a completely understandable way. Through the user friendly software, the results of the risk computation appear in the Bow-Tie diagram immediately and optically providing the user with a coloured scheme of the obtained risk values for top event and each consequence, after the introduction of the necessary prevention and mitigation barriers by the user.

ACKNOWLEDGEMENT

The authors confirm contribution to the paper as follows: study conception and design: P. Papageorgiou, Z. Dermatis; data collection: P. Papageorgiou; analysis and interpretation of results: P. Papageorgiou, P. Liargovas; draft manuscript preparation: P. Papageorgiou, Z. Dermatis. All authors reviewed the results and approved the final version of the manuscript.

REFERENCES

Agrawal, V. (2017). *A framework for the information classification in ISO 27005 standard* [Master's thesis, Norwegian University of Science and Technology]. NTNU.

Bichou, K. (2008). Security and risk-based models in shipping and ports: Review and critical analysis. *OECD International Transport Forum*, Discussion Paper No. 2008, 20.

Bradford, J. F. (2004). Japanese anti-piracy initiatives in Southeast Asia: Policy formulation and the coastal state responses. *Contemporary Southeast Asia*, *26*(3), 480–505. DOI: 10.1355/CS26-3E

Brooks, M., & Pelot, R. (2008). Port security: A risk-based approach. In Talley, W. (Ed.), *Maritime safety, security and piracy* (pp. xx–xx). Informa.

Chalk, P. (2008). *International security: Terrorism, piracy, and challenges for the United States*. RAND, Center for Terrorism Risk Management Policy.

DNV-GL. (2016, September). *Cyber security resilience management for ships and mobile offshore units in operation*. DNV-GL AS.

IMO. (2004, December 14). *Guidance to masters, companies and duly authorized officers on the requirements relating to the submission of security-related information prior to the entry of a ship into port* (MSC/Circ.1130).

IMO. (2009, June 9). *Revised guidance to masters, companies and duly authorized officers on the requirements relating to the submission of security-related information prior to the entry of a ship into port* (MSC.1/Circ.1305).

International Maritime Organization (IMO). (2003). *International ship and port facility security (ISPS) code and SOLAS amendments, adopted on 12 December 2002* (Electronic ed., Sales number E116E). London.

International Organization for Standardization (ISO). (2018). *Risk management—ISO 31000*. https://www.iso.org/files/live/sites/isoorg/files/store/en/PUB100426.pdf

ISO. (2007). *Ships and marine technology — Maritime port facility security assessments and security plan development (ISO 20858)*. https://www.iso.org/standard/46051.html

ISO. (2011). *Information technology—Security techniques—Information security risk management (ISO/IEC 27005)* [Technical report]. https://www.iso.org/iso/catalogue_detail?csnumber=56742

Jones, S. (2006). *Maritime security: A practical guide*. Nautical Institute.

Klein, N. (2011). *Maritime security and the law of the sea*. Oxford University Press. DOI: 10.1093/acprof:oso/9780199566532.001.0001

Kristiansen, S. (2004). *Maritime transportation: Safety management and risk analysis* (1st ed.). Elsevier-Butterworth-Heinemann.

Lambert, J. H. (2007). Risk-cost-benefit analysis for port environmental security investments. In Linkov, I. (Eds.), *Managing critical infrastructure risks* (pp. 299–307). Springer. DOI: 10.1007/978-1-4020-6385-5_15

Makrodimitris, G., Polemi, N., & Douligeris, C. (2014). Security risk assessment challenges in port information technology systems. In *Maritime security and privacy* (pp. xx–xx). Springer., DOI: 10.1007/978-3-319-11710-2_3

Ministry of Defence of Finland. (2011). *Security strategy for society: Government resolution 2010 Dec 16*. Ministry of Defence.

Nieminen, E., Salokorpi, M., & Rytkönen, J. (2012). *Maritime safety and security: Literature review*. Kymenlaakso University of Applied Sciences.

Nyman, T., Tuominen, R., Rytkönen, J., Kujala, P., & Ylitalo, T. (2010). *Preliminary study of developing a new method to control maritime safety and security risks together* [Electronic report, unpublished]. The Finnish Transport Agency, Ministry of Defence, The Finnish Border Guard, Aalto University, Kymenlaakso University of Applied Sciences, and Technical Research Centre of Finland. Presented by A. Fransas.

Organisation for Economic Co-operation and Development (OECD). (2003, July). *Security in maritime transport: Risk factors and economic impact*. OECD Maritime Transport Committee, Directorate for Science, Technology and Industry.

Parnell, G., Figueira, J. R., Bennett, S., Bobylev, N., Del Pup, M., Ganoulis, J., ... (2007, January). Decision analysis tools for safety, security, and sustainability of ports and harbors. In *Managing critical infrastructure risks* (pp. xx–xx). NATO Security through Science Series C: Environmental Security. DOI: 10.1007/978-1-4020-6385-5_13

Raymond, C. Z., & Morriën, A. (2008). Security in the maritime domain and its evolution since 9/11. In Rupert, H. B., Bateman, S., & Lehr, P. (Eds.), *Lloyd's MIU handbook of maritime security*. CRC Press. DOI: 10.1201/9781420054811.ch1

Reniers, G. (2011). Security of multimodal dangerous freights: The way forward. In Guarascio, M., Reniers, G., Brebbia, C. A., & Garzia, F. (Eds.), *Safety and security engineering* (pp. 327–335).

Risk Review. (2018). *Cyber security risk assessments: Using bowtie analysis with ISO 27000*. https://www.meercat.com.au/risk-management-blog/risk-management-news/cyber-risk-assessments-bowtie-analysis/

Skjong, R., Vanem, E., & Endresen, Ø. (2007). *Risk evaluation criteria* (Tech. Rep. SAFEDOR-D-4.5.2-2007-10-24-DNV-RiskEvaluationCriteria-rev-3.0). http://www.safedor.org/resources/SAFEDOR-D-04.05.02-2005-10-21-DNVRiskEvaluationCriteria-rev-3.pdf

UK Cabinet Office. (2005, May). *Regulatory impact assessments: 1st July to 31st December 2004*. Presented to Parliament by the Minister for the Cabinet Office.

United Nations. (2008). *General Assembly resolutions*.

U.S. Congress. (2002). *Maritime Transportation Security Act of 2002: Protecting America's ports*. Homeland Security.

US Navy, US Marine Corps, & US Coast Guard. (2007, July). *The commander's handbook on the law of naval operations*.

Yang, Z., Wang, J., & Ng, A. (2016). Toward robust management of maritime risk and security. In *Dynamic shipping and port development in the globalized economy* (pp. 122–149). Palgrave Macmillan. DOI: 10.1057/9781137514233_6

Chapter 10
Mapping AI's Influence in Politics Through Digital Authorship Analysis

Dimitrios Vagianos
https://orcid.org/0000-0001-6932-1513
University of Macedonia, Greece

Nikos Koutsoupias
https://orcid.org/0000-0003-1664-5404
University of Macedonia, Greece

ABSTRACT

Artificial intelligence has a considerable impact on national and international politics as well on international relations. It challenges geostrategic alliances, introduces new challenges to the international agenda, helps diplomats and negotiators in their work, and raise new questions and opportunities for human rights. Using R language tools, this study explores the digital presence of the scientific field of AI in national and international politics and provides a complete view of its authorship taking into account bibliometrical data including annual production rates, world-wide citation frequencies, word frequencies and word networking dynamics. The acquired datasets undergo additional analysis through the application of Multiple Correspondence Analysis (MCA) in conjunction with Hierarchical Clustering (HC). The goal of this analysis is to discern patterns and correlations among the diverse elements of these digital authorship datasets which can provide a clear semantic overview of the role of Artificial Intelligence in National and International Politics.

DOI: 10.4018/979-8-3693-9400-7.ch010

1. INTRODUCTION

AI has become involved in many aspects of modern life, given the proliferation of electronic media and networks. Its applications can be found in personal computers or smart phones, in the devices that make up the internet of things (smart TVs and watches, robotic vacuum cleaners, etc.), in online search engines, in various online machine translation applications, in targeted content delivery algorithms as well as in mechanisms to combat misinformation and fake news in social media, in online advertising in cybersecurity applications, etc. Its social and economic impact (Kokkoris et al, 2024), beneficial or not, is the subject of research from various perspectives. Its involvement in politics at national or international level is one of these perspectives.

The political use of AI is not clearly understood by a large proportion of citizens despite its wide range of applications. For example, machine learning algorithms are now able to automatically monitor social media, international news sources and other data streams in order to identify early indicators of impending conflicts or humanitarian emergencies, before which political decisions need to be taken at the national or international level. The international agenda can be significantly influenced by these processes. Additionally, specialised applications can play a key role in digital diplomacy and the corresponding decision-making process in the international political system. At the same time, bots that operate with AI algorithms successfully mimic human behaviour and can influence political situations such as conflicts or elections to the benefit of the person who deploys them.

For the above reasons, AI is among the most important sectors of concern for the competent institutions of the European Union in an effort to gain an edge over its competitors in this field. Efforts are being made to align its applications with legislation on the protection of human rights and international law.

Due to the importance of the aforementioned, the subject of the involvement of AI in politics has been an important topic of research and writing worldwide. In this study, an attempt is made to document the various important aspects of the relevant literature (Vagianos et al., 2023; Mikelis & Koutsoupias, 2023; Koutsoupias & Kokovidou, 2023) combined with statistical analysis of the relevant data (Bitzenis & Koutsoupias, 2024; Bitzenis et al.,2023; Drossos, 2006) and through this process the semantic understanding of the most important areas of AI involvement in National and International Politics.

2. LITERATURE REVIEW

Bibliometric analysis is a popular and accurate method for investigating and analysing large volumes of scientific data. It enables the evolutionary nuances of a particular field to be highlighted, while shedding light on its emerging areas. Donthu et al. presented an overview of bibliometric methodology, with particular emphasis on its various techniques, offering guidelines on which one can rely to perform rigorous and scientifically reliable bibliometric analyses (Donthu et al., 2021). Darvish investigated in 2020 the issue of bibliometric analysis using open source software. For this purpose he used Bibliometrix, an R package, for mining and analyzing available bibliometric metadata by performing a keyword specific search on the Web of Science database achieving the retrieval of 1155 scientific publications from 2000 to 2017 with the criterion of the country of origin of at least one author to be from his country (Darvish, 2020).

Passas argued that bibliometric analysis is an increasingly popular and thorough technique for examining and evaluating vast amounts of scientific data, which will now increasingly be used in research to identify patterns, trends and implications in a particular field. He identified the sub-steps of collecting data from relevant databases, cleaning and refining the data, and analysed them using various scientific methods, which he recognised as a necessary step in extracting meaningful information. For this reason, he thoroughly presented a bibliometric methodology, emphasizing its numerous pertinent techniques, providing a set of reliable, step-by-step instructions' guide for safely performing bibliometric analyses (Passas, 2024).

Bibliometrics is therefore very much related to the broad field of managing huge amounts of scientific data. Many factors, such as the development, accessibility and availability of bibliometric tools like R and Ω and scientific databases like Google Scholar, Scopus and Web of Science, are behind its increasing popularity. The interdisciplinary influence of bibliometric methodology from data science to operational research has also played an important role in its widespread adoption (Doulani, 2020). Scientists thus use bibliometric analysis for many other purposes, including revealing trends of specific topic-specific articles and journals, patterns of collaboration and research components, and exploring the build-up of a particular field within the existing literature (Abramo & D'Angelo, 2014).

The data involved in bibliometric analysis are often extensive but also objective, such as the number of citations and publications, keyword occurrences and topics. Through rigorous techniques for understanding and managing the vast unstructured data, bibliometrics, helps to map and understand the accumulated scientific knowledge and developments of specific subject areas. For example, Rodríguez et al. investigated global trends in scientific authorship on the rural population decline effect over time by analysing scientific sources and authors. The final sample contained 1150 articles published between 1979 and 2018 and was bibliometrically analysed using R's bibliometrix tool which retrieved and merged the metadata of two databases: the WoS and Scopus databases (Rodríguez et al., 2020). Similarly, Artyukhov et al. used R Studio Bibliometrix, VOSviewer and Connected Papers to structure the scholarly work on the adaptation of education systems during socioeconomic, political, migration, pandemic and climate crises in the 2020s. The research was based on more than 970 publications (articles, conference abstracts and monographs) indexed by Scopus in the period 2020-2024 through keyword searches. The analysis identified some important focus areas for the study of educational crises from a thematic perspective. These included education disruption and learning loss, as well as the broader impact of these issues on behavioural development as well as sustainable development. Among all the crises that took place in the world in the period 2020-2024, they cocame to the conclusion that COVID-19 and the full-scale war in Ukraine had the greatest impact on the transformation of education systems (Artyukhov et al., 2024). Within the same framework, Boiarskii et al. analysed and visualised scientific collaboration between Japan and Russia using a bibliometric analysis of collaborative publications from the Web of Science (WoS) database. They also used the R bibliometrix package for the analysis. The study covered a period of two decades, from 2000 to 2021, and highlighted the importance of promoting and maintaining scientific collaboration between the two countries (Boiarskii et al., 2024).

3. METHODOLOGY

The research conducted aimed to capture a detailed picture of the relevant digital authorship and popularity in the scientific field of AI applications in Politics, nationally or internationally. It was done using the open source software package bibliometrix of the statistical programming language R, which achieves comprehensive scientific bibliometric mapping analysis (Buyukkidik, 2022). This software package produces results by accessing journal articles, conference proceedings and book chapters as they appear in the following databases:

- Scopus (https://scopus.com)
- Clarivate Analytics Web of Science (https://www.webofknowledge.com)
- Digital Science Dimensions (https://www.dimensions.ai)
- Cochrane Library (https://www.cochranelibrary.com)
- Lens (https://lens.org)
- PubMed (https://pubmed.ncbi.nlm.nih.gov)

In this study, the Scopus search forms in the category "Social Science and Arts" were used as a data collection tool. As keywords the combinations "Artificial Intelligence" or " AI " were used in combination with the entries "International Relations" or "International Affairs" or "Global Affairs" or "World Affairs" or "World Politics" or "Foreign Relations" or "Global Politics" or "International Diplomacy" or "World Affairs" or "Internal Affairs" or "Internal Politics" or "Federal Politics" or "State Politics" or "Domestic Politics". The study covered the period from 1974 to 2024. Furthermore, the PRISMA 2020 guidelines (Preferred Reporting Items for Systematic Reviews and Meta-Analyses) were adhered to for the systematic organization and presentation of scientific papers, as illustrated in Figure 1 (Page, 2021).

The aim of the process was to draw conclusions of bibliometric analysis in the area under consideration based on annual levels of authorship, related scientific citations, their frequencies at local or global level, collaborative indicators, the distribution of word frequencies and the general dynamics of related networking (Fernández et al., 2022). In this study, bibliometrix's ability to create networks of co-citation coupling, recording of scientific collaborations and co-word analysis was exploited. The analysis outlined at a first level the authorship framework of AI in the field of Politics. With the help of this software, the thematic network (Cobo et al., 2011) was created, which represents the results of the analysis of the investigated two-word terms and created clusters, which can be considered as sub-topics or central ideas. This map gives a visual overview of the evolutionary dynamics of the words that appear most frequently next to keywords in the abstracts of the bibliographic database under examination.

Figure 1. Article selection based on the PRISMA 2020 statement

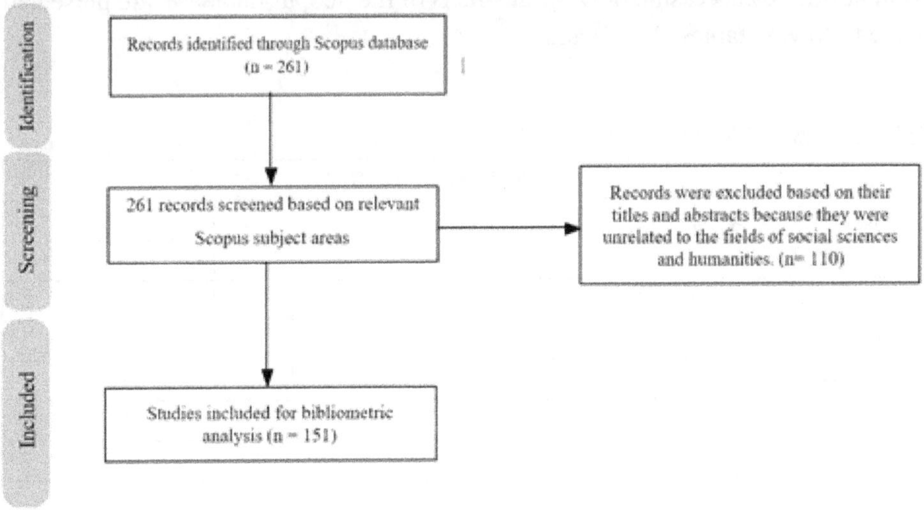

At a second level, the datasets produced by the bibliometrix can be further analysed using multivariate statistical methods (Papadimitriou, 2007).In the present research, Multiple Correspondence Analysis (MCA) was initially used in an attempt to capture underlying structures in the dataset that constitute the frequencies of occurrence of the searched terms in the abstracts of the relevant literature. Essentially, an attempt is made to apply a data analysis technique to nominal categorical data in order to identify thematic categories and explore them more deeply (Bechrakis, 1999) by identifying patterns and correlations of the authorial framework in the scientific field of AI in Politics. Subsequently, the Hierarchical Clustering (HC) method is also applied to the data through which it allows for the recording of the hierarchy of the formed thematic clusters.

The aim of the proposed combination of the use of the bibliometrix survey package in combination with multivariate statistics is to discern patterns and correlations between the digital authorship metrics of the subject under study and ultimately, through this process, the semantic understanding of the involvement of AI in national and international policy.

4. APPLICATION OF THE METHOD AND ANALYSIS OF THE RESULTS

4.1 First Level Bibliometrix Results

The first results of the initial keyword survey that were found and give a first picture of the volume of data collected by bibliometrics with keywords in the field of AI in Politics by accessing only the abstracts of the Scopus database, are presented in the following tables (1), (2) and (3).

Table 1. Basic Bibliometrix search data information.

BASIC DATA INFORMATION	
Sources (magazines, books, etc.)	120
Documents	151
Average number of years since publication	9,05
Reference period	1974-2024
Average number of reports per document	4,887
Annual growth rate	7,11%
Bibliographic references	1

Table 2. Types of documents found by Bibliometrix during the search.

TYPES OF DOCUMENTS	
Articles	72
Books	32
book chapters	30
conference articles	3
Reviews	7
Editorials	6

Table 3. Authors' data found by Bibliometrix during the search.

AUTHORS	
Authors	234
International author partnerships %	5,96%
Single-authored documents authors	76
Co-authors per document	1,83
Single-authored docs	86
Author's keywords	415
keywords related to the Scopus database	170

Figure 2 shows the annual authoring levels, where there is a sharp increase in the authorship of this subject from 2015 onwards, in the sample under consideration.

Figure 2. Annual levels of authorship in the scientific field under investigation

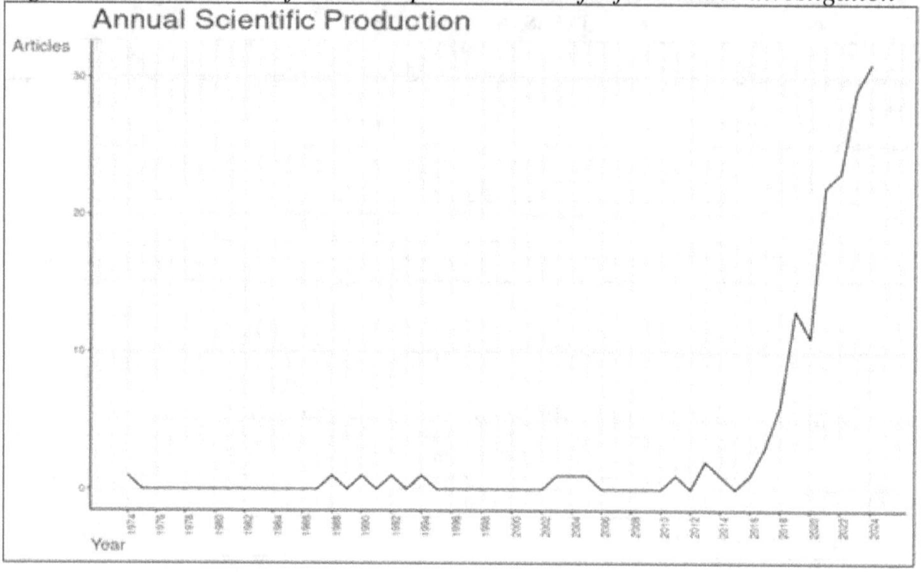

This observation indicates the date from which the applications of AI in various aspects of national and international politics are established. This was the time when the technology was mature enough to provide the necessary applications that could be used in the various fields of Politics, in particular those involving the Internet, such as e-diplomacy.

Table 4 lists the scientific sources with the largest collection of articles in the specific thematic area during the period under review.

Table 4. Scientific sources with the greatest collection of articles related to the searched terms in the given time period.

Name of Scientific Source	Articles
STUDIES IN SYSTEMS, DECISION AND CONTROL	4
ARTIFICIAL INTELLIGENCE AND DIGITAL DIPLOMACY: CHALLENGES AND OPPORTUNITIES	3
AUSTRALIAN JOURNAL OF INTERNATIONAL AFFAIRS	3
CONTRIBUTIONS TO INTERNATIONAL RELATIONS	3
EUROPEAN JOURNAL OF INTERNATIONAL SECURITY	3
THE ROUTLEDGE HANDBOOK OF US FOREIGN POLICY IN THE INDO- PACIFIC	3
BATTLESTAR GALACTICA AND INTERNATIONAL RELATIONS	2

continued on following page

Table 4. Continued

Name of Scientific Source	Articles
BULLETIN OF THE ATOMIC SCIENTISTS	2
COOPERATION AND CONFLICT	2
DEFENSE AND SECURITY ANALYSIS	2
INFORMACIOS TARSADALOM	2
NEW TECHNOLOGY IN SOCIOLOGY: PRACTICAL APPLICATIONS IN RESEARCH AND WORK	2
POLIS. POLITICAL STUDIES	2
RUSSIA IN GLOBAL AFFAIRS	2
SECURITY DIALOGUE	2
SOCIAL SCIENCE COMPUTER REVIEW	2
SOVREMENNAYA EVROPA	2
TECHNOLOGY AND INTERNATIONAL RELATIONS: THE NEW FRONTIER IN GLOBAL POWER	2

The drawn sample of literature examined revealed a significant number of authors active in the field of AI applications in Politics. Table 5 illustrates the most important authors of the period in terms of the number of sources that correspond to them, always in terms of finding the verbal associations of the subject under investigation in the abstracts of these sources.

Indicatively, some information on the professional profile and some details of the writing work of the top authors of table 5 are given below, always in the scientific field under investigation.

Table 5. Authors with the greatest collection of articles related to the searched terms in the given time period.

Author' s name	Articles	Author' s name	Articles
JOHNSON J	6	MARWALA T	2
HASSANIEN AE	4	MILLS WD	2
DARWISH A	3	NADIBAIDZE A	2
ERSKINE T	3	NEUMANN IB	2
SCHRODT PA	3	NYMALM N	2
SR	3	PAYNE K	2
ASLAM W	2	PÁLMA M	2
BENCE K	2	QIAO-FRANCO G	2

continued on following page

Table 5. Continued

Author' s name	Articles	Author' s name	Articles
BENCE S	2	SIENKNECHT M	2
BODE I	2	TAMÁS B	2
CUOMO S	2	TORKUNOV AV	2
ERMAN E	2	TURNER O	2
GRININ L	2	WATTS TFA	2
HAMDY W	2	WHYTE C	2
HUDSON VM	2	AALTOLA M	1
HUELSS H	2	ABD EL-AZIZ AA	1
KIERSEY NJ	2	AKAEV A	1
MALKOV S	2	AKILLI E	1

Dr James Johnson is Professor and Director of Strategic Studies in the Department of Politics and International Relations at the University of Aberdeen. He is an advisor to various departments of the US, UK and EU governments on AI and nuclear policy, including the US Department of Defense, the NATO Nuclear Planning Group, the UK Office for Artificial Intelligence and the Global Commission for Responsible AI in the Military. His research focuses on the relationship between AI and nuclear weapons, political psychology and strategy. For example, in his book "Artificial Intelligence and The Future of Warfare. The USA China and Strategic Stability" (Johnson, 2021), he emphasizes the challenges AI poses to strategic stability, nuclear deterrence, and how AI can influence nuclear weapons systems. He also highlights the implications and makes policy recommendations for states on how to manage the escalating risks posed by AI.

Dr. Aboul Ella Hassanein is the founder and head of the Egyptian Scientific Research Group (SRGE) and a professor of Computer Science at the Faculty of Computer and Information Science at the University of Cairo. His scientific interests cover diverse topics such as data mining, medical imaging, intelligent systems, social networks and smart environment. For example, in the chapter entitled "COVID-19 and Water Resources Nexus: Potential Routes for Virus Spread and Management Using Artificial Intelligence Techniques" included in the book "The Global Environmental Effects During and Beyond COVID-19" (Salem et al., 2021), he argues how the integration of methods using the Internet of Things (IoT) and Artificial Intelligence could effectively serve in outbreak monitoring and policy decision-making processes to limit the spread of the pandemic.

Professor Ashraf Darwish is Professor of Computer Science and former Head of the Department of Mathematics and Computer Science at the Faculty of Science, Helwan University, Egypt. Cognitively, he focuses on the use of Artificial Intelligence

(AI) and its applications, with expertise in areas such as machine learning, deep learning and computer vision. His expertise extends to cutting-edge technologies such as Explainable Artificial Intelligence and Generative Artificial Intelligence. A representative example of his writing in the field of AI is his contribution to the book "Artificial Intelligence for Environmental Sustainability and Green Initiatives" (Hassanien et al., 2024), which highlights the applications of AI in sustainability, exploring its potential in areas such as energy, healthcare, agriculture, transport and waste management. It talks about applications and innovations in green initiatives policies, including areas such as energy, finance and pharmaceutical research. The book is about the ethical challenges and benefits of integrating AI into the sustainability initiative. His contribution lies in the chapter entitled "Measuring Global Warming Effect by the Prediction of Climate Change on the Different Countries Using Machine Learning Approaches". Aboul Ella Hassanein and Ashraf Darwish have also co-authored the book "Artificial Intelligence for Sustainable Development: theory, practice and future applications" (Hassanien et al., 2021). This book presents the latest developments in the field of artificial intelligence and related technologies, with a particular focus on sustainable development and policy decisions related to environmentally friendly applications of AI. Analysing theory, applications and research, it covers all aspects of AI in the context of sustainable development policies.

Toni Erskine is Professor of International Politics at the Coral Bell School of Asia Pacific Affairs at the Australian National University (ANU) and a Fellow of the Leverhulme Centre for the Future of Intelligence at the University of Cambridge. She is currently a principal researcher on the two-year research project "Predicting the Future of War: Artificial Intelligence, Automated Systems and Decision Making for the Use of Force", funded by the Australian Government, Department of Defence. She has authored the article "AI and the future of IR: Disentangling flesh-and-blood, institutional, and synthetic moral agency in world politics" (Erskine, 2024), in which she analyses in detail the relationship between AI and global politics. More specifically, she argues that artificial intelligence (AI) is disrupting prevailing understandings in the field of international relations, explores what criteria intelligent machines must meet to qualify as moral agents in the field, and highlights the ways in which AI applications are eroding critical notions of responsibility in global politics. In addition, in her article "AI and the decision to go to war: future risks and opportunities" (Erskine & Miller, 2024) she comments on the current military AI landscape and turns to the neglected perspective of AI systems influencing state-level decision-making on the use of force. It argues that these systems will increasingly influence such deliberations in more straightforward ways - either as part of automated self-defence or through decision support systems that advise human deliberations. Furthermore, that the evolution of the arms race involving AI is inevitable and that it will probably happen in the near future with major consequences.

Philip Andrew Schrodt is a political scientist known for his work in automated coding of data and facts for political news. His main areas of research are formal models of political behavior, with a focus on international politics, and political methodology. His current research focuses on predicting political change using statistical and pattern recognition methods. For example, in his article "Artificial intelligence and formal models of international behavior" (Schrodt, 1988), he describes the ways in which, over the last five years, models and techniques developed in Artificial Intelligence (AI) have been applied to a wide range of issues in international relations. It identifies and analyzes three categories of applications in this area and emphasizes Natural Language (NL) approaches, which it considers to have significantly expanded the scope and complexity of formal models of international behavior. The same author has written a chapter entitled "Artificial Intelligence and International Relations: an Overview" which is included in the book "Artificial Intelligence And International Politics" (Schrodt, 1991). This chapter provides a brief history of the relevant developments in AI in order to provide an outline of the concepts used in the thematic axis of linking AI and International Relations. It underlines the problem of the overemphasis that traditional media and science fiction wrongly gives to AI and presents some AI and International Relations models with Rule Based systems being the most common.

Table 6 lists the total citations, i.e. the total number of total citations received by the sources of the sample under consideration during the period under investigation from "universal" sources, i.e. sources of the entire bibliographic database. The last column calculates the normalised citation score for documents, authors and sources using both global and local citations. In order to create a supervisory picture of the literature, a brief description of the scope of the five most important table's sources in terms of total citations is given below.

Table 6. The most cited bibliographic sources in the sample under consideration from the total literature database

Article	DOI	Total citations	Average annual reports per year	Normalised total reports
JOHNSON J, 2019, DEFENSE SECUR ANAL	10.1080/14751798.2019.1600800	92	15,34	6,8736
KIERSEY NJ, 2013, BATTLESTAR GALACTICA AND INTERNATIONAL RELATIONS	10.4324/9780203070055	39	3,25	1,69566
DER DERIAN J, 2020, SECUR DIALOGUE	10.1177/0967010620901905	35	7	3,9286
THAKUR R, 1994, J PEACE RES	10.1177/0022343394031002003	34	1,097	1
JENSEN BM, 2020, INT STUD REV	10.1093/isr/viz025	33	6,6	3,70409
JOHNSON J, 2022, J STRATEG STUD	10.1080/01402390.2020.1759038	28	9,334	6,376238
CERNY PG, 2003, GLOBALIZATION AND GOV	10.4324/9780203478790-15	28	1,27273	1
PAYNE K, 2018, STRATEGY, EVOL, AND WAR: FROM APES TO ARTIF INTELL		26	3,714	4,45714
MCKELVEY F, 2019, CAN J EDUC COMMUN	10.22230/cjc.2019v44n2a3509	25	4,1667	1,8678
BAZARKINA DY, 2019, RUSS GLOB AFF	10.31278/1810-6374-2019-17-1-147-170	22	3,667	1,64368
JOHNSON J, 2021, PAC REV	10.1080/09512748.2019.1676299	20	5	5,78948
GRININ L, 2023, WORLD-SYSTEMS EVOL GLOB FUTURES	10.1007/978-3-031-34999-7_10	19	9,5	7,5479
KARAKO K, 2020, BIOSCIENCE TRENDS	10.5582/bst.2020.03364	19	3,8	2,1326
CHOUCRI N, 1974, INT INTERACT	10.1080/03050627408434390	16	0,3138	1
TORRES P, 2019, BULL AT SCI	10.1080/00963402.2019.1604873	13	2,1667	0,97127

The most widely cited article addresses the implications of AI for future wars and international security, highlighting its decisive and potentially transformative impact on military power, strategic competition and global politics more broadly. It then further refines its analysis by arguing that if the uncertainties and vulnerabilities created by the rapid spread and diffusion of AI are not controlled, it could become a major source of instability and strategic competition among major powers. In general, several innovations and technological developments related to AI are highlighted, which can have a decisive impact on military applications from the perspective of the tactical battlefield to the strategic level (Johnson, 2019).

The second source in the order of total citations is a book that focuses on international relations through a science fiction story. More specifically, it analyses their most important issues in the present day and dedicates a significant part of it to analysing the philosophical implications of AI technologies which it contrasts with what it means to be a human being and with human rights in general (Kiersey & Neumann, 2013).

The third source is an article which, taking into account recent developments in quantum theory as well as innovations in quantum technology, focuses on the profound implications in international relations, especially in the field of international security. According to the authors, new scientific discoveries and technological applications suggest large-scale quantum phenomena with immediate implications for the creation of global entanglements. With the arrival of quantum computing, communications and the development and establishment of Artificial Intelligence, it describes the significant transformations that the nature, production and distribution of power and knowledge are undergoing (der Derian, & Wendt, 2020).

The fourth source is an article that discusses how rapid developments in Artificial Intelligence (AI) technologies could affect the creation and application of military power. Considering that there has been little empirical or even theoretical study from the perspective of international relations (IR) and security studies, it argues that, although the rapid adoption of AI technologies is set to transform the ways states fight on many fronts, the AI revolution brings with it new forms of risk that must be combined with simultaneous integration of algorithmic safety systems into all military functions related to decision-making and writing. More σpecifically, it highlights the need to develop accountability mechanisms and limit the irresponsible use of AI technologies (Jensen, et al. 2019).

The fifth article in the ranking has a similar theme, examining whether the use of AI in strategic decision making will be stabilising or destabilising, focusing on the risks and trade-offs of assigning military power to machines. It shows how non-nuclear states and non-state actors could use AI to their advantage to put pressure on nuclear states and devotes a significant part to how malicious use of AI-enhanced fake news, deepfakes, bots and other similar means by non-state actors or state agents can cause states to overreact to a threat from ambiguous or manipulated information, and therefore increasing instability (Johnson, 2020).

Figure 3 shows the country of origin of the bibliographic sources recorded in this thematic area. It can be observed that the United States of America, the United Kingdom and Australia are at the top of the list, all with English as the language of discourse. The presence of China in 6th place is significant, which can be related to that country's achievements in the broader field of AI.

Figure 3. Countries of origin of authors in the subject area under research both at local (SCO) and global (MCP) scale

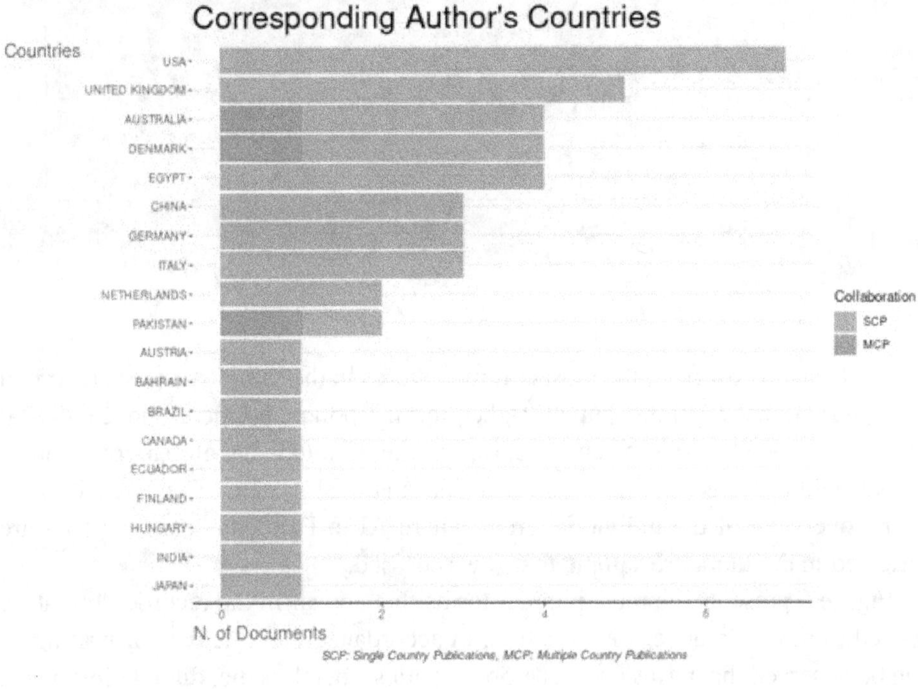

The particular political status of China, in combination with the responding sources dealing with the applications of AI in Politics, is now an interesting subject for conducting multifaceted research (Whiting, 2008). The presence of other Asian as well as Arab countries in the list reveals the global interest of this scientific field in recent years. It is noteworthy that most countries show independent authoring efforts with Australia, Denmark and Pakistan also showing a percentage of collaborative authoring with other countries.

The frequency of the words used in the relevant titles of the literature and in the available abstracts is presented in Figure 4.

Figure 4. Most frequent words in the abstracts of the recorded bibliography searched by keyword combinations related to AI in Politics.

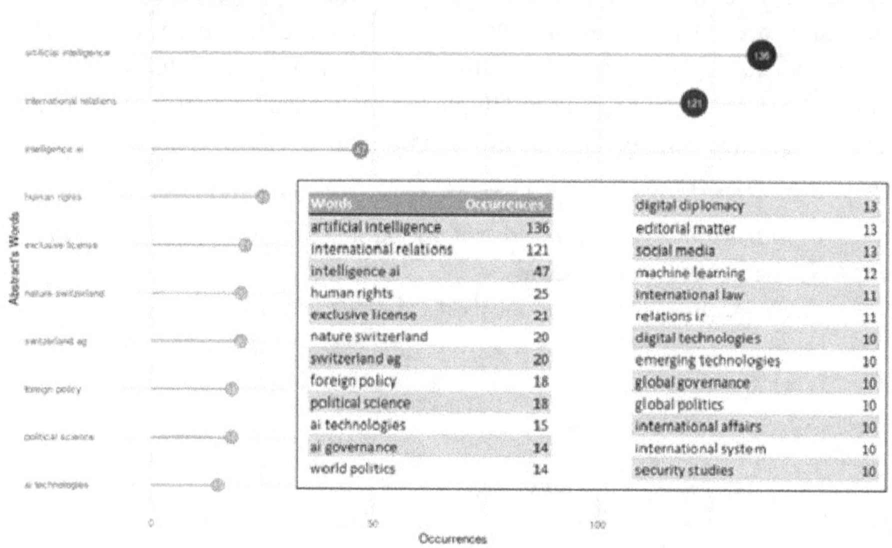

Words	Occurrences		
artificial intelligence	136	digital diplomacy	13
international relations	121	editorial matter	13
intelligence ai	47	social media	13
human rights	25	machine learning	12
exclusive license	21	international law	11
nature switzerland	20	relations ir	11
switzerland ag	20	digital technologies	10
foreign policy	18	emerging technologies	10
political science	18	global governance	10
ai technologies	15	global politics	10
ai governance	14	international affairs	10
world politics	14	international system	10
		security studies	10

As it can be observed, the most common words in the sample that was recorded are: international relations, human rights, foreign policy, political science, digital diplomacy, social media, machine learning, international law, global governance, global politics, international affairs and international system. These results give a first overview of the individual areas where AI in Politics is used, as they are reflected in the literature sample that was accessed.

Figure 5 presents the most popular terms that appear in the recorded literature related to the topic under investigation, in accordance with the date of writing. It can be observed that terms such as global politics, social media, digital diplomacy, international relations, human rights, and united nations appear with particularly increased frequency. The term human rights was the subject of a treatise at an earlier stage, while the fields of global politics and digital diplomacy came to the foreground at a later stage.

Figure 5. Countries of origin of authors in the subject area under research both at local (SCO) and international (MCP) scale

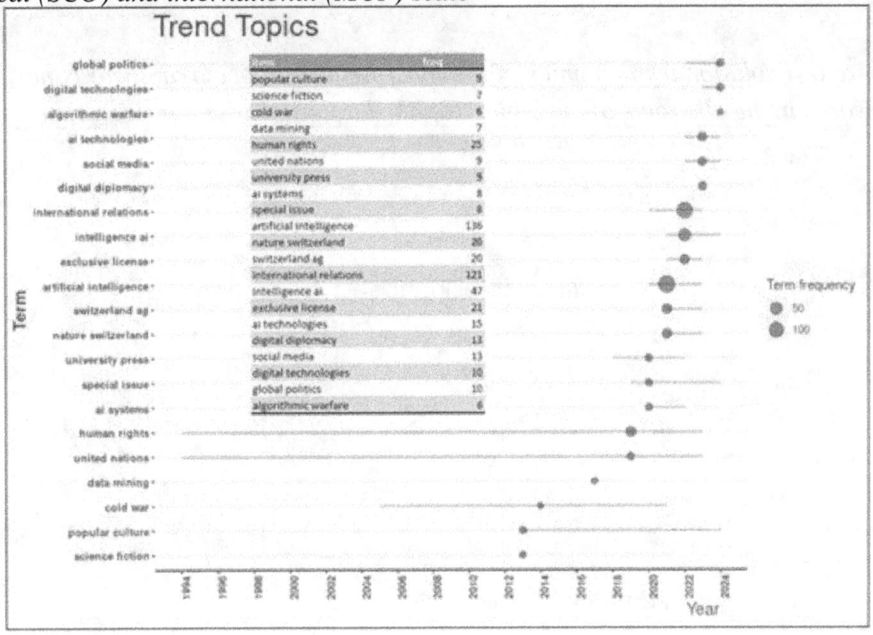

Additional information about these most common words is given in Figure 6, where the verbal connections are additionally marked. By observing it one can conclude that in the literature under consideration, there is a strong verbal connection between AI and International Relations with a variety of topics related to these two semantic poles: global politics, international law, international affairs, European Union, digital diplomacy, international system, human rights, global governance, and decision making in general. Moreover, AI seems to have influenced Political Science in general and its relation to data mining processes is also reflected. A separate semantic cluster includes global politics, the United Nations, digital transformation and the COVID pandemic, among which AI seems to be the common denominator. Finally, it verbally highlights the dominant role of China in the field of AI development, which is consistently a distinct thematic area of many research efforts due to its pioneering role in the field combined with its peculiar political characteristics (Mokry & Gurol, 2024). Many articles have referred to the competition for this type of technological dominance, pointing out that the dominant nation can influence global governance by establishing technical rules, shaping laws and restricting access to AI technologies. It is now widely known that AI can enhance productivity, innovation and competitiveness through economic growth (Maqsood et al., 2023).

Figure 6. Evolutionary dynamics of the most frequently occurring words next to keywords in the abstracts of the bibliographic database under consideration

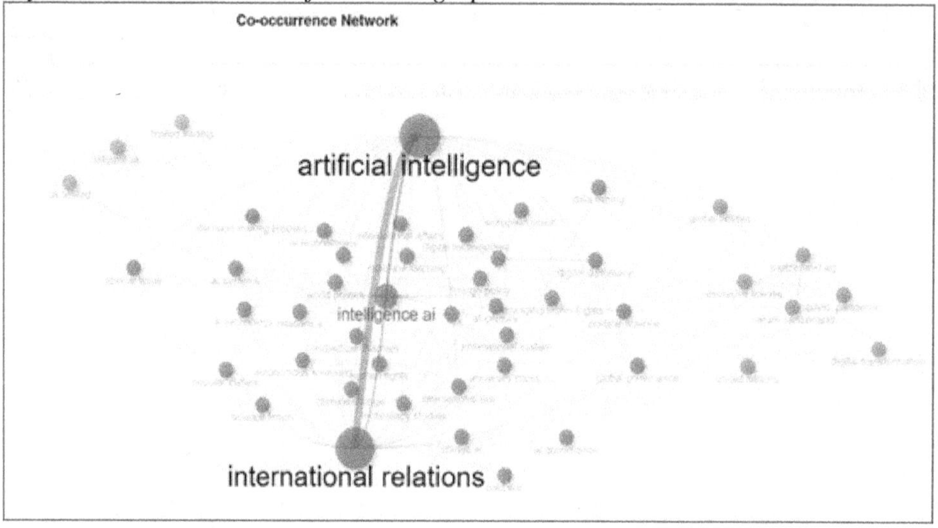

Figure 7 illustrates the thematic cluster map (Cobo et al., 2011) which represents the results of the analysis of the researched dialect terms and created clusters, which can be considered as individual themes or central ideas (themes). To do this, bi-word analysis is used in a longitudinal context in order to detect the different topics addressed by the research field during the specific time span. The performance analysis uses various bibliometric measures, including the h-index, to measure the impact of both the identified themes and topic areas. The presented approach essentially achieves a visualized overview of the thematic evolution of the field under study.

The two axes in figure 7 correspond to density (horizontal axis) and centrality (vertical axis). These two axes interpret the categorization of clusters according to the quadrant to which they belong (Greenacre, 1991). In the upper right quadrant, with higher density and centrality, appear the topics that are significantly related to the field under study and contribute to the construction of its conceptual framework. Similarly, in the bottom right-hand quadrant, are the important issues that cut across the various aspects of the study area but have a low level of analysis by the author community. The bottom left quadrant depicts the topics that have not been analyzed to a great extent and are of less interest, while the top left quadrant is for the topics that show a higher degree of analysis but little interest to the field of study. Essentially, centrality corresponds to the degree of authorial interest in the scientific community, while density corresponds to the degree to which the topic is analysed by it.

Figure 7. Thematic map of bi-words in the abstracts

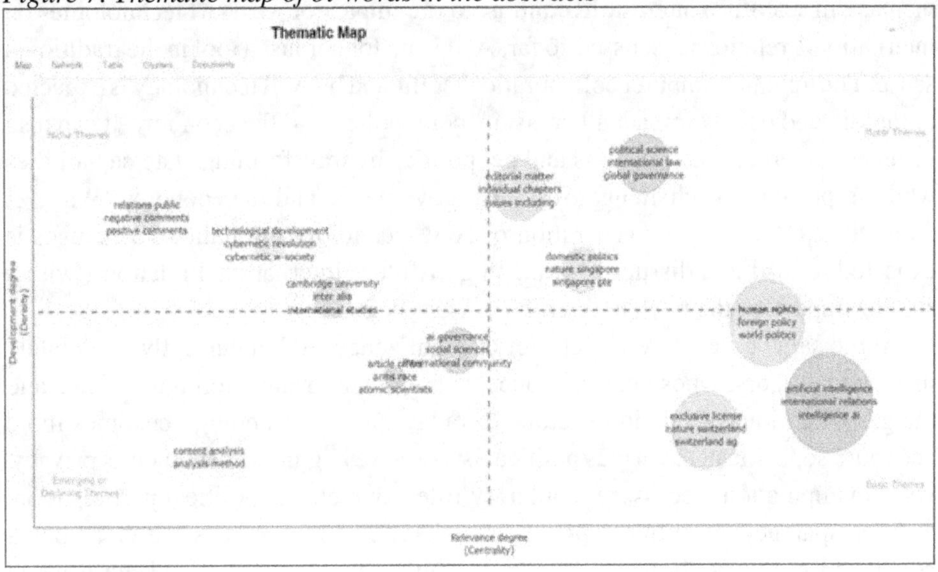

(Cobo et al., 2011)

Based on the above, it can be observed in Figure 7 that the coexistence of political science, international law and global governance in a cluster of high interest and engagement highlights the close connection between these fields, always in relation to AI. In the broad field of Political Science, research efforts are focused on revealing the political science background of AI, identifying its characteristics and potential in contemporary Political Theory and public administration practices in the context of the global experience of political and legal reality under conditions of digital transformation (Ilyina, 2022). As for the scientific field of International Law, the development of AI affects critical areas of the field. It is obvious that the future use of its technologies without a corresponding legal regulation may disturb the already fragile balance of the global order, which could be aggravated by state competition in AI technologies and its applications in the military sector, a grey area of international law (Kukeyeva et al., 2024). The relationship between AI and global governance is grounded in the literature that much of the existing global governance with AI elements lacks important prerequisites necessary for the necessary political legitimacy, which essentially does not exist if certain forms of decision-making are handed over to AI systems in blank (Erman & Furendal, 2024).

In the same graph, a thematic coherence between AI and international relations is also outlined, with a high level of authorship but low interest at the moment. The new generation of AI technologies (e.g. ChatGPT) can be seen as the beginning of the era of Artificial General Intelligence (AGI), which may lead to an even greater increase in scientific interest. In contrast to the impact of weak AI technologies on international relations discussed so far, AGI is no longer just a tool in the traditional sense. The human-computer collaboration facilitated by AGI technology is expected to rapidly and deeply impact all areas of social policy and the economy. It can also influence international relations and geopolitics by transforming state capabilities while imposing new challenges to global governance and the economy (Wang & Zvarych, 2024). The new generation of AGI technological qualitative changes is expected to lead to a disruptive reshaping of future international relations, with a similar impact on the relevant literature (Tianyu, 2024).

At the same time, there is a cluster of highly engaged but currently moderately interested authors, whose central concepts are human rights, foreign policies and the global political scene in general. Nowadays, as AI technology becomes more and more sophisticated, various political issues have begun to arise, such as privacy, voter manipulation, necessary regulatory rules, due ethics, political participation, and transparency of political processes (Li, 2024). Research is therefore being conducted in the field of taking modern politics to a new level and the ways in which the pathologies that technology can bring about can be combated, while it is increasingly understood that in addition to the protection of privacy and data, the way in which AI can be applied from the perspective of the protection of human rights in general must always be investigated (Lane, 2022). Regarding the existence of the subject of foreign policy in the same cluster, this can be justified by the fact that AI technology is now used in the decision-making mechanisms related to national foreign policies, which raises great scientific interest as to its rational use, its possible negative effects on the global political scene and the related ethics and ethics (Anderson & Thorson 1982).

It should be noted that all the above results were obtained by applying specific keyword combinations when using the bibliometrix package. The results, although at first level, succeed in giving a very good first picture of the authorial context under investigation by highlighting various aspects of it from different perspectives. The bibliometrix data can be further analysed using multivariate statistical methods. In the next section, the methods of Multiple Correspondence Analysis and Hierarchical Clustering are applied on the dataset.

4.2 Multiple Correlation Analysis and Hierarchical Clustering Application Results

In statistics, Multiple Correspondence Analysis (MCA) is a data analysis technique for nominal categorical data (Greenacre, 2007; Koutsoupias, 2024) used to identify and represent underlying structures in a dataset (Benzécri, 1992). It achieves this by representing data as points in a low-dimensional Euclidean space (Beh & Lombardo, 2021). The Multiple Multiple Correspondence Analysis can be seen as an extension of simple Correspondence Analysis based on the fact that it can be applied to a large set of categorical variables (Greenacre, 2006; Orús-Lacort & Jouis, 2019). Essentially, it can be seen as an extension of simple Correspondence Analysis (CA) in that it can be applied to a large set of categorical variables (Athanasiadis, 1995).

Hierarchical clustering (also called hierarchical cluster analysis or HCA) is a cluster analysis method that seeks to create a cluster hierarchy (Koutsoupias & Papadimitriou, 2001). By applying Multiple Correspondence Analysis followed by Hierarchical Clustering (HC), it is possible to identify the thematic clusters of the subject under consideration as well as to create their relative hierarchy (Athanasiadis, 1995).

The two methods mentioned above were applied to all the data obtained using the bibliometrix software, in order to capture the thematic sections and using other scientific approaches, based on the frequencies of occurrence of bi-word terms in the abstracts. The results of applying the Multiple Correspondence Analysis method are illustrated in Figure 8.

Figure 8. Multiple Correspondence Analysis for mapping thematic areas based on frequencies of occurrence of bi-word terms in the abstracts

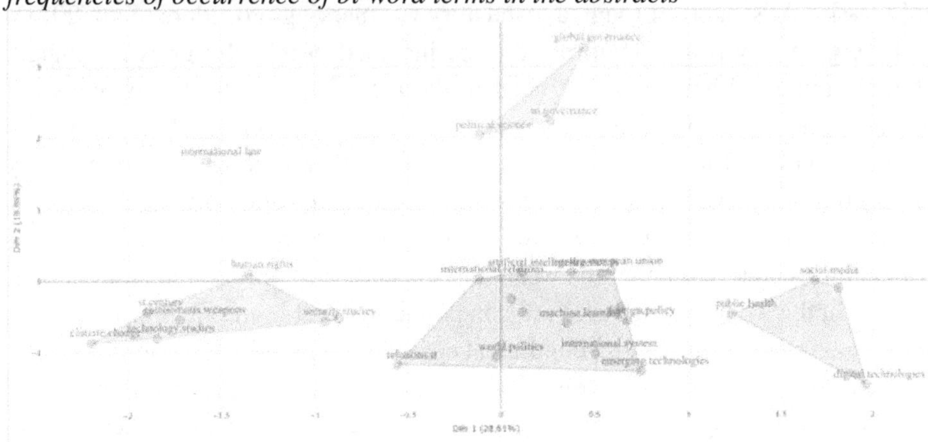

In Figure 8, the information represented is 28.1% on the first axis, 18.88% on the second axis and a total of about 47% on the first factorial plane. It can be observed that with this approach International Law forms an independent thematic cluster, while Political Science with International Governance belongs to a different one. These three thematic clusters belonged to the same thematic cluster according to the previous approach of Cobo et al. The classification of International Law into a thematically independent cluster can be justified by the fact that it is a very useful tool for Law and for Legal Science in general, as well as by the existing need to adapt International Law specifically in order to face and govern the AI revolution that is underway in the fields of automation, personalisation, weapon systems, control and standardisation. (Burri, 2017). The literature is also dealing with the attempt to answer the questions of what are the implications of these developments for collective security - both from a state-centric and human-centric perspective, as well as for legal systems worldwide. It is widely known that international law is the most suitable to create the framework that will legally cover the above developments worldwide (Tzimas, 2021).

In the MCA approach, the cluster containing AI with International Relations emerges enriched with additional topics such as the European Union, Foreign Policy, International Politics, International System and Machine Learning.

The inclusion of the European Union in this cluster is justified by its high awareness of advances in the field of AI. As early as 21 April 2021, the European Commission presented its Directive in this field (Artificial Intelligence Act). The Directive lays down basic horizontal rules for the development, trade and use of AI-based products, services and systems within the territory of the EU. The Directive introduced a sophisticated "product safety scheme" structured around a set of 4 risk categories. It imposes requirements for the market entry and certification of high-risk AI systems through a mandatory CE marking procedure (Kop, 2021).

Foreign policy is also legitimized to be included in this cluster as the impact of AI in this field has become crucial in the present times, as it transcends the borders of nations and represents a decisive change in international relations. The development of AI in this field is a new challenge that affects the foreign policy of countries, especially large countries, due to its connection with security issues and the international balance of power. Moreover, the entry of AI into the arms sectors raises concerns about the weapons race and its impact on the international system. In this context, countries are competing to excel in the field of AI, which has opened a new channel for geopolitical competition in the international system. It is therefore reasonable for foreign policy to be in this cluster since it is now a fact that AI is influencing the shaping of countries' foreign policy and how they can use it to pursue their interests and enhance their international influence.

This clearly affects the international political arena in general. Through targeted advertising algorithms and sentiment analysis techniques, it is possible to influence political processes such as e.g. manipulation of voters during electoral processes. The cases of the 2016 presidential election in the US and the 2019 general election in India have demonstrated how AI exerts significant influence on democratic political processes, leading to the conclusion that as AI becomes increasingly beneficial to human society, concerns about regulations, ethics, political participation and other issues emerge. The literature thus addresses the issue of maintaining democratic integrity in politics while promoting AI in ways that benefit contemporary international politics, suggesting regulatory policies that prevent the unethical engagement of AI in the international system.

It is also interesting to note the inclusion of machine learning in the cluster with the above thematic concepts. There are indeed many contexts relevant to the broader policy area in which this particular branch of AI is applied. Fake news is one of the biggest problems in today's technological world and has a huge impact on the decision making and the way of thinking of the global population. The main purpose of misinformation is to manipulate and confuse people in order to influence their opinion and bring advantages to specific stakeholders in areas such as politics, economy, etc.). Machine learning in this area is applied both to the creation of mechanisms that emulate human behaviour on the Internet in order to create and disseminate fake news, and to the development of mechanisms to detect fake news and to suppress the mechanisms for its production and dissemination.

In applying the Multiple Correlation Analysis method, the human rights issue appeared to belong to the thematic cluster that includes Climate Change and not Foreign Policy, as was the case in the prediction of Cobo et al. In the literature, this is based on the engagement of the scientific community with the view that climate change is a threat to human rights. In this view, climate change threatens the exercise of a wide range of human rights, while human rights law imposes duties on states with regard to climate change and necessitates international cooperation in this area. In the same thematic context, climate change should be addressed by international human rights bodies. Indeed, the UN Human Rights Council has already encouraged its competent sector to address climate change (Knox, 2009).

Finally, an independent cluster also appears that contains the concepts of Public Health and Social Media (Nahon, 2015). This particular finding can at this point be justified by the emergence of social media and AI algorithms involved in an innovative means of communication in the public health sector, providing unprecedented reach, interaction and real-time contact. As a result, they are now an important resource for public health professionals and organizations to communicate with people about health and healthcare (Kamanzi, 2024).

Next, the Hierarchical Clustering method was applied to the co-occurrence frequencies of the bi-words in the abstracts retrieved in the current study. The application of this method aimed to identify thematic patterns and correlations within the structures of the captured categorical authorship data in the scientific field under investigation. The results are illustrated in Figure 9.

Figure 9. Hierarchical clustering on co-occurrence frequencies of bi-word combinations in the abstracts

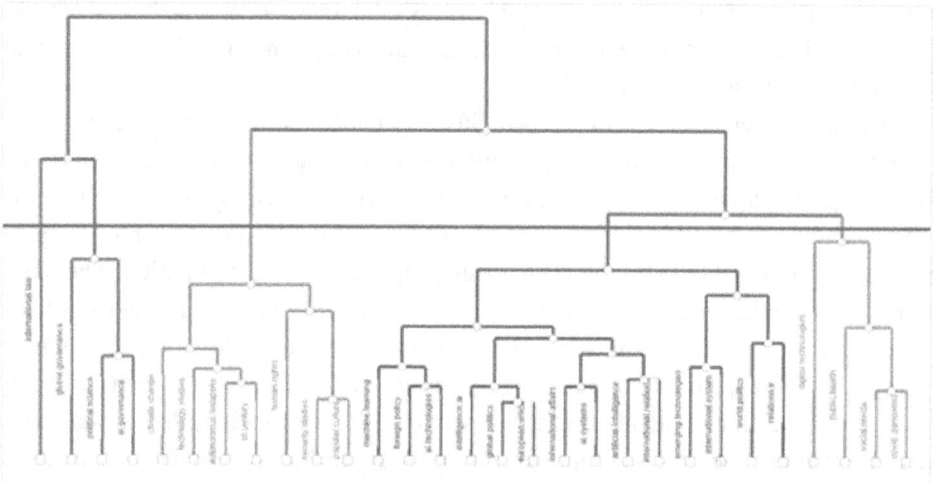

The following clusters with the following thematic objects are observed:

- International Law, (red)
- Political Science, Global Governance (blue)
- Human Rights, Climate Change (green)
- Technologies, Global Politics, International Relations, Foreign Policy, European Union, International Affairs (purple)
- Social media, Covid-19 pandemic, Public Health (ορανγε).

In terms of clustering, it can be observed that the results of this method are very similar to those of the Multiple Correlation Analysis. In the HC approach, the only difference with respect to the HCA approach is that the cluster containing the Public Health notion also contains that of the Covid-19 pandemic, highlighting the relevant studies included in the literature at the time of the Covid 19 pandemic and later. For example, Dmour et al. investigated the influence of social media platforms on public health protection during the Covid-19 pandemic through the mediating effects of public health awareness and behavioural change. Similarly, Neely et al.

explored how patients collected and evaluated health-related information online on platforms such as Facebook and Twitter during the pandemic, extracting useful information for healthcare providers and public health officials (Neely et al., 2021). The association of AI with this domain has to do with how content is distributed on these platforms. On the other hand, it is widely known that AI applications can now be used to tackle a future pandemic. They include early case detection, predictive modelling, healthcare management, risk communication and health surveillance. These applications, in particular prediction algorithms, have the potential to predict epidemics by integrating a variety of data sets, such as social media, meteorological data and mobile phone traffic data. AI-based tools can also optimize healthcare delivery by managing resource allocation and reducing healthcare workers' exposure to risk (McKee et al., 2024).

In terms of cluster hierarchy, the method highlighted two major themes of interest. The first included the clusters of International Law (red), and that of Political Remuneration and Global Governance (blue). The second contained an individual hierarchical combination of clusters which, lower down the hierarchy, had the cluster combining the cluster of Technologies, Global Politics, International Relations, Foreign Policy and European Union, International Affairs (purple) with the cluster of Social Media, Covid-19 Pandemic and Public Health (orange). The set of these two is combined at the next hierarchical level with the cluster containing Human Rights and Climate Change (green). At the lower hierarchical levels, one can identify interesting thematic literature associations such as Political Science with AI governance, AI technologies with foreign policies, the European Union with global politics, AI with international relations and Covid 19 pandemic with social media (Figure 9).

From the above it is concluded that the use of a combination of Multiple Correspondence Analysis and Hierarchical Clustering on the bibliometric data led to the classification of the thematic areas related to AI in clusters, a process that is largely assessed as successful in terms of capturing the related concepts to the field under investigation and eventually understanding how AI relates to national and international policy based on the literature data. The results ultimately encourage the application of this technique to larger literature datasets and other diverse subject areas (Shamsuddin et al., 2022).

5. CONCLUSIONS AND FURTHER WORK

In this study, R's bibliometrix was used to extract useful data from bibliographic databases based on keywords and their combinations related to the applications of AI in National and International Politics. The search forms of Scopus database were used and 150 documents were retrieved from a total of 234 authors. The exploitation of the package largely yielded a supervisory picture of the authoring context in the period from 1974 to 2024, providing useful and interesting data in terms of sources, authors, citations, relative frequencies, countries of origin and scientific collaborations. This particular thematic object appeared to be of major concern to the scientific community from 2014 onwards, which coincides with the maturing of the relevant technology not only in terms of Politics but also in terms of society in general. A significant number of authors come from countries in the Western world, with those of the United States, the United Kingdom and Australia predominating. It is noteworthy that scholars from China are active in this thematic area, as well as from other developing countries such as Egypt and Pakistan.

The bibliometrix identified as thematic areas of the field under investigation international relations, human rights, foreign policy, political science in general, digital diplomacy, social media, machine learning, international law, international law, global governance, global politics, international affairs and the international system. With the help of the software, the thematic network (Cobo et al., 2011) was created which classified the above areas into thematic clusters, indicating individual semantic associations in the scientific domain under investigation. The coexistence of political science, international law and global governance in a cluster of high interest and engagement was observed. Also, a thematic coherence was identified between AI and international relations of high but currently low interest. At the same time, the concepts of human rights, foreign policies and the global political scene were identified in a cluster of high authorial engagement but moderate interest.

A number of statistical methods can be further applied to the mined datasets (Karapistolis, 2011; Mavromatis, 1999) to build up a supervisory picture of the authorship over specific fields. In this case, Multiple Correspondence Analysis was used in combination with Hierarchical Clustering, which provided a supervisory picture from another perspective by classifying into clusters via another statistical approach. The categorical classification results of these two methods of multivariate statistical analysis showed an interesting differentiation both with respect to the thematic network and between them. Overall, it can be said that they added interesting complementary observations to the first level thematic network generated by the bibliometrix software itself. More specifically, in the MCA approach, the cluster containing AI with international relations appeared enriched with additional topics such as the European Union, foreign policy, international politics, international system and machine learning. In addition, the topic of human rights appeared to belong to the thematic cluster containing climate change and not foreign policy. In this approach, an independent cluster also appeared that contained the concepts of public health and social media. In addition, in the HC approach, the inclusion of the Covid-19 pandemic in the cluster containing the thematic concept of Public Health emerged.

Overall, with the above scientific approach and the analysis that followed, the relationship between AI technology and the field of Politics became absolutely clear and the individual areas in which it is applied were mapped, through the automated emergence of the corresponding authorship context on a global scale.

At this point it is considered appropriate to point out the limitations of this research. The overview formed is based solely on data from the abstracts derived from a specific literature database and using specific word combinations as described above. Also, only documents written in English were investigated. However, the encouraging results obtained from this research approach lead to the conclusion that this technique can be applied to more bibliographic databases and with more keywords or combinations of keywords. Non-English literature, which was not taken into account in this study, will certainly have useful information to offer. In such a case, however, the researcher will have at his disposal a multitude of data which will have to be carefully managed following a strictly scientific methodology (Donthu et al., 2021).

As additional suggestions for the expansion of the research, the individual periods, e.g. 1974-2014 and 2014-2024, can be carried out in detail in order to capture the individual aspects of the specific scientific field during its early maturation period and then during the period of its establishment. Also, networks of collaborations between authors - institutions and countries can be investigated in more depth. Finally, keyword co-occurrence networks could constitute an area that also deserves to be explored in greater depth.

REFERENCES

Abramo, G., & D'Angelo, C. A. (2014). How do you define and measure research productivity? *Scientometrics, 2014*(101), 1129–1144. DOI: 10.1007/s11192-014-1269-8

Abu Salem, H. S., Shams, M. Y., Hassanien, A. E., & Nosair, A. M. (2021). COVID-19 and Water Resources Nexus: Potential Routes for Virus Spread and Management Using Artificial Intelligence Techniques. In Hassanien, A. E., Darwish, A., Gyampoh, B., Abdel-Monaim, A. T., & Anter, A. M. (Eds.), *The Global Environmental Effects During and Beyond COVID-19. Studies in Systems, Decision and Control* (Vol. 369). Springer., DOI: 10.1007/978-3-030-72933-2_2

Anderson, P., & Thorson, S. (1982). Systems simulation artificial intelligence based simulations of foreign policy decision making. *Behavioral Science, 27*(2), 176–193. DOI: 10.1002/bs.3830270211

Artyukhov, A. & Lapidus, A. & Yeremenko, O. & Artyukhova, N. & Churikanova, O. (2024). An R Studio Bibliometrix Analysis of Global Research Trends of Educational Crises in 2020s. SocioEconomic Challenges. 8(2).

Athanasiadis, H. (1995). *Factor analysis of correspondences and hierarchical classification.* New Technologies Publications.

Bechrakis, Th. (1999). Multivariate Data Analysis: Methods and Applications. New Frontiers Publications. N.N. N.A., Athens, Greece.

Beh, E. & Lombardo, R. (2021). Multiple Correspondence Analysis. .DOI: 10.1002/9781119044482.ch6

Benzécri, J.-P. (1992). *Correspondence analysis handbook.* Marcel Dekker. DOI: 10.1201/9780585363035

Bitzenis, A., & Koutsoupias, N. (2024). Big Data in Economics Research. In Tsounis, N., & Vlachvei, A. (Eds.), *Applied Economic Research and Trends* (pp. 1063–1072). Springer Nature Switzerland., DOI: 10.1007/978-3-031-49105-4_61

Bitzenis, A., Koutsoupias, N., & Boutsiouki, S. (2023). Business Research and Data Mining: A Bibliometric Analysis. 2023 3rd International Conference on Electrical, Computer, Communications and Mechatronics Engineering (ICECCME), 1–6. DOI: 10.1109/ICECCME57830.2023.10252699

Boiarskii, B., Lyude, A., Anastasiia, B., Norikuni, O., & Hasegawa, H. (2024). Bibliometric analysis of the Japan-Russia scientific cooperation networks using R bibliometrix. Journal of Infrastructure. *Policy and Development.*, *8*(8), 6155. DOI: 10.24294/jipd.v8i8.6155

Burri, T. (2017). International Law and Artificial Intelligence. SSRN Electronic Journal. 60. DOI: 10.2139/ssrn.3060191

Buyukkidik, S. (2022). A Bibliometric Analysis: A Tutorial for the Bibliometrix Package in R Using IRT Literature. *Egitimde ve Psikolojide Ölçme ve Degerlendirme Dergisi*, *13*(3), 164–193. DOI: 10.21031/epod.1069307

Cobo, M. J., Lopez-Herrera, A. G., Herrera-Viedma, E., & Herrera, F. (2011). An approach for detecting, quantifying, and visualizing the evolution of a research field: A practical application to the fuzzy sets theory field. *Journal of Informetrics*, *5*(1), 146–166. DOI: 10.1016/j.joi.2010.10.002

Darvish, H. (2020). Bibliometric Analysis using Bibliometrix an R Package. *Journal of Scientometric Research.*, *8*(3), 156–160. DOI: 10.5530/jscires.8.3.32

der Derian, J., & Wendt, A. (2020). 'Quantizing international relations': The case for quantum approaches to international theory and security practice. *Security Dialogue*, *51*(5), 096701062090190. DOI: 10.1177/0967010620901905

Dmour, H., Masa'deh, R., Salman, P., & Abuhashesh, M. & Al-Dmour, Dr. (. (2020). The Influence of Social Media Platforms on Public Health Protection against Coronavirus (COVID-19) Pandemic Disease via the Mediating Effects of Public Health Awareness and Behavioral Change: An Integrated Model. *Journal of Medical Internet Research*, *22*. DOI: 10.2196/19996 PMID: 32750004

Donthu, N., Kumar, S., Mukherjee, D., & Lim, W. M. (2021). How to conduct a bibliometric analysis: An overview and guidelines. *Journal of Business Research*, *133*, 285–296. DOI: 10.1016/j.jbusres.2021.04.070

Doulani, A. (2020). A bibliometric analysis and science mapping of scientific publications of Alzahra University during 1986–2019. *Library Hi Tech*, *2020*(39), 915–935.

Drossos, G. (2006). *Statistics & Data Analysis*. Anikoula Publications.

Erman, E., & Furendal, M. (2024). Artificial Intelligence and the Political Legitimacy of Global Governance. *Political Studies*, *72*(2), 421–441. DOI: 10.1177/00323217221126665

Erskine, T. (2024). AI and the future of IR: Disentangling flesh-and-blood, institutional, and synthetic moral agency in world politics. *Review of International Studies*, *50*(3), 534–559. DOI: 10.1017/S0260210524000202

Erskine, T., & Miller, S. E. (2024). AI and the decision to go to war: Future risks and opportunities. *Australian Journal of International Affairs*, *78*(2), 135–147. DOI: 10.1080/10357718.2024.2349598

Fernández L., R. & Alonso, J. & Porraspita, D. & Sánchez, M. (2022). PROSPECTIVE SCENARIOS: A LITERATURE REVIEW USING THE R BIBLIOMETRIX PACKAGE. 18. 1-30.

Giordani, P. & Ferraro, M. & Martella, F. (2020). Hierarchical Clustering. .DOI: 10.1007/978-981-13-0553-5_2

Greenacre, M. (1991). Interpreting multiple correspondence analysis. *Applied Stochastic Models and Data Analysis*, *7*(2), 195–210. DOI: 10.1002/asm.3150070208

Greenacre, M., & Blasius, J. (Eds.). (2006). *Multiple correspondence analysis and related methods*. CRC press. DOI: 10.1201/9781420011319

Greenacre, M. J. (2007). *Correspondence analysis in practice*. Chapman & Hall/CRC. DOI: 10.1201/9781420011234

Hassanien A.E., Bhatnagar R., Darwish A. (2021). Artificial Intelligence for Sustainable Development: Theory, Practice and Future Applications, Part of the book series: Studies in Computational Intelligence (SCI, volume 912), Springer Cham. DOI: .DOI: 10.1007/978-3-030-51920-9

Hassanien A.E., Darwish A., Elghamrawy S.M. (2024). Artificial Intelligence for Environmental Sustainability and Green Initiatives, Springer Cham. DOI: .DOI: 10.1007/978-3-031-63451-2

Ilyina, E. (2022). Politics and Administration under Conditions of Digital Transformation: A Political Science Perspective of Artificial Intelligence. [Искусство управления]. *Ars Administrandi*, *14*(3), 403–421. DOI: 10.17072/2218-9173-2022-3-403-421

Johnson, J. (2019). Artificial intelligence & future warfare: Implications for international security. *Defense & Security Analysis.*, *35*(2), 1–23. DOI: 10.1080/14751798.2019.1600800

Johnson, J. (2020). Delegating strategic decision-making to machines: Dr. Strangelove Redux? *The Journal of Strategic Studies*, *45*(3), 439–477. DOI: 10.1080/01402390.2020.1759038

Johnson, J. (2021). Artificial Intelligence and The Future of Warfare. The USA China and Strategic Stability. Manchester: Manchester University Press, 240 S., ISBN: 978-1526145055 DOI: 10.7765/9781526145062

Kamanzi, N. & University III, Kampala International. (2024). *The Impact of Social Media on Public Health Communication. 5.* 9-12.

Karapistolis, D. (2011a). *Methods of Data Processing and Analysis.* Altintzis Publications.

Kiersey, N. J., & Neumann, I. B. (2013). Battlestar Galactica and International Relations by. Abingdon: Routledge, 2013. ISBN 978 0415632812 DOI: 10.4324/9780203070055

Knox, J. (2009). Linking Human Rights and Climate Change at the United Nations. The Harvard environmental law review: HELR.

Kokkoris, C., Koutsoupias, N., & Vassiliadis, N. (2024). Research on the economics of education: A multifaceted review (1993-2022). Education, Lifelong Learning, Research and Technological Development. *Innovation and Economy*, *3*, 1–16. DOI: 10.12681/elrie.7120

Kop, M. (2021). EU Artificial Intelligence Act: The European Approach to AI. SSRN Electronic Journal. Stanford - Vienna Transatlantic Technology Law Forum, Transatlantic Antitrust and IPR Developments, Stanford University, Issue No. 2/2021.

Koutsoupias, N. (2024). Multiple Correspondence Analysis in Social Sciences and Humanities Research: A Longitudinal Mapping. Data Analysis Bulletin, 20(1), 59–77. https://doi.org/DOI: 10.17605/OSF.IO/693DR

Koutsoupias, N., & Kokovidou, M. (2023). Mapping Research in International Law since 1896. 11th National Data Analysis Conference, GSDA, Univ. of W. Macedonia, GREECE, 35-42. https://doi.org/DOI: 10.17605/OSF.IO/XC89N

Koutsoupias, N., & Papadimitriou, I. (2001). Hierarchical Classification in Insurance Data Analysis. In C. Zopounidis, P. M. Pardalos, & G. Baourakis, Fuzzy Sets in Management, Economics and Marketing (pp. 69–81). World Scientific. DOI: 10.1142/9789812810892_0005

Kukeyeva, F. & Kurmangali, M. & Aktay, D. (2024). Theoretical and Methodological Approaches to Studying Artificial Intelligence in the Context of International Relations and International Law. Central Asia s Affairs. 1. 4-21.

Lane, L. (2022). CLARIFYING HUMAN RIGHTS STANDARDS THROUGH ARTIFICIAL INTELLIGENCE INITIATIVES. *The International and Comparative Law Quarterly*, *71*(4), 915–944. DOI: 10.1017/S0020589322000380

Li, J. (2024). Artificial Intelligence and Its Transformative Impact on Modern Politics. *Applied and Computational Engineering.*, *97*(1), 163–169. DOI: 10.54254/2755-2721/97/20241361

Maqsood A.A., & Khan, A. & Siddiqi, M.U. (2023). US-China Competition in Artificial Intelligence: Implications on Global Governance. .DOI: 10.62345/jads.2023.12.4.37

Mavromatis, G. (1999). *Statistical Models and Methods of Data Analysis*. University Studio Press.

McKee, M., Rosenbacke, R., & Stuckler, D. (2024). The power of artificial intelligence for managing pandemics: A primer for public health professionals. *The International Journal of Health Planning and Management*. DOI: 10.1002/hpm.3864 PMID: 39462894

Mikelis, K., & Koutsoupias, N. (2023). Mapping Research on Thucydides: a Longitudinal Bibliographical Analysis. 11th National Data Analysis Conference, GSDA, Univ. of W. Macedonia, 72-80. https://doi.org/DOI: 10.17605/OSF.IO/CVBS5

Mokry, S., & Gurol, J. (2024). Competing ambitions regarding the global governance of artificial intelligence: China, the US, and the EU. *Global Policy*, *15*(5), 955–968. DOI: 10.1111/1758-5899.13444

Nahon, K. (2015). Where There Is Social Media There Is Politics. .DOI: 10.4324/9781315716299-4

Neely, S., Eldredge, C., & Sanders, R. (2021). Health Information Seeking on Social Media During the Covid-19 Pandemic: A Survey of American SNS Users (Preprint). *Journal of Medical Internet Research*, *23*. DOI: 10.2196/29802 PMID: 34043526

Orús-Lacort, M., & Jouis, C. (2019). Simple and Multiple Correspondence Analysis. DOI: 10.13140/RG.2.2.16880.61448

Page, M. J., McKenzie, J. E., Bossuyt, P. M., Boutron, I., Hoffmann, T. C., Mulrow, C. D., Shamseer, L., Tetzlaff, J. M., Akl, E. A., Brennan, S. E., Chou, R., Glanville, J., Grimshaw, J. M., Hróbjartsson, A., Lalu, M. M., Li, T., Loder, E. W., Mayo-Wilson, E., McDonald, S., & Moher, D. (2021). The PRISMA 2020 statement: An updated guideline for reporting systematic reviews. *BMJ (Clinical Research Ed.)*, (71), n71. DOI: 10.1136/bmj.n71 PMID: 33782057

Papadimitriou, G. (2007). *Data Analysis*. Typothito Publications.

Passas, I. (2024). Bibliometric Analysis: The Main Steps. *Encyclopedia.*, *4*(2), 1014–1025. DOI: 10.3390/encyclopedia4020065

Rodríguez, S. (2020). Worldwide trends in the scientific production on rural depopulation, a bibliometric analysis using bibliometrix R-tool. *Land Use Policy*, *97*, 104787. DOI: 10.1016/j.landusepol.2020.104787

Schrodt, P.A. (1991). Artificial intelligence and formal models of international behavior. Am Soc 19, 71–85. DOI:DOI: 10.1007/BF02692375

Schrodt, P. A. (1991). Artificial Intelligence and International Politics (1st ed.). Routledge. ISBN DOI: 10.4324/9780429033575

Shamsuddin, S., Noriszura, I., & Roslan, N. (2022). A bibliometric analysis of insurance literacy using bibliometrix an R package. *AIP Conference Proceedings*, *2472*, 050023. DOI: 10.1063/5.0092721

Tianyu, X. (2024). Impact of next-generation Artificial General Intelligence (AGI) on international relations. *JOURNAL OF REGIONAL AND INTERNATIONAL COMPETITIVENESS.*, *5*(1), 64–69. DOI: 10.52957/2782-1927-2024-5-1-64-69

Tzimas, Th. (2021). Legal and Ethical Challenges of Artificial Intelligence from an International Law Perspective. .DOI: 10.1007/978-3-030-78585-7

Vagianos, D., Mpoitsis, I., & Koutsoupias, N. (2023). Politics of the Internet Digital Authorship Context: A Multivariate Statistical Analysis. 11th National Data Analysis Conference, GSDA, Univ. of W. Macedonia, DOI: 10.5281/ZENODO.10828143

Wang, Y., & Zvarych, R. (2024). Artificial Intelligence as Promoting Effect on International Economic Relations. *Journal of Intelligence and Knowledge Engineering.*, *2*(2), 1–5. DOI: 10.62517/jike.202404201

Whiting, M. (2008). The Great Firewall of China: A Critical Analysis. Graduate Research paper, AIR FORCE INSTITUTE OF TECHNOLOGY Wright-Patterson Air Force Base, Ohio, USA.

Chapter 11
Sentiment and Geopolitical Analysis in Sino-American Relations

Marios Nosios
https://orcid.org/0009-0007-6161-3854
University of Macedonia, Greece

ABSTRACT

This chapter investigates the integration of Sentiment Analysis, Multiple Correspondence Analysis, and Hierarchical Clustering analysis to advance the understanding of emotional dynamics in international relations discourse, with particular emphasis on Sino-American relations. Bibliometric analysis supplements this framework by offering a systematic overview of scholarly output. This integrated methodological approach enables the processing of large-scale textual data, the refinement of categorical distinctions through multiple correspondence analysis, and the identification of emotional patterns via hierarchical clustering. The findings indicate a predominance of positive sentiment and trust, while fear and anger underscore persistent geopolitical anxieties. Furthermore, the bibliometric results identify Australia, the United States, and China as principal contributors, alongside a growing pattern of global academic engagement. The synthesis of these methods provides a multidimensional perspective on emotional valence in geopolitical analysis.

INTRODUCTION

Sentiment Analysis (SA), also referred to as Opinion Mining (OM), is a technique encapsulated within the broader discipline of Natural Language Processing (NLP) that focuses on identifying, extracting, and analyzing subjective information

DOI: 10.4018/979-8-3693-9400-7.ch011

within textual data. As an emerging discipline in computational linguistics, SA aims to determine the emotional tone conveyed in a piece of text, categorizing it as positive, negative, or neutral. However, the scope of SA extends beyond simple polarity classification; it involves the complex task of understanding sentiment intensity, emotional states, and the subtleties embedded in textual expressions. This capability has become increasingly crucial in the contemporary, data-driven environment, where a large volume of textual data is generated daily across various domains, including social media interactions, official diplomatic communications, news articles, and scholarly discourse (Medhat et al., 2014).

While SA and OM are terms often used interchangeably in literature, they embody distinct aspects of text analysis depending on the context and focus of the research. SA is primarily engaged in detecting and classifying the emotional tone of a text, essentially determining whether it conveys positive, negative, or neutral sentiments. This comprehensive definition enables SA to be applied across a diverse range of contexts, including the assessment of the emotional tone in political speeches and the analysis of media portrayals of diplomatic events.

In contrast, OM focuses on extracting more specific opinions or attitudes articulated within the text. While SA is concerned with determining emotional polarity, OM seeks to identify viewpoints or stances regarding a subject, often aiming to capture the finer nuances of opinion, including its target or the subject, as well as the intensity or strength of the sentiment expressed. This distinction between the two is crucial when considering applications in disciplines such as International Relations (IR) where multiple aspects of sentiment need to be assessed in a structured manner (Sun et al., 2016).

Over time, the techniques utilized for SA have undergone significant development. Early approaches primarily relied on rule-based systems, where predefined dictionaries or lexicons were used to map words or phrases to sentiment scores. These lexicons consisted of collections of emotionally charged terms assigned fixed values (e.g., “happy” = +1, “sad” = -1). While such systems were adequate for basic SA tasks, they exhibited notable limitations in handling complex linguistic patterns or adapting to novel contexts. Furthermore, rule-based methods often struggled with ambiguous language, irony, and sarcasm, reducing their effectiveness in nuanced text interpretation (Berka, 2020).

As the field advanced, machine learning methods gained prominence, utilizing annotated datasets where text samples were labeled to train models for automated SA. Supervised learning emerged as a common methodology, enabling models to discern complex relationships between words, phrases, and emotion. This capability improved context sensitivity and more accurate classification across a broader range of textual inputs (Nalini et al., 2023; Patrick et al., 2023).

The introduction of deep learning techniques, including Recurrent Neural Networks (RNNs), transformer-based models such as BERT (Bidirectional Encoder Representations from Transformers), and GPT (Generative Pretrained Transformers), have further advanced SA. These models demonstrate proficiency in capturing intricate linguistic nuances, such as sarcasm, irony, and context-dependent sentiment shifts (Kim, 2014; Pang & Lee, 2008). For instance, in sentences such as *"The president's policies are a disaster"*, deep learning models are better equipped to identify the underlying negative tone, even when the text involves indirect or nuanced expressions.

Building on the imperative for advanced analytical clarity in IR textual data, Multiple Correspondence Analysis (MCA) offers a complementary approach for exploring patterns within categorical variables. It extends Correspondence Analysis (CA), a technique used to examine associations between categorical variables in contingency tables. MCA is particularly suited for dimensionality reduction and visualizing complex relationships within high-dimensional datasets. Unlike methods developed for continuous variables, MCA specifically addresses the extraction of patterns from categorical variables, making it particularly valuable in analyzing qualitative data such as survey responses, textual information, and social media content (Fávero et al., 2023; Heiser, 2001).

The primary function of MCA is to facilitate dimensionality reduction by constructing a geometric representation of categorical data in a lower-dimensional space. This process enhances data structuring, improving its suitability for subsequent analytical procedures. MCA achieves this by projecting categorical variables onto a spatial map, wherein categories exhibiting similarities are positioned in closer proximity to one another. One of the principal advantages of MCA lies in its ability to process multiple categorical variables simultaneously, thereby refining the structure of high-dimensional datasets and optimizing them for further analysis (Atkinson, 2023). However, in this study, MCA does not serve directly as an interpretative tool, but rather it functions as a preprocessing step that enhances data organization before Hierarchical Clustering Analysis (HCA). In the context of IR, MCA can be employed to refine categorical datasets, such as those containing diplomatic positions, geopolitical alignments, or thematic classifications, prior to the application of HCA. Its role in this context is exclusively to reduce noise and improve data structuring, ensuring that the process of HCA is applied to a more coherent and analytically robust dataset (Martin-Mazé, 2017).

Furthermore, HCA provides a valuable complement to the geometric framework of MCA by systematically organizing data into a hierarchy of nested clusters based on their degree of similarity or dissimilarity (Jafarzadegan et al., 2019). The methodology involves calculating pairwise distances between data points, often using metrics such as Euclidean or Manhattan distances, and progressively merging these points or clusters according to predefined linkage criteria. This iterative process culminates in the construction of a dendrogram, a tree-like diagram that visually represents the hierarchical relationships among clusters. The dendrogram enables researchers to explore the dataset at multiple levels of granularity, facilitating the identification of natural groupings and latent structures within the data (Almeida et al., 2007).

One of the key strengths of HCA lies in its flexibility, as it does not require prior specification of the number of clusters. This characteristic makes it particularly suitable for exploratory data analysis, where the objective is to uncover inherent patterns and relationships within complex datasets. By providing a hierarchical structure, HCA supports the examination of datasets with varying degrees of heterogeneity, allowing researchers to focus on the most meaningful clusters for their specific analytical goals. When employed alongside MCA, HCA enhances interpretative capacity by offering clear hierarchical visualizations and systematically identifying relationships among categories projected by MCA into a reduced-dimensional space. This sequential integration directly addresses the inherent complexity and multidimensionality of international relations datasets, thereby yielding deeper analytical insights that would be achievable through the independent application of these methodologies.

The integration of MCA and HCA in SA represents an innovative methodological approach to enhance the interpretation of complex, high-dimensional data. SA typically outputs a classification of text (positive, negative, neutral), which can be seen as a categorical variable. When combined with MCA and HCA, sentiment labels can be analyzed in conjunction with other categorical variables, such as geographical location, political affiliation, or media source. For instance, within an IR context, one might apply SA to a corpus of diplomatic cables, social media posts, or news articles to classify sentiment towards a specific country or political leader (Chambers et al., 2015). Once classification is complete, MCA can be employed as a preprocessing step to structure categorical data and reduce noise prior to applying HCA. The purpose of MCA in this context is not to interpret findings independently but to refine the dataset, enhancing the quality of input for subsequent clustering. HCA is then used to identify relationships between classified data and other categorical variables, such as the country of origin, political party, or international issues (e.g., trade, defense, environment). By integrating MCA before HCA, researchers can improve the clustering process, ensuring that sentiment correlations are more clearly delineated within the hierarchical structure.

Recent studies have shown the potential of SA, MCA, and HCA in handling complex categorical data in the field of IR research. Yiu et al. (2021) examined the influence of public sentiment on cross-border acquisitions, particularly in geopolitically sensitive situations. By analyzing over 100,000 blog posts, the study highlights how public attitudes toward the home countries of acquiring firms affect ownership decisions, revealing sentiment's pivotal role in shaping corporate strategies and legitimacy in international markets. Similarly, Garcia and Cunanan-Yabut (2022) apply SA to the Russo-Ukrainian conflict, analyzing nearly 27,894 tweets using the #UkraineRussia hashtag. The findings reveal a predominance of negative emotions, reflecting global condemnation and concern regarding the war, demonstrating SA's capacity to capture real-time emotional responses to international crises and providing valuable insights into the global public's stance on geopolitical events.

Abedin et al. (2018) also utilize SA to assess public reactions to the Iran nuclear deal by analyzing tweets through lexicon-based and machine-learning methods. The results reveal a positive sentiment toward the agreement, followed by negative and neutral responses, showcasing SA's potential to gauge public opinion on high-stakes international agreements. Georgiadou et al. (2019) integrates SA with Big Data Analytics to study public sentiment during the UK-EU Brexit negotiations. By examining over 13 million tweets, this research uncovers collective preferences regarding negotiation outcomes, showing how SA can enhance transparency and decision-making in international negotiations. In a similar vein, Shukla and Unger (2022) employ machine learning and deep learning techniques to analyze emotional tones in news articles, tracking these scores between countries over time to provide insights into the historical and potential impacts of policy decisions, thereby illustrating the role of artificial intelligence in forecasting IR dynamics.

Koutsoupias and Mikelis (2021) demonstrate the potential of MCA in conjunction with text mining and data analysis for examining scientific discourse. The application of MCA to a dataset of social science journal articles reveals underlying structural patterns in IR, offering a quantitative analysis of the evolution of scientific thought and uncovering latent relationships within academic literature. McCourt and Ruley (2023) employ MCA to analyze the structure of the US national security community's engagement with China-related issues. Their study identifies divisions within the community not based on ideological differences but on the degree of political engagement, distinguishing older think tanks and cultural organizations from business-oriented and politically connected groups. This application of MCA provides a nuanced understanding of the internal dynamics within the US national security establishment, offering a clearer picture of the diverse interests at play in US-China relations.

Kahma and Toikka (2012) apply MCA to the study of cultural consumption patterns in Finland, showcasing how MCA can be used to map cultural preferences without reference to socio-economic factors. The study identifies two key dimensions of cultural engagement: high engagement versus disengagement, and contemporary versus traditional culture. These findings are consistent with similar studies in other countries, illustrating how MCA can reveal underlying patterns of cultural differentiation that transcend socio-economic divisions. Koutsoupias (2024) presents a bibliometric analysis of MCA research, mapping its applications across various disciplines from 1983 to 2021. By examining over 1,200 articles, the study identifies trends, key authors, and significant publication channels, highlighting the growing influence of MCA in the social sciences, humanities, and arts. This meta-analysis offers a comprehensive overview of MCA's development and its increasing relevance as a methodological tool for research. This combined approach enriches traditional SA methods, offering a more sophisticated understanding of global power dynamics and providing a valuable tool for future research in both SA and IR.

Beyond the confines of IR, MCA has demonstrated remarkable versatility across a range of disciplinary contexts, underscoring its efficacy in managing categorical complexity. In the field of cultural economics, Koutsoupias and Boutsiouki (2019) applied MCA alongside HCA to reveal latent structures within the European trade of cultural goods, integrating economic and educational indicators with import-export data. Within the art market, Koutsoupias (2019) utilized MCA to facilitate decision-making among online art consumers by mapping leading sales platforms according to categorical variables such as visit frequency, prior purchasing experience, and platform navigation patterns. In a separate case study, Koutsoupias and Mikelis (2019) examined 58 IR articles published in *Perceptions* (2010–2012), employing Multiple Correspondence Analysis (MCA) in conjunction with Hierarchical Clustering Analysis (HCA) to detect structural patterns within the metadata, thereby linking multivariate techniques to the broader epistemological evolution of the IR field. These studies collectively highlight MCA's value as a tool for organizing data in contexts involving preference hierarchies or multidimensional rankings, conceptually reinforcing its applicability to sentiment clustering in geopolitical analyses.

Additional interdisciplinary evidence of MCA's adaptability emerges from a bibliometric study of painting-related PhD dissertations in Greece (Koutsoupias, 2021), where the technique was instrumental in visualizing institutional output, temporal trajectories, and thematic concentration across 156 records. Similarly, MCA has been employed to uncover structural patterns in IR journal articles, linking metadata and term-frequency distributions to discern thematic configurations within textual corpora (Koutsoupias & Mikelis, 2021). More broadly, the incorporation of ML techniques, including clustering algorithms and NLP, is gaining momentum across adjacent disciplinary landscapes. Concurrently, Bitzenis et al. (2025) conducted a systematic bibliometric analysis of such applications, elucidating emergent trajectories in natural language processing, clustering methodologies, and cross-disciplinary computational modeling. Their findings underscore the growing prominence of hierarchical techniques, particularly as these are adapted and extended across domains such as cultural analytics, historiography, and educational research. Parallel bibliometric research by Koutsoupias et al. (2024), focusing on Mount Athos and broader art and humanities domains, further attests to the expanding role of computational approaches in mapping knowledge production across cultural fields. Collectively, these studies affirm the relevance and cross-contextual transferability of MCA-based frameworks for refining large-scale, thematically heterogeneous datasets in both cultural and geopolitical spheres.

Furthermore, HCA enhances these methodological approaches by enabling the identification of intricate patterns and relationships within complex datasets. Samuelowicz and Bain (2001) employed this technique to analyze academics' beliefs about teaching and learning, distinguishing teaching-centered and learning-centered orientations through qualitative analysis combined with HCA. Similarly, Henry et al. (2005) demonstrated its versatility in family psychology, where it was applied to derive family typologies, examine longitudinal changes, and integrate qualitative and quantitative methods. Pişirgen and Peker (2024) utilized HCA in bibliometric research to categorize scholars based on features such as academic age, productivity, and impact, resulting in profiles like *"rising stars"* and *"superstars"* with practical implications for academic policy. Xu et al. (2020) advanced the methodology by introducing innovative validity indexes to address challenges associated with clustering imbalanced and complex datasets, improving the precision of results. Biresselioglu et al. (2020) applied semantic clustering in Arctic-related bibliometric research, identifying pivotal themes such as governance, climate change, and security. Collectively, these studies underscore the adaptability of HCA, demonstrating its potential to uncover multidimensional patterns and generate valuable insights across diverse disciplines.

In conclusion, the integration of SA with MCA and HCA provides a powerful methodological framework for examining complex issues in IR. By leveraging the strengths of both techniques, researchers can analyze the emotional tone of textual data and uncover deeper insights into how these emotions correlate with other critical factors, such as geopolitical relationships, political ideologies, or economic interests. This approach opens new avenues for understanding global political dynamics and equips scholars with tools to analyze large-scale, unstructured data in a meaningful way.

METHODOLOGY

This study employed a systematic methodological framework that integrates bibliometric analysis, SA, MCA, HCA, and visualization techniques to examine the discourse in USA-China relations. The dataset was compiled from scientific journals in the Scopus database, adhering to a well-defined search strategy to ensure a comprehensive and representative collection of relevant academic publications. The query used to extract the dataset is as follows:

TITLE-ABS-KEY(USA AND CHINA AND ("INTERNATIONAL RELATIONS" OR "INTERNATIONAL AFFAIRS" OR "GLOBAL AFFAIRS" OR "World Politics" OR "Foreign Relations" OR "Global Politics" OR "International Diplomacy" OR "World Affairs")) AND (LIMIT-TO (SUBJAREA,"SOCI") OR LIMIT-TO (SUBJAREA,"ARTS")) AND (LIMIT-TO (DOCTYPE,"ar"))

In addition to the above, the PRISMA 2020 guidelines were followed (Preferred Reporting Items for Systematic Reviews and Meta-Analyses) for the systematic organization and presentation of scientific papers as seen in Figure 1 (Page, 2021).

Figure 1. PRISMA 2020 flow diagram of the selection process

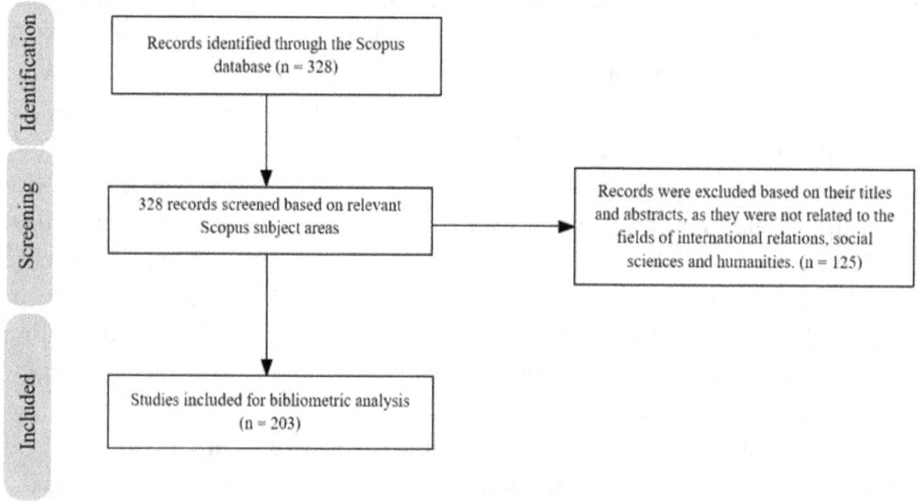

Following the construction of the dataset, the analysis was conducted in three stages: bibliometric analysis, SA, MCA and HCA on the extracted sentiment. These methods were applied sequentially to provide a comprehensive understanding of the academic discourse on USA-China relations. In the first stage, bibliometric techniques using the bibliometrix package were applied to identify key trends in authorship, publication patterns, international collaboration, and annual scientific production. In the second stage, SA was employed to examine the underlying tone and emotion conveyed within the dataset. Utilizing Natural Language Processing (NLP) techniques, the NRC lexicon was used to analyze the sentiment of titles, abstracts, and keywords, identifying dominant attitudes, emotions, and perspectives regarding USA-China relations (Mohammad & Turney, 2013). This analysis offered insights into temporal trends and fluctuations, illuminating how the discourse has evolved in response to geopolitical events. During the second stage, MCA was employed exclusively as a preprocessing step to structure categorical data and reduce noise, ensuring a more stable input for HCA. This refinement facilitated the application of hierarchical clustering in the final stage, where HCA was used to hierarchically categorize sentiment data and identify layered emotional and thematic clusters. This step enabled the visualization of relationships among emotional tones and metadata within the dataset. Together, MCA and HCA provided complementary perspectives, unveiling connections between academic topics and sentiments, and identifying critical intersections within the discourse to yield a deeper understanding of the interplay between emotional responses and thematic focus in scholarly publications.

The integration of these sequential stages directly addresses the methodological complexities inherent in analyzing IR textual data. While bibliometric analysis and SA each offer valuable insights into publication patterns and emotional dynamics, their isolated application falls short of capturing the structured relationships between emotional categories and contextual metadata. By incorporating MCA as an intermediate preprocessing step, the analysis achieves greater dataset coherence, effectively minimizing noise and enabling clearer hierarchical segmentation (Koutsoupias & Mikelis, 2021; McCourt & Ruley, 2023). Subsequently, HCA builds upon this refined dataset to uncover nuanced clusters, delivering an interpretative depth that exceeds what each methodology could offer independently (Koutsoupias, 2024). This structured combination proves particularly effective in dissecting the multifaceted discourse surrounding USA–China relations, yielding a more sophisticated understanding of how sentiment intersects with institutional, geopolitical, and thematic dimensions.

Finally, to facilitate the visualization of results, a range of graphical tools were employed, including network diagrams, thematic maps, and sentiment trend charts. These visualizations were instrumental in distilling complex analytical findings into accessible representations, enhancing the interpretability and impact of the research. Additionally, the results were summarized and presented through detailed tables and graphs, ensuring a clear and comprehensive depiction of the findings.

Results

The results section presents the findings from the bibliometric analysis, SA, MCA, and HCA offering an in-depth examination of key publication trends, international collaboration networks, emotional tones, as well as the relationships and latent patterns among critical variables.

Table 1. Descriptive statistics

Description	Results
MAIN INFORMATION ABOUT DATA	
Timespan	1980:2024
Sources (Journals, Books, etc)	102
Documents	203
Annual Growth Rate %	4.52
Document Average Age	8.56
Average citations per doc	6.793
References	0
DOCUMENT CONTENTS	
Keywords Plus (ID)	251
Author's Keywords (DE)	725
AUTHORS	
Authors	300
Authors of single-authored docs	126
AUTHORS COLLABORATION	
Single-authored docs	129
Co-Authors per Doc	1.52
International co-authorships %	8.867
DOCUMENT TYPES	
article	203

The descriptive statistics offer valuable insights into the research landscape, spanning over four decades from 1980 to 2024. The dataset includes 203 documents sourced from 102 distinct publications, with an annual growth rate of 4.52%, reflecting a consistent increase in scholarly output. The average document age of 8.56 years suggests a balance between foundational and contemporary works, while the average of 6.793 citations per document highlights a moderate impact within the field.

The thematic diversity of the dataset is evident in the inclusion of 251 *"Keywords Plus"* and 725 *"Author's Keywords"*. Authorship patterns reveal 300 distinct contributors, with 126 single-authored documents, indicating that individual scholarship plays a significant role, complemented by a modest level of collaboration. This is further evident in the 1.52 co-authors per document and 129 single-authored documents, emphasizing the prevalence of independent research. The international co-authorship rate of 8.867% points to some degree of global collaboration, though this remains an area with potential for growth to enhance the international scope of the field.

All documents in the dataset are journal articles, indicating a strong preference for peer-reviewed outlets as the primary medium of scholarly communication. The absence of books, book chapters, or conference proceedings reflects a focus on journal-based metrics, often considered the standard measure of academic productivity. Overall, the dataset represents a steadily growing field with moderate impact, diverse thematic interests, and a combination of independent and collaborative research efforts. The relatively low rate of international co-authorship and moderate citation impact suggest opportunities to expand global partnerships and enhance the field's scholarly influence (see Table 1).

Figure 2. Annual scientific production

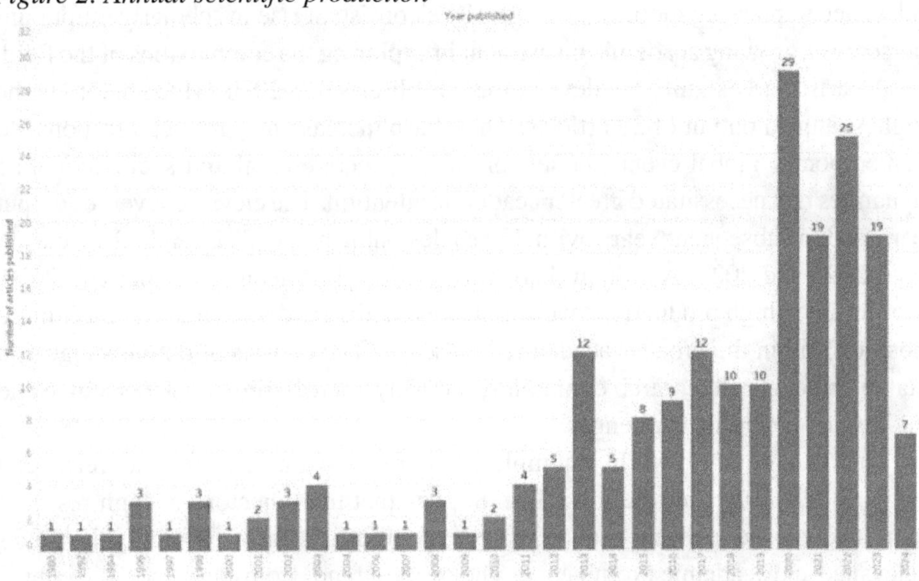

The chart illustrates the annual scientific production within the dataset, covering the period from 1980 to 2024, and provides an insightful representation of the evolution of research activity in the field. The data reveal a slow start in scientific output, with a minimal number of publications recorded during the earlier years. Between 1980 and 2002, annual production remained sporadic, often limited to a single article per year. This modest level of activity likely reflects the field's early developmental phase, characterized by isolated research efforts rather than a coordinated or widespread academic focus.

A gradual increase in research output becomes apparent in the early 2000s, with the number of articles beginning to rise incrementally. By 2003, the annual production reached four articles, signifying the emergence of a more structured engagement with the field. Between 2005 and 2012, annual output fluctuated between one and five articles, indicating a steady but incremental growth. This period can be interpreted as one of foundational consolidation, during which the groundwork for subsequent expansion was being laid. The relatively stable growth during these years reflects an initial broadening of interest and an expansion of the research community contributing to this domain.

The field witnessed a notable acceleration in research activity from 2013 onward. This period is marked by a sustained increase in annual publications, with the number of articles entering double digits in 2016 and maintaining that level thereafter. The steady growth during this phase suggests a maturation of the field, likely driven by its increasing recognition within academic and policy-making circles, as well as its relevance to pressing global challenges. The consistent rise in scholarly output underscores a growing academic investment in exploring the complexities of the field.

A particularly significant development is observed in 2020, which recorded the highest annual output of 29 articles. This sharp increase may reflect a response to contemporary global events, crises, or shifts in geopolitical and socio-economic dynamics that necessitated greater academic attention. The elevated levels of output persisted in subsequent years, with 25 articles published in 2021 and 19 articles in both 2022 and 2023. Although there was a slight decline after the peak in 2020, the sustained high production levels indicate that the field has achieved a solidified position within the broader academic landscape. The consistent output suggests a stable and engaged research community, actively contributing to the discourse and addressing emerging challenges.

The data for 2024, while incomplete, records seven articles to date, reflecting ongoing scholarly contributions. This implies that the trajectory of high research activity observed in recent years is likely to continue, further reinforcing the field's relevance and academic prominence. The overall trend, from early sporadic contributions to sustained and substantial outputs in recent years, illustrates the dynamic evolution of research activity in the field.

In conclusion, the chart provides a comprehensive overview of the growth trajectory of research output in this domain. The significant increase in annual publications, particularly over the last decade, underscores the field's expanding influence and its growing capacity to address contemporary global issues. The sharp rise in 2020 and the subsequent sustained production levels underscore the responsiveness of the academic community to evolving challenges, further cementing the field's importance within the broader landscape of international research (see Figure 2).

Figure 3. Most frequently used words

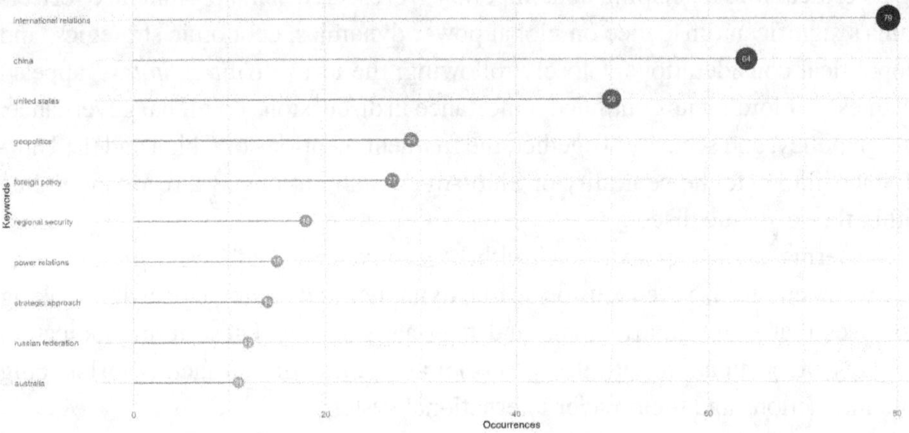

The analysis of the most frequently used keywords, as presented in Figure 3 and Table 2, offers valuable insights into the thematic focus of the research within the dataset. The frequency of specific terms reflects the prominence of recurring concepts and central themes, elucidating the intellectual priorities and dominant areas of inquiry in the field, highlighting the overall structure and orientation of the scholarly discourse.

Table 2. Words with the highest frequency

Words	Occurrences
international relations	79
china	64
united states	50
geopolitics	29
foreign policy	27
regional security	18
power relations	15
strategic approach	14
russian federation	12
australia	11

The term *"international relations"* dominates, with 79 occurrences, reaffirming its centrality as the foundational framework for the discourse. This prevalence highlights the overarching relevance of the field to the diverse topics explored in the dataset.

With 64 mentions, *"China"* emerges as the second most frequent term, underscoring its critical role in shaping contemporary IR research. This prominence reflects China's significant influence on global power dynamics, economic strategies, and geopolitical considerations. Closely following, the term *"United States"* appears 50 times, reinforcing its continued importance in discussions on global governance, foreign policy, and security. Together, the frequent mentions of China and the United States illustrate the centrality of Sino-American relations as a recurring theme within the academic discourse.

Key terms such as *"geopolitics"*, with 29 occurrences, and *"foreign policy"*, with 27, further emphasize the focus on spatial dynamics and strategic decision-making processes that shape international and regional stability. These terms indicate a scholarly preoccupation with the territorial and policy-driven factors influencing state interactions and the broader international system.

The term *"regional security"*, appearing 18 times, highlights the importance of examining localized dynamics within the context of broader IR. This focus suggests strong academic interest in understanding security issues specific to particular regions, often shaped by the strategic actions of major powers. Similarly, *"power relations"*, which appears 15 times, and *"strategic approach"*, mentioned 14 times, reflect a concerted focus on power dynamics and the calculated strategies states use to achieve their objectives on the global stage.

The inclusion of terms such as *"Russian Federation"*, with 12 occurrences, and *"Australia"*, with 11, points to their significance as case studies or key regional actors in IR. These mentions underscore the relevance of their roles in geopolitical tensions, strategic alliances, and regional security dynamics, contributing to a deeper understanding of their place in global discourse.

In summary, the distribution of these frequently used keywords demonstrates a nuanced interplay between global themes, such as IR, geopolitics, and power dynamics, and region-specific issues involving major actors like China and the United States. The prominence of both universal and localized themes reflects the complexity and breadth of academic inquiry within the field, offering a well-rounded perspective on the priorities and focus areas shaping contemporary IR research (see Figure 3 and Table 2).

Figure 4. Trend topics

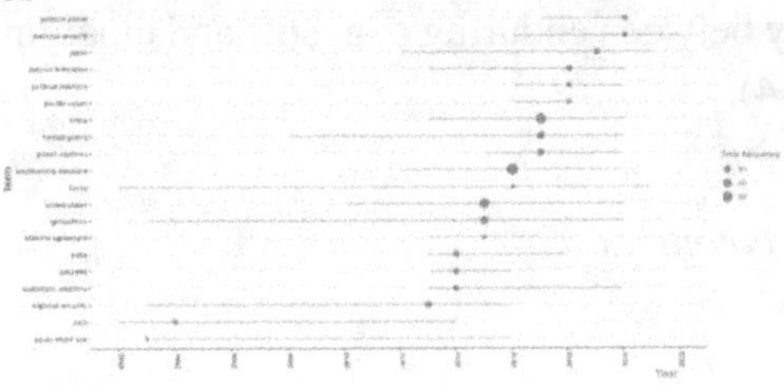

The trend topics graph provides a dynamic overview of the temporal evolution and prominence of key concepts within the dataset. The graph demonstrates how specific terms have gained or maintained relevance over time, highlighting the shifting priorities and focal areas of academic discourse in IR.

Notably, terms such as *"international relations"*, *"China"*, and *"United States"* show consistent appearances, indicating their enduring importance in the field. These terms dominate the dataset and reflect central themes of global governance, power dynamics, and strategic interactions. The term *"China"* exhibits a strong presence in recent years, aligning with its growing influence in international politics and its pivotal role in shaping regional and global affairs.

On the other hand, *"Geopolitics"* and *"foreign policy"* emerge as significant terms, with increasing frequency starting in the mid-2010s. This trend corresponds to heightened academic interest in spatial and strategic analyses, driven by geopolitical shifts and the redefinition of state interactions in the modern era. Similarly, *"regional security"* and *"power relations"* gain traction in this period, underscoring the scholarly focus on security challenges and the mechanisms of power distribution in both regional and international contexts.

Regional-specific terms, such as *"South China Sea"*, *"Asia"*, and *"Australia"*, also feature prominently, reflecting a growing emphasis on the geopolitical dynamics of the Indo-Pacific region. These terms illustrate the strategic importance of this area in contemporary IR and its implications for maritime security, territorial disputes, and multilateral engagements.

The visualization also highlights terms such as *"diplomatic relations"*, *"bilateral agreement"*, and *"national security"*, which gain relevance at different points, suggesting a diverse yet interconnected discourse. The presence of *"Japan"*, *"Korea"*, and *"Russian Federation"* as terms further points to the inclusion of key regional actors in discussions on global politics, demonstrating the complexity of IR as a field.

Overall, the chart unveils a clear progression in the academic discourse, where foundational themes such as IR and power dynamics are complemented by emerging regional and thematic issues. The evolving frequency of terms underscores the dynamic nature of IR research, shaped by historical events, geopolitical shifts, and changing global priorities. This visualization provides a comprehensive lens to examine the interplay between enduring concepts and emerging areas of interest in the field (see Figure 4).

Figure 5. Most cited countries

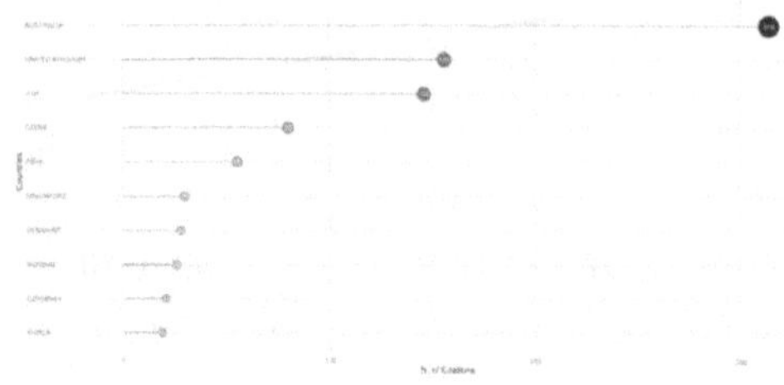

The graph and table illustrating the most cited countries offer a quantitative assessment of the global distribution of scholarly impact within the field, as measured by Total Citation (TC) counts. The dataset offers valuable insights into the geographic spread of influential research outputs, highlighting the disparities in citation impact across various nations. It is evident that certain countries exhibit a disproportionately high level of scholarly influence, with notable variations in their respective citation counts.

Table 3. Countries with the highest total citations

Country	TC
AUSTRALIA	314
UNITED KINGDOM	156
USA	146
CHINA	80
INDIA	55
SINGAPORE	30
DENMARK	28
NORWAY	26
GERMANY	21
KOREA	19

Australia emerges as the preeminent contributor in this dataset, with a total of 314 citations, significantly surpassing other countries. This total citation count highlights Australia's pivotal role in shaping academic research within the field, driven by its focus on producing high-quality work, robust academic networks, and strategic international partnerships. Australia's leading position underscores the significant influence of its academic community in shaping research priorities and advancing intellectual discourse within the discipline.

The United Kingdom and the United States follow as the second and third most cited countries, with citation counts of 156 and 146, respectively. These figures exemplify the enduring academic influence of research produced in these nations, characterized by advanced methodologies, diverse thematic contributions, and extensive collaborations within the global academic community. The prominence of these countries in terms of citation counts illustrates their ability to generate research of broad applicability and significance, consolidating their positions as key contributors to global scholarly discourse.

China, ranked fourth with 80 citations, demonstrates the growing prominence of emerging research contributors. The increasing citation impact of Chinese academic work indicates the country's strategic investments in academic development and its growing engagement in international research collaborations. China's rising influence is further bolstered by its focus on cutting-edge research areas, including data-driven methodologies and computational analysis, which align with contemporary trends in global research.

India, with a citation total of 55, also demonstrates a notable presence in the table. Although this figure is comparatively lower than those for Australia, the United Kingdom, and the United States, it reflects India's growing contributions to specific thematic areas within the field, particularly those related to regional issues and innovative approaches. Despite the relatively modest citation count, this data highlights the growing recognition of Indian research within global academic circles.

Smaller nations such as Singapore, Denmark, and Norway, with citation totals of 30, 28, and 26 respectively, exemplify the impactful contributions of countries with limited academic output on a global scale. Despite their relatively smaller research output, these countries maintain a significant level of citation influence, attributable to their focus on specialized research areas and their integration into established international academic networks. This concentrated expertise enables these nations to produce research that resonates within subfields, resulting in high citation counts within those domains.

Germany and Korea, with 21 and 19 citations respectively, represent the steady, consistent contributions of established academic systems. These nations, although not leading in citation impact, demonstrate the reliability and sustained nature of their scholarly output. The relatively modest citation counts illustrate the continued recognition of their research, which is often characterized by a focus on quality over quantity. Both nations' academic systems emphasize rigorous scholarship, leading to influential research that maintains a notable academic presence despite relatively lower citation counts.

In summary, the dataset reveals a pronounced unevenness in the distribution of citation impact across countries. While countries such as Australia, the United Kingdom, and the United States dominate the citation rankings, the presence of smaller or emerging contributors such as Singapore, Denmark, and India illustrate the diverse geographic origins of influential research. This analysis not only reflects the geographic reach of scholarly influence but also emphasizes the interconnected nature of modern academic networks, where research contributions from a wide range of nations collectively shape the intellectual trajectory of the field (see Figure 5 and Table 3).

Figure 6. Co-occurrence network

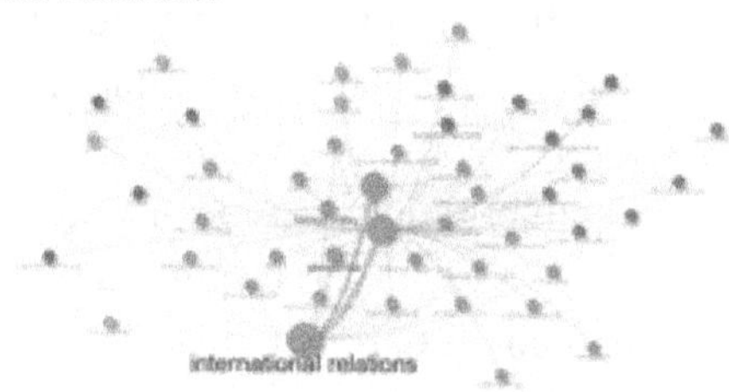

The co-occurrence network provides a detailed representation of how key topics within IR research are interconnected through shared scholarly attention. The network is constructed based on keyword co-occurrence, with nodes representing individual themes and the size of each node proportional to its frequency within the dataset. Larger nodes suggest higher prevalence, conveying the centrality of those themes in the academic discourse. The links between nodes depict co-occurrences, with thicker lines denoting stronger associations between themes. This structure enables the visualization of thematic relationships and the identification of prominent clusters within the research domain.

At the core of the network, the node for *"international relations"* emerges as the most prominent, signifying its foundational role in the field. Its large size and extensive connections underscore its centrality as a hub that links diverse research topics. Closely associated nodes, such as *"foreign policy"* and *"geopolitics"*, are also prominently represented, reflecting their significance as key areas of inquiry within the discipline. The strong links between these themes suggest frequent co-occurrence, implying their interconnected nature and the tendency of researchers to explore these topics in tandem.

Peripheral nodes, such as *"asean"* and *"political economy"*, are smaller in size and exhibit fewer connections to the core themes. These nodes represent secondary or niche areas of research that, while significant, do not dominate the broader field.

For instance, *"asean"* is linked to topics such as *"regional politics"*, indicating its role as a specific regional focus within the wider geopolitical discourse.

Lastly, the graph emphasizes both well-established research areas and emerging or specialized topics. Larger nodes and stronger connections signal areas of concentrated academic effort and widespread relevance, while smaller nodes and thinner connections suggest more fragmented or nascent research. This network thus serves as a valuable tool for understanding the intellectual structure of IR, identifying dominant themes, and uncovering areas that may warrant further exploration (see Figure 6).

Figure 7. Thematic map

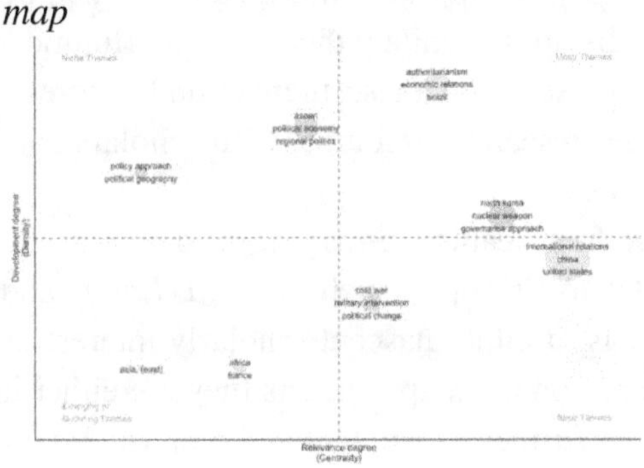

The thematic map provides an overview of the research landscape in IR by categorizing themes based on their relevance (*centrality*) and development (*density*). Centrality measures a theme's significance within the broader research network, indicating its integration and influence across various topics. Density reflects the internal coherence and maturity of a theme, signaling how developed the research in that area is. The map is divided into four quadrants, each representing a distinct category of thematic relevance: motor themes, niche themes, basic themes, and emerging or declining themes.

In the upper-right quadrant, motor themes are highlighted as highly central and well-developed. These themes represent the most influential and advanced areas of research, driving the intellectual discourse in IR. In the current analysis, topics such as *"authoritarianism"*, *"economic relations"*, and *"Brazil"* stand out in this quadrant. Their positioning reflects their dual role as pivotal to the research network and internally cohesive. likely due to its significance in understanding global governance challenges and shifts in political systems. Similarly, economic relations represent a substantial research stream exploring trade dynamics, financial interdependence, and their implications for international diplomacy. The inclusion of Brazil in this category suggests concentrated scholarly attention on the nation's geopolitical strategies and its influence in regional and global contexts.

The upper-left quadrant contains niche themes, which are well-developed but lack broader relevance within the research network. Themes such as *"policy approach"* and *"political geography"* are categorized here, reflecting their specialization and depth within narrower subfields. While these themes possess high internal coherence, their limited centrality implies they cater to more focused scholarly audiences or address specialized research questions.

Basic themes, found in the lower-right quadrant, are central to the field but underdeveloped. They serve as the foundation for future research and have the potential for further exploration and expansion. In this analysis, *"international relations"*, *"China"*, and *"United States"* appear in this quadrant. Their positioning highlights their centrality to the discipline while indicating opportunities for additional study and refinement. The presence of these themes underscores their importance as anchors for the broader research discourse, with scholars continually building on these foundational areas.

The lower-left quadrant features emerging or declining themes, characterized by low centrality and density. Topics such as *"Asia (East)"* and *"Africa"* are found here, representing areas of either nascent scholarly interest or waning focus. The position of these themes on the map suggests they are either in the early stages of development, requiring further conceptual and empirical work, or experiencing a decline in academic engagement.

Overall, the thematic map provides a comprehensive visualization of the intellectual structure within the field of IR. It reveals the dynamic interplay between well-established areas of research, emerging topics, and specialized niches, offering valuable insights into the evolving priorities of the scholarly community (see Figure 7).

Figure 8. Country collaboration map

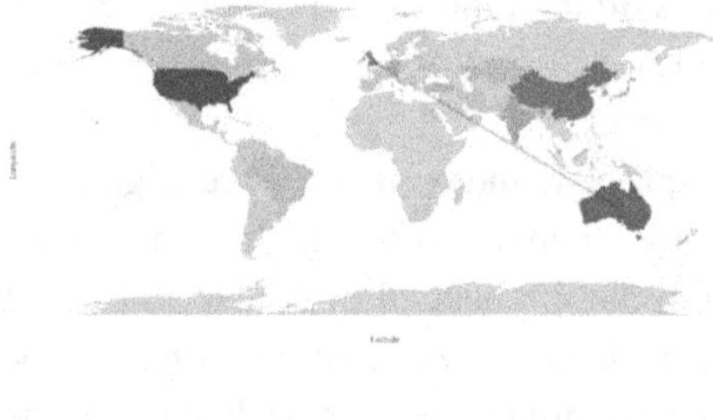

The country collaboration map and table provide insights into the interconnected nature of academic research in the field of IR. The map identifies key contributors, collaboration hubs, and patterns of interaction, offering a data-driven perspective on how research is distributed and conducted across nations.

Table 4. Country collaboration table

From	To	Frequency
Australia	United Kingdom	2
China	Australia	1
China	Kazakhstan	1
China	Singapore	1
China	United Kingdom	1
Italy	Switzerland	1
Japan	Austria	1
Japan	Norway	1
Korea	Bulgaria	1
Korea	Malaysia	1

The central positioning of the United States indicates its extensive involvement in international research collaboration. Table 4 illustrates significant connections with several countries, including the United Kingdom, Germany, and China. The density and breadth of these connections suggest a well-established infrastructure supporting transnational research initiatives. This pattern is likely influenced by the availability of funding resources, institutional capacity, and historical academic networks. Notably, the collaborative pathways involving the United States encompass both Western and Asian counterparts, signifying a wide-ranging scope of partnerships.

China's presence in the map reflects its substantial role in fostering international academic collaborations. Connections with neighboring countries such as Japan, South Korea, and India are prominent, as are linkages with major Western nations. The visualization suggests a dual focus on regional and global research networks. This positioning aligns with broader trends of increasing research output and expanding international engagement within China's academic community, as evidenced by the growing number of co-authored publications involving Chinese institutions.

In Europe, the clustering of nations such as Germany, France, and the United Kingdom is evident. These countries exhibit high levels of intra-regional collaboration, with strong ties extending to the United States and China. The map shows interconnectedness within the European research landscape, potentially reflecting shared funding mechanisms, collaborative frameworks, and policy priorities. Smaller but research-active nations, such as the Netherlands and Switzerland, also demonstrate significant connections, indicating their roles as contributors to broader academic networks.

Australia's position in the collaboration map depicts its strong affiliations with Western countries, particularly the United States and the United Kingdom. The map also shows partnerships extending to the Asia-Pacific region, indicating a dual orientation toward regional and international research. Similarly, Japan and South Korea appear as active participants, with visible connections to both regional and global academic hubs. These patterns are consistent with their historical emphasis on research in areas such as security studies and regional cooperation.

The map also provides data on underrepresented regions, such as Africa and Latin America, where the density of collaborations is comparatively lower. While countries like South Africa and Brazil appear as notable contributors, their networks are less extensive compared to those of the United States or Europe. The observed disparities may reflect differences in academic funding, infrastructure, and access to global research platforms. Emerging linkages in these regions, however, suggest increasing participation in international academic networks.

Collaboration between nations often involves multilateral frameworks or institutional partnerships. The map indicates the role of multinational entities in facilitating interactions among countries with differing research capacities. For instance, countries with fewer direct collaborations often show connections mediated through academic consortia or international organizations, underscoring the importance of such entities in bridging gaps in research output.

The bilateral linkages represented in the map also reveal key partnerships. For instance, the United States and Canada exhibit strong connections, likely influenced by geographic proximity and shared academic traditions. Similarly, ties between China and Russia are visible, suggesting mutual academic engagement. These bilateral networks may serve as conduits for addressing region-specific issues or shared research priorities.

In summary, the country collaboration map and table offer a detailed representation of global research networks. The map illustrates the distribution and intensity of academic partnerships, highlighting variations in collaborative capacity and regional focus. By presenting the results in a structured format, the map offers a clearer understanding of the dynamics driving international research efforts (see Figure 8 and Table 4).

Figure 9. Country scientific production map

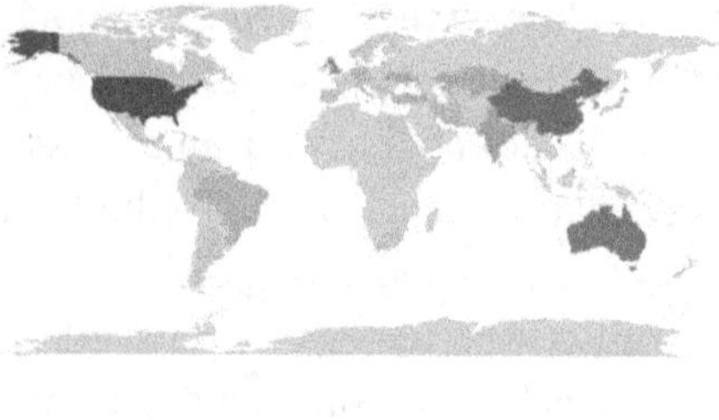

The country scientific production data, as summarized in Table 5 and shown in Figure 9, highlight the global distribution of academic contributions to the field. This distribution reflects the geographical concentration of research output and the varying degrees of scholarly engagement across regions, offering critical insights into the regional and institutional dynamics that shape the intellectual development of the discipline.

Table 5. Country scientific production table

Country	Frequency
USA	32
China	24
Australia	21
UK	18
South Korea	11
India	10
Germany	7
Kazakhstan	7
Ukraine	7
Brazil	6

The United States leads with 32 contributions, highlighting its position as a dominant player in IR studies. This prominence aligns with the robust infrastructure of U.S.-based academic institutions, which are well-funded and globally connected, fostering a consistent output of high-quality research.

China ranks second with 24 contributions, exhibiting its expanding role as a significant contributor to IR research. This growing academic output mirrors China's rising geopolitical and economic influence, as well as its increasing investment in academic development and international collaborations. Australia's third position, with 21 contributions, demonstrates its emphasis on research related to the Indo-Pacific region and strategic security, areas that align closely with its geopolitical interests.

The United Kingdom, with 18 contributions, continues to demonstrate its longstanding influence in the field, rooted in its strong academic tradition and global research networks. South Korea and India, with 11 and 10 contributions respectively, reflect the increasing scholarly engagement from Asian countries in addressing regional security and global governance issues. This involvement exemplifies the growing relevance of these nations in shaping contemporary IR discourse.

Germany, Kazakhstan, and Ukraine each contribute seven publications, highlighting the diversity of European and Central Asian engagement in the field. Germany's academic strength is well-established, while Kazakhstan and Ukraine's presence signals an emerging interest in leveraging regional perspectives to address broader international issues. Brazil, with six contributions, represents Latin America's input into the global dialogue on IR, particularly in areas related to regional security and multilateralism.

This geographic distribution underscores the field's multidimensional nature, with contributions spanning various regions and reflecting both global and localized academic priorities. The prominence of nations such as the United States, China, and Australia points to their leadership in shaping discourse on IR, while the involvement of countries like Kazakhstan, Ukraine, and Brazil illustrates the increasing diversification of scholarly voices. These findings highlight the interconnected nature of IR research, where global challenges are addressed through a multiplicity of perspectives and regional insights (see Figure 9 and Table 5).

Table 6. Most cited papers

Paper	Total Citations
Campbell H, 2007, Third World Q	115
Boylan Bm, 2020, Fudan J Humanities Soc Sci	82
Pan C, 2014, Aust J Int Aff	76
Hughes Cw, 2016, Chin J Int Pol	59
Medcalf R, 2014, Aust J Int Aff	57
Bisley N, 2013, Aust J Int Aff	43
Fiori A, 2015, Pac Rev	41
Schindler S, 2021, Trans Inst Br Geogr	37
Odgaard L, 2003, Secur Dialogue	28
Jacob Jt, 2020, China Rep	27
Mackerras C, 2015, East Asia	27
Lai Hh, 2001, Third World Q	27
Clarke M, 2016, Aust J Int Aff	26
He K, 2018, Aust J Int Aff	25
Layne C, 2017, Int Polit	23

The table with the 15 most cited papers provides critical insights into the thematic and geographic priorities of contemporary IR, addressing global power dynamics, regional security, and institutional strategies. Together, these studies form a foundation for understanding the complexities of international politics and diplomacy.

Campbell (2007) examines China's expanding influence in Africa, focusing on the country's strategic partnerships and their broader implications for global power competition. The study highlights the intersection of economic transformation, resource extraction, and challenges to U.S. dominance, offering valuable perspectives on China's evolving role in global governance. Boylan et al. (2020) explore the U.S.-China trade war, focusing on the nationalist motivations underlying protectionist policies. By addressing economic and political tensions between the two largest global economies, this work underscores the interplay of economic interdependence and political rivalry.

Pan (2014) critically evaluates the concept of the Indo-Pacific, challenging its naturalness and revealing its role in fostering great power competition and regional security dilemmas. Medcalf (2014) builds on this theme by analyzing the Indo-Pacific as a maritime super-region and its implications for Asian security, with particular attention to Australia's strategic policies in response to the region's dynamics. Bisley (2013) further examines Australia's alignment with the United States, emphasizing the domestic and international drivers behind its strong alliance in the context of China's rise.

Hughes (2016) investigates Japan's strategic transition from hedging to active balancing in its approach to China, with implications for regional security and U.S.-Japan relations. Fiori and Passeri (2015) address Southeast Asia's strategic responses to U.S.-China competition, using Myanmar as a case study to illustrate the hedging strategies employed by smaller states to navigate great power rivalries. Schindler et al. (2021) explore the geopolitical-economic competition between China and the United States, focusing on transnational infrastructure development as a key arena for global rivalry. The study highlights the opportunities and risks for smaller states engaging with competing powers.

Odgaard (2003) analyzes the South China Sea as a critical focal point for both cohesion and division within asean, emphasizing the pivotal role of regional consensus in addressing security challenges and mediating the strategic competition between the United States and China. Jacob (2020) examines China's external propaganda strategies during the COVID-19 pandemic, situating these efforts within a broader framework aimed at rehabilitating its global image and advancing its foreign policy objectives. Mackerras (2015) investigates the influence of the Xinjiang region on China's foreign relations, exploring the complex interplay between economic development facilitated by the Silk Road initiatives and the enduring challenges of regional instability and Islamist terrorism.

Lai (2001) provides a detailed exploration of China's accession to the WTO, offering insights into the domestic and international factors that facilitated this milestone and its broader implications for China's integration into the global economy. Clarke (2016) focuses on China's approach to Afghanistan, connecting Beijing's regional policies to its One Belt, One Road initiative and broader geopolitical ambitions. He (2018) introduces a theoretical framework incorporating institutional balancing and role theory to explain variations in state strategies during periods of order transition. Finally, Layne (2017) critiques the United States' grand strategy of primacy, advocating for offshore balancing as a means to better accommodate China's rise and maintain global stability.

Overall, all these highly cited papers collectively address key themes in IR, including the strategic significance of the Indo-Pacific, the dynamics of Sino-American competition, regional security challenges, and the evolving roles of economic and institutional frameworks. By integrating global and regional perspectives, these works provide essential theoretical and empirical contributions that continue to shape academic discourse and inform policy-making in an increasingly complex global order (see Table 6).

Table 7. Keywords and their sentiment with the highest frequency

Word	Sentiment	N times appeared
policy	trust	15
war	fear	14
war	negative	14
strategic	positive	12
foreign	negative	11
pacific	positive	11
diplomacy	anticipation	9
diplomacy	positive	9
diplomacy	trust	9
united	positive	8

Figure 10. Wordcloud of the most frequent words by the NRC lexicon

The SA conducted using the NRC lexicon provides a quantitative insight into the emotional dimensions of key terms within the field of IR. The association between specific words and their corresponding sentiments offers a more nuanced understanding of the prevailing tones in the discourse.

The analysis highlights the emotional dimensions of key terms within the discourse of IR. The term *"policy"*, linked to the sentiment *"trust"*, appears 15 times, emphasizing its role in discussions centered on trust-building and reliability. This suggests that policy serves as a mechanism to foster confidence and stability, reflecting trust as a recurring theme in diplomatic and strategic decision-making processes. Conversely, the term *"war"*, associated with both *"fear"* and *"negative"* sentiments and appearing 14 times, reveals anxiety, instability, and undesirable consequences often tied to conflict. These contrasting associations underscore the dual emotional underpinnings of trust and fear that shape the discourse, offering a nuanced perspective on the prevailing tones in IR.

The term *"strategic"*, associated with the sentiment *"positive"*, appears 12 times, implying that discussions of strategy are frequently framed in an optimistic or constructive light. This indicates that strategic considerations are often viewed in terms of opportunity, foresight, and the potential for favorable outcomes. The *"positive"* sentiment linked to *"strategic"* reinforces a forward-looking perspective, where planning and calculated actions are associated with achieving beneficial objectives in IR. In contrast, the term *"foreign"*, associated with a *"negative"* sentiment and appearing 11 times, highlights the cautious or uncertain tone often characterizing discussions of foreign relations. This negative sentiment likely conveys the complexities and risks inherent in international diplomacy, including the potential for discord or misunderstandings between states. The presence of *"foreign"* in a *"negative"* context accentuates concerns about the unpredictability and challenges posed by cross-border interactions.

The word *"pacific"*, associated with a *"positive"* sentiment and appearing 11 times, conveys an emphasis on peaceful relations and cooperation, particularly within the Pacific region. This association signifies the acknowledgment of diplomacy and stability as fundamental to fostering constructive relationships among states in this geopolitically significant area. The *"positive"* sentiment linked to *"pacific"* reinforces the focus on peacebuilding and conflict resolution within this context. Similarly, *"diplomacy"* associated with *"anticipation"*, *"positive"*, and *"trust"* sentiments and appearing nine times for each, emerges as a central term in the dataset. The range of sentiments connected to diplomacy highlights its multifaceted role in IR, with *"anticipation"* reflecting forward-looking expectations, *"positivity"* indicating hope and constructive engagement, and *"trust"* signifying its foundational importance. Together, these associations illustrate diplomacy as a critical mechanism for managing expectations, fostering cooperation, and ensuring stability in global politics.

Lastly, the term *"united"*, associated with *"positive"* sentiment and appearing eight times, underscores the value of unity and collective action in IR, emphasizing collaboration and mutual effort as vital for addressing global challenges and achieving shared goals. The SA of frequent terms in IR reveals a nuanced balance of emotional tones. Words such as *"policy"*, *"strategic"*, *"pacific"*, and *"united"*, are linked to *"positive"* sentiments, highlighting themes of cooperation, stability, and proactive planning. Conversely, terms like *"war"* and *"foreign"* carry *"negative"* connotations, reflecting concerns over conflict, instability, and the complexities of diplomacy. These findings offer insight into the emotional and rhetorical dimensions shaping the discourse on global political issues (see Table 7).

Table 8. Sentiment distribution across the dataset

Sentiment	Count
positive	134
trust	101
negative	95
fear	52
anger	44
anticipation	33
sadness	17
joy	13
surprise	11
disgust	5

Figure 11. NRC sentiment distribution

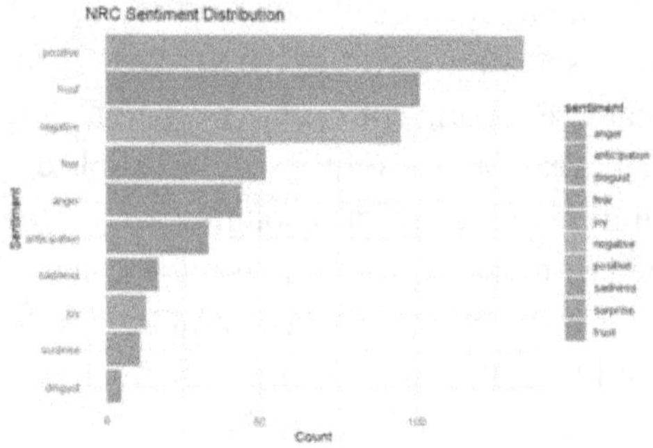

The sentiment distribution across the dataset highlights a multifaceted emotional landscape within IR discourse. Positive sentiments, particularly those related to trust, dominate the analysis, reflecting a scholarly emphasis on themes such as cooperation, diplomacy, and strategic optimism. Negative emotions, including fear, anger, and anticipation, highlight the inherent tension and uncertainty that characterize global political dynamics. These sentiments offer critical insights into rhetorical strategies and thematic underpinnings of the analyzed political discourse, revealing a complex interplay between constructive engagement and critical analysis that shapes the field of IR.

The *"positive"* sentiment emerges as the most frequent, appearing 134 times, highlighting the optimistic or favorable tone that predominates much of the discourse. This implies that discussions within the dataset tend to focus on constructive or forward-looking perspectives, often emphasizing the potential for cooperation, collaboration, and positive outcomes in the realm of IR. Similarly, the sentiment of *"trust"* is notably prominent, with 101 occurrences. This frequency indicates that the establishment of trust, both between states and within the broader geopolitical context, is a critical theme. Trust-related sentiments are indicative of the importance placed on building stable, reliable relationships, a core element in diplomatic and international political strategies.

The *"negative"* sentiment, recorded 95 times, underscores the challenges, uncertainties, and conflicts that pervade international politics. Its presence within the discourse highlights the skepticism, tension, and dissent that frequently shape global political interactions. This sentiment suggests that IR are often framed through the lens of caution, adversity, and contestation, reflecting competing interests and power structures. *"Fear"*, appearing 52 times, emerges as a significant sentiment, emphasizing the centrality of apprehension and perceived threats in shaping political narratives. This sentiment draws attention to the uncertainty and insecurity that often accompany discussions of global stability, security challenges, and military conflicts. The prominence of fear within the dataset highlights a heightened sensitivity to risks and potential disruptions in international order, illustrating its critical role in influencing diplomatic strategies and policymaking.

The sentiment *"anger"*, appearing 44 times, adds to the negative emotional undertones within the dataset, signaling frustration or indignation over perceived injustices or violations of international norms. This sentiment highlights the emotional intensity often present in diplomatic disputes or contentious geopolitical issues. In contrast, *"anticipation"*, with 33 occurrences, reflects the forward-looking nature of much political discourse. It captures expectations of future developments, decisions, or actions, emphasizing the anticipatory tone surrounding shifts in geopolitical dynamics, policy changes, and the resolution of ongoing conflicts. Together, these sentiments illustrate the interplay of emotional intensity and forward-thinking perspectives in the discourse on IR.

The sentiments of *"sadness"* occurring 17 times, and *"joy"* appearing 13 times, emphasize the affective responses to outcomes in IR. Sadness likely conveys disappointment or mourning over failed negotiations or conflicts, while joy represents moments of success, such as peace agreements or diplomatic achievements. The sentiment of *"surprise"*, with 11 occurrences, and *"disgust"*, appearing 5 times, further accentuate the emotional variability within the dataset. These sentiments suggest that IR often evoke unexpected or intense reactions, particularly in response to unforeseen events or controversial actions on the global stage.

The distribution of sentiments reveals a complex and multifaceted emotional landscape within the context of IR. While positive sentiments such as trust feature prominently in the discourse, negative emotions, particularly fear and anger, alongside anticipation, underscore the challenges and tensions inherent in global politics. This range of sentiments offers valuable insights into the emotional undercurrents shaping diplomatic discourse, highlighting the critical role of affective responses in the interpretation and analysis of international political communication (see Table 8).

Table 9. Cluster 1 characteristics based on the NRC lexicon

Description	Cluster 1
positive	10.6
trust	8.63
anticipation	6.82
joy	5.73
negative	4.78
surprise	4.15
LA=RUSSIAN	3.47
sadness	2.99
OA=ALL OPEN ACCESS; GOLD OPEN ACCESS	2.77
fear	2.7
anger	2.58
Affiliations=PRIMAKOV NATIONAL RESEARCH INSTITUTE OF WORLD ECONOMY AND INTERNATIONAL RELATIONS, RUSSIAN ACADEMY OF SCIENCES (IMEMO), 23, PROFSOYUZNAYA STR., MOSCOW, 117997, RUSSIAN FEDERATION	2.4
disgust	1.96

The results from the applied MCA and HCA on SA provide a detailed understanding of the relationships between emotional sentiments and associated metadata, including institutional affiliations and access types, within the dataset. Cluster 1, comprising 126 instances, highlights a strong association between specific emotional tones and particular institutional contexts and metadata characteristics, uncovering underlying trends in the corpus of IR research.

The sentiments in Cluster 1 are primarily characterized by *"positive"* tones, with *"trust"* and *"anticipation"* emerging as dominant themes. These sentiments suggest a prevailing tone of optimism, forward-thinking, and confidence in the diplomatic and geopolitical discussions analyzed. The emphasis on positive emotions reflects a scholarly focus on proactive IR strategies, emphasizing cooperation, trust-building, and anticipatory measures as central to the discourse. The presence of *"joy"*, though less frequent, further reinforces the positive emotional orientation, indicating that successful diplomatic engagements, peace-building efforts, and optimistic projections are recurring themes in the analyzed texts.

While positive sentiments dominate, *"negative"* emotions and *"surprise"* appear at lower frequencies. These sentiments likely reflect critical perspectives or unforeseen geopolitical developments. However, their relatively lower occurrence underscores that the overall tone of the research remains constructive, emphasizing potential solutions, strategic alliances, and diplomatic negotiations over challenges or conflicts in IR.

Institutional affiliations, particularly with Russian entities, play a significant role in this cluster. A strong connection emerges between Russian-based research and the positive emotional sentiments observed, highlighting the importance of institutions such as the Primakov National Research Institute of World Economy and International Relations (IMEMO). This finding underscores the contribution of Russian academic institutions to the discourse on trust, anticipation, and diplomatic strategies within IR.

The association with open-access publications further emphasizes the inclusive and transparent nature of the research. The prominence of open-access outputs highlights the broad availability and dissemination of these studies, fostering a more accessible and inclusive dialogue in the field. The alignment between open access and sentiments such as *"trust"* and *"anticipation"* suggests that the research aims to engage a wider audience and promote collaborative, forward-thinking discussions.

Although *"fear"* and *"anger"* appear in the dataset, they occur less frequently, indicating that while the research acknowledges security concerns, risks, and geopolitical tensions, these themes are not as central to the cluster. These sentiments likely arise in the context of geopolitical instability, military conflicts, or territorial disputes but are less prominent compared to the overarching emphasis on constructive engagement and strategic cooperation. Similarly, sentiments such as *"sadness"* and *"disgust"* reflect the acknowledgment of historical conflicts or perceived unethical actions in the geopolitical context, yet their lower frequencies confirm that these negative emotions are not the primary focus of the cluster.

In conclusion, the results for Cluster 1 reveal a significant alignment between positive emotions and strategic, forward-looking discussions in IR research. Thc association with Russian-affiliated institutions and the prevalence of open-access publications further emphasize the scholarly focus on transparency, cooperation, and constructive dialogue. Whilc critical and negative emotions are acknowledged, they appear less frequently, with trust, anticipation, and optimism defining the dominant tone of the research within this cluster (see Table 9).

Table 10. Cluster 2 characteristics based on the NRC lexicon

Description	Cluster 2
LA=ENGLISH	3.13
OA=NA	2.88

The analysis of Cluster 2 reveals an association between language of publication and access status, with English-language articles (3.13) and non-open access publications (2.88) being the defining characteristics of this cluster. This pattern reflects the broader trend in IR discipline, where English serves as the primary language of academic discourse, and a substantial portion of research output remains restricted behind paywalls.

In contrast to the other clusters, Cluster 2 does not exhibit a strong correlation with specific sentiment categories. Instead, it highlights the structural properties of the dataset, indicating that language and access type are relevant factors shaping the dissemination of research but do not, in this case, determine sentiment distribution. The lack of a distinct emotional pattern suggests that English-language, non-open access publications do not systematically align with sentiment expressions, at least within the dataset examined.

This finding underscores the importance of considering contextual publishing factors when analyzing sentiment trends in international relations literature. Future research could explore whether variations in journal policies, accessibility, and linguistic conventions contribute to differences in sentiment framing or rhetorical approaches in academic discourse (see Table 10).

Table 11. Cluster 3 characteristics based on the NRC lexicon

Description	Cluster 3
TC	11.4
Affiliations=DEPARTMENT OF POLITICAL SCIENCE, SYRACUSE UNIVERSITY, NEW YORK, NY, UNITED STATES	2.24
Affiliations=DEPARTMENT OF POLITICS AND INTERNATIONAL STUDIES, UNIVERSITY OF WARWICK, UNITED KINGDOM	2.24
Affiliations=INTERNATIONAL SECURITY PROGRAM, LOWY INSTITUTE, AUSTRALIA; BROOKINGS INSTITUTION, UNITED STATES; INDIAN STRATEGIC AFFAIRS AT THE UNIVERSITY OF NEW SOUTH WALES, AUSTRALIA	2.24
Affiliations=SCHOOL OF HUMANITIES AND SOCIAL SCIENCES, DEAKIN UNIVERSITY, AUSTRALIA	2.24
Affiliations=UNIVERSITY OF ALASKA FAIRBANKS (UAF), 603 C/D GRUENING BUILDING, 1747 SOUTH CHANDALAR DRIVE, FAIRBANKS, 99775, AK, UNITED STATES; EMERITUS OF POLITICAL SCIENCE, UNIVERSITY OF ALASKA FAIRBANKS (UAF), 1777 RED FOX DRIVE, FAIRBANKS, 99709, AK, UNITED STATES; SCHOOL OF INTERNATIONAL RELATIONS, CHINA UNIVERSITY OF INTERNATIONAL BUSINESS AND ECONOMICS (UIBE), MAIL BOX 100, NO. 10 HUIXIN EAST STREET, CHANGYANG DISTRICT, BEIJING, 100029, CHINA	2.24
surprise	1.94
LA=ENGLISH	1.17
fear	1.08
anger	0.969
joy	0.545
sadness	0.534
negative	0.438
positive	0.176
disgust	0.0811
trust	0.0723

Cluster 3, comprising 28 cases, provides significant insights into both the institutional affiliations and sentiment distribution associated with the dataset. The most notable metric within this cluster is the high value of *"TC"* (11.4), indicating a substantial citation impact. This suggests that the publications in this cluster hold considerable academic influence, reflecting their importance and relevance within the broader field of study.

The institutional affiliations associated with Cluster 3 highlight a diverse and globally distributed research base. Key institutions include *"Syracuse University's Department of Political Science (United States)"*, the *"Department of Politics and International Studies at the University of Warwick (United Kingdom)"*, and *"Deakin University's School of Humanities and Social Sciences (Australia)"*. Other notable contributors include the *"International Security Program at the Lowy Institute (Australia)"*, the *"Brookings Institution (United States)"*, and *"China University of International Business and Economics"*. This extensive geographic spread emphasizes the international collaboration underpinning the research in this cluster, underscoring its relevance to global academic discourse.

The sentiment distribution reveals *"surprise"* as the most prevalent sentiment within this cluster (1.94), followed by *"fear"* (1.08) and *"anger"* (0.969). This pattern suggests that the publications in this group often address topics characterized by unpredictability or significant geopolitical and strategic shifts. Such sentiments are commonly associated with studies examining international security, diplomatic strategies, or responses to crises. By contrast, sentiments such as *"trust"* (0.0723) and *"positive"* (0.176) are less pronounced, which could reflect the analytical and critical approach inherent in scholarship on complex international issues. The predominance of English as the language of publication (1.17) aligns with established trends in global academic communication, where English serves as the lingua franca for scholarly exchange. This reinforces the accessibility and dissemination of research findings across international audiences.

In summary, Cluster 3 illustrates a significant interplay between high citation impact, diverse institutional representation, and the nuanced emotional tones prevalent in the analyzed works. The presence of sentiments such as *"surprise"* and *"fear"* highlights the complex and often contentious nature of the topics addressed, while the broad geographic scope of institutional contributions underscores the global relevance of these scholarly efforts. These findings contribute to a deeper understanding of the intellectual and emotional dimensions shaping research in IR (see Table 11).

Comparative Analysis

The integration of MCA and HCA with SA constitutes a significant methodological advancement, enhancing the latter by providing a more nuanced and contextually enriched understanding of sentiment data. While SA classifies textual data into broad emotional categories, MCA and HCA enhance this process by elucidating the relationship between sentiment and various additional categorical variables, such as institutional affiliations, geopolitical contexts, or publication types. This complementary approach allows for a more detailed interpretation of sentimental data, facilitating a deeper comprehension of the underlying dynamics shaping sentiment expressions.

In the preliminary results of SA, sentiment is broadly classified into three primary categories: positive, neutral, and negative. Among these, positive sentiment is identified as the most prevalent (positive = 134, trust = 101, negative = 95). These results offer an initial overview of emotional trends within the dataset but are constrained in scope, as they do not explore the factors that may influence sentiment distribution or their correlation with specific contextual variables (see Table 8). While SA categorizes sentiments into general emotional classes, it does not reveal how sentiment relates to geopolitical regions, institutional affiliations, or temporal shifts in discourse.

MCA and HCA optimize SA by placing sentiment data within a multidimensional analytical framework. It offers a more comprehensive understanding by uncovering the relationships between sentiment and other categorical variables such as institutional affiliations, regional distinctions, or thematic focus in the research. For instance, in Cluster 1, a significant correlation between positive sentiment and institutions such as *MGIMO University* and *RUDN University* is observed. This association suggests that these institutions tend to publish content with a more optimistic tone, a pattern that would not be immediately apparent from the SA results alone (see Table 9). HCA enables the identification of relationships between sentiment classifications and institutional or geographical variables, thereby deepening the understanding of the emotional tone within the discourse. Furthermore, MCA contributes by structuring categorical data prior to clustering, facilitating the organization of sentiment classifications for improved hierarchical analysis.

Beyond identifying straightforward sentiment patterns, MCA and HCA facilitate the uncovering of more intricate associations between sentiment and thematic variables. In Cluster 3, the sentiment of *"fear"* is closely linked with *"surprise"* and primarily arises in the context of geopolitical instability. This relationship underscores the capacity of HCA to reveal how specific sentiments manifest in response to political or security events, a dimension that remains unexplored within a traditional SA framework (see Table 11). The *"fear"* sentiment, appearing 52 times in the SA results, is shown by HCA to correlate with specific geopolitical contexts, thereby providing a more detailed understanding of how such sentiments are framed within political and security discourses (see Table 8).

Furthermore, MCA and HCA enable the identification of relationships between sentiment and additional categorical variables, such as publication types. In Cluster 2, both *"positive"* and *"trust"* sentiments are associated with English-language, non-open access publications. This finding suggests that the accessibility and linguistic attributes of the publications may influence the emotional tone of the sentiment expressed. Such patterns would be difficult to detect through SA alone, as it classifies sentiment without accounting for the underlying contextual variables that could influence its distribution (see Table 10).

In conclusion, while SA serves as an essential foundation for detecting and categorizing emotional trends within textual data, the application of MCA and HCA significantly enhance this process by adding a contextual layer that reveals the complex relationships between sentiment and other categorical variables. MCA and HCA increase the interpretive power of SA by linking sentiment to key factors such as institutional affiliation, geopolitical context, and publication characteristics. This integration offers a more comprehensive, multidimensional approach to SA, yielding deeper insights into the emotional dynamics present in IR discourse. By providing a richer understanding of sentiment, MCA and HCA optimize the analytical potential of SA, enabling a more sophisticated examination of emotional and thematic patterns across the dataset.

CONCLUSIONS

This study has examined the integration of SA, MCA and HCA to enhance the understanding of emotional dynamics within IR discourse. The complexity of IR texts necessitates a methodological approach capable of capturing both emotional nuance and categorical structure, thus justifying the integration of these complementary analytical techniques. By combining these methodologies, the analysis has provided deeper insights into the categorization of sentiment and its relationships with key contextual factors, including institutional affiliations, geopolitical contexts, and publication characteristics. The principal strength of this mixed methodological approach lies in its ability to transcend the limitations of SA in isolation, enabling more nuanced and contextually grounded interpretations. The integration of MCA and HCA enabled a detailed examination of sentiment patterns, revealing nuanced associations between emotional tones and contextual metadata, including institutional affiliations, publication language, and accessibility. These intricate relationships, which extend beyond the scope of SA's traditional capabilities, underscore the added value of combining these methodologies for a more comprehensive analysis.

The findings from SA reveal that *"positive"* sentiment occurring 134 times, and *"trust"*, with 101 instances, were the most prevalent emotional categories in the dataset, indicating a predominantly optimistic tone in IR discussions. In contrast, negative sentiments, particularly *"fear"* with 52 instances and *"anger"* with 44, reveal a significant undercurrent of apprehension and frustration in the discourse (see Table 8). The integration of MCA and HCA facilitated a more nuanced understanding by linking these sentiments to specific institutional and geopolitical contexts. Cluster 1 revealed a strong association between positive sentiment and institutions such as *"MGIMO University"* and *"RUDN University"*, indicating a generally optimistic tone in publications emanating from Russian academic sources (see Table 9). In Cluster 2, English-language, non-open access publications were primarily associated with structural characteristics rather than distinct sentiment patterns. The results do not indicate a strong alignment between these publication attributes and specific sentiment expressions, suggesting that language and accessibility are not primary determinants of sentiment distribution within this dataset (see Table 10).

Furthermore, the co-occurrence network analysis revealed the interconnectedness of key terms in the discourse, with *"international relations"*, *"China"*, and *"United States"* emerging as central nodes within the network. This finding illustrates the global significance of these powers and their dominant role in shaping contemporary geopolitical discussions (see Figure 6). The co-occurrence of other critical terms, such as *"global security"* and *"power dynamics"*, further highlights the thematic focus of IR research, which often revolves around these central issues, thereby influencing the emotional tone observed in the discourse.

Cluster 3 revealed the prominence of sentiments such as *"fear"* and *"surprise"*, closely linked to themes of geopolitical instability and security challenges (see Table 11). The analysis demonstrated that *"fear"* was predominantly associated with concerns over security threats and territorial disputes, while *"surprise"* often corresponded to unforeseen developments in global politics. These findings underscore the critical role of geopolitical contexts in shaping the emotional layers of IR discourse. By integrating MCA and HCA, this study advanced the exploration of the relationships between sentiment and thematic variables, offering a deeper and more sophisticated perspective on how global events inform academic narratives.

Future research could extend the application of MCA and HCA to other geopolitical regions, such as the Middle East or Africa, to assess how sentiment dynamics may differ across diverse political contexts. The inclusion of alternative sentiment dictionaries, such as AFINN, VADER, or LIWC, could enhance the depth of the analysis by enabling comparative evaluations across diverse classification frameworks, thereby broadening the scope and interpretative potential of the findings. Longitudinal studies could also explore how sentiment evolves in response to major geopolitical shifts or crises, providing insight into the temporal patterns of emotional responses in IR. Moreover, the integration of deep learning techniques with MCA and HCA could offer greater precision and scalability, enabling more granular SA of large datasets and providing enhanced insights into the emotional dynamics of global political discourse.

In conclusion, integrating SA, MCA, and HCA offers a compelling and methodologically rigorous framework, well-suited to capturing the complex emotional and thematic dynamics that define contemporary international relations research. The integration of these methods offers a multidimensional framework for analyzing sentiment, uncovering complex interconnections between emotional tones and associated metadata, such as contextual and categorical variables. By revealing the complex relationships between sentiment, institutional affiliations, geopolitical context, and publication types, this approach offers a more comprehensive and multidimensional understanding of global political sentiment. The insights derived from this study contribute to the refinement of SA methodologies IR research, providing valuable tools for future exploration of the emotional and thematic dimensions of global diplomacy and policy.

REFERENCES

Abedin, E., Jafarzadeh, H., & Akhlaghpour, S. (2018). Opinion Mining on Twitter: A sentiment analysis of the Iran deal. *Pacific Asia Conference on Information Systems*, 220. https://aisel.aisnet.org/cgi/viewcontent.cgi?article=1219&context=pacis2018

Almeida, J., Barbosa, L., Pais, A., & Formosinho, S. (2007). Improving hierarchical cluster analysis: A new method with outlier detection and automatic clustering. *Chemometrics and Intelligent Laboratory Systems*, *87*(2), 208–217. DOI: 10.1016/j.chemolab.2007.01.005

Atkinson, W. (2023). Charting fields and spaces quantitatively: From multiple correspondence analysis to categorical principal components analysis. *Quality & Quantity*, *58*(1), 829–848. DOI: 10.1007/s11135-023-01669-w

Berka, P. (2020). Sentiment analysis using rule-based and case-based reasoning. *Journal of Intelligent Information Systems*, *55*(1), 51–66. DOI: 10.1007/s10844-019-00591-8

Biresselioglu, M. E., Demir, M. H., Solak, B., Kayacan, A., & Altinci, S. (2020). Investigating the trends in arctic research: The increasing role of social sciences and humanities. *The Science of the Total Environment*, *729*, 139027. DOI: 10.1016/j.scitotenv.2020.139027 PMID: 32498176

Bisley, N. (2013). 'An ally for all the years to come': Why Australia is not a conflicted US ally. *Australian Journal of International Affairs*, *67*(4), 403–418. DOI: 10.1080/10357718.2013.803029

Bitzenis, A., Koutsoupias, N., & Nosios, M. (2025). Machine learning in Social Sciences and Humanities research: A structured literature review. *ICIB 2024 Conferfence Proceedings, 1*(1), 1–17. DOI: 10.12681/icib.8050

Boylan, B. M., McBeath, J., & Wang, B. (2020). US–China relations: Nationalism, the trade war, and COVID-19. *Fudan Journal of the Humanities and Social Sciences*, *14*(1), 23–40. DOI: 10.1007/s40647-020-00302-6

Campbell, H. (2007). China in Africa: Challenging US global hegemony. *Third World Quarterly*, *29*(1), 89–105. DOI: 10.1080/01436590701726517

Chambers, N., Bowen, V., Genco, E., Tian, X., Young, E., Harihara, G., & Yang, E. (2015). Identifying Political Sentiment between Nation States with Social Media. *Proceedings of the 2021 Conference on Empirical Methods in Natural Language Processing*, 65–75. DOI: 10.18653/v1/D15-1007

Clarke, M. (2016). 'One Belt, One Road' and China's emerging Afghanistan dilemma. *Australian Journal of International Affairs, 70*(5), 563–579. DOI: 10.1080/10357718.2016.1183585

Fávero, L. P., Belfiore, P., & De Freitas Souza, R. (2023). Simple and multiple correspondence analysis. In *Elsevier eBooks* (pp. 215–234). DOI: 10.1016/B978-0-12-824271-1.00013-5

Fiori, A., & Passeri, A. (2015). Hedging in search of a new age of non-alignment: Myanmar between China and the USA. *The Pacific Review*, *28*(5), 679–702. DOI: 10.1080/09512748.2015.1012543

Garcia, M. B., & Cunanan-Yabut, A. (2022). Public Sentiment and Emotion Analyses of Twitter Data on the 2022 Russian Invasion of Ukraine. *9th International Conference on Information Technology, Computer, and Electrical Engineering (ICITACEE)*, 242–247. DOI: 10.1109/ICITACEE55701.2022.9924136

Georgiadou, E., Angelopoulos, S., & Drake, H. (2019). Big data analytics and international negotiations: Sentiment analysis of Brexit negotiating outcomes. *International Journal of Information Management, 51*, 102048. DOI: 10.1016/j.ijinfomgt.2019.102048

He, K. (2018). Role conceptions, order transition and institutional balancing in the Asia-Pacific: A new theoretical framework. *Australian Journal of International Affairs*, *72*(2), 92–109. DOI: 10.1080/10357718.2018.1437390

Heiser, W. (2001). Correspondence analysis. In *Elsevier eBooks* (pp. 2820–2824). DOI: 10.1016/B0-08-043076-7/00655-0

Henry, D. B., Tolan, P. H., & Gorman-Smith, D. (2005). Cluster analysis in family Psychology research. *Journal of Family Psychology*, *19*(1), 121–132. DOI: 10.1037/0893-3200.19.1.121 PMID: 15796658

Hughes, C. W. (2016). Japan's 'Resentful realism' and balancing China's rise. *The Chinese Journal of International Politics*, *9*(2), 109–150. DOI: 10.1093/cjip/pow004

Jacob, J. T. (2020). 'To Tell China's Story Well': China's International Messaging during the COVID-19 Pandemic. *China Report (New Delhi, India)*, *56*(3), 374–392. DOI: 10.1177/0009445520930395

Jafarzadegan, M., Safi-Esfahani, F., & Beheshti, Z. (2019). Combining hierarchical clustering approaches using the PCA method. *Expert Systems with Applications*, *137*, 1–10. DOI: 10.1016/j.eswa.2019.06.064

Kahma, N., & Toikka, A. (2012). Cultural map of Finland 2007: Analysing cultural differences using multiple correspondence analysis. *Cultural Trends*, *21*(2), 113–131. DOI: 10.1080/09548963.2012.674751

Kim, Y. (2014). Convolutional Neural Networks for Sentence Classification. *Proceedings of the 2014 Conference on Empirical Methods in Natural Language Processing (EMNLP)*. DOI: 10.3115/v1/D14-1181

Koutsoupias, D., Pavlos, S., & Vasileiadis, N. Kokovidou Marina, & Koutsoupias, N. (2024). Humanities and Arts Research on Mount Athos. In *Ant. Stamoulis Publishing* (pp. 477–491). OSF. https://osf.io/a7b2g/

Koutsoupias, N. (2019). *Online Art Buying Decision Support: A Multivariate Approach*. DOI: 10.5281/zenodo.2576123

Koutsoupias, N. (2021). Fine Art Analytics: Mining Doctoral Dissertations on Painting (Greece, 1986-2019). *Mediterranean Chronicle, 11*.

Koutsoupias, N. (2024). Multiple Correspondence Analysis in Social Sciences and Humanities Research: A Longitudinal Mapping, *Data Analysis Bulletin*, 20(1), pp. 59-77. DOI: 10.17605/osf.io/693dr

Koutsoupias, N., & Boutsiouki, S. (2019). Exploring European Cultural Goods Trade: A Multidimensional Perspective. *ICIB 2017-2018 Conference Proceedings*, 124–138. DOI: 10.5281/zenodo.1286354

Koutsoupias, N., & Mikelis, K. (2019). Exploring International Relations Journal Articles: A Multivariate Approach. In *Sage Research Methods Cases*. Sage., DOI: 10.4135/9781526486189

Koutsoupias, N., & Mikelis, K. (2021). Text, content and data analysis of journal articles: The Field of International Relations. In *Studies in classification, data analysis, and knowledge organization* (pp. 113–120). DOI: 10.1007/978-3-030-60104-1_13

Lai, H. H. (2001). Behind China's World Trade Organization agreement with the USA. *Third World Quarterly*, *22*(2), 237–255. DOI: 10.1080/01436590120037054

Layne, C. (2017). The US foreign policy establishment and grand strategy: How American elites obstruct strategic adjustment. *International Politics*, *54*(3), 260–275. DOI: 10.1057/s41311-017-0033-0

Mackerras, C. (2015). Xinjiang in China's foreign relations: Part of a new silk road or central Asian zone of conflict? *East Asia (Piscataway, N.J.)*, *32*(1), 25–42. DOI: 10.1007/s12140-015-9224-8

Martin-Mazé, M. (2017). Multiple correspondence analysis in international relations. In *Springer eBooks* (pp. 139–149). DOI: 10.1007/978-3-319-61979-8_10

McCourt, D. M., & Ruley, G. (2023). How is the American foreign policy establishment structured? A multiple correspondence analysis of the US China field. *Foreign Policy Analysis*, *19*(2), orad002. DOI: 10.1093/fpa/orad002

Medcalf, R. (2014). In defence of the Indo-Pacific: Australia's new strategic map. *Australian Journal of International Affairs*, *68*(4), 470–483. DOI: 10.1080/10357718.2014.911814

Medhat, W., Hassan, A., & Korashy, H. (2014). Sentiment analysis algorithms and applications: A survey. *Ain Shams Engineering Journal*, *5*(4), 1093–1113. DOI: 10.1016/j.asej.2014.04.011

Mohammad, S. M., & Turney, P. D. (2013). Nrc emotion lexicon. National Research Council, Canada, 2, 234. DOI: 10.4224/21270984

Nalini, C., Dharani, B., Baskar, T., & Shanthakumari, R. (2023). Review on sentiment analysis using supervised machine learning techniques. In *Lecture notes in networks and systems* (pp. 166–177). DOI: 10.1007/978-3-031-35507-3_17

Nielsen, F. (2016). Hierarchical clustering. In *Undergraduate topics in computer science* (pp. 195–211). DOI: 10.1007/978-3-319-21903-5_8

Odgaard, L. (2003). The South China Sea: ASEAN's security concerns about China. *Journal of Technological Advancements*, *34*(1), 11–24. DOI: 10.1177/0967010603034001672

Page, M. J., McKenzie, J. E., Bossuyt, P. M., Boutron, I., Hoffmann, T. C., Mulrow, C. D., Shamseer, L., Tetzlaff, J. M., Akl, E. A., Brennan, S. E., Chou, R., Glanville, J., Grimshaw, J. M., Hróbjartsson, A., Lalu, M. M., Li, T., Loder, E. W., Mayo-Wilson, E., McDonald, S., & Moher, D. (2021). The PRISMA 2020 statement: An updated guideline for reporting systematic reviews. *BMJ (Clinical Research Ed.)*, (71), n71. DOI: 10.1136/bmj.n71 PMID: 33782057

Pan, C. (2014). The 'Indo-Pacific' and geopolitical anxieties about China's rise in the Asian regional order. *Australian Journal of International Affairs*, *68*(4), 453–469. DOI: 10.1080/10357718.2014.884054

Pang, B., & Lee, L. (2008). *Opinion mining and sentiment analysis*. Now Publishers Inc. DOI: 10.1561/9781601981516

Patrick, H. A., J, P. G., Sharief, M. H., & Mukherjee, U. (2023). Sentiment Analysis Perspective using Supervised Machine Learning Method. *2023 Fifth International Conference on Electrical, Computer and Communication Technologies (ICECCT)*, 1–4. DOI: 10.1109/ICECCT56650.2023.10179807

Pişirgen, A., & Peker, S. (2024). A clustering approach for classifying scholars based on publication performance using bibliometric data. *Egyptian Informatics Journal*, *28*, 100537. DOI: 10.1016/j.eij.2024.100537

Samuelowicz, K., & Bain, J. D. (2001). Revisiting academics' beliefs about teaching and learning. *Higher Education*, *41*(3), 299–325. DOI: 10.1023/A:1004130031247

Schindler, S., DiCarlo, J., & Paudel, D. (2021). The new cold war and the rise of the 21st-century infrastructure state. *Transactions of the Institute of British Geographers*, *47*(2), 331–346. DOI: 10.1111/tran.12480

Shukla, D., & Unger, S. (2022). Sentiment Analysis of International Relations with Artificial Intelligence. *Athens Journal of Sciences*, *9*(2), 91–106. DOI: 10.30958/ajs.9-2-1

Sun, S., Luo, C., & Chen, J. (2016). A review of natural language processing techniques for opinion mining systems. *Information Fusion*, *36*, 10–25. DOI: 10.1016/j.inffus.2016.10.004

Xu, Q., Zhang, Q., Liu, J., & Luo, B. (2020). Efficient synthetical clustering validity indexes for hierarchical clustering. *Expert Systems with Applications*, *151*, 113367. DOI: 10.1016/j.eswa.2020.113367

Yiu, D. W., Wan, W. P., Chen, K. X., & Tian, X. (2021). Public sentiment is everything: Host-country public sentiment toward home country and acquisition ownership during institutional transition. *Journal of International Business Studies*, *53*(6), 1202–1227. DOI: 10.1057/s41267-020-00380-4

Chapter 12
Hierarchical Clustering in Social Sciences and Humanities: A Review

Nikos Koutsoupias
https://orcid.org/0000-0003-1664-5404
University of Macedonia, Greece

Marios Nosios
https://orcid.org/0009-0007-6161-3854
University of Macedonia, Greece

ABSTRACT

This chapter undertakes a systematic bibliometric analysis of hierarchical clustering applications within the social sciences and humanities, drawing upon publications indexed in the Scopus database from 1975 to 2024. The findings reveal the methodological versatility of clustering techniques in addressing complex research problems across a broad spectrum of academic disciplines. Thematic emphases on social network analysis and data management strategies underscore the centrality of clustering in elucidating intricate sociocultural systems. The United States, China, and the United Kingdom emerge as principal contributors, highlighting the global scope and interdisciplinary nature of this research domain. This analysis provides a comprehensive account of prevailing thematic trends, patterns of scholarly collaboration, and the intellectual architectures that underpin the field. It offers a robust foundation for both the methodological advancement of clustering techniques and their expanded application to pressing societal and academic challenges

DOI: 10.4018/979-8-3693-9400-7.ch012

INTRODUCTION

The rapid advancement of data collection technologies in the 21st century has led to an unprecedented proliferation of complex datasets. Researchers are increasingly faced with the challenge of analyzing large, multidimensional datasets to uncover underlying patterns, trends, and relationships. In response to these challenges, numerous clustering methodologies have been developed, with Hierarchical Clustering (HC) distinguished by its reliability and adaptability across a wide range of research contexts. This chapter builds upon a preliminary bibliometric analysis presented at the 12th Panhellenic Conference on Data Analytics and offers a systematic overview of the use and evolution of HC within the social sciences and humanities, providing insights into its methodological foundations, applications, limitations, and emerging trends.

As an unsupervised learning technique, HC is widely utilized in data analysis, particularly in disciplines dealing with large and complex datasets, to group data into clusters based on shared characteristics or similarities. Although this method was originally developed for the natural sciences, its ability to uncover complicated patterns and relationships has turned particularly popular within social sciences and humanities, in which datasets are often described by high heterogeneity and complexity (Jafarzadegan et al., 2019).

The assessment of similarity or distance between data points constitutes the foundation of HC, which is broadly classified into two primary types: agglomerative and divisive. Agglomerative Hierarchical Clustering (AHC) is the most common type following a bottom-up approach where each object starts in its own cluster, and pairs of clusters are merged as one moves up the hierarchy (Li et al., 2022; Murtagh & Contreras, 2011; Zhou et al., 2016). In contrast, Divisive Hierarchical Clustering (DHC) follows a top-down approach where all objects begin in a single cluster, which is recursively split into smaller ones. Both approaches involve calculating the distance between objects or clusters using a distance metric, such as Euclidean, or Manhattan distance, or cosine similarity (Roux, 2018). These distance metrics are crucial for quantifying the degree of similarity between data points, and the selection of an appropriate metric can profoundly impact the clustering results. Euclidean distance is commonly applied to numerical data, as it represents the straight-line distance between two points. In contrast, Manhattan distance is utilized when the absolute differences across dimensions are of importance, whereas cosine similarity is frequently employed in the analysis of text or vector data (Giordani et al., 2020).

The result of HC is typically represented as a tree-like diagram called a dendrogram, which visually depicts the nested clusters at different levels of granularity. An advantage of HC is that it does not require the user to specify the number of clusters in advance, unlike other clustering algorithms such as k-means. Instead, the number of clusters can be determined after the algorithm has been completed by cutting the dendrogram at the desired level. This flexibility is one of HC's major strengths, allowing researchers to explore hierarchical relationships within the dataset (Almeida et al., 2007).

Furthermore, the linkage criteria used in HC, such as single linkage, complete linkage, and average linkage, determine how the distance between clusters is calculated as the algorithm progresses. Single linkage measures the shortest distance between points in different clusters, complete linkage uses the maximum distance, and average linkage computes the mean distance between all points in the clusters. These linkage methods provide flexibility in how clusters are defined and can lead to different clustering results, depending on the characteristics of the dataset (Nielsen, 2016).

The ability of HC to handle various data types and its non-reliance on predefining the number of clusters makes it particularly useful in analyzing complex, multi-dimensional datasets typical in social sciences and humanities. These fields often deal with complex, multi-dimensional data, such as textual data, social networks, and historical data, which can be naturally grouped hierarchically, revealing deeper patterns and relationships. Several studies have applied HC, demonstrating its broad applicability across various disciplines and its ability to uncover meaningful patterns within complex datasets. Leydesdorff et al. (2017) employed HC to classify 11,359 journals from the *Journal Citation Reports*, mapping and grouping related academic disciplines, providing a clear visualization of the academic landscape and the nuanced boundaries between fields. Also, Chen and Deng (2019) applied HC to transportation data, identifying patterns of cooperation consciousness among urban travelers, highlighting the role of social networks and information use in shaping travel choices.

In the realm of cultural analysis, Louf et al. (2023) utilized HC to analyze social media data, identifying cultural regions in the United States based on regional variations in language use. This innovative approach showed HC's power in revealing cultural divisions without preconceived assumptions. Furthermore, Kong et al. (2024) used HC to assess the health of higher education systems by clustering various educational indicators, offering valuable insights for policy improvement. On the other hand, Zourna et al. (2024) combined HC with mixed methods to study students' experiences during the COVID-19 pandemic, using the technique to uncover distinct patterns in their responses. In the financial domain, Koutsoupias and Papadimitriou (2001) applied HC to analyze 1,234 life insurance contracts in Greece and the European region, using 145 variables to identify distinct and homogenous contract groups. This approach provided insights into the portfolio structure of life insurance companies without relying on prior assumptions, showcasing HC's effectiveness in handling complex datasets.

The applicability of HC extends beyond cultural and educational contexts, offering valuable insights across various domains in the social sciences and humanities. For instance, Liu (2023) applied this technique alongside other algorithms to segment consumer profiles based on data such as age, income, and consumption behavior. This analysis facilitated the identification of distinct consumer patterns, aiding in the development of targeted marketing strategies. Ortega et al. (2011) utilized clustering techniques to classify research institutes within the Spanish National Research Council by their scientific outputs, uncovering structural differences that informed more precise evaluation frameworks. Similarly, Lancho-Barrantes and Cantu-Ortiz (2021) examined research preferences among globally recognized universities using clustering approaches, identifying patterns of affinity and divergence, and providing insights into collaborative potential and the alignment of institutional research priorities. Pişirgen and Peker (2024) likewise introduced a comprehensive clustering framework to assess scholars' publication performance by analyzing bibliometric data and integrating dimensions such as academic age, productivity, impact, and recency, ultimately producing nuanced scholar profiles to guide decisions on recruitment, promotion, and resource allocation. These diverse applications accentuate the significant role of HC in various fields, emphasizing its effectiveness in identifying patterns, uncovering structure, and providing deeper insights into complex phenomena.

To sum up, HC is a versatile analytical method with significant applications in the social sciences and humanities. Its capacity to organize complex datasets into hierarchical structures and uncover latent relationships has made it a valuable tool in disciplines such as political science, sociology, cultural studies, and linguistics, and despite certain challenges, the continued research and development in these areas underline the method's enduring relevance.

This chapter is structured as follows: Section 1 introduces the research, Section 2 outlines the methodology, Section 3 presents the results, and Section 4 concludes with insights into existing research gaps and potential future directions.

METHODOLOGY

To analyze the application of HC within social sciences and humanities research, a systematic methodology integrating bibliometric analysis and advanced visualization techniques were employed. This approach aimed to provide a comprehensive overview of scholarly activity in the field, identify emerging trends, and explore collaborative and thematic patterns within literature.

The study commenced with the construction of a dataset comprising academic works that specifically addressed HC applications within these disciplines (Bitzenis & Koutsoupias, 2024A; Bitzenis et al., 2023). To ensure that the dataset was exhaustive and representative of the field, publications were systematically retrieved from the Scopus database, a widely recognized and authoritative repository of peer-reviewed academic literature (Baas et al., 2020). A well-defined and reproducible search strategy was employed, utilizing precise combinations of keywords and Boolean operators. This approach facilitated the retrieval of studies encompassing a broad range of HC applications across multiple subfields, ensuring the inclusion of diverse perspectives and research contributions. The exact query syntax utilized for this data extraction process is detailed below:

```
(TITLE-ABS-KEY("hierarchical cluster analysis" OR "hierarchical clustering") AND TITLE-ABS-KEY("social sciences" OR humanities OR "cultural studies" OR "social network analysis" OR "historical documents" OR "cultural artifacts") AND NOT (genomics OR biology OR medicine) AND TITLE-ABS-KEY("unsupervised learning")) AND (LIMIT-TO (SUBJAREA,"COMP") OR LIMIT-TO (SUBJAREA,"SOCI") OR LIMIT-TO (SUBJAREA,"ENGI") OR LIMIT-TO (SUBJAREA,"DECI") OR LIMIT-TO (SUBJAREA,"PSYC") OR LIMIT-TO (SUBJAREA,"BUSI") OR LIMIT-TO (SUBJAREA,"ECON") OR LIMIT-TO (SUBJAREA,"ARTS"))
```

To enhance the rigor and reproducibility of the study, the PRISMA (2020) guidelines (Preferred Reporting Items for Systematic Reviews and Meta-Analyses) were followed. These guidelines provided a structured framework for managing the review process, ensuring transparent and systematic identification, screening, eligibility assessment, and inclusion of manuscripts as depicted in Figure 1 (Page et al., 2021).

Figure 1. PRISMA 2020 flow diagram of the selection process

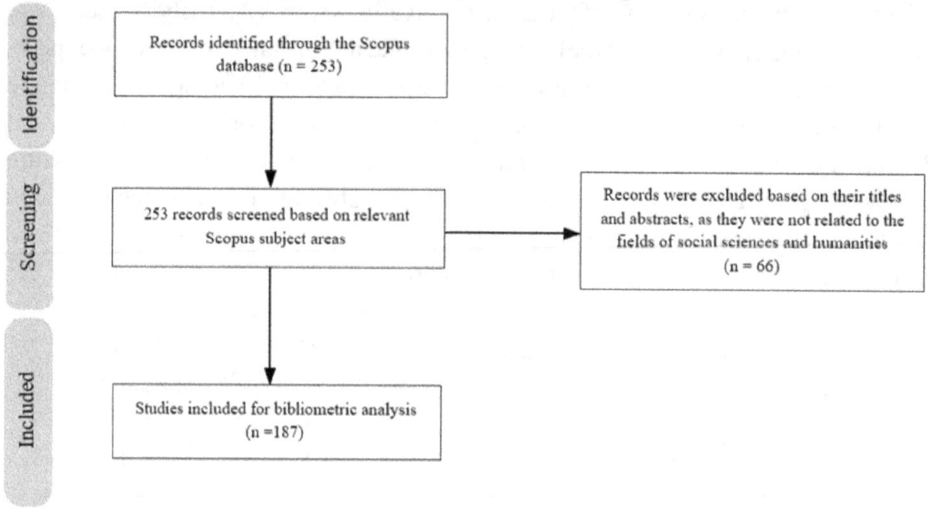

Following the construction of the dataset, a comprehensive bibliometric analysis was undertaken utilizing the R programming language, specifically employing the bibliometrix package. This analytical tool facilitated the extraction and examination of critical bibliometric indicators, such as citation metrics, co-authorship patterns, keyword co-occurrence networks, and journal impact measures. These indicators served to quantitatively delineate the structural and developmental dimensions of HC research within the social sciences and humanities (Aria & Cuccurullo, 2017; Bitzenis et al., 2025; Koutsoupias & Papadimitriou, 2020). Beyond the quantitative evaluation, advanced techniques such as thematic mapping and co-word analysis were applied to uncover the intellectual architecture of the field. These methodologies enabled the identification of dominant research themes and the detection of emerging topics, providing a deeper understanding of the interplay between established paradigms and novel applications of HC in these disciplines (Bitzenis & Koutsoupias, 2024B).

The results of the analysis were subsequently visualized through the creation of detailed graphical and tabular outputs. Network diagrams highlighted collaborative relationships among authors and institutions, citation maps traced the intellectual lineage and influence of key publications, and trend graphs illuminated temporal dynamics within the literature. These visualizations offered a comprehensive and intuitive portrayal of the dataset, revealing interconnected thematic clusters and the evolution of research trends over time (Bitzenis & Koutsoupias, 2023).

Ultimately, this systematic methodological framework not only ensured the rigorous examination of HC applications but also contributed to a holistic understanding of its role in advancing research across the social sciences and humanities. By integrating quantitative analysis with interpretive visualization, the study provided critical insights into the trajectory of this research domain and identified potential avenues for future inquiry.

RESULTS

This section presents the findings from the bibliometric analysis, offering a detailed overview of key trends, citation patterns, and thematic clusters within the literature, while highlighting leading countries in terms of citation impact, significant research topics, and influential papers, thereby providing a systematic view of the method's expanding global influence and interdisciplinary applications.

Table 1. Descriptive statistics

Description	Results
MAIN INFORMATION ABOUT DATA	
Timespan	1975:2024
Sources (Journals, Books, etc)	158
Documents	187
Annual Growth Rate %	4.81
Document Average Age	8.54
Average citations per doc	13.18
References	6837
DOCUMENT CONTENTS	
Keywords Plus (ID)	1098
Author's Keywords (DE)	605
AUTHORS	
Authors	555
Authors of single-authored docs	16
AUTHORS COLLABORATION	
Single-authored docs	16
Co-Authors per Doc	3.3
International co-authorships %	21.93
DOCUMENT TYPES	
article	111

continued on following page

Table 1. Continued

Description	Results
book	1
book chapter	5
conference paper	64
conference review	3
review	3

The dataset, spanning the years 1975 to 2024, represents a comprehensive longitudinal study of HC within the social sciences and humanities. It contains 187 works from 158 distinct sources, including journals, books, and conference proceedings, showcasing the extensive scholarly engagement with the methodology over the nearly five-decade timeframe. The breadth underlines the wide applicability of HC across various disciplines and contexts. The dataset's coverage across multiple publication formats, including 111 journal articles, 64 conference papers, 5 book chapters, 3 reviews, and 1 book, indicates that journal articles remain the primary medium for disseminating research, while conference proceedings contribute significantly to ongoing methodological discussions.

An annual growth rate of 4.81% signals a gradual but consistent expansion of interest in the topic. Such growth, while modest compared to rapidly emerging fields, indicates the enduring relevance of HC as a research tool. The average age of the documents, calculated at 8.54 years, demonstrates a balance between foundational works and more recent contributions. This equilibrium suggests a field rooted in established principles yet responsive to ongoing advancements and new applications. Also, the mean citation count of 13.18 per document underlines the significant academic influence of these studies, pointing to their relevance and utility within the scholarly community. Additionally, the content analysis highlights the presence of 1,098 Keywords Plus (automatically extracted keywords) and 605 Author's Keywords, indicating a diverse and specialized thematic focus. The difference in volume between these two categories suggests that automated indexing identifies a broader range of topics compared to the more targeted selections by authors.

The 6,837 references cited across the corpus exhibit the methodological rigor and intertextuality of the research. This high number of citations reflects a field deeply embedded in the academic ecosystem, relying on an extensive network of prior studies to inform its development. Regarding the collaboration metrics, these reveal the collective nature of scholarly inquiry into HC. The 555 contributing authors reflect a substantial pool of expertise, with a notable average of 3.3 co-authors per document. This collaboration intensity highlights the interdisciplinary and often international nature of the research, as evidenced by the 21.93% rate of international co-authorships. Single-authored works, numbering only 16, contrast with this trend, indicating that although individual contributions are present, most of the research derives from collaborative intellectual effort.

Overall, the dataset exemplifies a well-established, collaborative, and methodologically rigorous field. The steady growth rate, diverse publication formats, and extensive collaboration underscore the sustained scholarly interest in HC and its capacity to address complex, interdisciplinary research questions. The data suggest that HC remains a relevant methodological approach, with continued scholarly engagement, ongoing refinements, and applications across multiple disciplines.

Figure 2. Annual scientific production

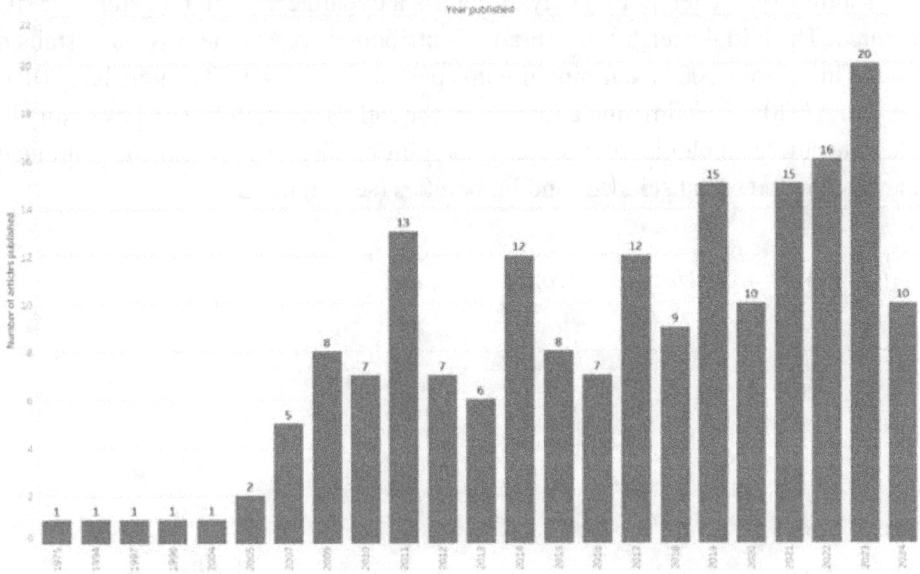

The annual scientific production data provides a chronological perspective on the development of HC research. The dataset begins in 1975 with a single publication, reflecting the nascent state of HC methods at the time. This early contribution likely laid the groundwork for subsequent studies, although no further publications are recorded until 1994, marking a nearly two-decade gap. This temporal discontinuity can be attributed to the limited computational resources and the relatively narrow scope of HC applications during this period.

From the mid-1990s onward, the field begins to show signs of growth. The publication of isolated articles in 1994, 1997, and 1998 indicates a gradual rekindling of interest, likely driven by advancements in computational power and the increasing availability of data. The real turning point occurs in the mid-2000s, with a marked increase in scientific production. The five articles published in 2007 represent a significant jump, signaling the field's transition into a period of accelerated growth. This trend continues into the late 2000s and early 2010s, with annual publication consistently exceeding five articles per year.

The most prolific years in the dataset are 2019 to 2023, with annual publication counts ranging from 15 to 20 articles. This peak reflects the maturity of HC as a research area and its widespread adoption across disciplines. The consistent output during these years indicates a stable and well-established field, underpinned by a growing research community and significant methodological advancements. The slight decline in 2024, with 10 publications, may be attributed to external factors, such as shifts in funding priorities or the inherent fluctuations in scholarly attention.

In summary, the temporal analysis reveals a dynamic growth trajectory for HC research. The initial decades of sporadic contributions paved the way for sustained growth in the mid-2000s, culminating in a period of peak activity in the late 2010s and early 2020s. This timeline underscores the field's adaptability and its ability to integrate new technologies and address emerging challenges, ensuring its continued relevance in the social sciences and humanities (see Figure 2).

Figure 3. Most frequently used words

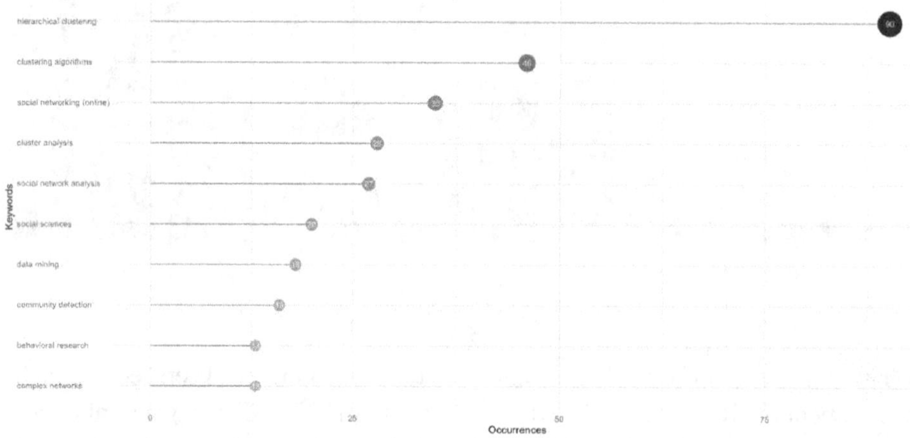

The frequency analysis of keywords within the dataset provides valuable insights into the intellectual landscape of HC research, revealing prevailing themes and areas of scholarly focus. The figure and table illustrate the most frequently used terms, highlighting the conceptual priorities and methodological emphases that characterize the field.

Table 2. Words with the highest frequency

Words	Occurrences
hierarchical clustering	90
clustering algorithms	46
social networking (online)	35
cluster analysis	28
social network analysis	27
social sciences	20
data mining	18
community detection	16
behavioral research	13
complex networks	13

As seen in Table 2 and Figure 3, at the core is the term *"hierarchical clustering"*, which appears 90 times. This prevalence underscores the centrality of this specific methodology in the corpus, highlighting its role as both a primary focus and a framing tool for numerous studies. Researchers' emphasis on this term indicates alignment between the research objectives and the methodological approaches employed, affirming its significance as an analytical framework.

The term *"clustering algorithms"*, occurring 46 times, extends this focus by situating HC within the broader context of algorithmic approaches. This frequency suggests that studies often compare hierarchical methods with alternative clustering techniques, such as k-means or density-based algorithms. Such comparative analyses are essential for refining methodological practices and assessing their suitability for various types of data and research objectives.

A notable thematic emphasis is revealed through terms such as *"social networking (online)"* with 35 occurrences and *"social network analysis"* with 27. These keywords reflect a prominent application of HC in understanding online relational structures and behaviors. Studies employing these terms likely explore phenomena such as community detection, influence dynamics, and interaction patterns within digital networks. The integration of HC with network analysis methodologies underlines its adaptability to contemporary challenges in data-intensive research.

Furthermore, *"cluster analysis"*, cited 28 times, serves as a broader conceptual category encompassing HC and related methodologies. This term often acts as a bridge between specific analytical approaches and their applications, positioning it within a wider spectrum of data partitioning techniques. Similarly, the occurrence of *"social sciences"*, cited 20 times, highlights the primary disciplinary focus of the dataset, underscoring the methodology's relevance to examining sociocultural phenomena.

Additional keywords, such as *"data mining"* with 18 occurrences, *"community detection"* with 16, and *"behavioral research"* with 13, illustrate the diverse range of applications addressed by HC. *"Data mining"* emphasizes the reliance on computational methods to extract patterns from large datasets, while *"community detection"* represents a specific focus on uncovering latent group structures within networks. The appearance of *"behavioral research"* signals the methodology's utility in studying human actions and decision-making processes, particularly in complex social or organizational contexts.

The term *"complex networks"*, appearing 13 times, underscores the field's engagement with highly interconnected systems. HC's ability to identify meaningful substructures within such networks makes it an indispensable tool for exploring phenomena ranging from social interactions to communication infrastructures. The consistent presence of terms related to networks, communities, and behaviors reflects the methodology's interdisciplinary reach and its capacity to address multifaceted research questions.

Ultimately, this analysis of keyword frequencies accentuates the intellectual diversity and methodological rigor of the research. By emphasizing core methodological terms alongside specific applications, the dataset captures the evolving interplay between HC techniques and the challenges they are designed to address (see Figure 3 and Table 2).

Figure 4. Trend topics

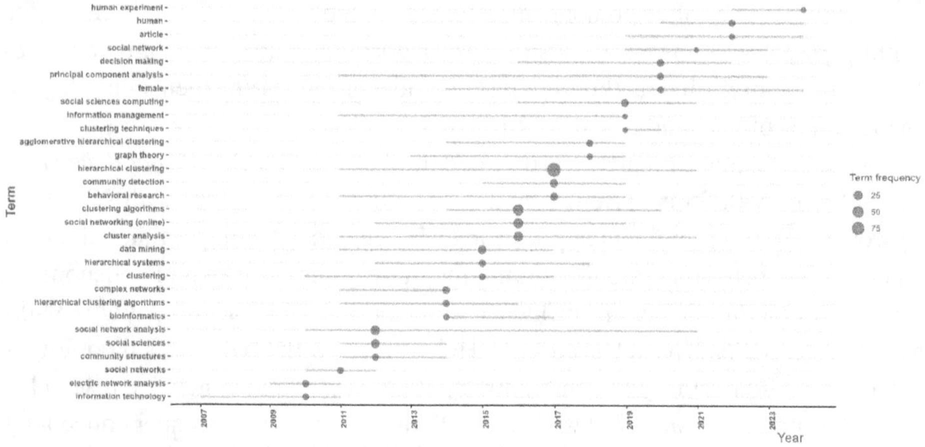

The trend topic analysis highlights the evolving focus within the domain of HC and its applications in the social sciences and humanities over the years. The graph illustrates the temporal distribution and frequency of key terms used in the corpus of articles analyzed, capturing shifts in research priorities and thematic relevance. The size of each node corresponds to the term frequency, with larger nodes indicating more prominent topics. In addition to the above, the timeline spanning from 2007 to 2023 reflects the trajectory of academic discourse, offering insights into how certain concepts gained prominence or diminished over time.

As seen in Figure 4, key terms such as *"human experiment"*, *"social network"*, and *"decision making"* emerge early in the timeline, suggesting foundational themes that shaped the field's initial inquiries. These terms were recurrently explored, signaling their central role in laying the groundwork for subsequent studies. By 2013, the focus diversifies, incorporating topics such as *"hierarchical clustering"*, *"community detection"*, and *"graph theory"*, signaling a growing interest in computational methods and their applicability in social structures. This phrase likely represents the convergence of algorithmic advancements with domain-specific applications, underscoring the interdisciplinarity of the field.

From 2015 onward, terms such as *"clustering algorithms"*, *"social networking (online)"*, and *"data mining"* gain prominence, reflecting the integration of big data analytics and the rise of digital platforms as significant areas of study. The sustained frequency of terms such as *"hierarchical systems"* and *"social sciences"* throughout the timeline points to their enduring relevance in the field. Notably, the emergence of *"bioinformatics"* and *"electric network analysis"* towards the later years suggests an expansion into adjacent disciplines, illustrating the adaptability and broad applicability of HC methodologies.

The temporal distribution of terms also reveals an underlying narrative about technological and methodological progress. For instance, the introduction of terms including *"principal component analysis"* and *"complex networks"* depicts a methodological evolution, aligning with the increasing computational capabilities available to researchers. The graph underscores the iterative nature of scientific inquiry, where earlier foundational topics pave the way for more specialized and complex investigations.

Additionally, the clustering of terms over time suggests the existence of distinct thematic periods in the research trajectory. Early phases focused on basic algorithmic development and application in well-defined social science scenarios, while later phases indicate a shift towards integrating HC with advanced machine learning techniques and interdisciplinary frameworks. Terms such as *"social dynamics" and "network theory"*, which appear towards the later years, point to the maturation of the field as researchers explore more complex phenomena and refine existing methodologies. This thematic evolution exhibits the adaptability of HC as a methodological tool capable of addressing diverse research questions across domains.

In summary, the trend topic analysis encapsulates the dynamic evolution of HC research within the social sciences and humanities, demonstrating the field's ability to adapt to emerging technologies and methodologies, while fostering a rich and diverse body of work that continues to expand its theoretical and practical horizons. The interplay between foundational topics and emerging trends reinforces the iterative and interdisciplinary nature of the field, ensuring its continued relevance and growth in the upcoming years (see Figure 4).

Figure 5. Co-occurrence network

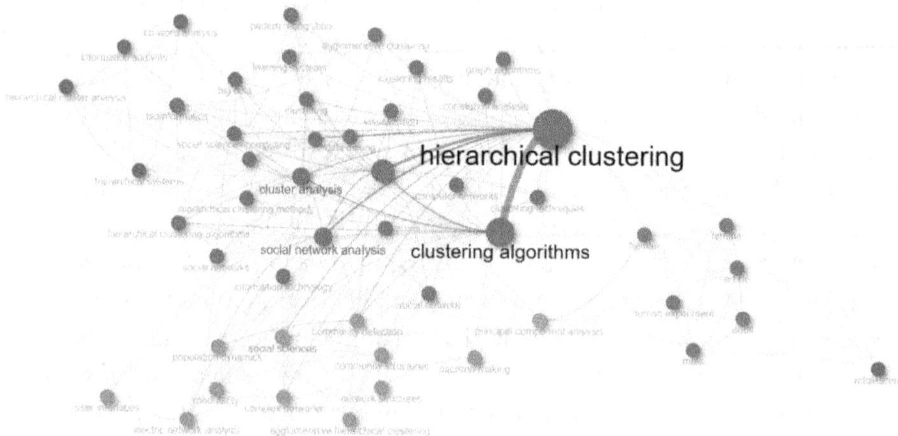

The network visualization depicted in Figure 5 illustrates the co-occurrence relationships between keywords in the dataset on HC research. Each node in the network represents a keyword, with its size proportional to the frequency of the term's occurrence, depicting its prominence within the field. The edges between nodes signify the frequency of co-occurrence, with thicker edges indicating stronger relationships. Additionally, the visualization employs distinct colors to group related keywords into thematic clusters, where nodes within the same color share close thematic connections, representing cohesive subfields of research.

As seen in Figure 5, the largest and most central node in the network is "hierarchical clustering", emphasizing its dominant role as the foundational concept in this research domain. Its dense connections to other key terms such as *"clustering algorithms"*, *"social network analysis"*, and *"data mining"* emphasize its extensive applicability and methodological significance. These terms form part of the dense central cluster, predominantly colored in red, which encompasses computational methodologies, including *"clustering techniques"*, *"visualization"*, and *"correlation analysis"*. This cluster represents the methodological core of HC research, showcasing the interplay of techniques and applications that underpin the advancement of the field.

Peripheral clusters provide insight into more specialized applications and emerging areas of interest. For instance, the purple cluster includes terms such as *"bioinformatics"* and *"hierarchical systems"*, reflecting the use of HC in biological and information systems contexts. Similarly, the green cluster comprises terms notably represented by *"community detection"*, *"network structures"*, and *"social sciences"*, highlighting applications in social and network science domains. Other terms including *"Population dynamics"* and *"electric network analysis"* appear on the outer periphery, representing niche or emerging topics with fewer connections to the central themes and pointing to areas of potential future exploration.

Overall, Figure 5 offers a comprehensive depiction of the intellectual structure of HC research, revealing the dominance of computational methodologies, the interdisciplinary reach of the field, and the presence of both well-established and emerging areas of interest.

Figure 6. Thematic map

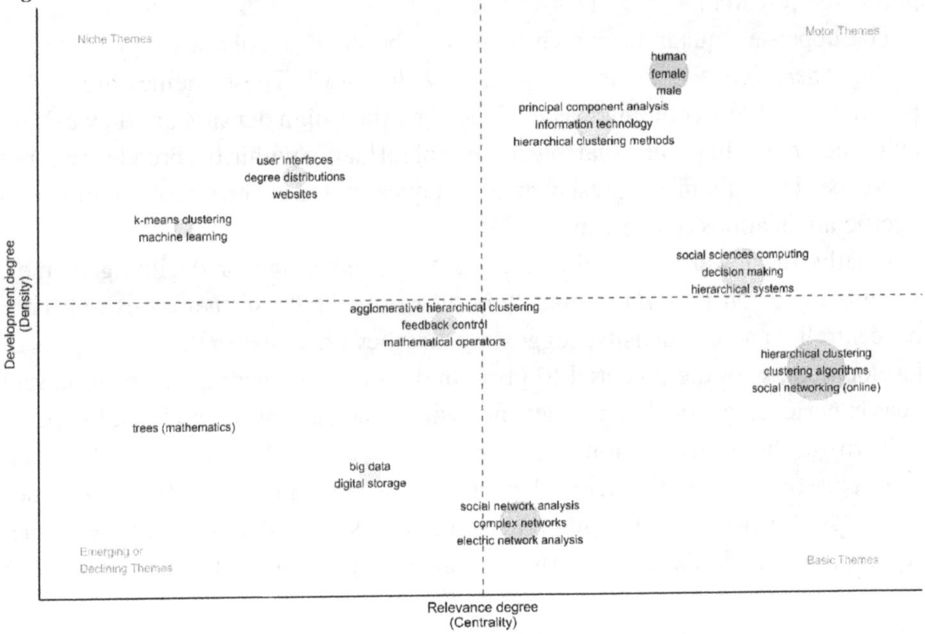

Figure 6 presents a thematic map that categorizes the identified research themes based on their centrality and density, offering a strategic overview of the field. Centrality, represented on the horizontal axis, measures the relevance of a theme within the broader research domain by assessing its connection to other themes. Density, depicted on the vertical axis, indicates the degree of internal development and cohesion within a theme. The map is divided into four quadrants, each representing distinct thematic characteristics.

The upper-right quadrant, which represents motor themes, includes topics such as *"human"*, *"hierarchical clustering methods"*, and *"principal component analysis"*. These themes are both highly central and well-developed, signifying their critical role in advancing the theoretical and methodological frameworks of HC. The high density of these themes reflects ongoing refinement and active research focus, while their centrality underscores their interconnectedness with other domains, establishing them as pivotal elements within the field.

Moreover, in the lower-right quadrant, labeled as basic themes, the core topics that appear are *"hierarchical clustering"*, *"clustering algorithms"*, and *"social networking"*. These terms exhibit high centrality, indicating their broad relevance and influence, but their relatively lower density suggests that they are less internally cohesive or still undergoing development. These themes form the backbone of the field, providing the foundational methodologies and concepts that underpin more specialized research.

The upper-left quadrant, which houses niche themes, contains topics encompassing *"user interfaces"* and *"degree distributions"*. These themes are highly specialized and self-contained, as evidenced by their high density, but they exhibit limited centrality, implying a narrower scope of influence within the broader research landscape. These themes represent areas of concentrated expertise, often limited to specific applications or contexts.

Finally, the lower-left quadrant, representing emerging or declining themes, contains topics on *"big data"* and *"digital storage"*. These themes exhibit both low centrality and low density, suggesting that they are either in the early stages of development, with the potential to grow into more prominent areas, or represent areas experiencing a decline in relevance within the current research landscape.

Through the categorization of themes into quadrants, Figure 6 provides a nuanced understanding of the field's thematic landscape, highlighting the coexistence of well-established and emerging topics, as well as areas that may require further exploration or revitalization, offering valuable insights for guiding future research efforts.

Figure 7. Country collaboration map

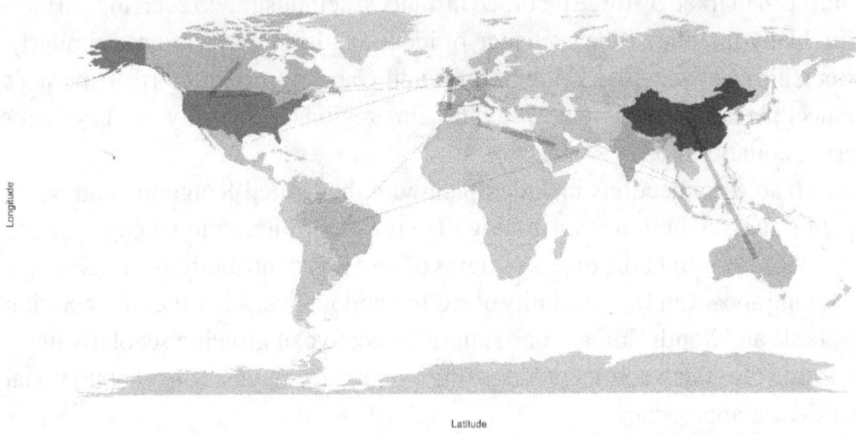

Figure 7 and Table 3 depict international research partnerships, providing insights into the global network of scholarly interactions within the domain of HC. Each connecting line between countries indicates a collaborative effort, with the thickness and frequency reflecting the strength and intensity of these partnerships. This analysis provides meaningful insights into the geographical distribution of academic contributions and the cross-border exchange of ideas and expertise.

Table 3. Country collaboration table

From	To	Frequency
China	Australia	3
Saudi Arabia	Tunisia	3
Spain	United Kingdom	3
USA	Canada	3
Brazil	Germany	2
China	Brazil	2
China	France	2
China	Japan	2
China	United Kingdom	2
India	Korea	2

China emerges as a pivotal player in the global research network, engaging in multiple collaborations with countries such as Australia, Brazil, France, Japan, and the United Kingdom. This dominance reinforces China's substantial investment in and contribution to HC research. The collaboration with Australia and the United Kingdom, each marked by three recorded instances, demonstrates recurring partnerships that likely facilitated the exchange of ideas and joint publications. Similarly, the partnerships between the United States and Canada, reflect North America's interconnected research ecosystem, emphasizing regional proximity as a key factor in fostering collaboration.

Other notable connections include Spain with the United Kingdom, and Saudi Arabia with Tunisia, both demonstrating the breadth of international engagement. These instances highlight the diverse origins of research contributions, suggesting a wide-reaching appeal and applicability of HC methodologies. Also, the collaboration between India and South Korea is particularly indicative of growing scholarly interactions within the Asian continent, depicting a regional emphasis on computational and data-driven approaches.

The map also demonstrates the uneven geographic distribution of partnerships, with regions like Africa and South America showing fewer linkages. This disparity may point to resource limitations or differing research priorities, which could be areas for future investigation. Nevertheless, the observed collaborations emphasize the inherently cooperative nature of academic research, where knowledge transcends national boundaries to address shared scientific challenges.

Further analysis reveals a strong alignment between research collaborations and global economic and technological hubs. Countries with established research infrastructures and robust funding mechanisms, such as the United States, China, and European nations, dominate the network. This trend emphasizes the importance of institutional support and resources in enabling collaborative research. At the same time, the limited representation of certain regions highlights the need for initiatives to promote global inclusivity and equitable access to research opportunities.

Regional collaborations likewise play a key role in knowledge exchange. For instance, partnerships within Europe, involving countries like France, Italy, and Spain, reflect a cohesive research environment supported by frameworks such as the European Union's Horizon 2020 program. Similarly, collaborations within Asia and between Asia and other regions underscore the growing emphasis on transcontinental research efforts, particularly in areas like computational social sciences and machine learning applications.

Overall, the country collaboration map captures the essence of global academic interconnectedness, showcasing key players and partnerships that propel the field forward while revealing both the strengths of existing networks and opportunities for enhancing inclusivity and diversification in research collaborations. By understanding the dynamics of international partnerships, researchers and policymakers can develop strategies to enhance global participation and address challenges of resource disparities, ensuring balanced and comprehensive progress in HC research.

Figure 8. Country scientific production map

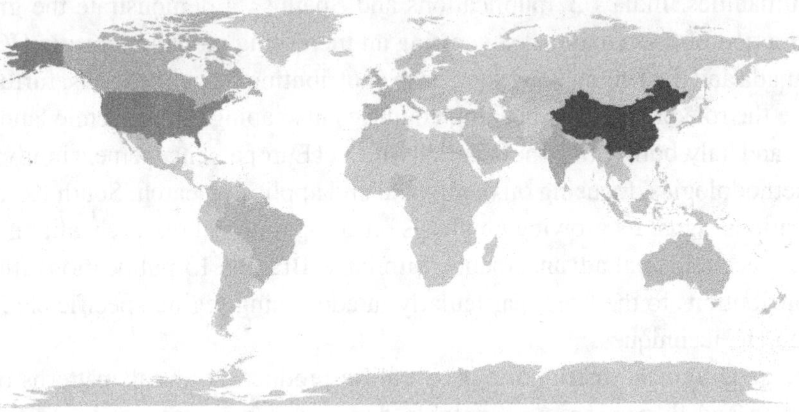

Figure 8 and Table 4 present a quantitative analysis of national contributions to HC research, offering a detailed perspective on the global distribution of scholarly activity. Together, they elucidate variations in research output, capturing disparities in the extent of participation across countries.

Table 4. Country scientific production table

Country	Frequency
China	162
USA	81
India	27
Spain	24
Canada	22
UK	22
France	19
Italy	19
South Korea	16
Brazil	15

As seen in Table 4, China leads with 162 publications, demonstrating its significant investment in research infrastructure and a strategic focus on computational and data-driven methodologies. The United States ranks second with 81 publications, reaffirming its position as a major contributor to global academic discourse. The significant output from the United States underscores its tradition of interdisciplinary research, bridging computational methods with applications in the social sciences and humanities. India's 27 publications and Spain's 24 demonstrate the growing participation of these nations, suggesting an increasing global interest in HC.

Canada and the United Kingdom, each contributing 22 publications, further emphasize the role of English-speaking countries in shaping the academic landscape. France and Italy, both with 19 publications, reflect Europe's engagement in advancing HC methodologies, focusing on theoretical and applied research. South Korea's 16 publications signal its growing emphasis on computational research, aligning with broader technological advancements. Similarly, Brazil's 15 publications illustrate its contributions to the field, particularly in addressing region-specific challenges through HC techniques.

The geographical distribution of scientific production reveals patterns of concentration and dispersion, with notable dominance by countries in Asia, North America, and Europe. This distribution illustrates the global appeal of HC as a research topic, while also indicating disparities in research output among regions. The lower contributions from African and South American countries, apart from Brazil, suggest potential areas for growth where targeted initiatives could foster greater participation and innovation.

Temporal analysis of publication output reveals trends in research activity across regions. The consistent increase in publications from China and the United States reflects sustained investment and interest in the field. The emergence of India and South Korea as notable contributors indicates a shift towards broader international participation, driven by advancements in computational resources and academic infrastructure. The steady output from European nations, particularly France, Italy, and Spain, underscores their established presence in the field, supported by collaborative frameworks and funding mechanisms.

In conclusion, the analysis of national scientific production offers a comprehensive overview of contributions to HC research, emphasizing the leadership of China, the United States, and other key players, while also identifying opportunities to enhance global inclusivity and diversification in academic research. By addressing regional disparities and fostering international collaboration, the field can continue to evolve, ensuring balanced and equitable representation of diverse perspectives and expertise.

Figure 9. Most cited countries

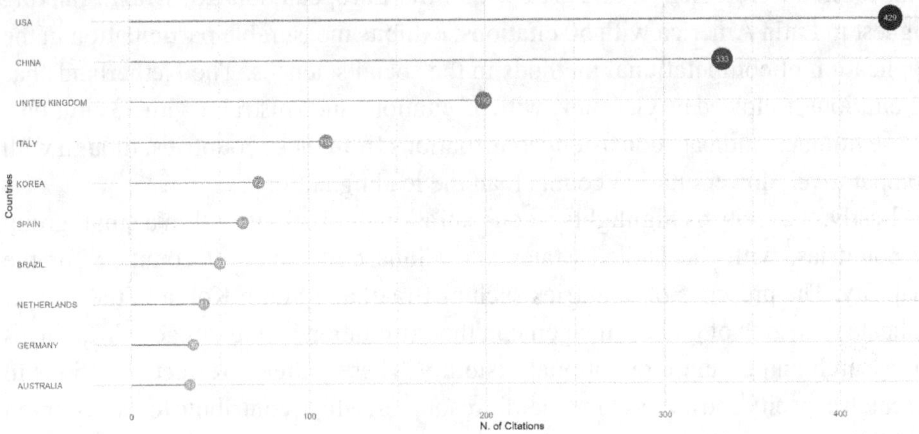

Table 5. Most cited countries table

Country	Total Citations
USA	429
CHINA	333
UNITED KINGDOM	199
ITALY	110
KOREA	72
SPAIN	63
BRAZIL	50
NETHERLANDS	41
GERMANY	35
AUSTRALIA	33

As seen in Figure 9, the United States lead with 429 citations, representing the highest impact globally, reflecting the country's extensive academic output. China follows as the second most cited country, accumulating 333 citations, underscoring the increasing prominence in academic research, driven by a growing emphasis on interdisciplinary methodologies and computational approaches. The United Kingdom ranks third with 199 citations, which can be attributed to its consistent contributions to the application of advanced clustering techniques in social science contexts.

Italy ranks fourth with 110 citations, representing the highest citation count among Southern European countries. South Korea, with 72 citations, records a significant presence in the Asia-Pacific region. Spain, with 63 citations, reflects a steady contribution to research in this area within the European context. Brazil, ranking highest in Latin America with 50 citations, exhibits measurable participation in the application of computational methods in the social sciences. The Netherlands has 41 citations, followed by Germany with 35 citations and Australia with 33 citations. These numbers indicate consistent contributions from these countries, though with comparatively lower citation counts than the leading nations.

Lastly, the analysis highlights a concentration of citations within a small group of countries, with the United States and China collectively accounting for the majority. The presence of countries such as Brazil and South Korea in the dataset indicates a degree of global engagement, though citation counts suggest differences in research output and international visibility. These patterns reflect variations in research capacity and access to academic resources, which contribute to the observed distribution of citations across regions (see Figure 9 and Table 5).

Table 6. Most cited papers

Paper	Total Citations
Breiger Rl, 1975, J Math Psychol	508
Xu Jj, 2005, Acm Trans Inf Syst	185
Kamis Nh, 2018, Appl Soft Comput J	84
Ge W, 2009, Workshop Appl Comput Vis, Wacv	78
Shin Jc, 2009, High Educ	63
Hjellbrekke J, 2018, Multiple Correspondence Analysis for the Soc Sciences	58
Gan C, 2015, Scientometrics	55
Waller Ng, 1998, Psychometrika	55
Xie W-B, 2020, Inf Sci	53
Qiu J-P, 2014, Scientometrics	47
Kamis Nh, 2019, Inf Fusion	43
Benites-Lazaro Ll, 2017, Energy Res Soc Sci	41
Sircova A, 2015, Time Perspective Theory; Review, Research and Application: Essays in Honor of Philip G Zimbardo	41
Leydesdorff L, 2017, Scientometrics	41
Didimo W, 2014, Inf Sci	40

Table 6 presents the most cited papers derived from the bibliometric analysis, illustrating the significant influence of HC methods across various disciplines. The papers highlighted are seminal works that contribute to the development and application of HC in fields such as social network analysis, criminology, decision-making, human behavior analysis, and education.

Breiger et al. (1975), whose paper has the highest TC of 508, introduced the CONCOR algorithm, a method for HC applied to relational data, which has since become a cornerstone in social network analysis. The algorithm was pivotal in the analysis of social structures, providing an effective means to identify clusters within relational data using correlation analysis. In a similar vein, Xu and Chen (2005), expanded the application of HC through the CrimeNet Explorer framework. This innovative framework integrates clustering with visualization and structured analysis to uncover patterns within criminal networks, significantly enhancing criminal investigations. By applying HC to crime data, they demonstrated the potential of clustering techniques to identify hidden relationships and structures within criminal organizations. Their work has been cited a total of 185 times, indicating the substantial impact it has had on criminology and law enforcement, particularly in the context of criminal network analysis.

Further advancing the application of HC, Kamis et al. (2018) introduced an innovative approach to decision-making using weighted preference networks. Their model employed HC to identify individuals with similar preferences, facilitating more efficient group consensus. This contribution, with 84 total citations, illustrates the utility of HC in enhancing collective decision processes. On the other hand, Ge et al. (2009) applied HC to analyze pedestrian behavior within crowds, identifying social groups through automated methods. Their study, which was cited 78 times, marked a significant advancement in the field of crowd analysis, offering real-time insights into group dynamics in public spaces. By employing HC to detect social groupings within large crowds, Ge et al. (2009) were able to provide deeper insights into collective human behavior, an area of significant interest for urban planners, security personnel, and event organizers.

Furthermore, Shin (2009) with 63 TC applied HC to the classification of higher education institutions based on research performance, providing an alternative to traditional university rankings. This paper offers a data-driven approach to evaluating academic institutions, enabling policymakers to assess universities' performance through objective metrics. By using HC, a novel classification system was proposed that classifies institutions based on their research output rather than subjective ranking criteria. Similarly, Hjellbrekke (2018) explored the application of HC in the context of Multiple Correspondence Analysis (MCA), a technique for analyzing complex categorical data. Their work, cited 58 times, contributes to the development of social science research methods by demonstrating how HC can complement MCA to reveal meaningful patterns within categorical data, emphasizing the adaptability of HC in understanding relationships between various categorical variables and their underlying structures.

Gan and Wang (2015) explored scientific research trends by focusing on the evaluation of research outputs in scientometrics. Their study with 55 TC identifies patterns in scientific production and collaboration, uncovering research areas and grouping within the scientific community, providing insights into the structural organization of research fields. Similarly, Waller et al. (1998), applied clustering techniques to psychometric data, identifying latent structures within psychological assessments, which refined methods for analyzing complex data, offering new interpretations of psychological test results and making a significant contribution to the field of psychometrics.

In the domain of information science, Xie et al. (2020) examined the relationships between scientific publications and research areas, utilizing clustering methods to understand how topics evolve and interconnect. Their study with 53 TC, underscores the growing importance of such techniques in bibliometrics and scientometrics. Likewise, Qiu et al. (2014) focused on evaluating scientific performance and collaboration by employing clustering techniques in scientometric analysis. Their research, cited 47 times, offered tools for assessing the development of academic productivity and research networks, contributing to the understanding of how scientific communities and collaborative networks evolve over time.

In addition to the above, Kamis et al. (2019) extended their previous work on preference networks by integrating clustering with information fusion to enhance decision-making processes. With 43 TC, their research demonstrates how this combination can synthesize diverse sources of information, producing cohesive groups that support more accurate decisions, especially in complex and uncertain contexts. In the environmental sciences, Benites-Lazaro et al. (2017) examined energy consumption patterns across various social and economic groups by grouping regions based on their consumption behaviors, with their study, which has 41 total citations, providing new insights that could guide the design of targeted energy efficiency programs.

Moreover, Sircova et al. (2015) investigated how time perspective theory applies to understanding individual and group behaviors. With 41 TC, their work showed how clustering techniques could identify distinct time perspectives within individuals and groups, contributing to behavioral sciences and offering a framework for future research and therapy. On the other hand, Leydesdorff et al. (2017) with 41 TC, employed HC to map the evolution of scientific knowledge and collaboration networks, revealing patterns in research areas and the interrelationships between different scientific communities, shedding light on the dynamics of academic research fields. Finally, Didimo and Montecchiani (2014) with 40 TC, examined the relationship between academic publications in information science, using clustering to explore the structure of knowledge. Their work introduced innovative methods for visualizing and interpreting academic collaboration networks, advancing the field of bibliometric analysis.

In conclusion, these studies collectively highlight the far-reaching influence of HC across diverse fields. By mapping relationships, uncovering hidden patterns, and providing deeper insights into complex data, HC has become a critical tool for understanding and interpreting the dynamics of research networks, scientific collaboration, and social phenomena (see Table 6).

Table 7. Most relevant affiliations

Affiliation	Articles
Southeast University	8
Wuhan University	7
Cavite State University	6
Huazhong University of Science and Technology	6
Jiangnan University	6
University of Leicester	6
Washington State University	6
Ewha Womans University	5
The University of Texas at Dallas	5
University of Toronto	5

The distribution of publications across various academic institutions provides important insights into the global engagement with HC in social sciences and humanities research. As seen in Table 7, Southeast University stands out as the most prolific contributor, with 8 articles. This high output indicates a significant institutional focus on the exploration and application of HC techniques, suggesting that Southeast University plays a pivotal role in advancing this analytical methodology within its academic community. Wuhan University follows closely with 7 articles, further reinforcing its substantial involvement in the development of HC applications. This reveals that Wuhan University is an influential research hub for scholars interested in data analysis methodologies, contributing extensively to the scholarly discourse on HC.

A cohort of institutions, such as Cavite State University, Huazhong University of Science and Technology, Jiangnan University, the University of Leicester, and Washington State University, each with 6 publications, demonstrates a broad and varied commitment to research utilizing this methodology. These universities, spanning diverse geographic regions and academic disciplines, collectively highlight the wide-ranging applicability of the technique. The studies originating from these institutions likely encompass fields such as engineering, social sciences, and information technology, with a shared emphasis on employing clustering methods to analyze complex data structures.

The 5 articles from Ewha Woman's University, the University of Texas at Dallas, and the University of Toronto, though fewer in number, continue to contribute to the growing body of literature. These institutions exemplify sustained engagement with clustering methods, producing studies that deepen understanding of their applications across different research contexts. Their contributions serve to refine existing techniques and expand their utility in various academic and scientific areas.

The diversity of institutions and geographical distribution within this dataset suggests that clustering techniques are widely recognized and utilized across a range of academic disciplines. This trend reflects the increasing recognition of the method as an essential tool in analyzing complex datasets, with institutions globally contributing to its continuous refinement and expansion into various research domains (see Table 7).

Table 8. Most relevant sources

Sources	
Lecture Notes in Computer Science (Including Subseries Lecture Notes in Artificial Intelligence and Lecture Notes in Bioinformatics)	9
Scientometrics	9
Information Sciences	4
Communications in Computer and Information Science	3
2011 International Conference on Internet Technology and Applications, ITAP 2011 - Proceedings	2
ACM International Conference Proceeding Series	2
Advances in Intelligent Systems and Computing	2
Energy Research and Social Science	2
Information Studies: Theory and Application	2
Proceedings of the International Conference on Information Technology Interfaces, ITI	2

The analysis of the most relevant sources in the field reveals which journals and conference proceedings have been most influential in publishing research on clustering techniques. As seen in Table 8, the sources *Lecture Notes in Computer Science (including subseries Lecture Notes in Artificial Intelligence and Lecture Notes in Bioinformatics)* and *Scientometrics*, each contributing 9 articles, stand out as the primary outlets for research on clustering techniques. These two journals are instrumental in advancing the understanding of complex data structures and the integration of clustering techniques within diverse academic fields. The prominence of *Lecture Notes in Computer Science* highlights the growing convergence of computational approaches with clustering methods, particularly in domains such as artificial intelligence, bioinformatics, and data science. The articles published in this series often tackle advanced topics, including machine learning, pattern recognition, and computational modeling, where clustering is essential for structuring and analyzing large, intricate datasets.

In addition to the above, several other journals have made noteworthy contributions to the literature on HC. The journal *Information Sciences*, contributing 4 articles, illustrates the growing importance of clustering in the management and retrieval of information. This source focuses on the use of such techniques to improve search algorithms, organize vast amounts of information, and optimize data retrieval processes. The research underscores how clustering methods can be leveraged to structure datasets and enhance the relevance and accessibility of information in a variety of research contexts.

Another relevant source, *Communications in Computer and Information Science* with a total of 3 articles, reflects the application of clustering within computer science and information systems, particularly in areas such as network analysis and data security. In these fields, clustering is essential for organizing large datasets, detecting patterns, and improving system performance. The research published in this source illustrates how the technique can be used to analyze and interpret complex information systems, thereby optimizing data processing and analytical efficiency.

Conference proceedings such as the *2011 International Conference on Internet Technology and Applications, ITAP 2011 – Proceedings* and the *ACM International Conference Proceeding Series*, each contributing 2 articles, reveal the emerging role of clustering methods in analyzing web and internet-related data. These sources emphasize the technique's utility in understanding relationships within online interactions, web traffic, and social networks, providing insights into user behavior and network structures.

Other sources, including *Advances in Intelligent Systems and Computing*, *Energy Research and Social Science*, and *Information Studies: Theory and Application*, each contributing 2 articles, highlight the diverse applications of clustering in interdisciplinary research. In intelligent systems, clustering is used to model complex behaviors and enhance decision-making processes. In energy research, it aids in identifying consumption patterns and understanding regional differences in energy use. Similarly, *Information Studies: Theory and Application* showcases the role of clustering in organizing and analyzing data related to human behavior, information flow, and knowledge management.

Lastly, the *Proceedings of the International Conference on Information Technology Interfaces (ITI)*, contributing 2 articles, further exemplifies the applications of clustering techniques in information technology. The studies in this source focus on clustering user data, optimizing system interfaces, and improving data visualization strategies, demonstrating the practical relevance of clustering in the development of technology systems.

In summary, the sources analyzed reveal the widespread application of clustering techniques in various academic and research fields. Whether in computational sciences, bibliometrics, energy research, or information systems, clustering plays a pivotal role in organizing complex datasets, detecting patterns, and enhancing decision-making processes. These sources not only contribute to the theoretical development of clustering methodologies but also illustrate their practical applications across multiple domains, reinforcing the technique's versatility and increasing importance in contemporary research.

CONCLUSIONS

The findings of this bibliometric analysis provide a meticulous and extensive examination of HC within the social sciences and humanities, elucidating its intellectual framework, thematic progression, and the collaborative networks underpinning this interdisciplinary field. The results reinforce HC's methodological significance, versatility, and broad applicability, while identifying avenues for further scholarly exploration and critical evaluation.

Furthermore, HC emerges as a cornerstone methodology, underpinning a vast array of research inquiries. Its prominence, as reflected in the frequent occurrence of terms such as *"hierarchical clustering"* and *"clustering algorithms"*, underscores its indispensable role in analyzing complex datasets and uncovering latent structures (see Figure 3 and Table 2). The intellectual centrality of HC is further evidenced by its dense connections to terms such as *"social network analysis"* and *"data mining"* in network visualizations. These linkages illustrate HC's adaptability to a broad spectrum of analytical contexts, ranging from computational techniques to applications in social sciences and digital humanities (see Figure 5). Moreover, the thematic map delineates HC's methodological core while simultaneously pointing to emerging and specialized areas of interest, such as *"big data"* and *"digital storage"* (see Figure 6).

The temporal evolution of HC research reveals a dynamic trajectory of growth and adaptation. From its sporadic early contributions to its more consistent output in recent decades, the field has demonstrated resilience and responsiveness to advances in computational resources and interdisciplinary collaborations. The marked growth from the mid-2000s onward, peaking in the late 2010s and early 2020s, coincides with the proliferation of big data and the increasing complexity of research questions in social and network sciences. While a slight decline in publication activity in 2024 may reflect natural fluctuations, it warrants attention as a potential signal of shifting academic priorities or funding dynamics (see Figure 2).

Collaboration and international engagement have significantly shaped the trajectory of HC research. The dataset's inclusion of 555 authors and a notable rate of international co-authorship (21.93%) attests to the collaborative ethos of the field. Country-specific analyses reveal disparities in research output, with China and the United States emerging as dominant contributors. In contrast, underrepresentation from regions such as Africa and parts of South America underscores the need for initiatives to foster global inclusivity (see Table 1). Strengthening international collaboration, particularly with underrepresented regions, could enrich the field with diverse perspectives and address localized challenges through HC methodologies.

A notable strength of HC lies in its interdisciplinary appeal, as evidenced by the diversity of its applications. From criminology and psychometrics to education and bioinformatics, HC has demonstrated its utility in addressing complex relational and categorical data. Seminal works, such as Breiger et al. (1975) CONCOR algorithm and Xu and Chen (2005) application in criminal network analysis, exemplify its transformative impact across domains. However, the iterative nature of HC's development suggests the need for continuous refinement and integration with emerging computational techniques, such as machine learning and artificial intelligence, to maintain its methodological rigor and expand its applicability (see Table 6).

Emerging themes identified in the thematic analysis, such as *"big data"* and *"digital storage"*, represent promising frontiers for HC research. These areas, while nascent, offer significant potential for advancing HC methodologies in the context of large-scale and heterogeneous datasets. Similarly, niche themes, characterized by their high density but limited centrality, present opportunities for deeper exploration and integration with mainstream research (see Figure 4). Collaborative efforts that bridge methodological expertise and domain-specific applications could unlock novel insights and theoretical advancements, further enriching the intellectual landscape of HC.

Despite its robust contributions, the field of HC is not without limitations. The geographic concentration of research output highlights systemic inequities in academic participation and resource allocation (see Figure 8 and Table 4). Addressing these disparities through targeted funding, capacity-building programs, and equitable collaboration networks is essential to democratizing access to HC research. Additionally, while the integration of HC with advanced computational tools has enhanced its analytical capabilities, future research could navigate ethical considerations, including data privacy and the reproducibility of results, to uphold methodological integrity.

In conclusion, HC remains a vital methodological pillar within the social sciences and humanities. Its capacity to address multifaceted research questions, adapt to evolving technological landscapes, and foster interdisciplinary collaboration underscores its enduring relevance. However, ensuring the field's continued impact necessitates proactive efforts to address geographic disparities, promote inclusivity, and explore emerging thematic directions. By leveraging the insights gleaned from this bibliometric analysis, future research can enhance the theoretical and practical dimensions of HC, positioning it to meet the challenges and opportunities of an increasingly data-driven academic environment.

REFERENCES

Almeida, J., Barbosa, L., Pais, A., & Formosinho, S. (2007). Improving hierarchical cluster analysis: A new method with outlier detection and automatic clustering. *Chemometrics and Intelligent Laboratory Systems*, *87*(2), 208–217. DOI: 10.1016/j.chemolab.2007.01.005

Aria, M., & Cuccurullo, C. (2017). bibliometrix : An R-tool for comprehensive science mapping analysis. *Journal of Informetrics*, *11*(4), 959–975. DOI: 10.1016/j.joi.2017.08.007

Baas, J., Schotten, M., Plume, A., Côté, G., & Karimi, R. (2020). Scopus as a curated, high-quality bibliometric data source for academic research in quantitative science studies. *Quantitative Science Studies*, *1*(1), 377–386. DOI: 10.1162/qss_a_00019

Benites-Lazaro, L., Mello-Théry, N., & Lahsen, M. (2017). Business storytelling about energy and climate change: The case of Brazil's ethanol industry. *Energy Research & Social Science*, *31*, 77–85. DOI: 10.1016/j.erss.2017.06.008

Bitzenis, A., & Koutsoupias, N. (2023, December). Visualizing Economics and Business in the field of Education: A Quantitative Analysis of Scholarly Literature. *Global Business & Economics Anthology. Volumes I, II*(II). DOI: 10.47341/GBEA.23126

Bitzenis, A., & Koutsoupias, N. (2024A). Big data in economics research. In *Springer proceedings in business and economics* (pp. 1063–1072). DOI: 10.1007/978-3-031-49105-4_61

Bitzenis, A., & Koutsoupias, N. (2024B). A Scientometric Review regarding FDI in SouthEast Europe (2005–2024). *Global Business & Economics Anthology, I&II* (2024), 117–128. DOI: 10.47341/GBEA.241210

Bitzenis, A., Koutsoupias, N., & Boutsiouki, S. (2023). Business Research and Data Mining: a Bibliometric Analysis. *2023 3rd International Conference on Electrical, Computer, Communications and Mechatronics Engineering (ICECCME)*, 1–6. DOI: 10.1109/ICECCME57830.2023.10252699

Bitzenis, A., Koutsoupias, N., & Nosios, M. (2025). Artificial intelligence and machine learning in production efficiency enhancement and sustainable development: A comprehensive bibliometric review. *Frontiers in Sustainability*, *5*, 1508647. DOI: 10.3389/frsus.2024.1508647

Breiger, R. L., Boorman, S. A., & Arabie, P. (1975). An algorithm for clustering relational data with applications to social network analysis and comparison with multidimensional scaling. *Journal of Mathematical Psychology, 12*(3), 328–383. DOI: 10.1016/0022-2496(75)90028-0

Chen, X., & Deng, H. (2019). A correlation analysis of information use, social networks and cooperation consciousness in travel behaviors. *Transportation Research Part F: Traffic Psychology and Behaviour*, *62*, 819–832. DOI: 10.1016/j.trf.2019.03.015

Didimo, W., & Montecchiani, F. (2014). Fast layout computation of clustered networks: Algorithmic advances and experimental analysis. *Information Sciences, 260*, 185–199. DOI: 10.1016/j.ins.2013.09.048

Gan, C., & Wang, W. (2015). Research characteristics and status on social media in China: A bibliometric and co-word analysis. *Scientometrics, 105*(2), 1167–1182. DOI: 10.1007/s11192-015-1723-2

Ge, W., Collins, R. T., & Ruback, B. (2009). Automatically detecting the small group structure of a crowd. *2009 Workshop on Applications of Computer Vision, WACV 2009*, 1–8. DOI: 10.1109/WACV.2009.5403123

Giordani, P., Ferraro, M. B., & Martella, F. (2020). Hierarchical clustering. In *Behaviormetrics* (pp. 9–73). DOI: 10.1007/978-981-13-0553-5_2

Hjellbrekke, J. (2018). *Multiple Correspondence Analysis for the Social Sciences* (1st ed.). Routledge., DOI: 10.4324/9781315516257

Jafarzadegan, M., Safi-Esfahani, F., & Beheshti, Z. (2019). Combining hierarchical clustering approaches using the PCA method. *Expert Systems with Applications, 137*, 1–10. DOI: 10.1016/j.eswa.2019.06.064

Jafarzadegan, M., Safi-Esfahani, F., & Beheshti, Z. (2019). Combining hierarchical clustering approaches using the PCA method. *Expert Systems with Applications, 137*, 1–10. DOI: 10.1016/j.eswa.2019.06.064

Kamis, N. H., Chiclana, F., & Levesley, J. (2018). Preference similarity network structural equivalence clustering based consensus group decision making model. *Applied Soft Computing*, *67*, 706–720. DOI: 10.1016/j.asoc.2017.11.022

Kamis, N. H., Chiclana, F., & Levesley, J. (2019). An influence-driven feedback system for preference similarity network clustering based consensus group decision making model. *Information Fusion*, *52*, 257–267. DOI: 10.1016/j.inffus.2019.03.004

Kong, J., Du, J., & Wu, C. (2024). Evaluating the Health of Higher Education: A Hierarchical Clustering and Multi-Model Approach. In *Atlantis highlights in social sciences, education and humanities/Atlantis Highlights in Social Sciences, Education and Humanities* (pp. 297–303). DOI: 10.2991/978-94-6463-574-4_35

Koutsoupias, N., & Papadimitriou, I. (2001). Hierarchical Classification In Insurance Data Analysis. In C. Zopounidis, P. M. Pardalos, & G. Baourakis, *Fuzzy Sets in Management, Economics and Marketing* (pp. 69–81). World Scientific. DOI: 10.1142/9789812810892_0005

Koutsoupias, N., & Papadimitriou, I. (2020). Multivariate Data Analysis in R - Μέθοδοι Πολυδιάστατης Ανάλυσης Δεδομένων με τη γλώσσα R. Zenodo *(CERN European Organization for Nuclear Research)*. DOI: 10.5281/zenodo.7557364

Lancho-Barrantes, B. S., & Cantu-Ortiz, F. J. (2021). Quantifying the publication preferences of leading research universities. *Scientometrics*, *126*(3), 2269–2310. DOI: 10.1007/s11192-020-03790-1 PMID: 33424051

Leydesdorff, L., Bornmann, L., & Wagner, C. S. (2017). Generating clustered journal maps: An automated system for hierarchical classification. *Scientometrics*, *110*(3), 1601–1614. DOI: 10.1007/s11192-016-2226-5 PMID: 28255188

Leydesdorff, L., Bornmann, L., & Wagner, C. S. (2017). Generating clustered journal maps: An automated system for hierarchical classification. *Scientometrics*, *110*(3), 1601–1614. DOI: 10.1007/s11192-016-2226-5 PMID: 28255188

Li, T., Rezaeipanah, A., & Din, E. M. T. E. (2022). An ensemble agglomerative hierarchical clustering algorithm based on clusters clustering technique and the novel similarity measurement. *Journal of King Saud University. Computer and Information Sciences*, *34*(6), 3828–3842. DOI: 10.1016/j.jksuci.2022.04.010

Liu, Y. (2023). Customer Segmentation in User Behavior Analysis: A Comparative Study of Clustering Algorithms. *Highlights in Business Economics and Management*, *21*, 758–764. DOI: 10.54097/hbem.v21i.14758

Louf, T., Gonçalves, B., Ramasco, J. J., Sánchez, D., & Grieve, J. (2023). American cultural regions mapped through the lexical analysis of social media. *Humanities & Social Sciences Communications*, *10*(1), 133. DOI: 10.1057/s41599-023-01611-3

Murtagh, F., & Contreras, P. (2011). Algorithms for hierarchical clustering: An overview. *Wiley Interdisciplinary Reviews. Data Mining and Knowledge Discovery*, *2*(1), 86–97. DOI: 10.1002/widm.53

Nielsen, F. (2016). Hierarchical clustering. In *Undergraduate topics in computer science* (pp. 195–211). DOI: 10.1007/978-3-319-21903-5_8

Ortega, J. L., López-Romero, E., & Fernández, I. (2011). Multivariate approach to classify research institutes according to their outputs: The case of the CSIC's institutes. *Journal of Informetrics*. DOI: 10.1016/j.joi.2011.01.004

Page, M. J., McKenzie, J. E., Bossuyt, P. M., Boutron, I., Hoffmann, T. C., Mulrow, C. D., Shamseer, L., Tetzlaff, J. M., Akl, E. A., Brennan, S. E., Chou, R., Glanville, J., Grimshaw, J. M., Hróbjartsson, A., Lalu, M. M., Li, T., Loder, E. W., Mayo-Wilson, E., McDonald, S., & Moher, D. (2021). The PRISMA 2020 statement: An updated guideline for reporting systematic reviews. *BMJ (Clinical Research Ed.)*, (71), n71. DOI: 10.1136/bmj.n71 PMID: 33782057

Pişirgen, A., & Peker, S. (2024). A clustering approach for classifying scholars based on publication performance using bibliometric data. *Egyptian Informatics Journal*, *28*, 100537. DOI: 10.1016/j.eij.2024.100537

Qiu, J., Dong, K., & Yu, H. (2014). Comparative study on structure and correlation among author co-occurrence networks in bibliometrics. *Scientometrics*, *101*(2), 1345–1360. DOI: 10.1007/s11192-014-1315-6

Roux, M. (2018). A comparative study of divisive and agglomerative hierarchical clustering algorithms. *Journal of Classification*, *35*(2), 345–366. DOI: 10.1007/s00357-018-9259-9

Shin, J. C. (2009). Classifying higher education institutions in Korea: A performance-based approach. *Higher Education*, *57*(2), 247–266. DOI: 10.1007/s10734-008-9150-4

Sircova, A.. (2015). Time Perspective Profiles of Cultures. In Stolarski, M., Fieulaine, N., & van Beek, W. (Eds.), *Time Perspective Theory; Review, Research and Application*. Springer., DOI: 10.1007/978-3-319-07368-2_11

Waller, N. G., Kaiser, H. A., Illian, J. B., & Manry, M. (1998). A comparison of the classification capabilities of the 1-dimensional kohonen neural network with two pratitioning and three hierarchical cluster analysis algorithms. *Psychometrika*, *63*(1), 5–22. DOI: 10.1007/BF02295433

Xie, W., Lee, Y., Wang, C., Chen, D., & Zhou, T. (2020). Hierarchical clustering supported by reciprocal nearest neighbors. *Information Sciences*, *527*, 279–292. DOI: 10.1016/j.ins.2020.04.016

Xu, J. J., & Chen, H. (2005). CrimeNet explorer: A framework for criminal network knowledge discovery. *ACM Transactions on Information Systems*, *23*(2), 201–226. DOI: 10.1145/1059981.1059984

Zhou, S., Xu, Z., & Liu, F. (2016). Method for determining the optimal number of clusters based on agglomerative hierarchical clustering. *IEEE Transactions on Neural Networks and Learning Systems*, *28*(12), 3007–3017. DOI: 10.1109/TNNLS.2016.2608001 PMID: 28113523

Zourna, C., Papavassiliou-Alexiou, I., & Koutsoupias, N. (2024). COVID-19 challenges: how ICT successfully allowed adaptation of Drama in Education experiential method during PhD research. In E. Karaiskou, A. Kokkinou, A. Mavrogianni, and G. Rentifis (Eds), Proceedings of the 3rd International Scientific Conference "Greece – Europe 2030: Education, Research, Innovation, New technologies, Institutions and Sustainable Development", Vol. 3, Heraklion Crete, 7-10 September 2023. Athens: Published by the Hellenic Scientific Institute of Economics of Education, Lifelong Learning, Research and Innovation, Electronic Publisher: Greek National Documentation Centre. DOI: 10.12681/elrie.7145

Chapter 13
Work-Related Stress Among Substitute Teachers in Greece

Lazaros Mitsou
https://orcid.org/0000-0002-6297-4738
Democritus University of Thrace, Greece

Theodora Vainterli
https://orcid.org/0009-0009-0363-7908
Democritus University of Thrace, Greece

ABSTRACT

This article examines the impact of work stress on substitute secondary school teachers in Greece, a group facing unique challenges and uncertainties. It analyzes the relationship between job stress and factors such as job performance and burnout. A questionnaire was designed with five factors: (1) job stress, (2) job performance, (3) burnout, (4) job satisfaction, and (5) coping strategies. Participants responded using a Likert scale, and data will be analyzed through regression to identify correlations and effects of stress. The study also explores strategies teachers use to manage stress and how these affect satisfaction and performance. Preliminary results suggest stress significantly impacts burnout and satisfaction. Coping strategies mitigate stress, while performance varies based on stress levels and strategies. This study highlights the need for interventions to support substitute teachers, aiming to improve mental health and performance. Findings could guide programs to enhance resilience and satisfaction in this critical educational sector.

DOI: 10.4018/979-8-3693-9400-7.ch013

INTRODUCTION

Work-related stress is a global phenomenon affecting nearly every professional sector, with the educational field being among the most vulnerable. Its implications extend beyond individual well-being, influencing the overall quality of education. Teachers often face intense pressures due to constant reforms, rising performance expectations, and the need to adapt to changing environments (Kyriacou, 2001; Johnson et al., 2005). The inherent complexity of the teaching profession, especially in secondary education, contributes to elevated levels of emotional exhaustion, reduced job performance, and lower job satisfaction (Skaalvik & Skaalvik, 2010).

Substitute teachers face additional stressors compared to their permanent counterparts, such as job insecurity, constant school changes, and limited support (Stavrou, 2017; Papadopoulou & Kaldi, 2010). Their inability to establish stable relationships with students and colleagues often leads to feelings of isolation, which may affect both their performance and mental health. This can result in a vicious cycle of stress and professional burnout. As defined by Maslach and Leiter (2016), burnout includes emotional exhaustion, depersonalization, and a diminished sense of efficacy—symptoms that are particularly common among substitutes (Van Droogenbroeck et al., 2014).

Stress levels are also closely linked to job satisfaction. Substitute teachers frequently report lower satisfaction due to the temporary nature of their roles and the lack of recognition and development opportunities (Skaalvik & Skaalvik, 2010). Reduced satisfaction can, in turn, diminish motivation and intensify stress.

Coping strategies play a critical role in stress management. Approaches such as social support, mindfulness, and the development of adaptive skills have been shown to alleviate the psychological burden of teaching (Lazarus & Folkman, 1984; Schaufeli & Bakker, 2004). Participation in support groups or resilience-building programs can serve as protective factors that improve both well-being and professional performance.

In Greece, systemic challenges such as limited resources, frequent transfers, and an underdeveloped support infrastructure exacerbate stress for substitute teachers (Papadopoulou & Kaldi, 2010; Stavrou, 2017). The demand for constant self-validation and the instability of appointments make it difficult for educators to establish professional identity and security. Additionally, cultural expectations—such as strong societal and familial pressure to secure permanent employment—further intensify emotional strain. Frequent relocations also disrupt social networks and hinder integration into school communities, leading to a greater sense of professional marginalization.

The present study explores the relationship between work-related stress and factors such as burnout, job satisfaction, and performance among Greek substitute secondary teachers. It also examines coping strategies used to manage stress. By investigating these dynamics, this research aims to inform the development of effective interventions that promote mental health and professional well-being among this often-overlooked segment of educators. Addressing their specific challenges is essential to strengthening the overall quality and sustainability of the educational system.

LITERATURE REVIEW

Work-related stress, particularly in the education sector, has been the focus of extensive research due to its significant implications for both individual well-being and organizational performance. Kyriacou (2001) emphasizes that teacher stress is a universal issue, stemming from factors such as role ambiguity, student behavior, and workload. However, recent studies have delved deeper into how these stressors manifest differently across cultural and socioeconomic contexts. For example, studies in Scandinavian countries have shown that despite robust social support systems, teachers still report significant stress due to high expectations for innovation and student outcomes (Skaalvik & Skaalvik, 2010). Comparatively, educators in developing nations face additional challenges such as resource scarcity and overcrowded classrooms, amplifying stress levels.

A pivotal aspect of understanding work-related stress lies in the coping strategies adopted by teachers. Lazarus and Folkman (1984) introduced the transactional model of stress and coping, which remains a cornerstone in this field. This model highlights that individuals appraise stressors and respond through problem-focused or emotion-focused strategies. In the context of education, problem-focused strategies might include seeking professional development opportunities to enhance teaching efficacy, while emotion-focused strategies could involve mindfulness practices or peer support groups.

Recent advancements have expanded on Lazarus and Folkman's foundational work. For instance, the role of resilience in mitigating teacher burnout has garnered considerable attention. Resilience, defined as the ability to recover from adversity, is increasingly recognized as a critical factor in sustaining educators' mental health. Research by Beltman et al. (2011) underscores that resilience-building interventions, such as mentorship programs and reflective practices, can significantly reduce stress and improve job satisfaction among teachers. These practices help teachers develop adaptive coping mechanisms, fostering a sense of control and accomplishment even in challenging environments.

Another growing area of interest is the integration of technology as a coping mechanism. Digital tools, such as online forums and stress management apps, have emerged as innovative solutions for teachers facing high workloads. A study by Johnson et al. (2020) found that educators who utilized apps designed for time management and mindfulness reported lower stress levels and improved productivity. This finding is particularly relevant in the post-pandemic era, where remote teaching has introduced new stressors but also new opportunities for technological support. Additionally, the increased use of collaborative tools and platforms during remote learning has emphasized the importance of digital proficiency in managing workload and fostering professional connections.

The Greek educational context offers a unique lens through which to examine work-related stress and coping strategies. Papadopoulou and Kaldi (2010) highlight that substitute teachers in Greece face heightened stress due to systemic issues such as frequent reassignments and lack of administrative support. Moreover, the economic crisis of the last decade has exacerbated these challenges, leading to reduced resources and increased job insecurity. Stavrou (2017) adds that these stressors are compounded by cultural expectations, where teachers are often perceived as role models but receive limited recognition and support. This cultural dynamic places additional emotional and professional strain on educators, requiring them to navigate societal pressures alongside their professional responsibilities.

Work-related stress has become a focal point of research in the field of educational sciences due to its increasing recognition as a significant factor impacting the physical and mental health of teachers. In education, work-related stress is influenced by factors such as a lack of resources, high demands, the complexity of school environments, and shifting policies (Montgomery & Rupp, 2005; Kyriacou, 2001). Teachers often face the dual challenge of managing their own expectations and stressors while simultaneously navigating the broader social and political contexts in which education operates (Travers & Cooper, 1996). Substitute teachers in secondary education, in particular, experience elevated levels of occupational stress compared to their permanent counterparts. Their work environment is characterized by uncertainty, as they rarely have stable placements or predictable working conditions (Antoniou, Ploumpi, & Ntalla, 2013; Stavrou, 2017). Additionally, frequent transitions to different schools each academic year create intense emotional and professional challenges (Bakker, Hakanen, Demerouti, & Xanthopoulou, 2007; Skaalvik & Skaalvik, 2010).

International literature clearly highlights the multifaceted effects of occupational stress on employees' mental health. Among teachers, chronic stress has been associated with higher levels of anxiety, depression, and other psychological difficulties (Hakanen, Bakker, & Schaufeli, 2006). Substitute teachers, due to the elevated stress levels they often experience, are more vulnerable to mental health disorders compared to their counterparts in stable positions (Pillay, Goddard, & Wilss, 2005). These impacts include emotional exhaustion, which is a key component of professional burnout. The demanding nature of their work, requiring quick adaptation to diverse school environments without adequate administrative or social support, further exacerbates their challenges (Schwarzer & Hallum, 2008; Johnson et al., 2005).

Burnout has been extensively studied in educational literature and is recognized as one of the primary risks associated with work-related stress. It is characterized by three core dimensions: emotional exhaustion, depersonalization, and a diminished sense of personal accomplishment (Maslach, Schaufeli, & Leiter, 2001; Schaufeli & Bakker, 2004). For substitute teachers, these symptoms occur more frequently due to the insecurity and instability of their positions (Kokkinos, 2007). Emotional exhaustion, in particular, lies at the heart of professional burnout. The constant struggle to meet daily professional demands, combined with the lack of support, often leads to reduced performance among substitute teachers. Teachers experiencing burnout frequently exhibit decreased enthusiasm and capacity for teaching, which, in turn, negatively impacts the quality of the educational process (Sutton & Wheatley, 2003).

Job satisfaction is another crucial factor closely linked to stress and burnout levels. Studies have shown that teachers who are more satisfied with their work report lower levels of stress and burnout (Klassen & Chiu, 2010). For substitute teachers, job satisfaction is significantly influenced by a sense of stability, recognition, and support within their work environment (Taris, Van Horn, Schaufeli, & Schreurs, 2004). Feelings of insecurity reduce their commitment to the profession and adversely affect their overall performance and psychological well-being (Zembylas & Papanastasiou, 2006; Papadopoulou & Kaldi, 2010). Teachers who feel undervalued often exhibit lower self-confidence and job satisfaction, which, in turn, increases their levels of occupational stress.

The ability to manage work-related stress varies among individuals and is influenced by factors such as personality, experience, and environmental support. Research has indicated that teachers with higher levels of self-efficacy are better equipped to handle occupational stress and are less likely to experience burnout (Bandura, 1997; Schwarzer & Hallum, 2008). Furthermore, the development of resilience skills through training programs can mitigate the effects of stress and emotional exhaustion (Jennett, Harris, & Mesibov, 2003; Day & Gu, 2009). Developing strong self-regulatory habits, such as setting achievable goals and prioritizing tasks, can empower teachers to maintain balance in their professional lives despite external pressures.

Coping strategies play a pivotal role in mitigating the negative effects of work-related stress among teachers. Specifically, the literature suggests that the use of active coping strategies, such as seeking social support, participating in professional development programs, and adopting self-care strategies, can reduce the impact of occupational stress and burnout (Skaalvik & Skaalvik, 2015; Folkman & Moskowitz, 2004). Social support from colleagues and school administration has been recognized as one of the most critical protective factors against burnout (Hobfoll, 1989). Teachers who receive support from their school environment experience a greater sense of belonging and increased job satisfaction, which leads to lower levels of stress and burnout (Day & Gu, 2009). Building collaborative communities within schools can also help foster shared goals and reduce the isolation often felt by substitute teachers.

The need for intervention and support programs aimed at reducing work-related stress and preventing burnout is becoming increasingly urgent. Research has shown that professional development programs focused on enhancing stress management skills and emotional resilience can provide significant benefits to teachers (Jennett, Harris, & Mesibov, 2003). Additionally, programs that promote professional collaboration and the development of positive social relationships within the school environment have proven particularly effective in mitigating the adverse effects of stress (Klusmann, Kunter, Trautwein, Lüdtke, & Baumert, 2008). Creating a supportive and collaborative school environment contributes to increased job satisfaction and psychological well-being among teachers, especially those in temporary positions.

Finally, policies aimed at enhancing job stability for substitute teachers can play a crucial role in improving their professional circumstances. Strengthening stability includes establishing predictable and fair placement processes as well as providing ample opportunities for professional development. These measures can reduce disparities and foster a sense of security and autonomy among teachers (Stavrou, 2017; Zembylas & Papanastasiou, 2006). Encouraging the implementation of well-structured mentorship programs and comprehensive wellness initiatives further supports educators in managing stress and cultivating resilience, ultimately benefiting both teachers and students.

The digital shift in education, particularly following the COVID-19 pandemic, has emerged as a significant factor influencing teacher stress levels. While remote teaching provided a temporary solution during crises, it introduced new stressors, especially for substitute teachers who often lacked sufficient training or access to digital infrastructure (Johnson et al., 2020). Many were forced to teach from inadequate home environments, with limited technological support and minimal guidance from school administrations.

Research indicates that insufficient digital proficiency is associated with elevated stress and feelings of professional inadequacy (Alghamdi et al., 2021). Additionally, the continuous reliance on digital platforms for communication with students, parents, and administrators led to workload overload and blurred boundaries between professional and personal life (Hertzog et al., 2022).

On the other hand, technology can also serve as a resilience-enhancing tool when used appropriately. Time-management apps, online support communities, and educational software can provide meaningful relief from stress by empowering teachers to manage their responsibilities more efficiently (Cheng & Lam, 2023).

FACTORS UNDER STUDY

Work-Related Stress

Work-related stress is defined as the psychological and physiological response of an individual to excessive work demands or pressures that exceed their available resources and capabilities (Lazarus & Folkman, 1984). In the educational sector, work-related stress often arises from factors such as a lack of resources, heightened performance expectations, and ongoing reforms in educational systems (Kyriacou, 2001). The sense of uncertainty that frequently characterizes substitute teachers is an additional source of stress, as it is associated with job instability and frequent changes in the work environment (Stavrou, 2017). Chronic work-related stress can lead to severe psychological and physical consequences, including depression, anxiety, cardiovascular problems, and diminished job performance (Hakanen, Bakker, & Schaufeli, 2006).

Job Performance

Job performance refers to an individual's ability to successfully fulfill their professional responsibilities and meet the demands of their role (Campbell, 1990). In the field of education, performance relates to the quality of teaching, interaction with students, and management of administrative tasks (Day & Gu, 2009). Job performance is heavily influenced by stress levels, with higher levels of stress linked to reduced effectiveness and professional burnout (Skaalvik & Skaalvik, 2015). Furthermore, support from colleagues and participation in professional development programs can enhance teachers' performance by providing tools for stress management and achieving professional goals (Hobfoll, 1989).

Professional Burnout

Professional burnout is a state of physical, emotional, and mental exhaustion caused by prolonged and excessive work-related pressure (Maslach, Schaufeli, & Leiter, 2001). It is characterized by three key components: emotional exhaustion, depersonalization, and a reduced sense of personal accomplishment. In the education sector, substitute teachers exhibit higher rates of burnout, primarily due to the uncertainty associated with their employment and the need for constant adaptation to different environments (Kokkinos, 2007). Burnout has severe consequences, such as reduced teaching quality and a deterioration in mental health, making interventions essential for addressing this issue (Schaufeli & Bakker, 2004).

Job Satisfaction

Job satisfaction refers to the level of contentment and fulfillment an individual feels in their professional life (Spector, 1997). Among educators, satisfaction is related to factors such as a sense of autonomy, recognition from management, relationships with colleagues, and the achievement of educational goals (Klassen & Chiu, 2010). For substitute teachers, the temporary nature of their positions negatively affects job satisfaction, as it is associated with increased insecurity and limited opportunities for professional growth (Taris, Van Horn, Schaufeli, & Schreurs, 2004). However, the presence of a supportive school environment and opportunities for personal development can improve job satisfaction and enhance their professional well-being (Zembylas & Papanastasiou, 2006).

Coping Strategies

Coping strategies encompass a set of behaviors and skills used by employees to manage stress and challenges in their professional environment (Lazarus & Folkman, 1984). For educators, these strategies include seeking social support, participating in professional development programs, managing time effectively, and taking care of their mental and physical health (Folkman & Moskowitz, 2004). Seeking social support from colleagues and administrators has been shown to reduce stress levels and improve job satisfaction (Hobfoll, 1989). Moreover, the use of active strategies, such as participating in support groups and developing resilience, can enhance teachers' ability to handle daily challenges and improve their overall well-being (Day & Gu, 2009).

EMPIRICAL INVESTIGATION: THE CASE OF SUBSTITUTE SECONDARY SCHOOL TEACHERS IN GREECE

This study was conducted over three months, with data collection facilitated through a Google Forms questionnaire. The questionnaire included closed-ended questions and utilized a five-point Likert scale to assess work-related stress, professional burnout, job satisfaction, and coping strategies employed by teachers.

A total of 105 substitute secondary school teachers from various regions of Greece participated in the study. Among them, 38 were male (36.19%), and 67 were female

(63.81%), reflecting the general gender distribution in the educational sector in the country. Regarding age, the majority of participants (75 individuals, 71.43%) were between 35 and 49 years old, 16 participants (15.24%) were aged 50 to 64, and 14 participants (13.33%) were between 20 and 34 years old.

In terms of educational attainment, the majority of participants held a postgraduate degree (79 individuals, 75.24%), while 19.05% (20 individuals) had completed undergraduate studies at universities or technical institutions. A smaller percentage (6 individuals, 5.71%) had earned a doctoral degree. This distribution highlights the high educational qualifications of substitute teachers, an important factor in understanding how education level impacts their ability to manage work-related stress.

Regarding professional experience, most of the sample (72 individuals, 68.57%) had between 1 and 5 years of experience, emphasizing the relatively recent entry of many substitutes into the educational system. Fifteen teachers (14.29%) had 6 to 15 years of experience, while 18 participants (17.14%) had over 16 years of experience. This indicates that professional experience among substitutes is not always correlated with job permanency.

The demographic characteristics of the participants are summarized in Table 1. Quantitative analysis of this data provided insight into the profile of the respondents and served as a foundation for the statistical processing of the study's broader findings. Furthermore, this analysis contributes to understanding the context in which substitute teachers in Greece operate. These educators often face challenges such as job insecurity, geographical mobility, and a lack of consistent support from administrative structures.

Table 1. Demographic characteristics of participants

Characteristic	**Participants**	
	Number of participants	**Percentage %**
Gender		
Male	38	36.19%
Female	67	63.81%
Age		
20 – 34 years	14	13.33%
35 – 49 years	75	71.43%
50 – 64 years	16	15.24%
Educational Level		
Bachelor degree	20	19.05%
Postgraduate degree	79	75.24%

continued on following page

Table 1. Continued

Characteristic	Participants	
	Number of participants	Percentage %
Doctorate degree	6	5.71%
Years of teaching experience		
1 - 5 years	72	68.57%
6 - 15 years	15	14.29%
16 years and above	18	17.14%

Scales Used: WSS, IWPQ, MBI-ES, BIAJS, CSI-SF

This study employed five distinct scales to fulfill its objective of analyzing the relationship between work-related stress and four other factors. The selection of these scales was based on their proven validity and reliability as established in previous research. Each factor in the study—(1) work-related stress, (2) job performance, (3) professional burnout, (4) job satisfaction, and (5) coping strategies—was measured through a set of questions rated using five-point or seven-point Likert scales.

Workplace Stress Scale (WSS)

The **WSS (Workplace Stress Scale)**, used to measure work-related stress, consists of 8 items that examine various dimensions of stress, including working conditions, role demands, uncertainty, relationships with colleagues and supervisors, and overall workplace pressure (American Institute of Stress, 2014). This scale is designed to capture both the frequency and intensity of workplace stress, allowing for an understanding of the factors that cause it. In the education sector, challenges such as managing large student groups, pressure for educational outcomes, and lack of resources are significant sources of stress. The WSS is widely accepted for its simplicity and effectiveness in evaluating stress across various professional domains.

Individual Work Performance Questionnaire (IWPQ)

The **IWPQ (Individual Work Performance Questionnaire)** assesses job performance through three main subscales: task performance, contextual performance, and counterproductive work behavior (Koopmans et al., 2014). Comprising 18 questions, this scale evaluates an employee's effectiveness in fulfilling their job responsibilities, interacting with their environment, and adhering to work demands. The counterproductive behavior dimension is particularly important as it examines how stress negatively impacts employee behavior. The IWPQ offers a comprehensive view of employee performance and has demonstrated reliability in numerous studies. In this research, it helped identify factors influencing teachers' productivity and the impact of work-related stress on their overall performance.

Maslach Burnout Inventory: Educators Survey (MBI-ES)

The **MBI-ES (Maslach Burnout Inventory – Educators Survey)** is specifically designed to capture and measure professional burnout among educators (Maslach & Jackson, 1981). It consists of 22 items divided into three primary dimensions: emotional exhaustion, depersonalization, and reduced personal accomplishment. This scale has been extensively used in studies to assess the level of burnout in teachers, highlighting the effects of chronic stress and high demands. For substitute teachers, the MBI-ES enabled the investigation of how job temporality and professional insecurity affect their mental health.

Brief Index of Affective Job Satisfaction (BIAJS)

The **BIAJS (Brief Index of Affective Job Satisfaction)** is a concise tool for measuring emotional job satisfaction (Thompson & Phua, 2012). It includes only 5 questions, focusing on evaluating an individual's general emotional state concerning their work. Unlike other job satisfaction scales, the BIAJS does not concentrate on specific aspects like salary or environment but instead assesses the overall sense of pleasure or dissatisfaction the employee feels about their job. In this study, the BIAJS was used to evaluate teachers' emotional satisfaction and its connection to work-related stress and professional burnout.

Coping Strategies Inventory: Short Form (CSI-SF)

The **CSI-SF (Coping Strategies Inventory – Short Form)** is used to evaluate stress coping strategies. The short version contains 16 items and is divided into two main dimensions: problem-focused strategies and emotion-focused strategies (Tobin, 2001). Problem-focused strategies include actions aimed at managing or solving the problem, while emotion-focused strategies involve efforts to manage emotional reactions. The CSI-SF was a valuable tool for examining the different ways teachers address work-related stress, contributing to understanding how these strategies influence their professional well-being.

To ensure the validity and reliability of the scales, Cronbach's Alpha was calculated for each. This index measures the internal consistency of the items within a scale, ensuring that all items evaluate the same characteristic (Cronbach, 1951).

The Cronbach's Alpha results for each scale are presented in Table 2, confirming the reliability of the measurements and their contribution to the data analysis.

Table 2. Cronbach's Alpha values for scales

Scale	Factor	Number of items	Cronbach's alpha value
WSS	Work-related stress	8	0.859
IWPQ	Job performance	18	0.772
MBI-ES	Burnout	22	0.864
BIAJS	Job satisfaction	5	0.755
CSI-SF	Coping Strategies	16	0.706

An analysis of Table 2 reveals that the Cronbach's Alpha values for all scales fall within satisfactory to high levels, indicating the reliability of the measurements. The highest reliability is observed in the WSS (Workplace Stress Scale) and MBI-ES (Maslach Burnout Inventory – Educators Survey), with values of 0.859 and 0.864, respectively. These findings confirm that these tools are suitable for assessing psychosocial factors related to the workplace environment. Although the Cronbach's Alpha value for the CSI-SF (Coping Strategies Inventory – Short Form) is relatively lower at 0.706 compared to the other scales, it is still considered acceptable given that it evaluates diverse coping strategies. The BIAJS (Brief Index of Affective Job Satisfaction), despite consisting of only five items, demonstrates satisfactory reliability (0.755), suggesting that even with a limited number of questions, it effectively measures job satisfaction. Overall, these results enhance the validity of the questionnaire, establishing it as an appropriate tool for the current study.

Table 3 outlines the factors evaluated by each measurement scale, as defined in the relevant literature. The scales used in this study encompass various aspects of work life, including work-related stress, professional burnout, job performance, job satisfaction, and coping strategies. The use of these scales has been highlighted in the literature as a fundamental approach for measuring psychological and organizational parameters, providing reliable tools to explore complex constructs associated with workplace experiences (Lazarus & Folkman, 1984; Maslach et al., 2001; Schaufeli et al., 2002).

Table 3. Factors by scale

Scale	Factor	Number of factors
WSS	Work-related stress	2
IWPQ	Job performance	3
MBI-ES	Burnout	3
BIAJS	Job satisfaction	1
CSI-SF	Coping Strategies	4

The WSS (Work Stress Scale) evaluates two key factors related to work-related stress, a critical parameter influencing employees' physical and psychological health, as well as their productivity (Cooper & Marshall, 1976). The IWPQ (Individual Work Performance Questionnaire) measures three dimensions of job performance, focusing on productivity, work quality, and innovation, offering a comprehensive approach to assessing professional performance (Koopmans et al., 2014).

The MBI-ES (Maslach Burnout Inventory – Educators Survey) examines three dimensions of professional burnout: emotional exhaustion, depersonalization, and reduced personal accomplishments, which are central concepts in the burnout literature (Maslach & Jackson, 1981). The BIAJS (Brief Index of Affective Job Satisfaction) focuses on job satisfaction as a unidimensional concept, highlighting employees' emotional reactions to their overall work experience (Judge & Klinger, 2007).

Finally, the CSI-SF (Coping Strategies Inventory – Short Form) assesses four coping strategies, including problem-solving, emotional management, avoidance, and positive reappraisal, which are essential tools for managing stress (Carver et al., 1989). The combined use of these scales provides an interdisciplinary and holistic approach to studying workplace conditions and the psychology of teaching professionals, contributing to the understanding of factors that influence their professional lives.

Factor Analysis – Validity Assessment

To conduct the factor analysis, it was necessary to verify the assumption that the data were suitable (KMO–Bartlett's Test). Beyond the theoretical classification, it was also essential to identify the number of factors within each scale based on participants' responses and to calculate the percentage contribution of each factor to the total variance of the data (Total Variance Explained). Table 4 presents the aforementioned information.

Table 4. KMO-Bartlett's Test results, number offsctors and explained total variance

Scale	KMO-Bartlett's Test Results		Factor analysis Results	
	KMO	Bartlett's	Number of factors	Explained Total Variance
WSS	0.758	0.000	2	64.35%
IWPQ	0.790	0.000	4	64.74%
MBI-ES	0.821	0.000	5	70.84%
BIAJS	0.755	0.000	2	74.59%
CSI-SF	0.656	0.000	5	65.32%

The data in Table 4 indicate sufficient correlation among the variables (KMO > 0.65) and confirm that the data are suitable for factor analysis (Bartlett's Test < 0.05). The highest KMO value is observed in the MBI-ES scale (0.821), indicating a high level of correlation among items evaluating professional burnout. Conversely, the lowest KMO value is found in the CSI-SF scale (0.656), which remains acceptable given the multidimensional nature of coping strategies.

The analysis of explained variance shows that the five factors of the MBI-ES account for 70.84% of the total variance, making this scale particularly robust in understanding burnout. Similarly, the BIAJS scale exhibits the highest explained variance (74.59%) despite its limited number of items, underscoring its effectiveness in assessing job satisfaction. The WSS scale is divided into two factors, explaining 64.35% of the variance, highlighting the multifaceted dimensions of work-related stress.

Based on these findings, the factors were redefined, non-significant items were removed, and the remaining items were categorized for inclusion in the analysis. Subsequently, mean scores for each factor's items were calculated and recorded as single variables per factor to facilitate regression analysis. Given that the aforementioned variables did not follow a normal distribution overall, Spearman's coefficient was used to identify potential correlations among these variables. Table 5 reports these correlation values and their significance levels.

Table 5. Correlations of variables (Spearman's Coefficient)

Factor	Work-related stress	Job performance	Burnout	Job satisfaction	Coping Strategies
Work-related stress	1.000	-	-	-	-
Job performance	0.200*	1.000	-	-	-
Burnout	0.600**	0.133	1.000	-	-

continued on following page

Table 5. Continued

Factor	Work-related stress	Job performance	Burnout	Job satisfaction	Coping Strategies
Job satisfaction	-0.551**	0.088	-0.542**	1.000	-
Coping Strategies	0.205*	0.056	0.256**	-0.216*	1.000

* Correlation significance level at 5%,
** Correlation significance level at 1%

From Table 5, it is evident that all variables significantly correlate with work stress. Specifically, the positive correlation between work stress and professional burnout ($r = 0.600$, $p < 0.01$) highlights the impact of stress on the emotional and psychological state of educators. Additionally, the negative correlation with job satisfaction ($r = -0.551$, $p < 0.01$) indicates that as stress increases, employees' satisfaction decreases. Professional burnout exhibits a moderate positive correlation with coping strategies ($r = 0.256$, $p < 0.01$), suggesting that educators experiencing burnout tend to employ more strategies to manage their condition. However, the negative correlation between job satisfaction and coping strategies ($r = -0.216$, $p < 0.05$) may imply that less satisfied employees seek more ways to handle stress. Finally, the correlations between work stress and job performance ($r = 0.200$, $p < 0.05$) and between coping strategies and professional burnout underscore the dynamic relationship among these factors, revealing how stress affects both educators' psychological state and performance. These findings emphasize the necessity for interventions aimed at reducing stress and enhancing professional satisfaction.

Regression Analysis

To implement linear regression and achieve the study's goal—analyzing the relationship between work stress and various factors such as job performance, professional burnout, job satisfaction, and coping strategies—certain assumptions needed to be verified (homoscedasticity, multicollinearity, independence of errors, etc.). The model met all these assumptions, enabling the analysis of linear regression. For the purposes of this study, linear regression was conducted with work stress as the dependent variable and performance, burnout, satisfaction, and coping strategies as independent variables.

The model's coefficient of determination (R^2) is 0.460, a moderate value indicating that only 46% of educators' work stress is explained by the factors used (satisfaction, burnout, performance, coping). This implies that the remaining 54% may be influenced by other factors not included in the model. The ANOVA test confirmed that the model is statistically significant (Sig.<0.05), indicating that the independent variables substantially contribute to predicting the dependent variable. Table 6 presents the results of the linear regression analysis.

Table 6. Linear regression results

Variables	Αποτελέσματα			
	Coefficient β	**Standardized Coefficient β**	**T-test**	**Significance level (Sig.)**
Constant	1.938	-	2.55	0.012
Job performance	0.284	0.161	2.10	0.039
Burnout	0.334	0.397	4.29	0.000
Job satisfaction	-0.349	-0.345	-3.70	0.000
Coping strategies	0.027	0.016	0.21	0.836

The Sig values below 0.05 indicate that all predictors, except for coping strategies (Sig=0.836), are statistically significant. Consequently, coping strategies were excluded from the predictive model. Analysis of the second column in Table 6 reveals the estimated values for each variable in the model. The predictive model derived from this analysis is as follows:

$$\boldsymbol{WRS} = \beta_0 + \beta_1 \times \boldsymbol{JP} + \beta_2 \times \boldsymbol{BO} + \beta_3 \times \boldsymbol{JS}$$

$$\boldsymbol{WRS} = 1{,}938 + 0{,}284 \times \boldsymbol{JP} + 0{,}334 \times \boldsymbol{BO} - 0{,}349 \times \boldsymbol{JS}$$

where:

WRS: Work-related Stress
JP: Job Performance
BO: Burnout
JS: Job Satisfaction

The model's constant (β_0) is 1.938, indicating that educators' work stress would be 1.938 if all independent variables were zero. The coefficient β_1=0.284 signifies a positive relationship between work stress and job performance. Specifically, a 1-unit increase in job performance would result in a 0.284-unit increase in educators' work stress, assuming other variables remain constant. Similarly, β_2=0.334 shows a positive association between professional burnout and stress; a 1-unit increase in burnout would lead to a 0.334-unit rise in stress. Finally, β_3=−0.349 demonstrates a negative relationship between job satisfaction and stress, implying that a 1-unit increase in satisfaction would reduce stress by 0.349 units, assuming other variables are unchanged.

The standardized beta coefficient, which compares the relative impact of each independent variable on the dependent variable, revealed that professional burnout has the strongest effect on educators' stress. This is expected, as physical and mental exhaustion often lead to stress crises or heightened tension. Conversely, job satisfaction has the strongest negative impact on stress, highlighting that individuals with higher satisfaction levels are less stressed. Job performance has a weaker influence on stress compared to the other variables, although the pressures educators face to achieve higher performance still elevate their stress levels. Notably, coping strategies were excluded from the model due to their lack of statistical significance. These findings underline the importance of addressing factors such as burnout and job satisfaction to mitigate educators' work stress while considering additional unexamined variables that might also influence stress levels.

APPLICATIONS: PRACTICAL INTERVENTIONS

The findings of this study highlight the importance of developing and implementing practical interventions aimed at reducing work-related stress and improving the mental health of educators. The need for such interventions is particularly evident among substitute teachers, who face heightened levels of job insecurity and professional burnout. Developing policies to support educators is critical to reducing stress and ensuring their professional well-being. A key step in this process is enhancing job stability by securing long-term contracts for substitute teachers, which could alleviate professional uncertainty and the associated stress (Antoniou et al., 2013). Additionally, establishing supportive structures, such as providing access to mental health counselors within schools, could enable educators to manage their stress effectively. Similar interventions have been successfully implemented in educational systems in other countries (Kyriacou, 2001).

Professional development and training represent another crucial area for strengthening educators' resilience to workplace stress. Training programs on stress management techniques, such as mindfulness, conflict resolution, and emotional intelligence development, can empower educators and reduce professional burnout (Jennett et al., 2003). Building educators' capacity to recognize and manage their emotions is a fundamental factor in reducing stress.

Moreover, recognition programs, such as awards or public ceremonies, can boost educators' self-esteem and reduce professional burnout (Hobfoll, 1989). Acknowledging educators' contributions creates a positive reinforcement environment, which enhances professional commitment and alleviates psychological pressure.

The implementation of pilot programs offers an opportunity to test and evaluate practical interventions on a small scale. For instance, a Teacher Mental Health Enhancement Program could include weekly counseling sessions, stress management workshops, and support groups. Participants would have the opportunity to share experiences and learn from their peers, fostering a culture of collaboration and solidarity. Pilot implementation of such programs has been proven effective in various countries and educational systems, significantly reducing participants' stress levels (Day & Gu, 2009).

Equally important is a Wellness and Work-Life Balance Program, offering activities such as yoga, meditation, or physical exercise within the school environment to reduce stress and improve overall health (Skaalvik & Skaalvik, 2015). These activities promote physical and mental well-being while also fostering a sense of community among participants. The program could also be expanded to include practices that address work-life balance, enhancing educators' overall well-being.

Pilot implementation of collaborative leadership is another promising approach to strengthen cooperation between administration and teaching staff. This approach involves educators in decision-making processes affecting the school environment, which fosters a sense of autonomy and recognition (Zembylas & Papanastasiou, 2006). Creating a participatory school environment where educators feel their voices are heard can reduce work stress and improve their overall psychological state.

Stress management strategies are also essential for coping with workplace stress and enhancing mental resilience. Mindfulness practices, for example, have been shown to effectively reduce anxiety and emotional exhaustion (Jennett et al., 2003). Educators can be trained in mindfulness techniques through specially designed seminars, equipping them with tools to effectively manage daily challenges.

In parallel, strengthening social relationships among colleagues can create a support network that reduces stress, as emphasized by Day and Gu (2009), who note that positive social relationships enhance resilience. A robust social support network helps educators cope more effectively with challenges while fostering a sense of belonging within the school community.

Psychological support and counseling on a regular basis is another strategy that can help educators identify the sources of their stress and develop more effective coping mechanisms. Providing counseling services within schools can ensure immediate and continuous support, significantly reducing the pressure educators experience during their professional lives.

The success of these interventions requires continuous evaluation and adaptation. Using quantitative and qualitative data can help understand the effectiveness of each strategy and improve interventions. Collaboration with specialists, such as psychologists and social scientists, can provide valuable insights for shaping more effective policies. Data from pilot programs can be used to develop broader strategies addressing the needs of diverse groups of educators.

In summary, practical interventions and policies to reduce workplace stress must be multidimensional and tailored to educators' specific needs. Creating a supportive environment, enhancing professional recognition, and providing tools for stress management are key pillars for promoting educators' well-being. The findings of this study provide the foundation for developing solutions that enhance the sustainability of the teaching profession and promote educators' overall well-being, thereby improving the quality of education and the broader school experience.

School leadership plays a pivotal role in shaping the psychosocial climate of the workplace and influencing teacher stress levels. Supportive and inclusive leadership practices can mitigate the impact of work-related stress, particularly for substitute teachers who often lack a stable support network. When school leaders foster a sense of trust, recognition, and open communication, teachers are more likely to feel valued and less isolated (Leithwood et al., 2006).

Participatory leadership models, which involve teachers in decision-making processes and school governance, can empower educators and enhance their sense of professional agency. This is especially important for substitute teachers, who may otherwise feel excluded from the school community due to their temporary status. Providing structured feedback, acknowledging their contributions, and including them in professional learning communities are essential strategies that can alleviate anxiety and promote a culture of collaboration (Day & Sammons, 2014).

Moreover, effective leadership can help bridge gaps in administrative support by implementing mentorship systems, organizing regular check-ins, and facilitating access to mental health resources. In environments where leadership is perceived as distant or authoritarian, substitute teachers often report increased levels of stress and depersonalization. Therefore, investing in leadership training that emphasizes emotional intelligence, empathy, and inclusive practices is a critical step toward reducing occupational stress and improving teacher retention.

Policy Recommendations

Addressing workplace stress among educators, particularly substitute teachers, demands a comprehensive and multi-dimensional approach. Enhancing job stability and career progression opportunities remains a cornerstone. Governments and educational institutions should develop policies that transition substitute teachers into permanent roles through structured pathways, such as tenure-track systems or long-term contracts. A centralized, digital system for managing teacher assignments can further enhance transparency and equity, allowing educators to select opportunities aligned with their expertise and preferences, thus reducing the uncertainty and anxiety caused by frequent relocations.

Expanding professional development opportunities is another essential strategy. Tailored CPD programs should incorporate advanced topics like emotional intelligence, inclusive teaching practices, and adaptive technologies. Workshops focusing on stress management techniques, including mindfulness and cognitive-behavioral strategies, can equip substitute teachers with tools to navigate high-stress environments. Additionally, incorporating resilience training into teacher induction programs can foster long-term coping mechanisms. Mentorship initiatives pairing novice substitute teachers with seasoned educators can bridge the experience gap and foster a sense of belonging.

Strengthening support systems within schools is critical. School administrators should provide regular performance reviews, constructive feedback, and avenues for teachers to voice their concerns. Recognizing substitute teachers' contributions through awards, career milestones, and public acknowledgment can significantly boost morale. Furthermore, fostering professional learning communities (PLCs) where teachers collaborate and share best practices can enhance peer support and camaraderie.

Leveraging technology to manage workplace stress offers untapped potential. Policymakers should invest in developing user-friendly apps for time management, interactive lesson planning, and mental health resources. Virtual peer support groups can serve as platforms for emotional and professional exchange. Training educators to integrate technology seamlessly into their workflow ensures these tools are used to their full potential.

Promoting work-life balance must be prioritized by reducing excessive administrative tasks and enforcing boundaries for after-hours communication. Offering periodic wellness days and access to mental health professionals within schools demonstrates institutional commitment to teacher well-being. Flexible schedules, particularly for substitute teachers, can help them manage personal and professional responsibilities more effectively.

Addressing cultural and societal perceptions of teaching is imperative. Governments should initiate public campaigns highlighting the importance of educators in societal development. Increased investment in education infrastructure and resources will not only ease workloads but also elevate the status of the teaching profession. Policymakers should collaborate with international organizations like the OECD to benchmark against global best practices, ensuring a tailored yet progressive framework for teacher support.

By implementing these enriched policy measures, stakeholders can create an environment that prioritizes the well-being of substitute teachers, thereby enhancing the quality of education and contributing to a more equitable and effective educational system.

Limitations

The limitations of the present study primarily concern the sample and the data collection method. The sample consists of substitute secondary school teachers, which may restrict the generalizability of the findings to other categories of teachers or educational levels. Additionally, the use of self-reported questionnaires may lead to subjective responses, as participants may present themselves in a socially desirable manner through their answers.

Another limitation lies in the cross-sectional nature of the study, which examines relationships between variables at a specific point in time. This limits the ability to explore causal relationships or the long-term impacts of workplace stress. Moreover, focusing exclusively on five specific factors (performance, burnout, satisfaction, coping, stress) may overlook other important variables that could influence the results, such as personality traits, cultural differences, or social support.

The use of online data collection through questionnaires (e.g., Google Forms) might also result in selection bias, as participants with access to technology and familiarity with online tools are more likely to participate. This could limit the representativeness of the sample. Additionally, the design of the questionnaire, including the selection of questions and the use of a Likert scale, may influence the precision of the responses and, consequently, the interpretation of the findings.

A further limitation is the scarcity of internationally comparable data specifically focused on substitute teachers. While there is abundant literature on teacher stress in general, most studies do not differentiate between permanent and substitute educators. This makes it difficult to position the findings of the present study within a broader international context or to benchmark them against policies and practices from other countries.

For future research, the use of mixed-method approaches, such as interviews or observations, is recommended to complement the data obtained from questionnaires. Additionally, conducting longitudinal studies could provide valuable insights into the dynamic relationships between factors and the long-term effects of workplace stress.

DISCUSSION

The findings of this study align with previous research examining the impact of workplace stress on educators. Specifically, studies have shown that professional burnout is directly associated with high levels of stress, as observed in this analysis, where professional burnout has the strongest positive impact on workplace stress. This aligns with the findings of Maslach, Schaufeli, and Leiter (2001), who argue that burnout leads to increased psychological burden and stress in professional educators. Similarly, Hakanen, Bakker, and Schaufeli (2006) emphasized that emotional exhaustion is the core of burnout and a significant predictor of workplace stress among teachers.

Furthermore, the negative relationship between job satisfaction and stress, also confirmed here, is consistent with the literature, where job satisfaction often acts as a protective factor against stress (Judge et al., 2001). This finding highlights the importance of supporting educators by improving working conditions, recognizing their efforts, and enhancing their involvement in decision-making processes. Klassen and Chiu (2010) similarly noted that teachers who are more satisfied with their work experience lower stress levels, a finding also supported by this study. The weaker impact of job performance on stress, while expected, requires further investigation, as some studies suggest that performance pressure can exacerbate stress (Skaalvik & Skaalvik, 2011). This discrepancy may be related to the different conditions faced by substitute versus permanent teachers.

However, the lack of statistical significance of coping strategies in the model requires further analysis. While previous research (Folkman & Lazarus, 1985) highlights their role in stress reduction, the present findings may reflect limitations in the selection and application of these strategies by participants. It is possible that the strategies employed by substitute teachers are not tailored to their specific needs or are insufficient to address the high levels of stress they experience. Additionally, some strategies may require more time to yield visible effects in stress reduction. Hobfoll (1989) and Day and Gu (2009) emphasize that social support and a collaborative school environment are critical for the effectiveness of coping strategies.

Future research could focus on exploring different coping strategies using more specialized measurement tools to determine whether specific strategies have a greater impact on workplace stress and educators' mental health. For example, strengthening social support, developing self-management skills, and introducing professional development programs could contribute positively to stress reduction. Skaalvik and Skaalvik (2015) found that problem-focused strategies are more effective in reducing stress compared to emotion-focused strategies, a topic worth further investigation.

Additionally, examining the role of organizational culture and the school environment in shaping stress levels and job satisfaction could provide significant insights. Zembylas and Papanastasiou (2006) highlighted that a supportive school environment and participation in decision-making are directly linked to job satisfaction and stress reduction.

Expanding the sample and examining results over a longer timeframe could reveal long-term trends not evident in short-duration studies. For example, the use of longitudinal data could capture fluctuations in workplace stress during different periods of the school year or various phases of educators' professional lives. Including qualitative methods, such as interviews or focus groups, could also provide a deeper understanding of educators' experiences and uncover hidden factors influencing stress and professional well-being.

FUTURE RESEARCH DIRECTIONS

While this study provides valuable insights into the stressors affecting substitute teachers and the potential strategies for alleviation, further research is necessary to address the dynamic and evolving nature of workplace stress in education. One promising avenue for future research involves exploring the long-term impacts of stress on teachers' career trajectories. Longitudinal studies could provide a clearer picture of how chronic stress influences job satisfaction, retention rates, and overall mental health over time.

Additionally, there is a need for more cross-cultural studies that compare the experiences of substitute teachers in different educational systems. Such research could identify best practices from various countries and provide a comprehensive understanding of how cultural, economic, and policy factors influence workplace stress.

The role of technology in managing stress warrants further exploration. While existing studies have highlighted the potential benefits of digital tools, there is limited research on their long-term effectiveness and accessibility across different demographics. Investigating how technology can be tailored to meet the specific needs of substitute teachers, particularly in resource-constrained settings, could yield practical solutions.

Finally, future studies should examine the intersectionality of stressors, considering how factors such as gender, age, and socioeconomic status interact to shape the experiences of substitute teachers. Although the present study did not find statistically significant differences in stress levels between male and female participants, this absence of variation warrants further investigation. Future research could explore whether gender-related stress patterns emerge under different institutional, regional, or cultural conditions. By adopting an intersectional approach, researchers can uncover nuanced insights that inform targeted interventions and policies.

CONCLUSION: RESULTS

This study aimed to examine the impact of workplace stress on substitute secondary school teachers in Greece by analyzing its relationship with four key factors: job performance, professional burnout, job satisfaction, and coping strategies. The findings underscore the complex nature of workplace stress, its diverse sources, and how various factors interact to shape it.

Firstly, professional burnout emerged as the factor with the strongest influence on workplace stress. This finding is particularly significant as it indicates that substitute teachers, who face frequent changes in their work environment and lack stability, are highly vulnerable to psychological exhaustion. Emotional exhaustion, as a core component of burnout, appears to be at the center of the issue, affecting educators' overall mental health and their ability to meet professional demands. This highlights the need for interventions targeting burnout reduction, such as the provision of supportive structures and the enhancement of psychological resilience.

Equally important is the negative relationship between job satisfaction and stress. Teachers who are more satisfied with their jobs appear to experience lower stress levels. This finding aligns with existing literature that identifies job satisfaction as one of the most important protective factors against stress. Improving job satisfaction can be achieved by creating a supportive work environment, recognizing teachers' contributions, and providing opportunities for professional development. The data indicate that job satisfaction not only affects educators' mental health but also their overall performance.

Job performance, although related to workplace stress, appears to have a smaller impact compared to burnout and satisfaction. This finding reflects the complex relationship between performance and stress, as the pressure to perform can act either as a motivator or as a source of anxiety, depending on the individual and the context. In any case, providing resources and support to facilitate the achievement of professional goals can help reduce the pressure experienced by educators.

In contrast, coping strategies were not statistically significant as a factor influencing workplace stress. This may reflect either the insufficient application of strategies by participants or the need for more time for these strategies to produce results. It is also possible that the strategies employed by teachers are not well-suited to the specific demands and challenges of their profession. This result underscores the need for training teachers in effective stress-coping strategies, such as time management, self-awareness enhancement, and emotional regulation skills development. For instance, mindfulness practices and social support reinforcement have proven to be effective tools for reducing anxiety and increasing resilience.

The study also highlights the importance of job stability and a supportive work environment. Substitute teachers, who often work under uncertain conditions, face a higher risk of stress and burnout. Enhancing job security through long-term contracts and fostering a positive school environment can significantly reduce the psychological burden experienced by educators. Additionally, implementing professional development and wellness programs, such as yoga and meditation, can improve work-life balance, contributing to educators' overall well-being.

Another important conclusion is the need for targeted policies and practices that promote teachers' professional well-being. Developing professional development programs, strengthening communication among school staff, and fostering positive social relationships can play a key role in reducing stress. Schools can also implement policies that promote work-life balance, thereby alleviating the pressure on teachers. Specifically, adopting collaborative leadership, which involves teachers in decision-making processes, can enhance their sense of autonomy and professional recognition while reducing stress.

The significance of professional recognition emerges as a critical factor for reducing workplace stress. Teachers who feel their efforts are valued experience higher job satisfaction and lower stress levels. This recognition can take various forms, from providing constructive feedback and professional development opportunities to enhancing autonomy in decision-making. Recognition programs, such as awards or acknowledgment events, can boost educators' commitment and psychological well-being.

While this research provides valuable insights, it has certain limitations that should be acknowledged. The limited sample size and the exclusive focus on substitute secondary school teachers may restrict the generalizability of the findings. Furthermore, using questionnaires for data collection may be subject to subjective responses and the potential for socially desirable behavior. Future research could expand to larger samples and different categories of teachers, while also employing qualitative data collection methods to gain a more comprehensive understanding of the phenomena.

In summary, this study demonstrates that workplace stress among substitute teachers is a complex phenomenon influenced by factors such as professional burnout and job satisfaction. Enhancing support for teachers, improving working conditions, and implementing targeted interventions can contribute to reducing stress and improving their professional well-being. The results of this study represent a significant step toward understanding the challenges faced by teachers and developing solutions that enhance the sustainability of the profession while improving the quality of education and the overall school experience.

REFERENCES

Alghamdi, A. K. H., (2021). Teachers' stress in online learning: Challenges of digital transformation. *Journal of Educational Technology Systems*, *50*(1), 102–118.

American Institute of Stress. (2014). Workplace stress scale. Retrieved from www.stress.org

Antoniou, A. S. G., Ploumpi, A., & Ntalla, M. (2013). Occupational stress and professional burnout in teachers of primary and secondary education: The role of coping strategies. *Psychology (Irvine, Calif.)*, *4*(3), 349–355. DOI: 10.4236/psych.2013.43A051

Bakker, A. B., Hakanen, J. J., Demerouti, E., & Xanthopoulou, D. (2007). Job resources boost work engagement, particularly when job demands are high. *Journal of Educational Psychology*, *99*(2), 274–284. DOI: 10.1037/0022-0663.99.2.274

Bandura, A. (1997). *Self-efficacy: The exercise of control*. W. H. Freeman and Company.

Beltman, S., Mansfield, C., & Price, A. (2011). Thriving not just surviving: A review of research on teacher resilience. *Educational Research Review*, *6*(3), 185–207. DOI: 10.1016/j.edurev.2011.09.001

Campbell, J. P. (1990). Modeling the performance prediction problem in industrial and organizational psychology. In Dunnette, M. D., & Hough, L. M. (Eds.), *Handbook of industrial and organizational psychology* (Vol. 1, pp. 687–732). Consulting Psychologists Press.

Carver, C. S., Scheier, M. F., & Weintraub, J. K. (1989). Assessing coping strategies: A theoretically based approach. *Journal of Personality and Social Psychology*, *56*(2), 267–283. DOI: 10.1037/0022-3514.56.2.267 PMID: 2926629

Cheng, X., & Lam, S. F. (2023). Technology-assisted resilience: Coping with teacher stress through digital tools. *Computers & Education*, *197*, 104709.

Cooper, C. L., & Marshall, J. (1976). Occupational sources of stress: A review of the literature relating to coronary heart disease and mental ill health. *Journal of Occupational Psychology*, *49*(1), 11–28. DOI: 10.1111/j.2044-8325.1976.tb00325.x

Cronbach, L. J. (1951). Coefficient alpha and the internal structure of tests. *Psychometrika*, *16*(3), 297–334. DOI: 10.1007/BF02310555

Day, C., & Gu, Q. (2009). Teacher emotions: Wellbeing and effectiveness. In Schutz, P. A., & Zembylas, M. (Eds.), *Advances in teacher emotion research* (pp. 15–31). Springer., DOI: 10.1007/978-1-4419-0564-2_2

Day, C., & Gu, Q. (2009). Veteran teachers: Commitment, resilience and quality retention. *Teachers and Teaching*, *15*(4), 441–457. DOI: 10.1080/13540600903057211

Day, C., & Sammons, P. (2014). *Successful School Leadership*. Education Development Trust.

Folkman, S., & Lazarus, R. S. (1985). If it changes it must be a process: Study of emotion and coping during three stages of a college examination. *Journal of Personality and Social Psychology*, *48*(1), 150–170. DOI: 10.1037/0022-3514.48.1.150 PMID: 2980281

Folkman, S., & Moskowitz, J. T. (2004). Coping: Pitfalls and promise. *Annual Review of Psychology*, *55*(1), 745–774. DOI: 10.1146/annurev.psych.55.090902.141456 PMID: 14744233

Hakanen, J. J., Bakker, A. B., & Schaufeli, W. B. (2006). Burnout and work engagement among teachers. *Journal of School Psychology*, *43*(6), 495–513. DOI: 10.1016/j.jsp.2005.11.001

Hertzog, H., Johnson, A., & Shapiro, L. (2022). Digital fatigue and teacher burnout in pandemic-era education. *Teaching and Teacher Education*, *117*, 103814.

Hobfoll, S. E. (1989). Conservation of resources: A new attempt at conceptualizing stress. *The American Psychologist*, *44*(3), 513–524. DOI: 10.1037/0003-066X.44.3.513 PMID: 2648906

Hofstede, G. (2001). *Culture's Consequences: Comparing Values*. Behaviors, Institutions and Organizations Across Nations.

Ingersoll, R. M., & Strong, M. (2011). The impact of induction and mentoring programs for beginning teachers: A critical review of the research. *Review of Educational Research*, *81*(2), 201–233. DOI: 10.3102/0034654311403323

Jennett, H. K., Harris, S. L., & Mesibov, G. B. (2003). Commitment to philosophy, teacher efficacy, and burnout among teachers of children with autism. *Journal of Autism and Developmental Disorders*, *33*(6), 583–593. DOI: 10.1023/B:JADD.0000005996.19417.57 PMID: 14714928

Johnson, S., Cooper, C., Cartwright, S., Donald, I., Taylor, P., & Millet, C. (2005). The experience of work-related stress across occupations. *Journal of Managerial Psychology*, *20*(2), 178–187. DOI: 10.1108/02683940510579803

Johnson, S., Smith, R., & Williams, K. (2020). The role of technology in mitigating teacher stress. *Teaching and Teacher Education*, *90*, 103–120. DOI: 10.1016/j.tate.2020.103120

Judge, T. A., & Klinger, R. (2007). Job satisfaction: Subjective well-being at work. In M. Eid & R. J. Larsen (Eds.), *The science of subjective well-being* (pp. 393–413). Guilford Press.DOI: 10.1037/a0019237

Judge, T. A., Thoresen, C. J., Bono, J. E., & Patton, G. K. (2001). The job satisfaction-job performance relationship: A qualitative and quantitative review. *Psychological Bulletin*, *127*(3), 376–407. DOI: 10.1037/0033-2909.127.3.376 PMID: 11393302

Karakatsani, D. (2014). Education and society in Greece: Cultural traditions and social transformations. *Journal of Modern Greek Studies*, *32*(2), 207–224.

Klusmann, U., Kunter, M., Trautwein, U., Lüdtke, O., & Baumert, J. (2008). Teachers' occupational well-being and quality of instruction: The important role of self-regulatory patterns. *Journal of Educational Psychology*, *100*(3), 702–715. DOI: 10.1037/0022-0663.100.3.702

Kokkinos, C. M. (2007). Job stressors, personality and burnout in primary school teachers. *The British Journal of Educational Psychology*, *77*(1), 229–243. DOI: 10.1348/000709905X90344 PMID: 17411497

Koopmans, L., Bernaards, C. M., Hildebrandt, V. H., Schaufeli, W. B., de Vet, H. C. W., & van der Beek, A. J. (2014). Conceptual frameworks of individual work performance: A systematic review. *Journal of Occupational and Environmental Medicine*, *56*(3), 210–219. DOI: 10.1097/JOM.0000000000000130 PMID: 24561507

Kyriacou, C. (2001). Teacher stress: Directions for future research. *Educational Review*, *53*(1), 27–35. DOI: 10.1080/00131910120033628

Lazarus, R. S., & Folkman, S. (1984). *Stress, appraisal, and coping*. Springer Publishing Company.

Leithwood, K., Day, C., Sammons, P., Harris, A., & Hopkins, D. (2006). *Successful School Leadership: What It Is and How It Influences Pupil Learning*. Department for Education and Skills Maslach, C., & Jackson, S. E. (1981). The measurement of experienced burnout. *Journal of Organizational Behavior*, *2*(2), 99–113. DOI: 10.1002/job.4030020205

Maslach, C., & Leiter, M. P. (2016). *Burnout: A critical overview*. In The handbook of stress and health: A guide to research and practice (pp. 101-120). Wiley Blackwell.

Maslach, C., Schaufeli, W. B., & Leiter, M. P. (2001). Job burnout. *Annual Review of Psychology*, *52*(1), 397–422. DOI: 10.1146/annurev.psych.52.1.397 PMID: 11148311

Montgomery, C., & Rupp, A. A. (2005). A meta-analysis for exploring the diverse causes and effects of stress in teachers. *Canadian Journal of Education*, *28*(3), 458–486. DOI: 10.2307/4126479

Papadopoulou, M., & Kaldi, S. (2010). Investigating the job satisfaction of primary and secondary substitute teachers in Greece. *Educational Studies*, *36*(5), 493–502. DOI: 10.1080/03055690903424722

Pillay, H., Goddard, R., & Wilss, L. (2005). Well-being, burnout and competence: Implications for teachers. *The Australian Journal of Teacher Education*, *30*(2), 22–33. DOI: 10.14221/ajte.2005v30n2.3

Schaufeli, W. B., & Bakker, A. B. (2004). Job demands, job resources, and their relationship with burnout and engagement: A multi-sample study. *Journal of Organizational Behavior*, *25*(3), 293–315. DOI: 10.1002/job.248

Schaufeli, W. B., Bakker, A. B., & Salanova, M. (2006). The measurement of work engagement with a short questionnaire: A cross-national study. *Educational and Psychological Measurement*, *66*(4), 701–716. DOI: 10.1177/0013164405282471

Schaufeli, W. B., & Enzmann, D. (1998). *The burnout companion to study and practice: A critical analysis*. Taylor & Francis.

Schwarzer, R., & Hallum, S. (2008). Perceived teacher self-efficacy as a predictor of job stress and burnout: Mediation analyses. *Applied Psychology*, *57*(s1, S1), 152–171. DOI: 10.1111/j.1464-0597.2008.00359.x

Skaalvik, E. M., & Skaalvik, S. (2010). Teacher self-efficacy and teacher burnout: A study of relations. *Teaching and Teacher Education*, *26*(4), 1059–1069. DOI: 10.1016/j.tate.2009.11.001

Skaalvik, E. M., & Skaalvik, S. (2011). Teacher job satisfaction and motivation to leave the teaching profession: Relations with school context, feeling of belonging, and emotional exhaustion. *Teaching and Teacher Education*, *27*(6), 1029–1038. DOI: 10.1016/j.tate.2011.04.001

Skaalvik, E. M., & Skaalvik, S. (2015). Job demands and job resources as predictors of teacher motivation and well-being. *Social Psychology of Education*, *18*(1), 17–39. DOI: 10.1007/s11218-014-9283-7

Skaalvik, E. M., & Skaalvik, S. (2015). Job demands and job resources as predictors of teacher motivation and well-being. *Social Psychology of Education, 18*(1), 29–44. DOI: 10.1007/s11218-014-9283-8

Spector, P. E. (1997). *Job satisfaction: Application, assessment, causes, and consequences*. Sage Publications. DOI: 10.4135/9781452231549

Stavrou, N. (2017). Substitute teachers in Greece: Challenges and prospects. *Educational Policy and Development Journal, 12*(2), 45–58.

Sutton, R. E., & Wheatley, K. F. (2003). Teachers' emotions and teaching: A review of the literature and directions for future research. *Educational Psychology Review, 15*(4), 327–358. DOI: 10.1023/A:1026131715856

Taris, T. W., Van Horn, J. E., Schaufeli, W. B., & Schreurs, P. J. (2004). Inequity, burnout and psychological withdrawal among teachers: A dynamic exchange model. *Anxiety, Stress, and Coping, 17*(1), 103–122. DOI: 10.1080/10615800310001516 20

Thompson, E. R., & Phua, F. T. T. (2012). A brief index of affective job satisfaction. *Group & Organization Management, 37*(3), 275–307. DOI: 10.1177/1059601111434201

Tobin, D. L. (2001). User manual for the Coping Strategies Inventory. Retrieved from www.uccs.edu

Travers, C. J., & Cooper, C. L. (1996). *Teachers under pressure: Stress in the teaching profession*. Routledge.

Van Droogenbroeck, F., Spruyt, B., & Vanroelen, C. (2014). Burnout among senior teachers: Investigating the role of workload and interpersonal relationships at work. *Teaching and Teacher Education, 43*, 99–109. DOI: 10.1016/j.tate.2014.07.005

Zembylas, M. (2005). *Teaching with emotion: A postmodern enactment*. Information Age Publishing.

Zembylas, M., & Papanastasiou, E. (2006). Sources of teacher job satisfaction and dissatisfaction in Cyprus. *Compare: A Journal of Comparative Education, 36*(2), 229–247. DOI: 10.1080/03057920600741289

Chapter 14
Gender-Based Violence in the Workplace:
Psychological and Professional Impacts

Theodora Vainterli
https://orcid.org/0009-0009-0363-7908
Democritus University of Thrace, Greece

Lazaros Mitsou
https://orcid.org/0000-0002-6297-4738
Democritus University of Thrace, Greece

ABSTRACT

Gender-based violence in the workplace is a complex and multidimensional issue that affects employees' psychological and professional lives daily. This phenomenon can manifest in various forms, such as sexual harassment, verbal abuse, psychological abuse, and bullying. The aim of this research is to identify the forms of gender-based violence in the workplace, analyze the risk factors, and evaluate the effectiveness of existing policies and practices aimed at addressing it. The research, conducted through the collection of electronic questionnaires, seeks to document the experiences of employees across various sectors and levels regarding gender-based violence in the workplace. Understanding and mapping the factors that increase or reduce the risk of gender-based violence will help develop more effective prevention and response strategies. The research will employ a quantitative approach using questionnaires distributed to employees from various sectors and different professional levels within the Region of Eastern Macedonia and Thrace.

DOI: 10.4018/979-8-3693-9400-7.ch014

INTRODUCTION

Gender-based violence in the workplace is a widely prevalent problem that affects individuals across all businesses and sectors. It includes various harmful behaviors, such as sexual harassment, discrimination, and abuse, all of which arise from unbalanced power dynamics and entrenched gender norms (Ali et al., 2024). Despite progress in gender equality, the workplace remains an environment where such violence undermines safety, dignity, and professional development. Addressing gender-based violence is essential for human well-being and for promoting healthy, productive work environments. This essay will examine the causes, effects, and solutions to this critical issue.

Definition

Gender-based violence in the workplace refers to any harmful action or behavior targeting an individual due to their gender, resulting in physical, emotional, or psychological harm. This violence can take various forms, such as sexual harassment, bullying, physical assault, and discrimination. Sexual harassment, arguably the most recognized type, includes inappropriate remarks and actions, as well as unsolicited sexual advances or coercion (Blindow et al., 2024). Gender-based violence also encompasses less overt manifestations, such as gender-based discrimination, where individuals experience unequal treatment or are denied opportunities solely because of their gender. In some cases, this includes pay inequities, limited advancement opportunities, or gender-biased evaluations.

Power dynamics significantly contribute to the perpetuation of gender-based violence in the workplace (Ali et al., 2024). Individuals in positions of power often misuse their influence to exploit or dominate employees, fostering an environment of fear and subjugation. This phenomenon is often evident in male-dominated industries or organizations where traditional gender roles are deeply ingrained in the culture. However, victimization is not limited to women (Latcheva, 2017). Gender-based violence affects individuals of all genders, although women and gender minorities are disproportionately impacted. Additionally, workplace gender-based violence intersects with other identities, including race, ethnicity, and sexual orientation, exacerbating the vulnerabilities of marginalized populations.

Understanding gender-based violence in the workplace requires recognizing its various manifestations, ranging from explicit physical assault to more insidious but equally harmful behaviors, such as exclusion or emotional manipulation (Latcheva, 2017). This broad spectrum highlights the complexity of addressing the issue and the need for comprehensive strategies to create safe and respectful workplaces for all employees.

CAUSES AND CONTRIBUTING FACTORS

The origins of gender-based violence in the workplace stem from a convergence of social, cultural, and organizational elements that perpetuate unequal power dynamics and biased attitudes (Nyberg et al., 2021). The enduring presence of patriarchal norms is the primary culprit, where traditional gender roles grant control and authority to men, while women and gender minorities are often relegated to subordinate or inferior positions. These entrenched ideas can foster an environment where gender-based violence, particularly sexual harassment and discrimination, is normalized or overlooked.

Hostile workplace cultures significantly contribute to the perpetuation of gender-based violence. In some organizations, particularly those with hierarchical structures, power imbalances are exacerbated, leading to atmospheres where individuals in positions of authority perceive themselves as entitled to mistreat subordinates (Nyberg et al., 2021). This is particularly evident in male-dominated industries, such as construction, banking, or technology, where women and gender minorities may experience marginalization or harassment solely due to their minority status. Furthermore, workplaces that prioritize profit or efficiency over employee well-being can foster cultures of silence, where victims of gender-based violence fear retaliation or professional consequences if they report abuses.

Another contributing factor is the absence of effective policies and enforcement mechanisms within organizations (Agbaje et al., 2021). Even in cases where policies to address gender-based violence are in place, their implementation is often inadequate, or enforcement is weak. Victims may encounter barriers to reporting incidents due to concerns about distrust, stigma, or the belief that their complaints will be ignored. Management may occasionally overlook misconduct, especially when the perpetrator holds a senior position or is considered an asset to the organization. Additionally, cultural disparities and legal gaps between nations may exacerbate the issue, as many regions lack robust legislation or social support mechanisms to effectively address gender-based violence. The interplay of these factors creates a multifaceted and widespread problem in global workplaces.

THE ROLE OF ORGANIZATIONAL LEADERSHIP AND TRAINING

The job of group leaders in dealing with gender-based harm at work is very important. Chiefs, especially those in strong roles, have the duty to form the way of life of the group and set a clear style on what is seen as okay behavior. Leaders promise to not allow bullying or unfair treatment shows a loud word to all workers that these actions will not be accepted under any case. By putting in hard rules and backing no-nonsense met hods, chiefs can build a safer and kinder place to work.

Also, the group's leader need to make sure that these rules are not just set but actually followed. Clear talk about what is expected, what happens if someone breaks a rule, and an easy way to report issues is key for making sure workers feel safe to share problems without worry of payback or bad outcomes. This means giving different ways to report, like secret phone lines

Teaching plans on variety, fairness, and including can help a lo t in boosting knowledge and giving useful tools for workers and leaders. These plans must talk about things like seeing hidden biases, knowing power flows, and fixing wrong behaviors. By adding DEI teaching into normal job growth, groups can make sure all workers any way of gender, race have the know-how and skills to build a more welcome space.

Additionally, organizational leaders must actively promote and support employee resource groups, particularly those focused on gender and diversity. These groups serve as platforms for employees to share experiences, seek support, and provide feedback on workplace policies. By involving employees in the decision-making process, leadership can better understand the unique challenges faced by different groups and create targeted solutions that address their needs.

A comprehensive approach to preventing and responding to gender-based violence also requires that organizations integrate these values into their recruitment, hiring, and promotion practices. Ensuring equal opportunities for all employees and addressing gender biases in performance evaluations are key to creating an environment where all individuals have an equal chance to succeed. Moreover, organizations should be proactive in evaluating their work culture and addressing any underlying issues that may contribute to discriminatory practices.

Ultimately, fostering an inclusive and respectful work environment has significant benefits for both employees and the organization as a whole. When employees feel valued, safe, and supported, their job satisfaction and productivity improve. Additionally, organizations that prioritize gender equality and a respectful workplace are more likely to attract and retain top talent, enhancing their reputation in the industry and contributing to their long-term success.

IMPACT ON VICTIMS AND THE WORKPLACE

Gender-based violence in the workplace has far-reaching consequences, affecting both the individuals involved and the overall organizational atmosphere. The effects on victims can be profound and long-lasting, often extending beyond the initial trauma of the incident. Many victims experience significant emotional and psychological turmoil, including anxiety, despair, and post-traumatic stress disorder (PTSD) (Nelson et al., 2024). These mental health issues may lead to decreased focus, challenges in completing work tasks, and diminished confidence in professional abilities. Over time, victims may become disillusioned with their careers, resulting in reduced job satisfaction and performance. In severe cases, chronic stress and emotional strain can compel individuals to resign from their positions, either voluntarily or as a result of retaliation or ostracism.

The professional consequences for victims are equally damaging. Career progression may be hindered when victims, especially women and gender minorities, are often excluded from opportunities, promotions, or key projects due to discrimination or retaliation they face after reporting gender-based violence (Nelson et al., 2024). This can lead to a cycle of exclusion that obstructs their career growth and potential. Furthermore, the concern about further victimization or damage to their reputation may discourage individuals from voicing their experiences, thereby maintaining a culture of silence that emboldens perpetrators.

In addition to individual consequences, gender-based violence negatively impacts the entire workplace. Organizations that fail to address or prevent such violence often face reduced productivity, as staff members operate in an atmosphere of fear, distrust, or hostility (Nyberg et al., 2021). Increased turnover rates prevail, as employees distance themselves from unsafe environments, and the process of replacement can incur significant costs. Moreover, the company's reputation may deteriorate, particularly if incidents of violence become public or lead to legal disputes. Potential litigation, sanctions, and settlements drain financial resources and may also generate adverse media coverage, tarnishing the public image of the organization.

LEGAL FRAMEWORK AND WORKPLACE POLICIES

The legal framework governing gender-based violence in the workplace varies by region and country. However, international entities such as the International Labour Organization (ILO) have played a crucial role in establishing global standards. The ILO Convention on Violence and Harassment (No. 190), adopted in 2019, was a significant step forward in defining and addressing violence and harassment in the workplace (O'Connor et al., 2021). This convention underscores the right of all individuals to a workplace free from violence and harassment, including gender-based violence, and advocates for states to adopt measures that protect workers from such threats. However, not all nations have ratified this convention, and even when legal frameworks are established, implementation can be inconsistent.

National legislation regarding gender-based violence in the workplace varies widely, from comprehensive to nearly nonexistent, depending on the country. In many developed nations, legislation to address sexual harassment and discrimination has been firmly established, giving workers the power to report such violations and seek legal recourse. However, legislative gaps still exist in many contexts, particularly concerning the protection of gender minorities, informal workers, and individuals in precarious employment (Agbaje et al., 2021). In low-income countries, the lack of strong legal protections and cultural stigmas associated with reporting abuse hinder victims from seeking justice. In these cases, victims often face retaliation or ostracism, perpetuating the cycle of abuse.

At the organizational level, measures designed to prevent and address gender-based violence are crucial. Many companies have implemented policies that include clear definitions of inappropriate behavior, reporting protocols, and punitive measures for offenders. The effectiveness of these rules often depends on the organizational culture and the leadership's commitment to enforcing them. In many instances, policies are merely theoretical and have minimal impact due to insufficient enforcement or a lack of employee understanding (O'Connor et al., 2021). Training programs, awareness initiatives, and the establishment of safe reporting mechanisms can significantly improve the effectiveness of these rules.

The legal and organisational approach to combating gender-based violence in the workplace is crucial for its prevention and acceptance in the social and professional environment. Despite the existence of international agreements and national legislative frameworks, their implementation often remains problematic, as organisations and governments often do not invest sufficiently in the mechanisms needed to effectively implement these policies. The prevailing culture in many companies, particularly those with hierarchical structures, can create barriers to the implementation of anti-gender violence policies, encouraging an atmosphere of impunity for perpetrators and support for victims.

Furthermore, creating safe and confidential channels for reporting incidents of violence is a vital priority to ensure the well-being of workers. Victims must be able to report incidents without fear of retaliation or disapproval. Workers, particularly in workplaces where traditional gender perceptions remain strong, need proper training and awareness of their rights and the options available to them to address these issues.

Developing policies and practices that incorporate a commitment to gender equality and the prevention of gender-based violence in the workplace is critical to changing the culture and creating a safe and equitable work environment for all workers. Organisations need to ensure that there is clarity around the concept of gender-based violence.

STRATEGIES FOR PREVENTION AND MITIGATION

Gender-based violence in the workplace is a widespread issue that requires comprehensive solutions for prevention and relief. Addressing it necessitates a multi-faceted strategy that incorporates awareness, training, policy enforcement, and cultural reform (Agbaje et al., 2021). Employers, employees, and governments must collaborate to create environments where such violence is recognized, prevented, and effectively addressed when it occurs. This discussion highlights various tactics, emphasizing their interactivity and the necessity of sustained commitment.

An essential approach to preventing gender-based violence is cultivating awareness and understanding among all members of the organization. Workplace violence often stems from deep-rooted biases, stereotypes, and unclear power inequalities (Nelson et al., 2024). Organizations should frequently implement training programs that educate employees about the different manifestations of gender-based violence, which include harassment, discrimination, and both physical and emotional abuse. These programs should prioritize identifying inappropriate behavior and understanding its harmful impact on individuals and the workplace dynamic. Awareness efforts should also consider the intersectionality of gender-based violence, emphasizing how factors such as race, class, and sexual orientation exacerbate risks and consequences for underrepresented populations.

Establishing a culture of respect and inclusion is equally necessary. The workplace culture significantly influences the behaviors and attitudes of employees. When businesses prioritize equality and respect, they lay a foundation for constructive relationships and reduce the likelihood of abusive behavior (Agbaje et al., 2021). Leadership should set an example by exhibiting behavior that reflects a clear intolerance for gender-based violence. This involves addressing inappropriate behavior promptly and transparently, ensuring that employees feel safe reporting issues without fear of retaliation. Leaders must actively engage in discussions on gender equality, signaling that these issues are a priority for the company, not merely compliance obligations.

Well-developed workplace policies are fundamental for preventing and alleviating gender-based violence. Clearly articulated policies outline inappropriate behavior, reporting protocols, and consequences for violations. These policies should be communicated to all employees during onboarding and periodically reviewed through training sessions. The reporting mechanisms outlined in these policies must be accessible and confidential (Nelson et al., 2024). Many victims of workplace violence remain reluctant due to stigma or professional consequences. Ensuring confidentiality and offering various reporting methods can allow victims to come forward and seek help. Additionally, policies should prioritize restorative approaches when appropriate, facilitating solutions that focus on the victim's needs while also addressing the perpetrator's behavior.

Organizations must not only establish policies but also enforce them rigorously. A common drawback in managing workplace violence is the inconsistency between policy and enforcement. When violations occur, employers are obligated to conduct thorough and impartial investigations, applying measures commensurate with the severity of the violation. Consistent policy enforcement not only ensures accountability for the perpetrator but also reinforces the organization's commitment to safety and equality. It is crucial that any enforcement process incorporates external oversight when necessary to ensure impartiality, especially in cases involving senior executives or prominent figures (Nelson et al., 2024).

Addressing the root causes of gender-based violence is essential for its prevention. Power dynamics often underpin workplace violence, as perpetrators exploit their influence to threaten or harm others. Organizations should aim to mitigate power imbalances by strengthening diversity in leadership, establishing equal opportunities for advancement, and ensuring transparency in decision-making processes. Moreover, promoting teamwork and mutual respect among employees can reduce hierarchical structures that enable abusive behavior (Oliveira et al., 2018). Facilitating team-based problem-solving initiatives and mentoring helps dismantle hierarchies among staff and cultivates trust within the organization.

A frequently overlooked approach is involving bystanders in preventive initiatives. Numerous incidents of workplace violence occur in the presence of bystanders who feel unable or reluctant to intervene. Organizations can empower staff to identify harmful behavior and respond effectively by providing bystander intervention training. This training may include techniques for de-escalating high-tension scenarios, offering assistance to victims, and documenting incidents (Oliveira et al., 2018). Promoting shared responsibility for ensuring a safe workplace reduces the burden on victims and fosters an environment where all individuals are accountable for maintaining behavioral standards.

At the same time, organizations must implement support networks for victims of gender-based violence. Prevention and mitigation are inadequate without effective ways to support those affected. Employee Assistance Programs (EAPs), counseling services, and access to legal or medical resources provide vital support (Oliveira et al., 2018). Employers should widely disseminate information regarding the availability of these resources and ensure they are offered at no or minimal cost to employees. Additionally, fostering peer support networks or confidential support groups can offer victims a sense of solidarity and mitigate feelings of isolation. This holistic strategy for victim care facilitates healing and strengthens trust in the organization's commitment to employee well-being.

Ultimately, compliance with legislation and collaboration with external stakeholders is critical to preventing and mitigating workplace violence. Employers must be educated about their legal responsibilities regarding gender-based violence and ensure their practices align with national and international standards. Engaging with advocacy groups, trade unions, and community organizations can provide valuable insights and resources to tackle workplace violence. External relationships signal an organization's seriousness in addressing gender-based violence, thereby enhancing its reputation and commitment to transformation.

Implementing these methods requires continuous effort and evaluation. The dynamics of the workplace are constantly evolving, and new challenges may arise, requiring flexibility in prevention and mitigation strategies. Regular evaluation of policies, soliciting employee input, and analyzing the effectiveness of training programs can help businesses identify gaps and strengthen their strategies. Moreover, embedding these methods into the corporate culture guarantees their longevity, preventing them from becoming ephemeral efforts.

CASE STUDIES AND REAL-WORLD EXAMPLES

Gender-based violence in the workplace is a significant issue that disrupts lives, careers, and organizational dynamics. Analyzing real-world examples and case studies provides valuable insights into the complex nature of this issue and highlights lessons that can inform prevention and mitigation strategies (Mirance et al., 2024). These narratives clarify the underlying causes, expose institutional gaps, and offer viable approaches to combating gender-based violence in various professional environments.

A prominent example is the widely reported case of sexual harassment in the entertainment industry, exemplified by the allegations against Harvey Weinstein. The Weinstein case, which sparked the global #MeToo movement, exposed the pervasive culture of secrecy and fear that enabled powerful individuals to exploit their authority. Many women reported being harassed or assaulted by Weinstein, detailing experiences ranging from unsolicited advances to coercion. These reports highlighted how gender-based violence often thrives in sectors with strong hierarchical power dynamics, where victims fear retaliation or professional failure if they reveal their experiences (Mirance et al., 2024). The #MeToo movement not only amplified the voices of survivors but also underscored the need for institutional reform to address workplace abuse, including the implementation of clearer reporting protocols and strengthened legal safeguards.

An alternative but equally enlightening scenario arose in the technology sector, a historically criticized industry for gender inequalities and its exclusionary ethics.

The 2017 Uber scandal serves as a prominent illustration. Susan Fowler, a former engineer at the company, wrote a comprehensive blog post revealing a culture of pervasive sexism and harassment. Her allegations included being propositioned by a manager and experiencing professional retaliation when she reported the incident to human resources. Uber's failure to implement appropriate measures highlighted how organizational policies, even if well-designed, can falter when leadership neglects enforcement or accountability (Mirance et al., 2024). The fallout from Fowler's revelations was significant, resulting in public scrutiny, the resignation of Uber's CEO, and a transformation of its workplace culture. This case demonstrated that transparency and whistleblowing can trigger change while also emphasizing the need for proactive measures to prevent such issues in the first place.

A study in South Africa shed light on the levels of gender-based violence faced by female nurses and other healthcare professionals. The cases involved verbal abuse and harassment from male colleagues as well as physical assault from patients (Rezwana, 2023). One particularly painful incident involved a nurse who was attacked by a male supervisor after rejecting his advances. When she reported the incident, her concerns were ignored, and she subsequently faced ostracism from colleagues who perceived her as a troublemaker. This case illustrates how workplace violence can escalate in environments where gender norms and cultural stigmas prevent women from voicing their concerns. It underscores the need for industry-specific solutions, especially in healthcare, including enhanced security protocols and gender-sensitive training, to protect staff from violence.

Corporate environments are also susceptible to these challenges. A multinational financial firm faced criticism after a female employee revealed harassment by her manager during a corporate retreat. Despite evidence supporting her claims, the company initially downplayed the incident, citing concerns about reputational damage. The inertia led other employees to disclose similar incidents, uncovering a broader pattern of inappropriate behavior across the organization (Rezwana, 2023). The ensuing public relations disaster resulted in the dismissal of several senior executives and a thorough review of the company's workplace standards. This example underscores the need to prioritize employee well-being over institutional reputation, illustrating the reputational and financial consequences of neglecting the effective handling of workplace violence.

There are positive case studies that provide insights into effective strategies for combating gender-based violence. A construction company in India developed an innovative program to address harassment and violence against its female employees. Recognizing that many cases went unreported due to stigma, the company partnered with local NGOs to create anonymous reporting mechanisms and implemented comprehensive gender sensitivity training for all employees, including senior management (Cerdán-Torregrosa et al., 2023). The initiative included a mentorship program designed to inspire female employees to take on leadership roles. Ultimately, these efforts led to a significant reduction in reported incidents and boosted morale and productivity in the workplace. This case demonstrates how a proactive and inclusive strategy can enhance workplace culture and create a safer environment for all workers.

Similarly, a Canadian university serves as an example of institutional commitment to preventing and addressing gender-based violence. In response to allegations involving a distinguished faculty member, the institution launched an independent investigation and implemented a series of measures (Cerdán-Torregrosa et al., 2023). These included revising the sexual harassment policy, establishing a specialized office for handling complaints, and requiring consent and bystander intervention training for all staff and students. The university's honest handling of the situation and willingness to take responsibility not only restored trust within its community but also set a standard for other academic institutions facing similar challenges.

The garment industry offers an alternative perspective on gender-based violence, especially in underdeveloped countries (Cerdán-Torregrosa et al., 2023). Factory reports from Bangladesh, Cambodia, and Vietnam have shown widespread harassment and mistreatment of female employees by male supervisors. One worker who rejected her supervisor's advances was fired, making her economically vulnerable. International organizations and advocacy groups have intervened, working with factory owners to establish anti-harassment committees and provide training on workplace rights. Although these actions are ongoing, they demonstrate the potential for cross-sector collaboration to address systemic concerns.

ROLE OF LEADERSHIP AND ORGANIZATIONAL CULTURE

Leadership and organizational culture play a critical role in combating gender-based violence in the workplace. Leaders set the tone for behavior within the workplace, and the culture they cultivate dictates whether harmful activities are addressed or perpetuated. Collectively, they can either act as catalysts for transformation or facilitators of systemic exploitation (Cayir et al., 2021). Addressing gender-based violence requires leaders to demonstrate commitment, accountability, and a vision for fostering an inclusive and respectful environment. Similarly, cultivating a culture that emphasizes safety, equality, and empowerment is vital for the prevention and treatment of such violence.

Effective leadership starts with recognition and accountability. Leaders must acknowledge that gender-based violence is not just an individual or isolated concern but a systemic issue often embedded throughout business hierarchies and practices (O'Connor et al., 2021). Ignoring or downplaying its existence not only perpetuates harm but also undermines employee trust and organizational credibility. Leaders must recognize the painful realities within their organizations and commit to resolving them. This involves openly declaring a zero-tolerance policy for harassment, discrimination, or any form of violence and ensuring that all employees understand the authenticity and stability of this commitment.

Leaders play a pivotal role in modeling appropriate behavior. Employees often look to authority figures for guidance, and leadership that exemplifies respect, equality, and empathy sets an important precedent. When leaders adhere to the same standards they impose on their teams, it sends the message that no one is exempt from the rules (O'Connor et al., 2021). This consistency is essential for building trust and ensuring that measures to address gender-based violence are not seen as cosmetic or selective. Conversely, when leaders fail to model these ideals or ignore misconduct, they foster an environment where harmful behavior is tacitly accepted.

Transparency is a foundational trait of good leadership in mitigating workplace violence. Leaders must establish transparent and accessible communication channels that allow employees to express concerns without fear of retaliation. This involves implementing confidential reporting systems, regularly disseminating policies, and proactively encouraging staff to provide feedback (Gardiner & Finn, 2022). Leaders must maintain transparency regarding the handling of complaints, ensuring that investigations are thorough, impartial, and consistent. When employees observe that gender-based violence complaints are taken seriously and addressed equitably, they are more likely to trust the system and feel confident in coming forward when necessary.

Organizational culture, intrinsically linked to leadership, provides the broader framework for workplace behavior. A culture that tolerates or minimizes gender-based violence—through humor, dismissive behaviors, or inadequate consequences—fosters an environment conducive to abuse (Gardiner & Finn, 2022). In contrast, a culture that prioritizes respect and equality serves as a protective measure, reducing the likelihood of violence and empowering employees to voice concerns when they arise. Transforming a corporate culture that traditionally ignores or facilitates violence requires deliberate action and sustained commitment.

Promoting diversity is essential for this cultural shift. Organizations characterized by diverse leadership and a commitment to equality are less likely to sustain power imbalances that often contribute to gender-based violence (Cayir et al., 2021). Leaders must diligently work to diversify their teams, ensuring that women and other underrepresented groups have fair access to opportunities and representation (Gardiner & Finn, 2022). Inclusion beyond numerical representation entails fostering an environment where every employee feels valued and recognized. This can be achieved through initiatives like mentorship programs, employee resource groups, and ongoing discussions about diversity and inclusion.

Training and education are key tools for building organizational culture. While policies are necessary, they are insufficient without initiatives to transform mindsets and behaviors (Cayir et al., 2021). Comprehensive educational programs can help staff at all levels understand the nature of gender-based violence, recognize its impact, and acquire the skills to intervene or report incidents effectively. This training should be ongoing rather than a one-off event, reinforcing the organization's commitment to continuous learning and improvement (Gardiner & Finn, 2022). Most importantly, leaders must participate in these initiatives, demonstrating that they are not exempt from the standards they impose on others.

Leadership and culture are vital for creating accountability. Organizations must establish processes that hold individuals accountable, regardless of their status or position (Mirance et al., 2024). This involves not only responding to reported incidents but also proactively recognizing and mitigating risks. Organizations can conduct regular audits to assess workplace dynamics, evaluate the effectiveness of current regulations, and identify behavioral patterns that may signal systemic issues. Ensuring accountability at the leadership level is crucial, as the behavior of individuals in senior positions often sets the organizational culture. When leaders face consequences for misconduct, it reinforces the idea that no one is exempt from scrutiny.

Establishing a culture of trust is critical in combating gender-based violence. Employees must be confident that their concerns will be addressed with empathy and action, not skepticism or dismissal. Trust in leadership allows workers to feel safe in reporting incidents and participating in efforts to reduce workplace violence. Organizational culture thus serves as the foundation for creating a work environment where safety, equality, and respect are not just policies, but everyday realities.

Gender-based violence in the workplace is a pervasive and complex issue that requires multi-dimensional solutions. Effective prevention and response demand strong leadership, a commitment to accountability, and an organizational culture that prioritizes safety, equality, and respect. By adopting transparent, inclusive, and proactive strategies, organizations can create a safer environment for all employees. Ultimately, addressing gender-based violence is not only a moral imperative but also a critical aspect of maintaining a productive, supportive, and thriving workplace.

RESEARCH PURPOSE

The purpose of this research is to understand and analyze the impact of gender-based violence in the workplace, as well as to assess the existing policies within organizations for addressing such incidents in organizations and institutions in the Eastern Macedonia and Thrace region. Furthermore, the research focuses on recording the participants' views on how psychological and social conditions in cases of gender-based violence could be improved, proposing better training and support from the organizations and institutions themselves, whether private or public.

An important aspect of this research is the investigation of the existing legal framework for gender-based violence in the workplace. Greek legislation includes a series of regulations aimed at protecting victims and addressing the phenomenon. However, the research revealed that many employees are not adequately informed about their rights, indicating the need for further information and awareness. Familiarity with legislative processes could enhance employees' trust in seeking help and reporting incidents.

The research revealed that gender-based violence significantly impacts the psychological well-being of employees, often leading to isolation, lack of self-confidence, and increased levels of stress. The social consequences include the disruption of relationships between colleagues and the creation of a toxic work environment. In this context, it is clear that psychosocial support from organizations is crucial for restoring balance in the work environment.

The research question on which this study is based is what forms and frequencies of gender-based violence employees face in their workplace, and how these affect both their psychological state and their professional performance. Furthermore, the perceptions of participants regarding the effectiveness of organizational policies for addressing gender-based violence are explored. The second research question concerns how an incident of gender-based violence affects the psychological well-being and productivity of employees.

The research collects basic demographic data from the participants, such as gender, age, education level, and professional sector. The composition of the sample is as follows: 86.67% of the participants are women and 13.33% are men. Participants represent a broad age range, including categories such as 18-24, 25-34, 35-44, 45-54, 55-64, and 65 and older. Specifically, 39% of the sample is between the ages of 35-44, 27% belong to the 25-34 range, while the remaining categories account for less than 15%. Next, the education level was examined; 34 out of 75 participants hold a master's degree, 19 are university graduates, 1 holds a doctoral degree, while the rest are secondary education or vocational school graduates. Additionally, participants come from various professional sectors, including the public sector, private sector, self-employment, and non-profit organizations. Specifically, 46% of the sample works in the private sector, 44% in the public sector, while the remaining percentage are self-employed or work in NGOs.

Continuing with the main data of the sample, the research also examines the impact on employee productivity and whether it is influenced by whether they have experienced gender-based violence in the workplace or not. For this correlation, an independent samples t-test was used, which is typically used to test if there are statistically significant differences in the mean values of two independent groups, in this case, employees who have experienced gender-based violence and those who have not. Subsequently, Levene's Test is used to determine if the variances (dispersion) between the two groups are equal. If the result of Levene's Test is statistically significant (Sig. < 0.05), then the variances between the two groups are different.

The test results for employee performance indicate that the variances between the two groups are not equal. Furthermore, there is no statistically significant difference between the groups concerning the examined variable (Sig. = 0.416). The mean difference is small (0.295), and it is not statistically significant, as confirmed by the 95% confidence interval. Therefore, employee performance does not appear to differ based on whether the employee has experienced gender-based violence or not.

Regarding the t-test results for emotional well-being, it is evident that the variances between the two groups are homogeneous, as indicated by the Sig. of Levene's Test. At the same time, there is a statistically significant difference between the two groups regarding how often they feel stress or tension due to gender-based violence (Sig. = 0.004). The mean difference (-0.496) suggests that one group feels significantly less stress or tension than the other.

Furthermore, to detect potential statistically significant differences between the means (M.O.) of employee performance and emotional well-being based on the forms of violence, an analysis of variance (One-way ANOVA) was used, with the results summarized in Tables 3 and 4.

At this point, it must be verified whether the assumption of homogeneity of variances is met, a prerequisite for conducting the ANOVA analysis. The rule followed is as follows:

> If the Asymptotic Significance value is < 0.05, the null hypothesis (H0) is rejected, and the alternative hypothesis (H1) is accepted, meaning the variances of the observations are not equal.
> If the Asymptotic Significance value is > 0.05, the null hypothesis (H0) is accepted, meaning the variances of the observations are equal.

In the case of emotional well-being, the sample meets the above condition, meaning the variances between the groups are homogeneous (Sig. > 0.05), allowing for the ANOVA analysis, as there is no violation of the assumption of homogeneity of variances. In contrast, in the case of performance, the sample does not meet the above condition, as the variances between the groups are not homogeneous (Sig. < 0.05). However, for the purposes of the study, the analysis will be conducted regardless.

From the above analysis of the variance tables, it is concluded that the values of Significance are 0.034 and 0.035, both less than 0.05, leading to the rejection of the null hypothesis (H0). Thus, the alternative hypothesis (H1) is accepted, indicating the presence of statistically significant differences between certain categories of the qualitative variable (forms of violence). However, this analysis is not sufficient to determine which forms of violence differ statistically from each other.

Beyond these correlations, this research highlighted other critical issues that need immediate resolution. Initially, the data reveal that the roles in such incidents are distributed between observers or victims, or both. A significant proportion of the participants, over 25%, have been in both positions.

One of the key proposals that emerged from the research is the development of specialized training programs for the prevention and addressing of gender-based violence. These programs could include training on recognizing forms of violence, intervention methods, and victim support. It is also suggested that such actions be included annually to ensure continuous information and training of staff.

The research could benefit from comparing the policies implemented in Eastern Macedonia and Thrace with best practices applied internationally. For example, many countries promote anonymous reporting platforms for incidents and specialized intervention teams within organizations. Such a comparison could lead to substantial recommendations for improving the local framework.

Furthermore, a very critical part of the answers concerns the organization's policies. Participants seem to suggest improvements such as better staff training and awareness, more immediate intervention and support from supervisors, and the creation of a more structured framework for supporting victims and those involved. Immediate intervention by organizations is seen as crucial by participants, especially in cases where they have experienced or witnessed gender-based violence incidents. Staff awareness and training are cited as key factors that can improve the psychological state of individuals involved in such incidents.

Additionally, a significant finding from the results of the study is that many participants (up to 65%) are not familiar with the existing legal framework in Greece for gender-based violence. This lack of awareness may burden victims and discourage them from seeking legal help or reporting incidents, as they are unaware of the process or how to proceed. Therefore, providing information on the legal framework and employees' rights in cases of gender-based violence is considered particularly crucial. The 75 responses analyzed in this study highlight that both practical and emotional support can contribute to improving the working conditions within an organization, creating an environment where individuals feel safe and protected.

Finally, this analysis examines the emotional consequences experienced by individuals after gender-based violence incidents, based on the collected responses. The findings reveal that after an incident of gender-based violence in the workplace, 28 out of 75 participants felt intense anxiety, 24 tended to feel anger, while 15 expressed disappointment as their primary emotion. Furthermore, 13 participants felt isolation, mainly those who were victims. The feeling of shame occurred 12 times, and it is noteworthy, as it may relate to the social stigma surrounding gender-based violence incidents. Subsequently, the feeling of fear was reported 11 times, indicating that individuals felt their safety was threatened after such incidents. Lastly, in addition to the above, sleep disturbances were mentioned 8 times as an immediate consequence, combined with eating disorders reported by 5 individuals who were victims.

LIMITATIONS OF THE RESEARCH AND DISCUSSION

Regarding the limitations identified in the research, it should be noted that despite the anonymity maintained in the questionnaires, the responses may not be entirely honest due to the social embarrassment participants may have felt regarding the phenomenon of gender-based violence. This particular behavior can often underestimate the extent of the phenomenon. Secondly, many employees may not be aware of the exact policies established by the organization they work for to address gender-based violence or may not have direct experience, resulting in difficulties in providing informed answers. This may limit the accuracy of the data related to each organization's policies regarding the phenomenon. Furthermore, the research relies on self-reports to measure the psychological impact of gender-based violence. This means that accurately measuring stress or other emotions and psychological consequences may be difficult, as participants may not have the ability to assess their situation precisely or may hesitate to share it. Even though the research is still ongoing, the sample may not be fully representative, and therefore the conclusions may not reflect the actual situation of the general population. Finally, another limitation identified during the research is the subjectivity in interpreting the responses. This means that the questions focusing on participants' perceptions of the intensity of violence or the support they receive may be subjective, as they may interpret the same situation differently, a characteristic that complicates data analysis. The high reports of stress, anger, and frustration suggest that the existing policies are insufficient to meet the needs of victims, which may be due to the lack of supportive structures and services. Furthermore, the connection between psychological and physical effects, such as sleep disturbances and eating disorders, confirms the need for comprehensive interventions that will, in turn, improve the mental and physical health of individuals. Generally, proposals for intervention strategies should include psychological support programs and educational seminars to raise awareness about the phenomenon, even within the workplace context. The combination of the research findings and the data emerging from the literature is crucial for understanding the complexity of the issue and the need for a holistic approach aimed at combating gender-based violence both within and outside the workplace.

CONCLUSIONS

In conclusion, gender-based violence in the workplace is a widespread issue that affects human well-being and harms organizational effectiveness. It is driven by deeply rooted social norms, power inequalities, and inadequate policies, resulting in significant emotional, psychological, and professional consequences for victims. Legal reforms and enhanced workplace policies are needed, but they require a transformation of the culture that emphasizes safety, respect, and equality for all workers. By addressing these challenges through thorough legislative safeguards and effective organizational initiatives, we can begin to create workplaces where gender-based violence is categorically rejected.However, this analysis also suggests that, while gender-based violence may not directly affect performance, it has a significant impact on the psychological state of employees, which could have indirect long-term effects on their professional well-being. It also highlights that the emotional responses of individuals who have experienced gender-based violence are multifaceted, encompassing both negative emotions and psychological difficulties that require investigation and immediate support. In particular, frequent reports of stress, anger, and frustration, along with generally negative emotions, demonstrate the need for better support and intervention from organizational structures.

Therefore, it is accepted that organizational policies and procedures must focus on psychological support and employee education for the prevention and timely handling of this phenomenon.

REFERENCES

Agbaje, O. S., Arua, C. K., Umeifekwem, J. E., Umoke, P. C. I., Igbokwe, C. C., Iwuagwu, T. E., Iweama, C. N., Ozoemena, E. L., & Obande-Ogbuinya, E. N. (2021). Workplace gender-based violence and associated factors among university women in Enugu, South-East Nigeria: An institutional-based cross-sectional study. *BMC Women's Health, 21*(1), 124. DOI: 10.1186/s12905-021-01273-w PMID: 33757498

Ali, M. A., Semahegn, A., Ashenafi, W., & Legesse, H. (2024). Workplace gender-based violence among female staff in public higher education institutions in eastern Ethiopia: Institution-based cross-sectional study. *PLOS Global Public Health*, *4*(7), e0003489. DOI: 10.1371/journal.pgph.0003489 PMID: 39058720

Blindow, K. J., Cedstrand, E., Elling, D. L., Hagland, M., & Bodin, T. (2024). Gender-based violence and harassment at work and health and occupational outcomes. A systematic review of prospective studies. *BMC Public Health*, *24*(1), 1788. DOI: 10.1186/s12889-024-19304-0 PMID: 38965519

Cayir, E., Spencer, M., Billings, D., Hilfinger Messias, D. K., Robillard, A., & Cunningham, T. (2021). "The only way we'll be successful": Organizational factors that influence psychosocial well-being and self-care among advocates working to address gender-based violence. *Journal of Interpersonal Violence*, *36*(23–24), 11327–11355. DOI: 10.1177/0886260519897340 PMID: 31928126

Cerdán-Torregrosa, A., Nardini, K., & Vives-Cases, C. (2023). "I reject it, but that's what normally happens": Grey zones of gender-based violence and gender roles in young people. *Journal of Interpersonal Violence*, *38*(11–12), 7656–7677. DOI: 10.1177/08862605221147070 PMID: 36710522

Gardiner, R. A., & Finn, H. (2022). Implementing gender-based violence policies in the neoliberal university: Challenges and contradictions. *Gender in Management*, *38*(2), 215–229. DOI: 10.1108/GM-07-2022-0228

Latcheva, R. (2017). Sexual harassment in the european union: A pervasive but still hidden form of gender-based violence. *Journal of Interpersonal Violence*, *32*(12), 1821–1852. DOI: 10.1177/0886260517698948 PMID: 30156991

Mirance, K., Foltz, K. E., Hattery, A. J., Kiss, M., & Smith, E. (2024). Theorizing feminist abolitionist approaches to gender-based violence: A descriptive case study of gender-based violence in sportsworld. *Gender & Society*, *38*(4), 557–585. DOI: 10.1177/08912432241266271

Nelson, S., Ayaz, B., Baumann, A. L., & Dozois, G. (2024). A gender-based review of workplace violence amongst the global health workforce—A scoping review of the literature. *PLOS Global Public Health*, *4*(7), e0003336. DOI: 10.1371/journal.pgph.0003336 PMID: 38954697

Nyberg, A., Kecklund, G., Hanson, L. M., & Rajaleid, K. (2021). Workplace violence and health in human service industries: A systematic review of prospective and longitudinal studies. *Occupational and Environmental Medicine*, *78*(2), 69–81. DOI: 10.1136/oemed-2020-106450 PMID: 32414952

O'Connor, P., Hodgins, M., Woods, D. R., Wallwaey, E., Palmen, R., Van Den Brink, M., & Schmidt, E. K. (2021). Organisational characteristics that facilitate gender-based violence and harassment in higher education? *Administrative Sciences*, *11*(4), 138. DOI: 10.3390/admsci11040138

Oliveira, C., Keygnaert, I., Oliveira Martins, M., & Dias, S. (2018). Assessing reported cases of sexual and gender-based violence, causes and preventive strategies, in European asylum reception facilities. *Globalization and Health*, *14*(1), 48. DOI: 10.1186/s12992-018-0365-6 PMID: 29743084

Rezwana, N. (2023). Risks and challenges in fieldwork on gender-based violence: Gendered identity, social taboo and culture. In Uddin, N., & Paul, A. (Eds.), *The Palgrave Handbook of Social Fieldwork* (pp. 237–251). Springer International Publishing., DOI: 10.1007/978-3-031-13615-3_14

Chapter 15
Greek Men's Knowledge of Osteoporosis

Stylianos Papalexandris
Democritus University of Thrace, Greece

Sofia Anastasiadou
https://orcid.org/0000-0001-6404-5003
University of Western Macedonia, Greece

ABSTRACT

Osteoporosis is a skeletal disease, characterised by reduced bone density, disrupted bone microarchitecture and susceptibility to fracture. Its epidemiology and pathophysiology justify the tendency to be considered as a women's disease. The study aims at determining the level of knowledge of Greek men on osteoporosis. Data were collected with the FOOQ and the MOKQ. It comprises 20 statements, assessing the knowledge on bone development, osteoporosis risk factors, prevention and treatment. 330 men participated in the study. SIA' results showed that the hardest statement of the FOOQ was the one referring to whether postmenopausal hormone replacement therapy can slow down bone loss. The easiest statement refers to the presentation of an osteoporotic fracture in 20% of women older than 50, if no preventative measures are taken. The most difficult statement of the MOKQ is the one that assesses whether hormonal treatment for prostate cancer does not reduce the risk of osteoporosis, whereas the easiest statement assesses whether bone loss increases in men after the age of 70.

DOI: 10.4018/979-8-3693-9400-7.ch015

1. INTRODUCTION

Aging and the changing epidemiological profile of the population have rendered degenerative diseases clinical entities of particular concern for public health (Omran, 2005; Harper & Armelagos, 2010). Within this context, osteoporosis is included—a systemic skeletal disease characterized by reduced bone density and disrupted bone microarchitecture, which increases bone fragility and susceptibility to fractures (Anonymous, 1993; Kanis et al., 1994). Osteoporosis develops and progresses silently, typically becoming apparent only after the occurrence of fractures, which are often the first reason for clinical evaluation, laboratory investigation, and patient awareness (Slavkin, 1996; Hightower, 2000). This absence of symptoms contributes to a general complacency and is one of the reasons for the low awareness within the general population regarding primary prevention of the disease (Levine, 2011). Statistical data on the incidence of osteoporosis, the frequency of osteoporotic fractures, and their resulting morbidity and mortality confirm the extent of the issue and the need for preventive measures well before the onset of the disease and its complications, particularly among men (Hightower, 2000; Johnell & Kanis, 2006; Singer et al., 1998; Morin et al., 2011; Barrett-Connor et al., 2010; Freitas et al., 2008; Paspati et al., 1998; Sterling, 2011). Additionally, the widespread belief that osteoporosis is a 'women's disease' exposes men to increased risk, as prevention is neglected, diagnosis is delayed or missed entirely, and treatment is often inadequate (Madeo et al., 2007; Kiebzak et al., 2002).

Knowledge of osteoporosis, both among healthy young individuals and older adults, remains lower than expected (Spencer, 2007; Gaines, Narrett, & Parrish, 2010; Ford, Bass, & Keathley, 2007; Edwards & Fraser, 1997; Fok, Leung, & Lee, 2008; Kutsal et al., 2005; Johnson et al., 2008; Anderson, Chad, & Spink, 2005). Most studies of this nature focus on women, while those involving mixed populations conclude that men generally believe the disease does not concern them, and they appear to lack sufficient knowledge and awareness. Studies examining knowledge of osteoporosis exclusively in male populations reveal a lack of awareness or limited understanding of the disease and its consequences, highlighting the need for the development and implementation of health education programs (Gaines & Marx, 2011). In the Greek literature, such studies are extremely limited and mainly refer to female populations (Alexandraki et al., 2008).

Human behavior related to health or illness can be interpreted—and even predicted—through neuropsychological, cognitive, psychoanalytic, and behavioral theories. These theories are based on organic processes or disorders, awareness and learning mechanisms, and subconscious responses (National Cancer Institute, 2005; Gras et al., 1997). Each theory consists of a set of concepts, definitions, and propositions that explain or predict phenomena by describing the relationships between examined variables (National Cancer Institute, 2005). Social cognitive behavior models assume that stimuli are cognitively processed before a protective or health-risk behavior is expressed. Cognitive factors serve as a convergence point for external influences and mediate between stimuli and final behavior. They also represent the most modifiable parameter for influencing behavior (Koulierakis, 2000; National Cancer Institute, 2005).

In this context, the interpretation of men's low levels of osteoporosis knowledge and the attempt to challenge the prevailing notion that it is a "women's disease" can be supported by applying social cognitive behavior models.

Objective of the Chapter: The objective of this chapter is to assess the level of knowledge and awareness of Greek men regarding osteoporosis and to explore their beliefs about the disease. The following research questions were formulated to guide the study:

- What percentage of Greek men have received at least some information about osteoporosis, and what is their knowledge level?
- What proportion of men exhibit risk factors for developing the disease?
- What are men's beliefs about the severity of osteoporosis and its complications?
- To what extent do they believe they are at risk of developing the disease?
- What do they know about preventive measures, their effectiveness, and their ability to implement them?
- Which parameters of the Health Belief Model can and should be modified to change Greek men's attitudes toward osteoporosis?

Purpose of the Chapter: The chapter aims to examine the degree to which the male population in Greece is informed about osteoporosis and to highlight any knowledge gaps. Simultaneously, it records men's beliefs about the disease, as interpreted through the Health Belief Model.

1.1 Significance of the Study

- It collects, analyzes, and interprets data on osteoporosis in a population group that has not previously been studied in Greece, and only to a limited extent internationally.
- It identifies specific educational and informational needs regarding osteoporosis within the male population.
- It supports the design, organization, and implementation of public health and education programs.

- It increases awareness among healthcare professionals and public health practitioners, guiding them in tailoring interventions to the identified needs of men concerning osteoporosis prevention.

2. THEORETICAL FRAMWORK

2.1. Bone Composition

The human skeleton provides structure and shape to the body, plays a central role in movement, contributes to the protection of internal organs, and facilitates the storage and transport of minerals—primarily calcium and phosphorus. The red bone marrow carries out hematopoietic functions by producing the formed elements of blood.

Bone tissue is composed of an organic and an inorganic component. This composition accounts for both the elasticity and strength of bones. The organic part constitutes approximately 30% of the bone mass and is composed of cells and extracellular matrix. The matrix consists of a ground substance and inorganic salts, with 95% of the ground substance composed of collagen.

The cellular population of bone includes osteoblasts, osteoclasts, and osteocytes. Osteoblasts are cells that produce the ground substance, into which osteocytes will later embed. They also secrete regulatory substances important for various bone functions. Osteoblasts play a key role in calcium homeostasis and in regulating the activity of osteoclasts. Their function is mainly regulated by hormonal factors, and their numbers decline with age.

Osteoclasts are primarily responsible for the resorption and removal of damaged bone tissue. They also appear to participate in immune-type responses and self-regulation during inflammatory musculoskeletal diseases, such as rheumatoid arthritis. Their activity is mainly regulated by hormones (e.g., parathyroid hormone, calcitonin), vitamin D, and cytokines, especially interleukins 1, 6, and 11. Osteoclasts function in close connection with osteoblasts through a receptor system on their surface known as the RANK/RANKL/OPG system (Receptor Activator of Nuclear Factor-kB / its Ligand / Osteoprotegerin).

Osteocytes are essentially former osteoblasts that, after completing their role in producing the ground substance, become trapped within it following mineral deposition. Osteocytes contribute to calcium homeostasis but have a limited role in bone remodeling. They maintain functional connections with each other and with osteoblasts to influence osteoblast activity depending on mechanical stimuli or bone injury.

The inorganic part of bone is mainly composed of hydroxyapatite crystals, which are in fact calcium phosphate salts [$Ca_{10}(PO_4)_6(OH)_2$]. Besides calcium salts, bone also contains salts of magnesium, sodium, and potassium, though these do not form a structured crystalline network like hydroxyapatite. These inorganic salts are tightly bound to the collagen matrix, which is why a reduction in their overall quantity is always accompanied by resorption of the organic matrix.

2.2 Bone Remodeling

Unlike other components of the musculoskeletal system—such as articular cartilage and tendons—bones undergo a continuous process of renewal and remodeling. This process involves the removal of damaged or stressed bone tissue and its replacement with new bone. Through bone remodeling, the following are achieved:

- The ongoing replacement of stressed or fatigued bone, reducing the risk of fractures due to repetitive mechanical loading on the skeleton
- The maintenance of bone shape during skeletal development in childhood and adolescence
- The alignment of trabeculae along lines of mechanical stress
- The correction of angular deformities, under certain conditions, in cases of fractures
- The release of calcium into the bloodstream as part of the body's mechanism for maintaining calcium homeostasis

2.3 Definition of Osteoporosis

Osteoporosis is a systemic skeletal disorder characterized by reduced bone mass and disruption of the microarchitectural structure of bone tissue. These changes result in decreased bone strength and resilience, making bones more susceptible to fractures.

Osteoporosis arises from a dysfunction in the coordinated activity of osteoblasts and osteoclasts, leading to increased bone resorption, decreased bone formation, or a combination of both.

2.4 Pathophysiology of Osteoporosis in Men

Bone mass loss in men is an inevitable consequence of aging, just as it is in women. However, in men, the rate of loss typically accelerates after the age of 70. Osteoporosis in men can also be associated with diseases or medications that disrupt bone metabolism, in which case it is classified as secondary osteoporosis. Approximately 85% of secondary osteoporosis cases in men are attributed to corticosteroid use, excessive alcohol consumption, and hypogonadism. The causes of secondary osteoporosis in men are outlined in Table 1.

The skeletal changes that occur with aging significantly determine fracture risk, based on the following triad:

- Failure to achieve peak bone mass
- Excessive bone resorption, leading to reduced bone mass and disruption of the bone's microarchitecture
- Inadequate bone formation mechanisms that are unable to compensate for the resorption

2.5 Risk Factors for Osteoporosis

Risk factors for osteoporosis share a common denominator: the disruption of the balanced function between osteoblasts and osteoclasts. These factors can be viewed through the lens of the two main mechanisms previously mentioned—excessive bone resorption and/or inadequate bone formation.

The most important and common risk factors for osteoporosis include[1]:

- Heredity (genetic predisposition)
- Race
- Sex
- Loss of the protective role of estrogens after menopause in women and hypogonadism in men
- Inadequate intake of calcium and vitamin D
- Low physical activity
- Smoking
- Constitution (body type)
- Excessive alcohol consumption
- Excessive consumption of coffee and tea
- High protein intake
- Use of medications (e.g., corticosteroids)

Table 1. Causes of secondary osteoporosis.

Nutritional causes	**Endocrine causes**
Malabsorption syndromes Malnutrition Post-gastrectomy Cirrhosis Scurvy	Hypogonadism[2] Hyperparathyroidism Hyperthyroidism Cushing's syndrome Diabetes mellitus
Inflammatory diseases	**Malignant diseases**
Rheumatic diseases Ankylosing spondylitis Tuberculosis	Carcinomatosis Multiple myeloma Leukemia
Medications	**Various / Other Factors**
Corticosteroids *Alcohol* Antiepileptic drugs Heparin Immunosuppressants Non-steroidal anti-inflammatory drugs Thyroxine	Smoking Chronic Obstructive Pulmonary Disease Chronic Kidney Disease Cystic fibrosis Sickle cell anemia Neuromuscular disorders Osteogenesis imperfecta

3. RESEARCH METHODOLOGY

3.1. Sample and Sampling Procedure

The study was conducted at the 424 General Military Training Hospital in Thessaloniki, following approval by the hospital's scientific ethics committee. The population from which the sample was drawn consisted of individuals attending the hospital's Orthopedic Outpatient Clinics. Data collection took place every second working day, to ensure representation of patients from both orthopedic departments. Therefore, systematic sampling was employed in terms of timing, and convenience sampling in terms of the study population.

Participation was voluntary, and informed written consent was obtained after participants were briefed on the purpose and procedures of the study.

A total of 330 men aged 18 to 65 years participated in the study.

Exclusion criteria included:
Prior studies in health sciences
Employment in a healthcare profession

Although the study was conducted in a military hospital, the expected participation of conscripted soldiers and civilian personnel was deemed sufficient to give the sample the characteristics of a general male population rather than a strictly military one.

The mean age of participants was 28.33 years (SD = 8.18), and the median age was 26 years. The Kolmogorov–Smirnov test indicated that the age variable did not follow a normal distribution, a finding attributed to the underrepresentation of participants over the age of 40, who made up only 8% of the sample.

The Body Mass Index (BMI) was calculated by dividing weight by the square of height (measured in kilograms and centimeters, respectively, in accordance with SI units). BMI categories were determined based on the World Health Organization (WHO) criteria, as presented in Table 2

Table 2. Body type categories based on body mass index (BMI)

BMI Range (kg/m^2)	Weight Category
<18.5	Underweight
18.5-24.9	Normal weight
25.0-29.9	Overweight
30.0-34.9	Obesity Class I
35.0-39.9	Obesity Class II
>40.0	Obesity Class III

The distribution of participants across BMI categories, according to WHO criteria, is presented in Table 3. Out of the 330 respondents, 3 individuals (0.9%) did not provide sufficient information for BMI calculation. Among those who provided complete data, 43.7% had a normal weight, while 45.9% were overweight.

Due to the large number of different responses regarding participants' place of residence, the locations were grouped into the following categories:

- Residents living abroad
- Residents of urban centers with a population greater than 100,000
- Residents of residential areas with a population less than 100,000

The distribution of respondents by place of residence is shown in Table 4.

Table 3. Distribution of the statistical sample by body mass index (BMI)

Weight Category	Frequency (N)	Percent (%)	Valid Percent (%)
Underweight	1	0.3	0.3
Normal weight	143	43.3	43.7
Overweight	150	45.5	45.9
Obesity Class I	29	8.8	8.9
Obesity Class II	3	0.9	0.9
Obesity Class III	1	0.3	0.3
Total	327	99.1	100.0
Missing System	3	0.9	
Total	330	100.0	

Table 4. Classification of study population by place of residence

Place of Residence	Frequency (N)	Percent (%)	Valid Percent (%)
Residents living abroad	12	3.6	4.4
Residents of urban areas with >100,000 inhabitants	142	43.0	52.2
Residents of residential areas with <100,000 inhabitants	118	35.8	43.4
Total	250	75.8	100.0
Missing System	80	24.2	
Total	330	100.0	

Occupational diversity was also observed among the respondents, with a high proportion of permanent military personnel included in the sample. Personnel serving in the armed forces and security bodies are, according to regulations, subject to regular clinical and laboratory health screening, a factor that may positively influence their awareness and knowledge on health issues, including osteoporosis. These considerations led to the categorization of the sample into the following groups:

- Employees in security services (military personnel, police officers, firefighters)
- Other professions, with proportions presented in Table 5.

Table 5. Classification of participants based on occupational status

Occupational Status	Frequency (N)	Percent (%)	Valid Percent (%)
Employees in security services	113	34.2	45.2
Other professions	137	41,5	54.8
Total	250	75.8	100.0
Missing System	80	24.2	
Total	330	100.0	

Educational level is also considered a factor that may influence both the level of knowledge and, potentially, attitudes toward health-related issues, thereby affecting the adoption of preventive behaviors. The overwhelming majority (93.6%) reported having completed more than the mandatory nine years of basic education. The educational composition of the sample is presented in detail in Table 6.

Table 6. Composition of the sample according to educational level

Occupational Status	Frequency (N)	Percent (%)	Valid Percent (%)
Higher education	148	44.8	44.8
Lyceum	161	48.8	54.8
High school	11	3.2	
Elementary education	7	2.1	
Total	327	100	100.0
Missing System	3	0.9	
Total	330	100.0	

The association between marital status and the factors of the OHBS (Osteoporosis Health Belief Scale), as well as the total scores on the FOOQ (Facts on Osteoporosis Quiz) and MOKQ (Male Osteoporosis Knowledge Questionnaire), was only statisti-cally viable for single and married participants (75% and 23.1%, respectively), as the proportion of divorced (1.5%) and widowed (0.3%) participants was too small to yield statistically reliable results. The marital status of respondents is shown in Table 7.

Table 7. Distribution of respondents according to marital status

Occupational Status	Frequency (N)	Percent (%)	Valid Percent (%)
Single	243	73.6	75.0
Married	75	22.7	23.1
Divorced	5	1.5	1.5
Widowed	1	0.3	0.3
Total	324	98.2	100.0
Missing System	6	1.8	
Total	330	100.0	

Alcohol abuse and smoking are well-documented risk factors for the development of osteoporosis. Out of the 330 participants, 316 (95.8%) responded to the question: "Do you smoke? If yes, how many cigarettes do you smoke per day?". Among them, 123 (38.92%) identified themselves as smokers. Of the smokers, 22% reported smoking 1–10 cigarettes per day, 56% reported smoking 10–20 cigarettes per day, and 22% stated that they smoke more than 20 cigarettes daily.

Tables 8. Percentages of participants informed about, examined for, and diagnosed with osteoporosis

Table 8a. Have you ever been informed about osteoporosis?

Diagnosis	Frequency (N)	Percent (%)	Valid Percent (%)
Yes	43	13.0	13.2
No	283	85.8	86.8
Total	326	98.8	100.0
Missing System	4	1.2	
Total	330	100.0	

Table 8b. Have you ever had your bone mass measured?

Diagnosis	Frequency (N)	Percent (%)	Valid Percent (%)
Yes	6	1.8	1.8
No	320	97	98.2
Total	326	98.8	100.0
Missing System	4	2.2	
Total	330	100.0	

Table 8c. Have you ever been diagnosed with osteoporosis?

Diagnosis	Frequency (N)	Percent (%)	Valid Percent (%)
Yes	4	1.2	1.2
No	325	98.5	98.8
Total	329	99.7	100.0
Missing System	1	0.3	
Total	330	100.0	

Regarding the question on alcohol consumption, 303 participants provided a response. Of these, 95% reported consuming fewer than 6 units of alcohol per week, while 42.2% stated that they consume alcohol occasionally and in very small quantities. In response to the question about whether they watch their diet, 55.9% of the respondents answered affirmatively, while 194 (58.8%) stated that they exercise regularly (Table 9).

Tables 9. Percentages of respondents who watch their diet and exercise regularly

Table 9a. Do you watch your diet?

Diagnosis	Frequency (N)	Percent (%)	Valid Percent (%)
Yes	184	55.8	55.9
No	145	43.9	98.8
Total	329	99.7	100.0
Missing System	1	0.3	
Total	330	100.0	

Table 9b. Do you exercise regularly?

Diagnosis	Frequency (N)	Percent (%)	Valid Percent (%)
Yes	194	58.8	55.9
No	136	41.2	98.8
Total	330	100.0	100.0
Missing System	0	0	
Total	330	100.0	

3.2. Statistical Methodology

The Implicative Analysis of Régis Gras: (Gras et al., 1997) emphasizes the need for a data analysis method that constitutes "an accurate mechanism for the collection and processing of data suitable for supporting or refuting hypotheses and for drawing conclusions." A characteristic example is a method that hierarchizes and connects factors. The method proposed by (Gras et al., 1997) is considered suitable when the goal is to identify:

(a) the main discriminant factors within a population through variables,
(b) a partitioning of the variables,
(c) a typology or classification – i.e., a hierarchical classification of similarities, and
(d) implications among variables or classes of variables – in the form of an implication tree or a hierarchical implication structure.

The implicative method allows for tracking the development of a capability and identifying invariants or stable elements in the thinking of social subjects. It does not imply causal relationships, but rather serves as a quality indicator, allowing one to assert that success in one task implies success in another task to which it is linked. Conversely, failure in one item implies failure in another linked item (Fotiadis & Anastasiadou, 2019; Anastasiadou et al., 2010; Anastasiadou, 2019; Anastasiadou, 2020).

The approach includes three main graphical representations:

(a) the Implicative Graph,
(b) the Similarity Diagram, and
(c) the Hierarchy Diagram.

In the Implicative Graph, various implication relationships among variables are illustrated. These implications may be significant at the 99% confidence level (bold arrow) or the 95% level (thin arrow).

The Similarity Diagram illustrates similarity relationships among different items. Items for which social subjects behave similarly when responding are grouped together. Horizontal black lines represent similarity at the 99% confidence level.

Finally, the Hierarchy Diagram displays implication relationships among all variables, in order of priority. Implications represented by bold red lines indicate significance at the 99% level.

3.3. Research Instruments

3.3.1 Facts on Osteoporosis Quiz (FOOQ)

The Facts on Osteoporosis Quiz (FOOQ) is the only research instrument for measuring knowledge about the disease that is based on a well-defined conceptual framework (Appendix G3). It is grounded in Orem's Self-Care Theory, which posits that an individual's knowledge about health problems is a prerequisite for adopting self-care and preventive behaviors.

The revised version of the FOOQ consists of 20 statements/questions designed to assess knowledge about bone development, risk factors for osteoporosis, and its prevention and treatment. Each item offers three possible responses: "true", "false", and "don't know". A correct answer is scored as 1 point, yielding a possible total score ranging from 0 to 20. The "don't know" option is intended to differentiate ignorance from misinformation or random guessing and is scored 0, as are incorrect answers. The revised FOOQ has a validity coefficient of 0.87 and a Cronbach's alpha coefficient of 0.76, indicating good internal consistency.

3.3.2 Male Osteoporosis Knowledge Quiz (MOKQ)

The Male Osteoporosis Knowledge Quiz (MOKQ) is essentially a set of six questions specifically addressing male osteoporosis, and it is used as a supplement to the Facts on Osteoporosis Quiz (FOOQ) (Appendix G4) [152]. Each item offers the response options "true," "false," and "don't know." A correct answer is awarded 1 point, while incorrect answers and "don't know" responses are scored 0.

The difficulty index of the MOKQ was found to be high by its original developers only for Question 1, which states: *"Men with small body frames and low weight are more likely to have osteoporosis than men with large body frames and higher weight"*, with fewer than 20% of respondents answering correctly.

The convergent validity of the MOKQ with the FOOQ was assessed using the Pearson correlation coefficient, which was calculated at $r = 0.76$ ($p < 0.001$).

Cronbach's alpha reliability coefficient was 0.72 for the MOKQ alone, and 0.90 for the combined set of 26 items derived from integrating the MOKQ with the FOOQ.

4. RESULTS

4.1. Facts on Osteoporosis Quiz (FOOQ)

The Similarity Diagram: The similarity diagram (Figure 1) illustrates the clustering of projects based on respondents' answers to the FOOQf questionnaire. The observed similarities are statistically significant at the 99% confidence level

Figure 1. Similarity diagram FOOQ

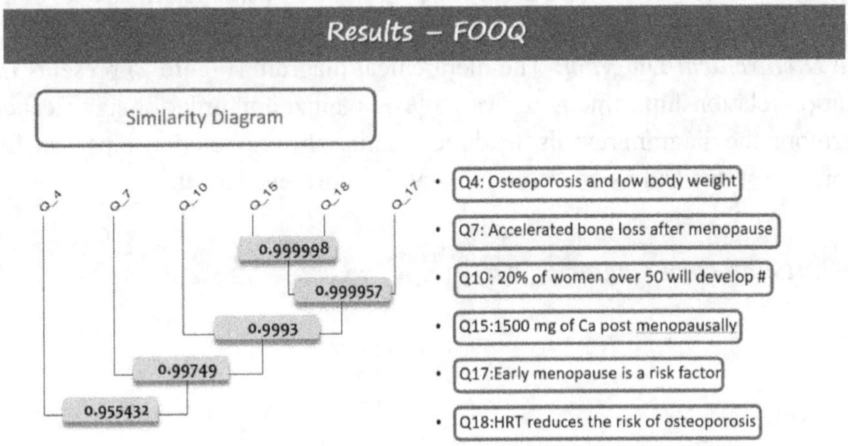

The Similarity Diagram: In the similarity diagram (Figure 1), a distinct cluster of similarity (Group A) is identified. This similarity cluster (Group A) refers to similarity relationships among variables Q_4, Q_7, Q_10, Q_15, Q_18, and Q_17 (similarity: 0.955432). The strongest similarity within this first cluster, Group A, is observed between variables Q_15 and Q_18 (similarity: 0.999998), which is nearly perfect. These items refer to the knowledge that postmenopausal women who do not take estrogen require approximately 1500 mg of calcium daily (e.g., five glasses of milk) — question Q_15 — and the knowledge that hormone replacement therapy after menopause can slow bone loss — question Q_18.

Specifically, the similarity between Q_15 and Q_18 indicates a consistent pattern among respondents in how they perceive and approach the role of estrogen and hormones in relation to the physiological impacts of menopause. Despite this pattern, the success rates for these questions remain low, at 21.8% and 8.2%, respectively. Similarly high similarity is found between variables Q_15, Q_18, and Q_17 (similarity: 0.999957). Another strong similarity is observed across variables Q_7,

Q_10, Q_15, Q_18, and Q_17 (similarity: 0.99749), reaffirming that respondents interpret in a consistent manner the concepts of early menopause as a risk factor for osteoporosis, the preventive role of interventions in predicting fracture incidence after the age of 50, the replacement of estrogen with approximately 1500 mg of calcium per day for postmenopausal women, and the slowing of bone loss through hormone replacement therapy.

Finally, a significant similarity (similarity: 0.955432) is noted across variables Q_4, Q_7, Q_10, Q_15, Q_18, and Q_17, indicating that respondents associate the question linking low body weight in women with a higher incidence of osteoporosis — in comparison to women with higher body weight — with the broader set of the aforementioned knowledge areas.

The Hierarchical Diagram: The hierarchical diagram (Figure 2) presents the implication relationships among the variables, organized in order of significance. Furthermore, the diagram reveals the directionality of these relationships. The implications are statistically significant at the 99% confidence level.

Figure 2. Hierarchical diagram FOOQ

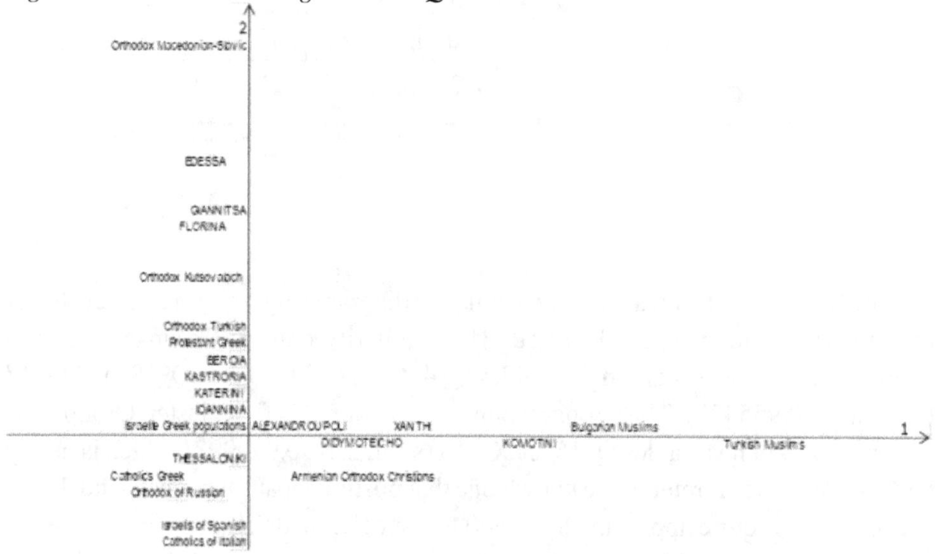

With regard to a single hierarchical group, it involves the items Q_7, Q_4, Q_18, Q_17, Q_15, and Q_10 (cohesion: 0.517). Success in understanding that bone loss naturally accelerates after menopause (Q_7) implies success in recognizing that women with low body weight are more likely to develop osteoporosis than women with higher body weight (Q_4). This, in turn, implies success in understanding that hormone replacement therapy (HRT) after menopause can slow bone loss (Q_18), which subsequently implies understanding that early menopause is a risk factor for osteoporosis (Q_17). This leads to success in questions Q_15 — concerning the need for 1500 mg of calcium per day (e.g., five glasses of milk) for postmenopausal women not taking estrogen — and Q_10, which refers to the role of preventive measures and the statistic that 20% of women over the age of 50 may suffer a fracture due to osteoporosis in the absence of such measures.

The highest cohesion is observed between variables Q_15 and Q_10 (cohesion: 1), where success in Q_15 — understanding the calcium requirements for postmenopausal women not receiving estrogen — implies success in Q_10, which highlights that without preventive measures, 20% of women over 50 will experience a fracture due to osteoporosis. Also noteworthy is the high cohesion among variables Q_17, Q_15, and Q_10 (cohesion: 0.997), as well as among Q_18, Q_17, Q_15, and Q_10 (cohesion: 0.892). Significant cohesion is also found among variables Q_4, Q_18, Q_17, Q_15, and Q_10 (cohesion: 0.769).

The Implication Graph: The implication diagram (Figure 3) depicts the logical implication relationships among the variables, highlighting how the understanding of certain items may lead to the understanding of others.

Figure 3. The implication graph FOOQ

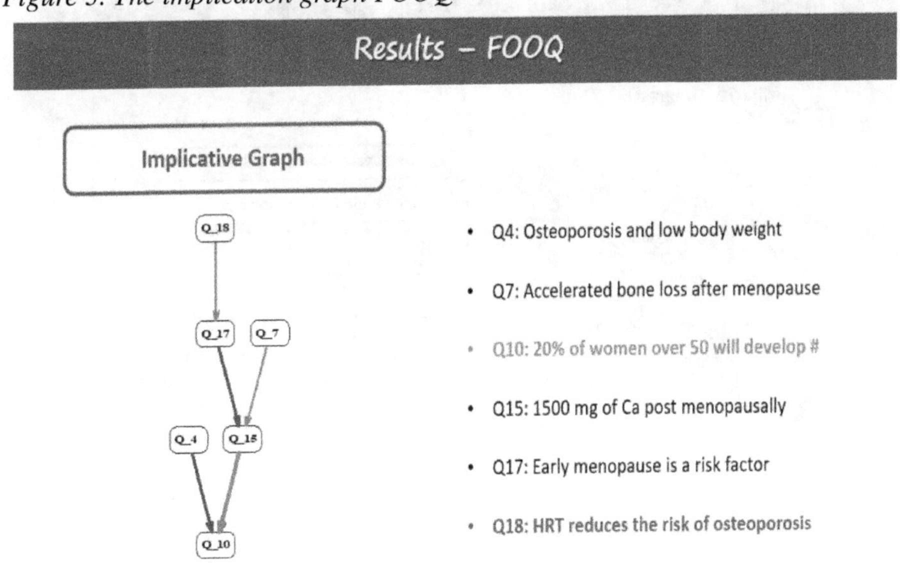

More specifically, in the first implication chain (Q_18 → Q_17, Q_17, Q_7 → Q_15 → Q_10, Q_4 → Q_10), knowledge that hormone replacement therapy after menopause can slow bone loss (Q_18) implies an understanding that early menopause is a risk factor for osteoporosis (Q_17). Furthermore, knowledge of early menopause as a risk factor (Q_17), combined with the understanding that bone loss accelerates naturally after menopause (Q_7), together imply a correct understanding that postmenopausal women who do not take estrogen require approximately 1500 mg of calcium daily (e.g., five glasses of milk) — as reflected in Q_15.

In turn, this leads to the successful recognition that, in the absence of preventive measures, 20% of women over the age of 50 will experience a fracture due to osteoporosis (Q_10). Finally, the implication Q_4 → Q_10 indicates that knowledge of the higher prevalence of osteoporosis among women with low body weight compared to those with higher body weight (Q_4) leads to the understanding that lack of preventive measures results in a 20% fracture incidence among women over 50 (Q_10).

4.2. Male Osteoporosis Knowledge Quiz (MOKQ)

The Similarity Diagram: The similarity diagram (Figure 4) presents the clustering of items based on the respondents' answers to the MOKQ. The observed similarities are statistically significant at the 99% confidence level.

Figure 4. Similarity diagram MOKQ

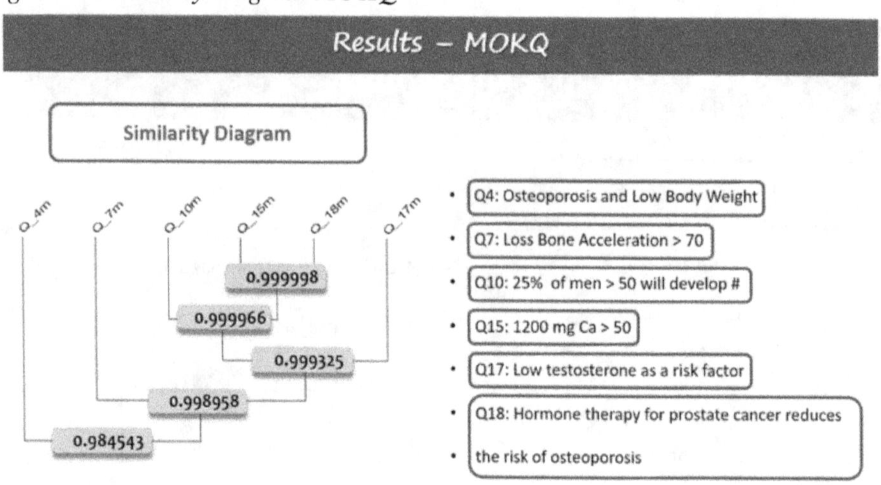

In the similarity diagram (Figure 5), a distinct similarity group (Group A) is identified. Group A refers to similarity relationships among the variables Q_4m, Q_7m, Q_10m, Q_15m, Q_18m, and Q_17m of the MOKQ, with a high overall similarity score (similarity: 0.984543), all relating to knowledge about male osteoporosis.

The strongest similarity within Group A, nearly equal to 1, is observed between Q_15m and Q_18m (similarity: 0.999998), highlighting the respondents' similar understanding of whether hormone therapy for prostate cancer reduces the risk of osteoporosis, and whether men over the age of 50 require approximately 1200 mg of calcium per day.

Equally strong is the similarity among Q_10m, Q_15m, and Q_18m (similarity: 0.999966), indicating a shared pattern in how respondents perceive the relationship between preventive measures and the fact that, without them, 25% of men over 50 may suffer a fracture due to osteoporosis, in relation to the previous two items (Q_15m, Q_18m).

Notable similarity is also found among Q_10m, Q_15m, Q_18m, and Q_17m (similarity: 0.999325), as well as among Q_7m, Q_10m, Q_15m, Q_18m, and Q_17m (similarity: 0.998958), both approaching unity. Similarly, a high degree of similarity exists among Q_4m, Q_7m, Q_10m, Q_15m, Q_18m, and Q_17m (similarity: 0.984543) of MOKQ.

This pattern suggests that respondents tend to respond in similar ways to the following items, Q_4m, related to whether men of smaller body size and lower weight are more likely to develop osteoporosis than men of larger body size and higher weight; Q_7m, related to whether bone loss increases in men after the age of 70; Q_10m, related to whether, without preventive measures, 25% of men over 50 are likely to suffer a fracture due to osteoporosis; Q_15m, related to whether men over 50 need approximately 1200 mg of calcium per day; Q_18m, whether hormone therapy for prostate cancer reduces the risk of osteoporosis; Q_17m, whether low testosterone levels are not a risk factor for osteoporosis.

Therefore, respondents exhibit a similar pattern in their responses to questions concerning: the role of small/large body size and low/high body weight in the development of osteoporosis, the increase in bone loss in men over the age of 70, the influence of the presence or absence of preventive measures on the incidence of fractures in men over the age of 50—estimated at approximately 25%, the required daily calcium intake, the role of hormone therapy for prostate cancer in reducing the risk of osteoporosis, and, finally, the role of low testosterone levels as a potential risk factor for osteoporosis.

Thus, the respondents appear to demonstrate a consistent pattern in how they understand the role of body size and weight in osteoporosis risk, age-related bone loss, fracture risk in the absence of preventive measures, calcium intake needs, the effect of hormone therapy, and low testosterone as a potential risk factor for osteoporosis.

The hierarchical diagram: In the hierarchical diagram (Figure 5) the causal relationships between the variables are presented in order of significance. Furthermore, in the hierarchical diagram we can see the direction of these relationships. The causal relationships are significant at the 99% significance level.

Figure 5. Hierarchical diagram MOOQ

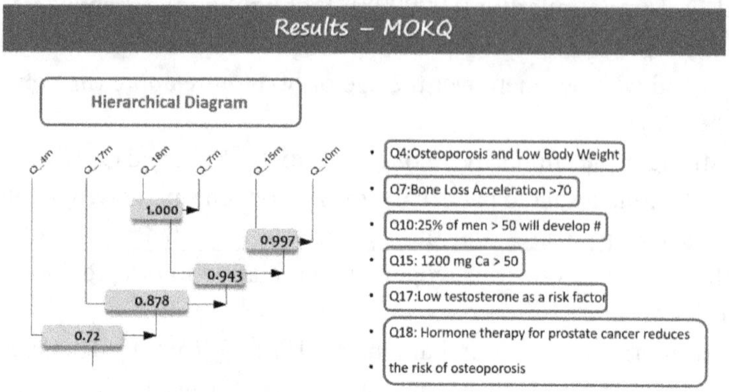

Specifically, the diagram illustrates a large hierarchical group that includes all variables. Perfect cohesion is observed between variables Q_18m and Q_7m (*cohesion: 1*), indicating that success in question Q_18m — which examines whether hormone therapy for prostate cancer reduces the risk of osteoporosis — implies success in question Q_7m, which addresses whether bone loss increases in men after the age of 70. Noteworthy cohesion (*cohesion: 0.997*) is also found between Q_15m and Q_10m, where success in Q_15m — assessing whether men over 50 require approximately 1200 mg of calcium daily — implies success in Q_10m, which investigates whether, without preventive measures, 25% of men over 50 will experience a fracture due to osteoporosis. Strong cohesion is also evident among Q_18m, Q_7m, Q_15m, and Q_10m (*cohesion: 0.943*), where success in Q_18m and Q_7m implies success in Q_15m and Q_10m. Moreover, success in Q_17m — regarding whether low testosterone levels are a risk factor for osteoporosis — implies success across the set of Q_18m, Q_7m, Q_15m, and Q_10m.

Finally, cohesion between the variables Q_4m, Q_18m, Q_7m, Q_15m, and Q_10m is around 72% (*cohesion: 0.72*). Success in Q_4m, which investigates whether men of smaller body size and lower weight are more likely to have osteoporosis than those of larger size and higher weight, implies success in Q_17m (concerning the role of low testosterone as a risk factor) and, consequently, success in Q_18m, Q_7m, Q_15m, and Q_10m.

The implication diagram: The implication diagram shows the inferential relationships between the variables (Figure 6).

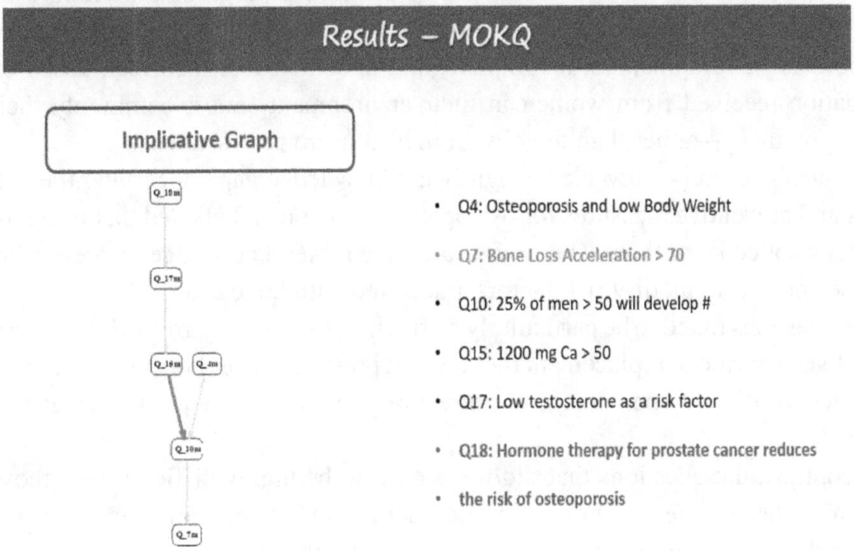

In the inferential chain Q_18m → Q_17m → Q_15m → Q_10m → Q_7m, knowledge that hormone therapy for prostate cancer does not reduce the risk of osteoporosis (Q_18m) implies awareness that low testosterone levels are a risk factor for osteoporosis (Q_17m). This, in turn, implies correct knowledge that men over the age of 50 require approximately 1200 mg of calcium per day (Q_15m), which leads to the understanding that, without preventive measures, 25% of men over the age of 50 will suffer a fracture due to osteoporosis (Q_10m). Subsequently, this knowledge implies a correct response to Q_7m, which explores whether bone loss increases in men over the age of 70.

Additionally, in the inferential sequence Q_4m → Q_10m → Q_7m, a correct response to Q_4m—which investigates whether men with smaller body size and lower weight are more likely to develop osteoporosis than those with larger body size and higher weight—implies correct knowledge of Q_10m, regarding the 25% fracture risk in men over 50 without preventive action, which in turn implies a correct response to Q_7m about age-related bone loss beyond 70.

Notably, Q_18m, which examines whether hormone therapy for prostate cancer does not reduce the risk of osteoporosis, appears to be the most difficult question, while Q_7m, assessing whether bone loss increases in men after age 70, seems to be the easiest for respondents to answer correctly.

CONCLUSIONS

The central conclusion of the study, which confirmed the authors' initial research hypothesis, was the limited awareness among men regarding osteoporosis, many of whom characterize it as a “women's disease.” Most men's awareness of osteoporosis was fragmented, incomplete, and in some cases incorrect, based on information received from women in their environment—most commonly their wives or partners—rather than directly from healthcare professionals.

The men's responses revealed a significant knowledge gap concerning the risk factors and preventive measures for osteoporosis. Most men believed that osteoporosis does not concern them, and as a result, they lacked knowledge of prevention strategies or how to modify risk factors associated with bone health.

The questions found to be particularly difficult were those regarding the relationship between hormone replacement therapy and prostate cancer hormone treatment with osteoporosis. The percentage of correct responses to these questions was less than 10%.

In contrast, the questions that did not appear to be highly difficult were those concerning the occurrence of osteoporotic fractures in 20% of women over the age of 50, and the increase in bone loss rate in men after the age of 70.

REFERENCES

Anastasiadis, L., Anastasiadou, S.D., Vandikas, I., & Angeletos, T. (2011). Implicative Statistical Analysis and Principal Components Analysis in Recording Students' Attitudes to Electronics and Electrical Construction Subjects. The International Journal of Technology, Knowledge, and Society: Annual Review, 7, 63-78.

Anastasiadou, S. (2019). Comparison of contemporary advanced statistical methods regarding construct validity evaluation of TEIque-SF instrument: Statistical Implicative Analysis vs. Principal Components. In the proceedings of 9ème Colloque International sur Analyse Statistique Implicative (ASI 10). (pp. 148-163). Belfort, France.

Anastasiadou, S. (2020). Pre-service teachers' performance on the learning of probability distributions and the role of projects: A multilevel statistical analysis. In J-C Regnier, F. Spagnolo, B. Di Paola, & R. Gras (Eds), Proceedings of the 5th International Conference on Implicative Statistic Analysis (pp. 73-84). Palermo, Italy.

Alexandraki, K. I., Syriou, V., Ziakas, P. D., Apostolopoulos, N. V., Alexandrakis, A. I., Piperi, C., Kavoulaki, E., Myriokefalitakis, I., Korres, G., & Diamanti-Kandarakis, E. (2008). The knowledge of osteoporosis risk factors in a Greek female population. *Maturitas*, *59*(1), 38–45. DOI: 10.1016/j.maturitas.2007.10.008 PMID: 18079073

Anderson, K. D., Chad, K. E., & Spink, K. S. (2005). Osteoporosis knowledge, beliefs, and practices among adolescent females. *The Journal of Adolescent Health*, *36*(4), 305–312. DOI: 10.1016/j.jadohealth.2004.03.007 PMID: 15780785

Anonymous, . (1993). Consensus development conference. Diagnosis, prophylaxis and treatment of osteoporosis. *The American Journal of Medicine*, *94*(6), 646–650. DOI: 10.1016/0002-9343(93)90218-E PMID: 8506892

Barrett-Connor, E., Nielson, C. M., Orwoll, E., Bauer, D. C., & Cauley, J. A. (2010). Epidemiology of rib fractures in older men: Osteoporotic Fractures in Men (MrOS) prospective cohort study. [Electronic Version]. *BMJ (Clinical Research Ed.)*, *340*(mar15 1), c1069. DOI: 10.1136/bmj.c1069 PMID: 20231246

Berger, C., Goltzman, D., Langsetmo, L., Joseph, L., Jackson, S., Kreiger, N., Tenenhouse, A., Davison, K. S., Josse, R. G., Prior, J. C., & Hanley, D. A. (2010). Peak bone mass from longitudinal data: Implications for the prevalence, pathophysiology, and diagnosis of osteoporosis. *Journal of Bone and Mineral Research : the Official Journal of the American Society for Bone and Mineral Research*, *25*(9), 1948–1957. DOI: 10.1002/jbmr.95 PMID: 20499378

Bonjour JP, Chevalley T, Ferrari S, Rizzoli R. (2009). The importance and relevance of peak bone mass in the prevalence of osteoporosis. Salud Publica Mex. 51(1):Suppl 5 17.

Clarke, B. L., & Khosla, S. (2010). Physiology of bone loss. *Radiologic Clinics of North America, 48*(3), 483–495. DOI: 10.1016/j.rcl.2010.02.014 PMID: 20609887

Edwards, L., & Fraser, M. (1997). How do we increase awareness of osteoporosis? *Bailliere's Clinical Rheumatology, 11*(3), 631–644. DOI: 10.1016/S0950-3579(97)80024-6 PMID: 9367041

Fok, M. W., Leung, H. B., & Lee, W. M. (2008). Osteoporosis: Public awareness, commitment, and perspectives. *Hong Kong Medical Journal, 14*(3), 203–208. PMID: 18525089

Ford, M. A., Bass, M. A., & Keathley, R. (2007). Osteoporosis knowledge and attitudes: A cross sectional study among college-age students. *Journal of American College Health, 56*(1), 43–47. DOI: 10.3200/JACH.56.1.43-48 PMID: 17711825

Fotiadis, Th. & Anastasiadou S. (2029). Contemporary advanced statistical methods for the science of marketing: Implicative Statistical Analysis vs Principal Components Analysis. International Journal of Entrepreneurship and Innovative Competitiveness – IJEIC, (2019), Vol 1, Issue 1

Freitas, S. S., Barrett-Connor, E., Ensrud, K. E., Fink, H. A., Bauer, D. C., Cawthon, P. M., Lambert, L. C., & Orwoll, E. S. (2008). Rate and circumstances of clinical vertebral fractures in older men. *Osteoporosis International, 19*(5), 615–623. DOI: 10.1007/s00198-007-0510-1 PMID: 18038109

Gaines, J. M., & Marx, K. A. (2011). Older men's knowledge about osteoporosis and educational interventions to increase osteoporosis knowledge in older men: A systematic review. *Maturitas, 68*(1), 5–12. DOI: 10.1016/j.maturitas.2010.08.013 PMID: 20950969

Gaines, J. M., Narrett, M., & Parrish, J. M. (2010). The effect of the addition of osteoporosis education to a bone health screening program for older adults. *Geriatric Nursing (New York, N.Y.), 31*(5), 348–360. DOI: 10.1016/j.gerinurse.2010.04.011 PMID: 20933147

Gras, R., P. Peter, H. Briand, & J. Philippé. (1997). Implicative Statistical Analysis. In C. Hayashi, N. Ohsumi, N. Yajima, Y. Tanaka, H. Bock, Y. Baba (Eds.), Proceedings of the 5th Conference of the International Federation of Classification Societies. (Vol. 2, pp. 412-419). Springer-Verlag: Tokyo, Berlin, Heidelberg, & New York.

Harper, K., & Armelagos, G. (2010). The Changing Disease-Scape in the Third Epidemiological Transition. *International Journal of Environmental Research and Public Health*, *7*(2), 675–697. DOI: 10.3390/ijerph7020675 PMID: 20616997

Hightower, L. (2000). Osteoporosis: Pediatric disease with Geriatric Consequences. *Orthopedic Nursing*, *19*(5), 59–62. DOI: 10.1097/00006416-200019050-00010 PMID: 11153388

Johnell, O., & Kanis, J. A. (2006). An estimate of the worldwide prevalence and disability associated with osteoporotic fractures. *Osteoporosis International*, *17*(12), 1726–1733. DOI: 10.1007/s00198-006-0172-4 PMID: 16983459

Johnson, S. C., McLeod, W., Kennedy, L., & McLeod, K. (2008). Osteoporosis health beliefs among younger and older men and women. *Health Education & Behavior*, *35*(5), 721–733. DOI: 10.1177/1090198107301331 PMID: 17602100

Kanis, J. A., Melton, L. J., Christiansen, C., Johnston, C. C., & Khaltaev, N. (1994). The Diagnosis of Osteoporosis. *Journal of Bone and Mineral Research : the Official Journal of the American Society for Bone and Mineral Research*, *9*(8), 1137–1141. DOI: 10.1002/jbmr.5650090802 PMID: 7976495

Kiebzak, G. M., Beinart, G. A., Perser, K., Ambrose, C. G., Siff, S. J., & Heggeness, M. H. (2002). Undertreatment of Osteoporosis in Men with Hip Fracture. *Archives of Internal Medicine*, *162*(19), 2217–2222. DOI: 10.1001/archinte.162.19.2217 PMID: 12390065

Koulierakis, G. (2000). Socio-psychological models of behavior in relation to health and illness. In Koulierakis, G., Metallinou, O., & Pantzou, P. (Eds.), *Sociological and psychological approach of hospitals/health services* (pp. 15–82). Hellenic Open University.

Kutsal, Y. G., Atalay, A., Arslan, S., Başaran, A., Cantürk, F., Cindaş, A., Eryavuz, M., İrdesel, J., Karadavut, K. İ., Kirazlı, Y., Sindel, D., Şenel, K., Güler-Uysal, F., & Yıldırım, K. (2005). Awareness of osteoporotic patients. *Osteoporosis International*, *16*(2), 128–133. DOI: 10.1007/s00198-004-1678-2 PMID: 15197545

Levine JP. Identification, Diagnosis, and Prevention of Osteoporosis. Am J Manag Care 201; 17(6):Suppl 170-176.

Madeo, B., Zirilli, L., Caffagni, G., Diazzi, C., Sanguanini, A., & Pignatti, E.. (2007). The osteoporotic male: Overlooked and undermanaged? *Clinical Interventions in Aging*, *2*(3), 305–312. PMID: 18044181

Matkovic, V., Jelic, T., Wardlaw, G. M., Ilich, J. Z., Goel, P. K., Wright, J. K., Andon, M. B., Smith, K. T., & Heaney, R. P. (1994). Timing of peak bone mass in Caucasian females and its implication for the prevention of osteoporosis. Inference from a cross-sectional model. *The Journal of Clinical Investigation*, *93*(2), 799–808. DOI: 10.1172/JCI117034 PMID: 8113412

Morin, S., Lix, L. M., Azimaee, M., Metge, C., Caetano, P., & Leslie, W. D. (2011). Mortality rates after incident non-traumatic fractures in older men and women. *Osteoporosis International*, *22*(9), 2439–2448. DOI: 10.1007/s00198-010-1480-2 PMID: 21161507

National Cancer Institute. (2005). Theory at a glance. A guide for health promotion practice (2nd edition). Washington, DC.

Omran, A. (2005). The Epidemiologic Transition: A Theory of the Epidemiology of Population Change. *The Milbank Quarterly*, *83*(4), 731–757. DOI: 10.1111/j.1468-0009.2005.00398.x PMID: 16279965

Paspati, I., Galanos, A., & Lyritis, G. P. (1998). Hip Fracture Epidemiology in Greece during 1977 1992. *Calcified Tissue International*, *62*(6), 542–547. DOI: 10.1007/s002239900476 PMID: 9576984

Singer, B. R., McLauchlan, G. J., Robinson, C. M., & Christie, J. (1998). Epidemiology of fractures in 15,000 adults: The influence of age and gender. *The Journal of Bone and Joint Surgery. British Volume*, *80*(2), 243–248. DOI: 10.1302/0301-620X.80B2.0800243 PMID: 9546453

Slavkin, H. C. (1996). Notes on a “silent disease”. *The Journal of the American Dental Association*, *127*(6), 801–803. DOI: 10.14219/jada.archive.1996.0318 PMID: 8708284

Spencer, S. J. (2007). Lack of knowledge of osteoporosis: A multi-centre, observational study. *Scottish Medical Journal*, *52*(1), 13–16. DOI: 10.1258/rsmsmj.52.1.13 PMID: 17373418

Sterling, R. S. (2011). Gender and Race/Ethnicity Differences in Hip Fracture Incidence, Morbidity, Mortality, and Function. *Clinical Orthopaedics and Related Research*, *469*(7), 1913–1918. DOI: 10.1007/s11999-010-1736-3 PMID: 21161737

Wren, T. A., Kim, P. S., Janicka, A., Sanchez, M., & Gilsanz, V. (2007). Timing of peak bone mass: Discrepancies between CT and DXA. *The Journal of Clinical Endocrinology and Metabolism*, *92*(3), 938–941. DOI: 10.1210/jc.2006-1570 PMID: 17164299

ENDNOTES

1 Modifiable risk factors are indicated in italics.

2 The use of corticosteroids, excessive alcohol consumption, and hypogonadism are the three most significant and common causes of secondary osteoporosis in men.

Chapter 16
Exploring AI Ethics Through Factorial Correspondence Analysis

Sofia D. Anastasiadou
https://orcid.org/0000-0001-6404-5003
University of Western Macedonia, Greece

ABSTRACT

Artificial intelligence (AI) has made its presence felt in almost all sciences with a multitude of applications. Big data or massive data and AI algorithms make the application possibilities of machine learning and deep learning evident. Ethical considerations focus on issues of privacy, confidentiality, privacy, personal data protection, justice, human rights, integrity, resilience, and robustness of algorithms. The present research approaches, records, analyzes and deepens issues of AI ethics. 239 male and female students participated in the final survey. The processing of the data was carried out with the help of the factorial analysis of correspondences. Two research instruments were used. The first is consisted of the conceptual constructs named "Human Rights", "Autonomy", "Bias", "Justice", "Safety". The second is consisted of the conceptual constructs named "Data Protection", "Privacy". The results of empirical research showed that social subjects are largely aware of the ethical issues raised, such as those of Autonomy, Privacy, Bias, Privacy, Justice, Human Rights, etc.

1. INTRODUCTION

Artificial Intelligence (AI) represents today's digital revolution. Quantum computers, the internet, big data collection, processing, and utilization, artificial

DOI: 10.4018/979-8-3693-9400-7.ch016

intelligence algorithms and quantum algorithms, artificial neural networks (ANN), the Internet of Things (IoT), Health Internet of Things (Health-IoT, H-IoT), and the Internet of Everything (IoE), cloud computing, blockchain technologies, GAFAMI platforms (Google, Amazon, Facebook, Apple, Microsoft), BHATX platforms (Baidu, Huawei, Alibaba, Tencent, Xiaomi), and 5G technology, among others, are shaping new economic, political, strategic, geopolitical, and military dynamics.

The term Artificial Intelligence (AI) is now widely recognized. Ordinary people and scientists from all over the world refer to it on a daily basis. AI applications have become integrated at all levels of daily human life (Sindermann et al., 2022). Applications of AI enable image, sound, voice, face, and video recognition; disease diagnosis (e.g., IBM's Watson Healthcare system, facilitating early diagnosis of various forms of cancer); remote disorder diagnosis (e.g., Morpheo system for remote sleep disorder diagnosis); patient monitoring; analysis of clinical trials; development of clinical decision support and management systems; identification of epidemics and tracking infectious diseases; creation of autonomous vehicles; development of transportation systems; transaction automation; crop monitoring; combating climate change; support for electric power generation plants; renewable energy systems management and energy storage; monitoring and prediction of hazardous weather conditions; analysis of legal documents; development of judicial and administrative decision-making systems; and the development of intelligent weapon systems, missiles, satellites, and drones that, with the assistance of quantum computers and 5G technology, may even become autonomous, among many others. While the benefits of artificial intelligence development are numerous, undeniable too are the negative impacts and risks it poses to human life and rights.

2. PROBLEMATIC OF THE TOPIC

Artificial Intelligence (AI) technologies give rise to ethical considerations, as well as ethical and bioethical dilemmas related to law, human rights, and ethics.

These dilemmas include issues concerning justice, human rights, transparency and governance, accountability, autonomy, human liberties (freedom of expression, freedom of assembly and association), privacy, confidentiality, consent, bias, fairness/equity, security, protection of sensitive personal data, informed consent, and ownership of data—particularly biomedical and medical data. Additionally, ethical and bioethical challenges center around data and metadata ownership, particularly regarding how such data are used, who controls the platforms and algorithms, and who has authority over them. Special attention has also been given to the role of AI in patient tracking during the coronavirus pandemic, aimed at containing virus spread (Kanellos, 2014; Kanellos, 2021, p. 225). Further critical bioethical issues have been raised regarding individual self-determination, autonomy, and consequently human dignity, free will, freedom of assembly and association, privacy, and human rights (Kanellos, 2021, p. 228). Moreover, in the future, significant risks will arise in relation to DNA (Deoxyribonucleic Acid) analysis, the identification of potential diseases throughout the human lifespan, disease prevention, and human enhancement (Ruenda, 2023). This clearly raises issues concerning which social groups will be able to benefit from such services.

Moreover, issues regarding responsibility attribution based on impact require particular study and attention. Artificial intelligence applications are difficult to comprehend, not only by ordinary people but even by distinguished scientists. These applications potentially harbor risks that may not be easily recognized by the average citizen due to the complexity of artificial intelligence systems, algorithm architectures, cloud computing, and the capabilities these systems develop (Panagopoulou-Koutnatzi, 2023, p. 52), particularly when we refer to applications involving biomedical or medical data.

Artificial intelligence may even threaten an individual's free will. This can occur when decisions are made with the help of algorithms through automated processes (Panagopoulou-Koutnatzi, 2023, p. 54). AI can also pose threats to human life, as in the cases of autonomous vehicles and weapon systems. Managing big data through IoT (Internet of Things) and H-IoT (Health-Internet of Things), machine learning, deep learning, and neural networks will bring new dynamics to warfare and terrorist activities, possibly placing civilian populations and others at great risk (Roumeliotis, 2021, pp. 328-333; Girasa, 2020, p. 270). Autonomous vehicles and weapons can make decisions regarding human life without any possibility of human intervention (Roumeliotis, 2021, p. 333). Another potential threat is patient protection (physical, mental, psychological health) and rights in cases of medical errors directly linked to artificial intelligence decisions. Attribution of responsibility is crucial. It remains unclear who bears responsibility if a robotic medical device causes severe harm or even patient death—does liability fall upon the doctor, the hospital, the creator of the robotic device, the maintenance technician, or the manufacturing company? Kanellos characteristically questions this (Kanellos, 2021, p. 225).

3. RESEARCH METHODOLOGY

In this paragraph, the research questions will be presented first, followed by a presentation of the statistical methodology and research instruments. The section will conclude with an evaluation of the reliability and validity of the research instruments.

3.1. Research Questions

The research questions of this study are the following:

RQ1: What is the attitude towards the general use of Artificial Intelligence (AI)?
RQ2: What are the attitudes toward AI and the implications of its development regarding issues related to Human Rights, Autonomy, Bias, Fairness, and Security?
RQ3: What are the attitudes toward AI and the implications of its development regarding issues related to Personal Data Protection and Privacy?
RQ4: Is there a statistically significant correlation among the ethical dimensions arising from the use of AI algorithms and Big Data concerning Privacy, Autonomy, Confidentiality, Bias, Fairness, Security, and Human Rights?
RQ5: Do attitudes towards the use of AI algorithms, Big Data, and ethical issues differ according to the demographic characteristics of the respondents?

3.2. Statistical Methodology

Data analysis was conducted using Descriptive Statistics, Principal Component Analysis (PCA), Confirmatory Factor Analysis (CFA), and Correspondence Factor Analysis (AFC).

Principal Component Analysis (PCA) was employed to reveal the Greek conceptual constructs-dimensions (factors) of the research instruments. Specifically, these instruments include the tool *"Human Rights, Autonomy, Bias, Fairness, Security and AI"* developed by Anastasiadou (2020), and the tool *"Data Protection, Privacy and AI"* also developed by Anastasiadou (2020). PCA contributed to variable reduction for easier interpretation and established the construct validity of the questionnaire.

Confirmatory Factor Analysis (CFA) was used to verify whether the empirical data confirmed the previously hypothesized theoretical structure. Its contribution was essential in confirming the validity of the theoretical model, verifying construct validity, and assessing the reliability of individual conceptual constructs within the theoretical model underlying the research instrument's development.

Correspondence Factor Analysis (Analyse Factorielle des Correspondances - AFC), a powerful multidimensional statistical tool suitable for processing qualitative categorical data (Papadimitriou, 2007), was utilized to highlight significant associations among complex variables/conceptual constructs named *"Human Rights," "Autonomy," "Bias," "Fairness," "Security," "Data Protection," "Privacy and AI," "Data Protection,"* and *"Privacy and AI."*

Correspondence Factor Analysis (AFC, Analyse Factorielle des Correspondances/ Factorial Correspondence Analysis) was employed in this doctoral dissertation as the primary method for processing and visualizing qualitative (categorical) data derived from responses to research instruments, specifically the tools titled *"Human Rights, Autonomy, Bias, Fairness, Security and AI"* and *"Data Protection, Privacy and AI"*, both developed by Anastasiadou (2020).

Additionally, this statistical methodology facilitated the visual and statistical detection of relationships between categories concerning perceptions, including ethical issues associated with Artificial Intelligence, providing valuable insights into underlying structures of social perceptions.

Finally, employing a combination of statistical methodologies—Principal Component Analysis (PCA), Confirmatory Factor Analysis (CFA), and Analyse Factorielle des Correspondances/ Factorial Correspondence Analysis (AFC)—enhanced the validity, reliability, and multidimensional understanding of the dissertation's research topic. The application of these methods was not fragmented but complementary, covering various aspects of the research approach. This combination contributed significantly to methodological precision, as PCA revealed underlying dimensions of attitudes, simplifying data complexity; CFA confirmed the construct validity of factors measured according to theoretical models; and AFC mapped associations between categorical variables, uncovering relationships that might remain undetected by other statistical techniques, effectively providing a multidimensional "scan" of the data dimensions, both theoretically and through the lens of social representations.

3.3. Presentation of Research Instruments

This section will present the two research instruments used in this study, specifically the research instrument titled *"Human Rights, Autonomy, Bias, Fairness, Security and AI,"* and the research instrument titled *"Data Protection, Privacy and AI."*

3.3.1 The Research Instrument Titled "Human Rights, Autonomy, Bias, Fairness, Security and AI"

It was developed by Anastasiadou (2020) and consists of 68 items organized into six conceptual constructs: The first construct, titled "Human Rights," comprises 21 items (e.g., Hur1: AI algorithms violate human rights). The second construct, titled "Autonomy," comprises 5 items (e.g., Aut1: Machine learning algorithms do not undermine human autonomy). The third construct, titled "Bias," comprises 16 items (e.g., Bias1: There are bias issues in AI). The fourth construct, titled "Fairness," comprises 8 items (e.g., Fair3: Machine learning algorithms do not produce fair decision-making models). The fifth construct, titled "Transparency," comprises 12 items (e.g., Tran11: The ability to control AI algorithm training is a prerequisite for accountability). The sixth construct, titled "Security," comprises 6 items (e.g., AI technologies are vulnerable to cyber-attacks).

Table 1. Research Instrument "Human Rights, Autonomy, Bias, Fairness, Security and AI"

Conceptual Construct	Code	Items
Human Rights	Hur1:	AI algorithms violate human rights.
	Hur2:	AI systems instrumentalize individuals.
	Hur3:	The use of social media algorithms leads to the proliferation of prejudices.
	Hur4:	AI algorithms on social media platforms amplify extreme political beliefs.
	Hur5:	AI algorithms on social media platforms allow the spread of homophobic prejudices.
	Hur6:	AI algorithms on social media platforms reproduce gender stereotypes.
	Hur7:	The representation of women in search engines reproduces societal stereotypes.
	Hur8:	Social media platforms introduce discrimination.
	Hur9:	AI algorithms on social media platforms allow hate speech.
	Hur10:	AI algorithms are used for mass surveillance of individuals.
	Hur11:	Women are equally represented in the design of AI systems.
	Hur12:	Women are equally represented in the development of AI algorithms.
	Hur13:	AI algorithms introduce social discrimination.
	Hur14:	AI algorithms introduce racist stereotypes.
	Hur15:	AI systems introduce racial stereotypes.
	Hur16:	Since algorithms are trained on real data containing inequalities, there is a risk they will reproduce these inequalities.
	Hur17:	Since algorithms are trained on real data containing inequalities, there is a risk they will reproduce representations of these data.
	Hur18:	Since algorithms are trained on real data containing social biases, there is a risk they will reproduce these biases.
	Hur19:	Since AI algorithms are trained on real data containing racial biases, there is a risk they will reproduce these biases.
	Hur20:	Human dignity is undermined by AI systems.
	Hur21:	Social media platforms disseminate fake news.
Autonomy	Aut1:	Machine learning algorithms do not undermine human autonomy.
	Aut2:	Machine learning algorithms do not undermine human freedom.
	Aut3:	Machine learning algorithms do not undermine human rights.
	Aut4:	AI systems do not harm humans.

continued on following page

Table 1. Continued

Conceptual Construct	Code	Items
	Aut5:	AI systems may become autonomous themselves and undermine human autonomy.
Bias	Bias1:	There are bias issues in AI.
	Bias2:	Social stereotypes can be embedded into AI algorithms.
	Bias3:	Training data for natural language processing algorithms may reproduce gender stereotypes.
	Bias4:	Training natural language processing algorithms can reproduce racial stereotypes.
	Bias5:	Training data for criminal behavior algorithms may reproduce stereotypes about various social groups.
	Bias6:	Training data for algorithms may reproduce biases related to sensitive variables (gender, ethnicity, race, sexual orientation, skin color).
	Bias7:	Training data for algorithms may perpetuate social inequalities.
	Bias8:	Training algorithms based solely on medical data from men produces biased results.
	Bias9:	Facial recognition algorithms reproduce biases concerning social minorities.
	Bias10:	Facial recognition neural networks reproduce biases regarding skin color or shades.
	Bias11:	AI algorithm training data may not come from representative samples regarding group demographics, thus reproducing biased results.
	Bias12:	AI algorithms can detect and minimize human biases.
	Bias13:	Machine learning algorithms do not replicate user preferences.
	Bias14:	AI systems do not produce biased outcomes.
	Bias15:	AI algorithms can identify the location of users.
	Bias16:	Bias in machine learning is difficult to detect.
Fairness	Fair1:	There is fairness concerning the accuracy of facial recognition.
	Fair2:	Unfairness in machine learning algorithms exists.
	Fair3:	Machine learning algorithms do not produce fair decision-making models.
	Fair4:	Machine learning algorithms reproduce unfair decisions through their training models.
	Fair5:	Loan decision-support systems are more favorable toward certain social groups.
	Fair6:	Neural networks that perform text representation reproduce social stereotypes.
	Fair7:	Neural networks that perform text representation reproduce biases against minorities
	Fair8:	Neural networks that perform text representation reproduce biases concerning people's sexual identity.

continued on following page

Table 1. Continued

Conceptual Construct	Code	Items
Transparency	Tran1:	There are transparency issues on social media platforms concerning who contributes to these platforms.
	Tran2:	There are issues related to the lack of transparency regarding who works behind social media platforms.
	Tran3:	There are opacity issues concerning ownership of social media platforms, leading to transparency concerns.
	Tran4:	The way social networking algorithms are developed and managed is unclear.
	Tran5:	It is unclear how personal information and behaviors are used for analyzing consumer preferences.
	Tran6:	The manner in which personal information and behaviors are used for targeted advertising is not clear.
	Tran7:	I understand how information provided through search engines is reproduced.
	Tran8:	Standards and policies regarding data usage and exploitation on social media platforms are unknown.
	Tran9:	The user is aware of the standards and usage policies of social media platforms.
	Tran10:	The user is aware of cookie standards and policies.
	Tran11:	The ability to control AI algorithms is a prerequisite for transparency.
	Tran12:	The ability to monitor AI algorithm training is a prerequisite for accountability.
	Tran13:	Transparency in the use of AI systems is an ethical obligation
Security	Sec1:	AI technologies are vulnerable to cyber-attacks.
	Sec2:	Οι οικονομικές συναλλαγές στα τραπεζικά συστήματα Financial transactions in banking systems are not secure.
	Sec3:	Online purchases are monitored to create consumer profiles of users.
	Sec4:	Intellectual property rights issues arise in AI systems.
	Sec5:	Intellectual property rights violations are relatively easy in AI systems.
	Sec6:	AI systems are susceptible to attacks.

3.3.2 The Research Instrument "Data Protection, Privacy and AI"

It was developed by Anastasiadou (2020) and consists of 29 items organized into two conceptual constructs. The first construct, named "Data Protection," includes 14 items (e.g., DaP3: AI systems ensure privacy protection). The second construct, named "Privacy," comprises 15 items (e.g., Priv2: Cookies can monitor user privacy). All negatively phrased items were reverse-coded for analysis purposes (Table 2).

Table 2. Data Protection, Privacy and AI

Conceptual Construct	Code	Items
Data Protection	DaP1:	AI systems ensure personal data protection.
	DaP2:	Machine learning algorithms guarantee information accuracy.
	DaP3:	AI systems ensure privacy protection.
	DaP4:	Personal data collection is provided voluntarily by users on social media.
	DaP5:	Personal data collection is provided voluntarily by users through online shopping.
	DaP6:	Personal data collection is provided voluntarily by users through the use of search engines.
	DaP7:	Personal data collection is provided voluntarily by users through collecting likes on various posts (products, movies, music preferences, online friends, etc.).
	DaP8:	Machine learning systems have appropriate data protection frameworks.
	DaP9:	Consumers' personal data become vulnerable through AI platforms.
	DaP10:	Users do not read the privacy policies on AI platforms regarding their personal data protection.
	DaP11:	Users consent easily in order to download desired web pages and do not read privacy policies on AI platforms.
	DaP12:	Personal data collected through AI platforms create user profiles.
	DaP13:	Cookies exacerbate personal data security issues.
	DaP14:	Search engine users become vulnerable through the use of cookies.
Privacy	Priv1:	Privacy violations occur in cloud computing technologies.
	Priv2:	Cookies can monitor user privacy.
	Priv3:	Cookies are a technique for collecting personal data, through which user autonomy is violated.
	Priv4:	AI systems can monitor people's habits.
	Priv5:	AI systems can control people's lives.
	Priv6:	AI systems may lead to infringements on human personality.
	Priv7:	Privacy issues arising from AI systems restrict individuals' personal freedom.
	Priv8:	Privacy issues arising from AI systems restrict individual autonomy.
	Priv9:	Large-scale data or big data include elements from an individual's private life that are difficult to protect.
	Priv10:	Large-scale data or big data gather personal data from specific social groups whose privacy should be inviolable.
	Priv11:	The use of large-scale data or big data can lead to violations of individuals' private lives.
	Priv12:	GPS data from mobile phones can collect information about a user's private life.
	Priv13:	Data from medical records can be compromised.

continued on following page

Table 2. Continued

Conceptual Construct	Code	Items
	Priv14:	Data from tax declarations can be compromised.
	Priv15:	Privacy concerning personal data on social media platforms is not protected.

4. PILOT STUDY RESULTS

4.1. Pilot Sample

The pilot sample consisted of 111 respondents, of whom 41 (36.9%) were women and 70 (63.1%) were men. Regarding respondents' age, 57 (51.4%) were aged 18–20 years, 27 (24.3%) were 20–22 years, 13 (11.7%) were 22–24 years, and finally, 14 (12.6%) were over 24 years old.

Concerning the place of residence, 86 out of the 111 respondents (77.5%) reported living in an urban area, 22 respondents (19.8%) in a semi-urban area, and finally, 3 respondents (2.7%) in a rural area.

With respect to the school of study, 69 respondents (62.2%) were from the School of Engineering, 30 (27%) from the School of Natural Sciences, 7 (6.3%) from the School of Health Sciences, 3 (2.7%) from the School of Economics and Social Sciences, and finally, 2 (1.8%) from the School of Humanities.

Regarding the year of studies, 42 respondents (37.8%) were in the 1st year, 26 (23.4%) in the 2nd year, 36 (32.4%) in the 3rd year, 13 (11.7%) in the 4th year, 7 (6.3%) in the 5th year, 7 (6.3%) in the 6th year, 2 (1.8%) were graduating students, and 3 respondents (2.7%) were enrolled in postgraduate or doctoral studies (Table 3).

Table 3. Demographic Characteristics of the Pilot Sample

Demographic Characteristics	Category	Frequency N	Relative Frequency (%)
Gender	Male	70	63.1
	Female	41	36.9
Age	18-20	57	51.4
	20-22	27	24.3
	22-24	13	11.7
	>24	14	12.6

continued on following page

Table 3. Continued

Demographic Characteristics	Category	Frequency N	Relative Frequency (%)
Place of Residence	Urban area	86	77.5
	Semi-urban area	22	19.8
	Rural area	3	2.7
School of Study	Engineering	69	62.2
	Natural Sciences	30	27
	Health Sciences	7	6.3
	Economics and Social Sciences	3	2.7
	Humanities	2	1.8
Year of Studies	1st year	42	37.8
	2nd year	26	23.4
	3rd year	13	11.7
	4th year	11	9.9
	5th year	7	6.3
	6th year	7	6.3
	Graduating (final year)	2	1.8
	Postgraduate or doctoral study program	3	2.7

4.2 Evaluation of Reliability and Validity of the Research Instruments

The following section presents the evaluation of reliability and validity for the research instrument "Human Rights, Autonomy, Bias, Fairness, Security, and AI" as well as the research instrument "Data Protection, Privacy and AI."

4.2.1 Evaluation of Reliability and Validity of the Instrument "Human Rights, Autonomy, Bias, Fairness, Security, and AI"

Results of Principal Component Analysis (PCA): According to the Kaiser-Meyer-Olkin (KMO) statistical index, which is KMO=0.709 and exceeds the acceptable threshold value of 0.60, it is demonstrated that correlations among variables are high, and thus, the data are suitable for processing using Principal Component Analysis. Additionally, Bartlett's test of sphericity rejects the null hypothesis that the correlation matrix is an identity matrix (χ^2=3949.776, df=2145, p<0.001), indicating that the analysis can proceed.

Table 4 presents the results of Principal Component Analysis. The eigenvalue criterion (eigenvalue ≥1) is satisfied for the six factors identified. The first factor/component has an eigenvalue of 3.240, explaining 21.382% of the total inertia or variance, which is a highly satisfactory percentage. The first factor includes items Hur1, Hur2, Hur3, Hur4, Hur5, Hur6, Hur7, Hur8, Hur9, Hur10, Hur11, Hur12, Hur13, Hur14, Hur15, Hur16, Hur17, Hur18, and Hur19, with respective loadings of 0.723, 0.587, 0.578, 0.508, 0.854, 0.631, 0.617, 0.861, 0.621, 0.653, 0.604, 0.653, 0.631, 0.748, 0.748, 0.681, 0.872, 0.878, 0.891, 0.816, and 0.804. These loadings exceed the threshold value of 0.40, confirming the satisfactory quality of the items. This particular factor, which essentially constitutes a conceptual construct, was named the "Human Rights" factor.

The second factor has an eigenvalue of 3.154, explaining 17.003% of the total variance, a very satisfactory percentage. It comprises items Aut1, Aut2, Aut3, Aut4, and Aut5, with loadings of 0.838, 0.678, 0.825, 0.783, and 0.642 respectively, verifying good item quality. This factor, which represents a conceptual construct, was named "Autonomy."

The third factor has an eigenvalue of 2.942, explaining 15.459% of the total variance, also a very satisfactory percentage. It includes items Bias1, Bias2, Bias3, Bias4, Bias5, Bias6, Bias7, Bias8, Bias9, Bias10, Bias11, Bias12, Bias13, Bias14, Bias15, and Bias16, with respective loadings of 0.761, 0.866, 0.769, 0.798, 0.626, 0.784, 0.674, 0.673, 0.733, 0.852, 0.644, 0.651, 0.644, 0.609, 0.527, and 0.772. It becomes evident that these loadings exceed the threshold value of 0.40, confirming the satisfactory quality of the items. This factor, representing a conceptual construct, was named "Bias."

The fourth factor has an eigenvalue of 2.129, explaining 14.782% of the total variance, a highly satisfactory percentage. It gathers items Fair1, Fair2, Fair3, Fair4, Fair5, Fair6, Fair7, and Fair8, with respective loadings of 0.757, 0.759, 0.676, 0.729, 0.689, 0.748, 0.688, and 0.603, verifying good item quality. This factor, representing a conceptual construct, was named "Fairness."

The fifth factor/component has an eigenvalue of 1.782, explaining 10.582% of the total variance, a very satisfactory percentage. This factor includes items Tran1, Tran2, Tran3, Tran4, Tran5, Tran6, Tran7, Tran8, Tran9, Tran10, Tran11, Tran12, and Tran13, with loadings of 0.774, 0.797, 0.785, 0.668, 0.546, 0.732, 0.717, 0.802, 0.718, 0.716, 0.748, 0.678, and 0.700, respectively. These loadings exceed the threshold value of 0.40, indicating satisfactory item quality. This conceptual construct was named the factor of "Transparency."

The sixth factor has an eigenvalue of 1.863, explaining 11.525% of the total variance, also a very satisfactory percentage. This factor consists of items Sec1,

Sec2, Sec3, Sec4, Sec5, and Sec6, with loadings of 0.894, 0.706, 0.765, 0.755, 0.867, and 0.797, respectively, verifying item quality. This conceptual construct was named the factor of "Security."

Clearly, all loadings have high values greater than the threshold of 0.50, confirming convergent validity. Additionally, all six factors have eigenvalues greater than unity, and items load onto one factor with loadings above 0.50, further confirming both convergent validity and construct validity.

The six factors cumulatively explain 90.733% of the total variance. The Average Variances Extracted (AVE) for factors one through six are 0.506, 0.574, 0.515, 0.501, 0.525, and 0.640, respectively.

These AVE values exceed the recommended threshold of 0.50, demonstrating that more than half of the variance observed in the items was attributed to their respective theoretical constructs, thereby confirming convergent validity.

The reliability of the instrument "Human Rights, Autonomy, Bias, Fairness, Transparency, Security, and AI" was evaluated using Cronbach's Alpha, resulting in a value of 0.768 (α=0.768). Specifically, Cronbach's Alpha reliability values for individual factors were 0.741 for "Human Rights," 0.749 for "Autonomy," 0.751 for "Bias," 0.632 for "Fairness," 0.854 for "Transparency," and 0.851 for "Security" and thay are considered to be satisfactory (Anastasiadou & Dimitriadou 2011; Anastasiadou & Tiliakou, 2016; Anastasiadou, & Taraza, 2019;).

Composite Reliability (CR) values were 0.950, 0.869, 0.943, 0.889, 0.934, and 0.914, for the factors "Human Rights," "Autonomy," "Bias," "Fairness," "Transparency," and "Security," respectively. These Cronbach's Alpha and Composite Reliability (CR) values indicate reliability and internal consistency of the measurement tool's factors (Fornell & Larcker, 1981; Nunnally, 1978).

Moreover, the goodness-of-fit indices of the six-factor conceptual model confirmed its validity: (χ^2/df=1.88, CFI=0.95, GFI=0.97, AGFI=0.96, TLI=0.96, RMSR=0.04, RMSEA=0.06, IFI=0.94, CD=0.95) (Table 4).

Table 4. "Human Rights, Autonomy, Bias, Fairness, Transparency, Security, and AI"

Conceptual Construct	Code	Eigen Value	%Variance	Loadings	Cronbrach's Alpha	CR	AVE
Human Rights, Autonomy, Bias, Fairness, Transparency, Security, and AI					0.768		
Human Rights		3.240	21.382		0.741	0.950	0.506
	Hur1			0.723			
	Hur2			0.587			
	Hur3			0.578			

continued on following page

Table 4. Continued

Conceptual Construct	Code	Eigen Value	%Variance	Loadings	Cronbrach's Alpha	CR	AVE
	Hur4			0.508			
	Hur5			0.854			
	Hur6			0.631			
	Hur7			0.617			
	Hur8			0.861			
	Hur9			0.621			
	Hur10			0.653			
	Hur11			0.604			
	Hur12			0.653			
	Hur13			0.631			
	Hur14			0.748			
	Hur15			0.748			
	Hur16			0.681			
	Hur17			0.872			
	Hur18			0.878			
	Hur19			0.891			
	Hur20			0.816			
	Hur21			0.804			
Autonomy		3.154	17.003		0.749	0.869	0.574
	Aut1			0.838			
	Aut2			0.678			
	Aut3			0.825			
	Aut4			0.783			
	Aut5			0.642			
Bias		2.943	15.459		0.751	0.943	0.515
	Bias1			0.761			
	Bias2			0.866			
	Bias3			0.769			
	Bias4			0.798			
	Bias5			0.626			
	Bias6			0.784			
	Bias7			0.674			
	Bias8			0.673			
	Bias9			0.733			
	Bias10			0.852			
	Bias11			0.644			

continued on following page

Table 4. Continued

Conceptual Construct	Code	Eigen Value	%Variance	Loadings	Cronbrach's Alpha	CR	AVE
	Bias12			0.651			
	Bias13			0.644			
	Bias14			0.609			
	Bias15			0.527			
	Bias16			0.772			
Fairness		2.129	14.782		0.632	0.889	0.501
	Fair1			0.757			
	Fair2			0.759			
	Fair3			0.676			
	Fair4			0.729			
	Fair5			0.689			
	Fair6			0.748			
	Fair7			0.688			
	Fair8			0.603			
Transparen-cy		1.863	11.525		0.854	0.934	0.525
	Tran1			0.774			
	Tran2			0.797			
	Tran3			0.785			
	Tran4			0.668			
	Tran5			0.546			
	Tran6			0.732			
	Tran7			0.717			
	Tran8			0.802			
	Tran9			0.718			
	Tran10			0.716			
	Tran11			0.748			
	Tran12			0.678			
	Tran13			0.700			
Security		1.782	10.582		0.851	0.914	0.640
	Sec1:			0.894			
	Sec2:			0.706			
	Sec3:			0.765			
	Sec4:			0.755			
	Sec5:			0.867			
	Sec6:			0.797			

x^2/df=1.88, CFI=0.95, GFI=0.97, AGFI=0.96, TLI=0.96, RMSR=0.04, RMSEA=0.06, IFI=0.94, CD=0.95

4.2.2 Evaluation of Reliability and Validity of the Instrument "Data Protection, Privacy and AI"

Results of Principal Component Analysis (PCA): According to the Kaiser-Meyer-Olkin (KMO) statistic, specifically KMO = 0.796, which exceeds the acceptable threshold of 0.60, it is demonstrated that correlations among variables are high, and thus the data are suitable for Principal Component Analysis. Moreover, Bartlett's test of sphericity rejects the null hypothesis that the correlation matrix is an identity matrix ($\chi^2 = 1663.199$, df = 378, $p < 0.001$), allowing the analysis to proceed using PCA.

As evident from Table 5 and based on the analysis results for all items of the instrument "Data Protection, Privacy and AI," the eigenvalue criterion (eigenvalue ≥ 1) is satisfied for two factors. The first factor refers to the extent to which respondents respond to the construct titled "Data Protection," with an eigenvalue of 5.403. This factor explains 41.562% of the total variance, an acceptable percentage for this methodology. The items loading onto this factor are DaP1, DaP2, DaP3, DaP4, DaP5, DaP6, DaP7, DaP8, DaP9, DaP10, DaP11, DaP12, DaP13, and DaP14, with notably high loadings of 0.745, 0.685, 0.708, 0.665, 0.653, 0.749, 0.723, 0.669, 0.862, 0.768, 0.674, 0.601, 0.726, and 0.723 respectively. According to the eigenvalue criterion (≥1), it is verified that items DaP1, DaP2, DaP3, DaP4, DaP5, DaP6, DaP7, DaP8, DaP9, DaP10, DaP11, DaP12, and DaP13 constitute a single conceptual construct.

The second factor refers to the extent to which respondents respond to the construct titled "Privacy," which has an eigenvalue of 3.893. This factor explains 25.954% of the total variance, an acceptable percentage for this methodology. The items loading onto this factor are Priv1, Priv2, Priv3, Priv4, Priv5, Priv6, Priv7, Priv8, Priv9, Priv10, Priv11, Priv12, Priv13, Priv14, and Priv15, with notably high loadings of 0.718, 0.688, 0.765, 0.823, 0.652, 0.692, 0.698, 0.656, 0.623, 0.785, 0.695, 0.664, 0.758, 0.765, and 0.615, respectively. According to the eigenvalue criterion (≥1), it is verified that items Priv1, Priv2, Priv3, Priv4, Priv5, Priv6, Priv7, Priv8, Priv9, Priv10, Priv11, Priv12, Priv13, Priv14, and Priv15 constitute a single conceptual construct.

It is noteworthy that all loadings exhibit high values, all exceeding even 0.50, demonstrating convergent validity. Additionally, both factors have eigenvalues greater than unity, and items load onto one factor with loadings greater than 0.50, confirming both the convergent validity and construct validity of the instrument.

Together, the two factors explain 67.516% of the total variance. The Average Variances Extracted (AVE) values for the first and second factors are 0.506 and 0.501, respectively. These AVE values exceed the recommended threshold of 0.50, indicating that more than half of the variance observed in the items is attributed to their corresponding theoretical constructs, confirming convergent validity.

The reliability of the instrument *"Data Protection, Privacy, and AI"* was assessed using Cronbach's Alpha, yielding an overall value of 0.815 (α=0.815) indicated a strong reliability (Taraza & Anastasiadou, 2019; Anastasiadou et al., 2014; 2022; Anastasiadou et al., 2023). Specifically, the reliability for the factor "Data Protection" is 0.738 (α=0.738) and for the factor "Privacy" is 0.783 (α=0.783). The Composite Reliability (CR) values for the factors "Data Protection" and "Privacy" are 0.930 and 0.938, respectively indicated an acceptable reliability (Fotiadis & Anastasiadou, 2018a). Both Cronbach's Alpha and Composite Reliability (CR) values indicate strong reliability and internal consistency of the instrument's factors (Fornell & Larcker, 1981; Nunnally, 1978).

Additionally, the goodness-of-fit indices of the two-factor conceptual model of the *"Data Protection, Privacy, and AI"* instrument confirmed its validity:

(χ^2/df = 1.96, CFI = 0.96, GFI = 0.97, AGFI = 0.95, TLI = 0.98, RMSR = 0.06, RMSEA = 0.07, IFI = 0.93, CD = 0.96) (Table 5).

Table 5. "Data Protection, Privacy, and AI"

Conceptual Construct	Code	Eigen Value	%Variance	Loadings	Cronbrach's Alpha	CR	AVE
«Data Protection, Privacy and TN					0.815		
		5.403	41.562		0.738	0.930	0.508
	DaP1			0.745			
	DaP2			0.685			
	DaP3			0.708			
	DaP4			0.665			
	DaP5			0.653			
	DaP6			0.749			
	DaP7			0.723			
	DaP8			0.669			
	DaP9			0.862			
	DaP10			0.768			
	DaP11			0.674			
	DaP12			0.601			
	DaP13			0.726			
	DaP14			0.723			
Privacy		3.893	25.954		0.783	0.938	0.503

continued on following page

Table 5. Continued

Conceptual Construct	Code	Eigen Value	%Variance	Loadings	Cronbrach's Alpha	CR	AVE
	Priv1			0.718			
	Priv2			0.688			
	Priv3			0.765			
	Priv4			0.823			
	Priv5			0.652			
	Priv6			0.692			
	Priv7			0.698			
	Priv8			0.656			
	Priv9			0.623			
	Priv10			0.785			
	Priv11			0.695			
	Priv12			0.664			
	Priv13			0.758			
	Priv14			0.765			
	Priv15			0.615			

x^2/df=1.96, CFI=0.96, GFI=0.97, AGFI=0.95, TLI=0.98, RMSR=0.06, RMSEA=0.07, IFI=0.93, CD=0.96

5. FINAL RESEARCH RESULTS

5.1 Final Research Sample

The final research sample consisted of 239 participants, of whom 114 (47.7%) were women and 125 (52.3%) were men. Regarding participants' age groups, 59 (24.7%) were aged 18–20, 65 (27.2%) were 20–22, 61 (25.5%) were 22–24, and finally, 54 (22.6%) were older than 24 years.

Concerning the place of residence, 169 out of 239 respondents (70.7%) reported living in an urban area, 66 respondents (27.6%) in a semi-urban area, and 4 respondents (1.7%) in a rural area.

Regarding their school of study, 62 (25.9%) were from the School of Engineering, 65 (27.2%) from the School of Natural Sciences, 47 (19.7%) from the School of Health Sciences, 40 (16.7%) from the School of Economics and Social Sciences, and finally, 25 (10.5%) from the School of Humanities.

Regarding the year of studies, 44 respondents (18.4%) were in the 1st year, 48 (20.1%) in the 2nd year, 46 (19.2%) in the 3rd year, 45 (18.8%) in the 4th year, 19

(7.9%) in the 5th year, 16 (6.7%) in the 6th year, 14 (5.9%) were graduating students, and 7 respondents (2.9%) were enrolled in postgraduate or doctoral studies (Table 6).

Table 6. Demographic Characteristics of the Final Sample

Demographic Characteristics	Category	Frequency N	Relative frequency. (%)
Gender	Male	114	47.7
	Female	125	52.3
Age	18-20	59	24.7
	20-22	65	27.2
	22-24	61	25.5
	>24	54	22.6
Place of Residence	Urban area	169	70.7
	Semi-urban area	66	27.6
	Rural area	4	1.7
School of Study	Engineering	62	25.9
	Natural Sciences	65	27.2
	Health Sciences	47	19.7
	Economics and Social Sciences	40	16.7
	Humanities	25	10.5
Year of Studies	1st year	44	18.4
	2nd year	48	20.1
	3rd year	46	19.2
	4th year	45	18.8
	5th year	19	7.9
	6th year	16	6.7
	Graduating (final year)	14	5.9
	Postgraduate or doctoral study program	7	2.9

5.2. Attitudes Toward Artificial Intelligence

By applying Analyse Factorielle des Corespondance (AFC) to the entire research instrument *"Human Rights, Autonomy, Bias, Transparency, Fairness, Security, and AI,"* a Chi-square (X^2) test of independence was conducted among the categories of variables, using Cramer's V criterion, and inertia was calculated. The interpretation of the Analyse Factorielle des Corespondance (AFC) results was based on the indicators of inertia and contribution (Benzécri, 1973; Papadimitriou, 2007). These indicators allow for the identification of the most significant and decisive variables or the categories of variables that contribute to the construction of factorial axes (Anastasiadou, 2013; 2016). Specifically, the interpretation of Analyse Factorielle des Corespondance (AFC) results was based on the inertia explained by each factorial axis, as well as on the indicators of correlation and contribution (Anastasiadou, 2019; Anastasiadou & Karakos, 2011).

Analysis of the initial data matrix using Analyse Factorielle des Corespondance (AFC) provides us with Figure 1, which includes the eigenvalues of the Burt table and also the percentages of inertia corresponding to each factorial axis. The table enables us to identify the number of most significant factorial axes, upon which the interpretation of results is based. The percentage of inertia for each factorial axis corresponds to the degree of significance represented by that specific factorial axis.

Based on cumulative frequency, the first two factorial axes explain 25.50% of the total inertia, variance, or total information, a percentage considered satisfactory (Karapistolis, 2015). Subsequently, by following Analyse Factorielle des Corespondance (AFC) procedures and based on criteria (inertia, correlation, and contribution), the variables or variable categories contributing to the formation of factorial axes and factorial planes were identified. Data processing was conducted using the MAD software (Karapistolis, 2000). The variables or variable categories highlighted based on two criteria—correlation (Cor ≥ 200) (Criterion 2) and contribution (Ctr ≥ 1000/226 = 4.484 ≈ 5) (Criterion 3)—are presented (Karapistolis, 2000; 2015).

Table 7.

PROJECTION OF EINGENVALUES				
TOTAL INERTIA 0.22761				
AXIS	INERTIA	%VARIANCE	COMMULATIVE	HISTOGRAM OF CHARACTERISTIC ROOTS
1	0.0146270	13.47	13.17	***
2	0.0130662	12.03	25.50	***
3	0.0097589	8.99	34.49	*************************************
4	0.0070441	6.49	40.98	************************
5	0.0059056	5.44	46.11	*******************
6	0.0051036	4.70	51.11	***************
7	0.0044029	4.05	55.17	**********
8	0.0041538	3.83	58.99	*********
9	0.0035857	3.30	62.30	*******
10	0.0032667	3.01	65.30	******
11	0.0030729	2.83	68.13	*****
12	0.0028171	2.59	70.73	****

The first factorial plane e_1 x e_2: The categories of variables deemed most significant based on the criteria of inertia, correlation, and contribution, specifically those with the highest Cor and Ctr values, are presented, studied, and analyzed below.

The first factorial plane e_1 x e_2 (Figure 1) explains 29.40% of the total inertia, variance, or total information, a percentage evaluated as satisfactory (Fotiadis & Anastasiadou, 2018b; Papadaki & Anastsasiadou, 2019) (Table 7).

In the first factorial plane e_1 x e_2, specifically in the first quadrant (e_1+,e_2+), the group of respondents stands out who believe that the user is aware of cookie standards and policies (Tran103), and that the ability to monitor AI algorithms is a prerequisite for transparency (Tran113). Consequently, this group also considers that the ability to control AI algorithm training is a prerequisite for accountability (Tran123). Furthermore, respondents in this group believe that AI systems are vulnerable to attacks (Sec63), and as a result, financial transactions in banking systems are not secure (Sec23), AI systems raise intellectual property rights issues (Sec43), and finally, intellectual property rights violations in AI systems are relatively easy (Sec53), producing biased results (Bias141). Thus, in the first factorial plane e_1 x e_2, specifically in the first quadrant (e_1+,e_2+), the group highlighting issues of Transparency, Security, and Bias clearly emerges.

Figure 1. First Factorial Plane e1 × e2 – Human Rights, Autonomy, Bias, Transparency, Fairness, Security, and AI

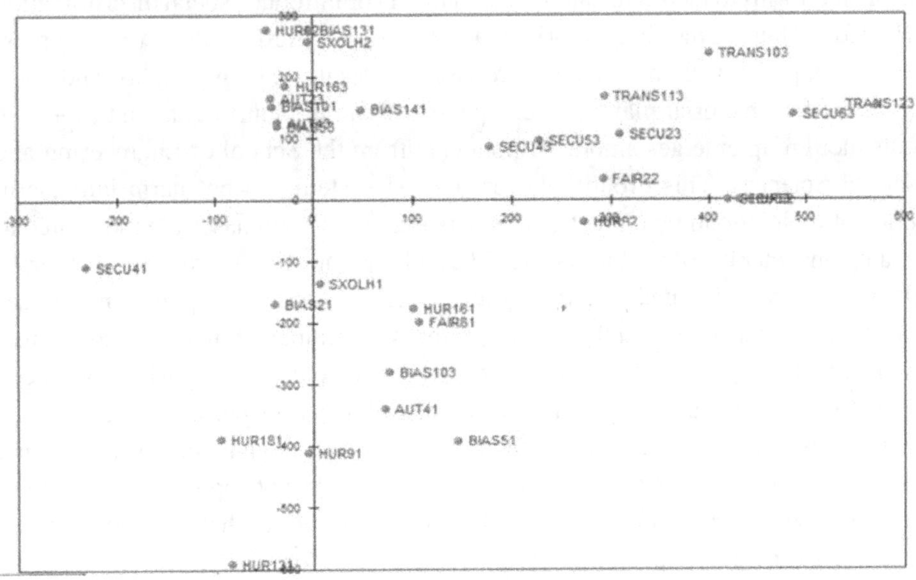

In the first factorial plane e_1 x e_2, and specifically in the second quadrant (e_1-,e_2+), the group of respondents enrolled in the School of Natural Sciences stands out. This group believes that algorithm training data related to criminal behavior might reproduce stereotypes about various social groups (Bias53), asserts that machine learning algorithms replicate users' preferences (Bias131), and claims that neural networks used in facial recognition reproduce biases concerning skin color or shades (Bias103). Additionally, this group believes that AI algorithms used through social media platforms reproduce gender stereotypes (Hur63), and, precisely because algorithms are trained on real data containing inequalities, they strongly believe there is a risk of reproducing these inequalities (Hur163). Despite these beliefs, this particular group considers that machine learning algorithms do not undermine human freedom (Aut23) and concludes that AI systems do not harm humans (Aut43). Thus, in the first factorial plane (e1 × e2), the second quadrant (e_1-,e_2+) represents a group positively oriented toward autonomy issues, yet negatively oriented toward issues related to human rights and bias.

In the first factorial plane e_1 x e_2, and specifically in the third quadrant (e_1-,e_2-), a different group of respondents emerges. This group believes that AI systems do not raise intellectual property rights issues (Sec41), that social stereotypes cannot be embedded into AI algorithms (Bias21), and that despite algorithms being trained on real data containing social biases, there is no risk of reproducing these biases (Hur18). Moreover, they argue that it is impossible for AI algorithms used on social media platforms to facilitate hate speech (Hur91) or introduce social discrimination (Hur131). Thus, in the first factorial plane e_1 x e_2, the third quadrant (e_1-,e_2-) presents a group positively oriented toward issues of security, human rights, and bias.

In the first factorial plane e_1 x e_2, specifically in the fourth quadrant (e_1+,e_2-), a distinct group emerges among respondents from the School of Engineering and Natural Sciences. This group believes that AI systems do not harm humans in terms of their autonomy (Aut41), considers that online purchases are monitored to create consumer profiles of users (Sec33), and does not clearly express whether AI algorithms on social media platforms allow the spread of homophobic prejudices (Hur53). Furthermore, even though algorithms are trained on real data containing inequalities, this group perceives no risk that these inequalities might be reproduced (Hur161). Additionally, they consider that neural networks performing text representation do not reproduce biases concerning people's sexual identity (Fair81), and that training data for criminal behavior algorithms do not reproduce stereotypes about various social groups (Bias51). However, they believe that facial recognition neural networks reproduce biases related to skin color or shades (Bias103).

In the first factorial plane e_1 x e_2, the fourth quadrant (e_1+,e_2-) thus identifies a group positively oriented toward issues of autonomy, security, fairness, human rights, and bias.

5.3. The Instrument "Data Protection, Privacy and AI"

The interpretation of the Analyse Factorielle des Corespondance (AFC) results was based on the indices of "inertia" and "contribution." These indices highlight the most significant and decisive variables or categories of variables contributing to the construction of factorial axes. Specifically, the interpretation relied on the inertia explained by each factorial axis, as well as the correlation and contribution indicators.

The analysis of the original data table using Analyse Factorielle des Corespondance (AFC) provides the Burt table of eigenvalues, along with the inertia percentages corresponding to each factorial axis (Benzécri, 1973). This allows identification of the number of significant factorial axes upon which the interpretation of results is based. The percentage of inertia for each factorial axis corresponds to the significance expressed by each respective axis (Table 8).

According to the values derived from the histogram (Table 8), the significance percentage of the first factorial axis amounts to 35.07%, the second factorial axis is 14.06%, the third factorial axis is 7.49%, the fourth factorial axis is 6.29%, etc. The total information explained by all factorial axes is 87.04%, as explicitly presented in Table 8.

Table 8.

PROJECTION OF EINGENVALUES				
TOTAL INERTIA 0.22761				
AXIS	INERTIA	%VARIANCE	COMMULATIVE	HISTOGRAM OF CHARACTERISTIC ROOTS
1	0.0798148	35.07	35.07	***
2	0.0320060	14.04	49.13	*****************
3	0.0170475	7.49	56.62	*********
4	0.0143123	6.29	62.90	********
5	0.0113342	4.98	67.88	******
6	0.0085564	3.76	71.64	*****
7	0.0080078	3.52	75.16	*****
8	0.0070744	3.11	78.27	****
9	0.0062836	2.76	81.03	****
10	0.0052294	2.30	83.33	***
11	0.0046991	2.06	85.39	***
12	0.0037529	1.65	87.04	**

Based on cumulative frequency, the first two factorial axes explain 49.13% of the total inertia, variance, or total information—a percentage considered satisfactory (Karapistolis, 2015).

Subsequently, through the application of Correspondence Factor Analysis (CFA), and according to the criteria (inertia, correlation, and contribution), we highlight the variables or categories of variables contributing to the formation of factorial axes and factorial planes ((Benzécri, 1973; Papadimitriou, 2007). Data analysis was conducted using the MAD software (Karapistolis, 2000).

Variables or categories of variables identified based on the two criteria—Correlation (Cor $\geq$ 200) (Criterion 2) and Contribution (Ctr $\geq$ 1000/109 = 9.1743 $\approx$ 10) (Criterion 3)—are highlighted (Karapistolis, 2015).

The first factorial plane e_1 x e_2: The variable categories that are most significant according to the criteria of inertia, correlation, and contribution—specifically, those showing the highest Cor and Ctr values—are presented, studied, and analyzed below.

The first factorial plane e_1 x e_2 (Diagram 2) explains 49.13% of the total inertia, variance, or total information, a percentage evaluated as satisfactory (Table 8)

In the first factorial plane e_1 x e_2, specifically in the first quadrant (e_1+,e_2+), a distinct group of respondents emerges who believe that cookies are a technique for collecting personal data, thereby violating user autonomy (Priv33). This group also believes that cookies can monitor user privacy (Priv23), considers that AI systems can monitor human habits (Priv43), and strongly supports that cloud computing technologies are associated with privacy violations (Priv13).

Figure 2. Data Protection, Privacy, and AI

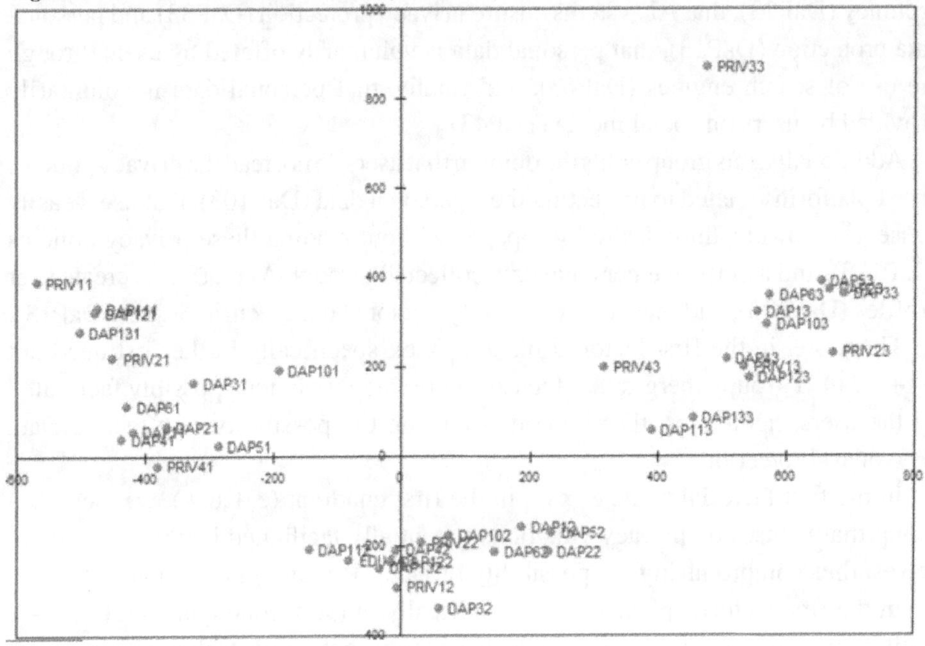

Moreover, this group strongly believes that personal data collection is voluntarily offered by users through online shopping (DaP53), machine learning algorithms guarantee information accuracy (DaP23), AI systems ensure privacy protection (DaP33) and personal data protection (DaP13), personal data is voluntarily provided by users through search engines (DaP63), and finally, personal data is voluntarily offered by users on social media platforms (DaP43).

Additionally, this group believes users do not read privacy policies on AI platforms concerning the protection of their personal data (DaP103), that users easily consent to download desired web pages without reading these privacy policies (DaP113), and that personal data collected via AI platforms create user profiles (DaP123). They also assert that cookies amplify personal data security problems (DaP13).

In the first factorial plane e_1 x e_2, specifically in the first quadrant (e1+, e2+), a distinct group of respondents emerges who believe that cookies are a technique for collecting personal data through which user autonomy is violated (Priv33). This group also believes that cookies can monitor user privacy (Priv23), considers that AI systems can monitor human habits (Priv43), and strongly maintains that privacy violations occur in cloud computing technologies (Priv13). Moreover, this group strongly believes that personal data collection is voluntarily offered by users through online shopping (DaP53), that machine learning algorithms guarantee information accuracy (DaP23), that AI systems ensure privacy protection (DaP33) and personal data protection (DaP13), that personal data is voluntarily offered by users through the use of search engines (DaP63), and finally, that personal data is voluntarily provided by users on social media (DaP43).

Additionally, this group holds the opinion that users do not read the privacy policies on AI platforms related to protecting their personal data (DaP103), that users easily consent to downloading desired webpages without reading these privacy policies (DaP113), and asserts that personal data collected through AI platforms create user profiles (DaP123), and that cookies amplify personal data security issues (DaP13).

Therefore, in the first factorial plane e_1 x e_2, specifically in the first quadrant (e1+,e2+), a group emerges that focuses on privacy violations possibly facilitated by the users themselves, thereby compromising the possibility of effective Data (Personal) Protection.

In the first factorial plane e_1 x e_2, in the first quadrant (e_1+,e_2+), we identify a group that focuses on privacy violations potentially facilitated by the users themselves, thus compromising the possibility of Data (Personal) Protection.

In the first factorial plane e_1 x e_2, specifically in the second quadrant (e_1-,e_2+), a different group emerges, believing that cloud computing technologies do not present privacy violations (Priv11) and that cookies cannot control user privacy (Priv21). Moreover, this group believes that users do read the privacy policies on AI platforms concerning personal data protection (DaP101), and do not easily consent to download web pages without reading privacy policies (DaP111). They further believe personal data collected via AI platforms do not create user profiles (DaP121) and assert that cookies do not amplify personal data security problems (DaP131).

Additionally, this group argues that personal data collection is not voluntarily offered by users on social media platforms (DaP41), not voluntarily offered by users through online shopping (DaP51), and not voluntarily offered through the use of search engines (DaP61). Lastly, this group strongly believes that machine learning algorithms do not guarantee information accuracy (DaP21).

Thus, in the first factorial plane e_1 x e_2, in the second quadrant (e_1-,e_2+), a group emerges that does not recognize issues related to privacy violations and does not attribute responsibility to users for voluntarily providing personal data through AI systems.

In the first factorial plane e_1 x e_2, specifically in the third quadrant (e_1-,e_2-), a group of respondents from the School of Health Sciences emerges, who believe that AI systems cannot monitor people's habits (Priv41) and that cookies cannot monitor user privacy (Priv21). This particular group does not hold a clear stance regarding whether personal data is voluntarily provided by users through social media (DaP42), whether users consent effortlessly to downloading desired webpages without reading privacy policies on AI platforms (DaP112), and finally, whether cookies amplify the security issues related to personal data (DaP132).

Thus, in the first factorial plane e_1 x e_2, and specifically in the third quadrant(e_1-,e_2-) we identify a group characterized primarily by a lack of a specific stance regarding certain issues of Data (Personal) Protection raised by AI technologies.

In the first factorial plane e_1 x e_2, specifically in the fourth quadrant (e_1+,e_2-), a group of respondents emerges who lack a clearly defined stance regarding whether AI systems ensure personal data protection (DaP12), whether personal data collection is voluntarily provided by users through online shopping (DaP52), whether machine-learning algorithms guarantee information accuracy (DaP22), whether cookies can monitor user privacy (Priv22), whether AI systems ensure privacy protection (DaP32), whether users neglect to read privacy policies on AI platforms concerning their personal data protection (DaP102), and finally, whether personal data collection is voluntarily provided by users through the use of search engines (DaP62).

Therefore, in the first factorial plane e_1 x e_2, specifically in the fourth quadrant (e_1+,e_2), a group emerges characterized mainly by a lack of a specific stance, maintaining neutrality regarding crucial issues related to Data (Personal) Protection raised by AI technologies.

6. DILEMMAS

The use of AI has raised numerous ethical dilemmas. Some of these involve questions such as:

- Are there informed consent issues, for instance, regarding users of AI systems for secondary research?
- Does informed consent by AI system users serve as a safeguard for protecting sensitive personal data?
- Should data protection by AI systems yield to the interest of public health protection in pandemic situations?
- Should personal confidentiality related to AI systems yield to the interest of public health protection during pandemics?
- Can AI systems restrict a patient's or citizen's general freedom to make life choices?
- Does using data from AI systems for secondary analysis undermine user autonomy?
- Does opacity in the collection, analysis, and use of AI system data undermine citizens' self-determination?

- Is there a risk of harm from citizens' overexposure to Health-IoT systems?
- Who inherits the governance of data collected by Health-IoT systems?

These questions require extensive study and investigation. By studying the dimensions of artificial intelligence ethics using fact8orial correspondence analysis, we attempted to shed some light on these issues. Understanding students' behaviors and attitudes toward artificial intelligence (AI) requires further research, particularly leveraging methods involving artificial intelligence (Souravlas & Anastasiadou, 2022; Souravlas et al., 2020), cloud computing technologies (Souravlas et al., 2021a; 2021b; 2021c), advanced technological solutions, and novel educational methods. Conducting this research is essential due to the importance of thoroughly exploring the emotional and cognitive dimensions related to AI (Souravlas et al., 2022a; 2022b; 2023).However, the discussion remains extensive, as AI development is still in its early stages.

7. CONCLUSIONS

The development of technology, along with increased use of sensors, Internet of Things (IoT) servers, electronic transactions, etc., generate enormous volumes of data globally, which are stored on cloud computing platforms. This increase in information and its subsequent use, management, and storage alone can raise significant research questions and challenges. Artificial intelligence (AI) is applied across nearly all scientific fields. The analysis of large-scale data (Big Data) and AI algorithms demonstrate virtually unlimited capabilities in handling increasingly larger data streams.

Contemporary society is experiencing the discovery of vast applications of cutting-edge technologies. Machine learning, deep learning, and artificial intelligence offer significant and innovative applications in fields such as justice, medicine, biomedicine, genetics, neuroscience, and robotics, significantly enhancing human life. On one hand, the benefits appear immense, while on the other hand, there are substantial disadvantages, risks, and ethical and bioethical dilemmas they bring about.

Neuroscience relies heavily on collecting large volumes of structured data, machine learning, deep learning, and the use of intelligent algorithms for processing and analysis. Artificial intelligence maximizes opportunities for new discoveries concerning human brain functions, simultaneously raising methodological and ethical challenges. These include, but are not limited to, issues of privacy, confidentiality, trust, data protection, human rights, freedom, as well as fairness, algorithmic integrity, resilience, and robustness. Bioethics, based on four fundamental principles—autonomy, beneficence, non-maleficence, and justice—provides the framework for the proper development and use of cutting-edge technologies, as well as for managing significant ethical and bioethical challenges that arise.

These issues, related to autonomy, human rights, bias, fairness, transparency, security, data protection, privacy, and confidentiality, were thoroughly analyzed in the theoretical part of this thesis. Particular attention must be given to autonomy, and consequently, the dignity and moral status of individuals, and informed consent, particularly regarding medical, biomedical, and genetic data processed by machine learning algorithms. The processing of these large-scale datasets poses risks of opacity, bias, and potential exploitation, which may lead to violations of personal dignity. Protecting the confidentiality of donors of biological and genetic data requires exceptional attention. Ensuring confidentiality and privacy safeguards the individuals' autonomy, freedom, and fundamental rights. Any attempt to violate such sensitive personal data is ethically unacceptable, restricting individual freedoms and fundamental rights, and undermining personal dignity and autonomy.

Any attempt to manipulate genetic, biological, medical, or personal data generally is ethically unacceptable concerning human value. Exploiting sensitive and socially vulnerable groups under the guise of scientific advancement violates consent rules and poses significant risks, creating inequalities and undermining human rights. Instrumentalizing humans is ethically unacceptable, highlighting the critical issue of accountability. However, the ability to process genetic data can lead to significant advancements in disease prevention and public health protection. Here, the bioethical principles of beneficence and solidarity are paramount. Only by potentially balancing the two primary bioethical pillars—autonomy and social solidarity—and limiting possible harm, can contributions to humanity and the common good be maximized. Bioethics principles provide the framework for our actions, with respect for the autonomy of every rational being offering critical guidance, as emphasized by Onora O'Neill (2011).

Therefore, digital literacy is essential, particularly among young people. Although empirical analysis revealed significant awareness of the risks associated with AI's extensive development, cloud computing, and large-scale data use, it is crucial to educate individuals further on the threats, explicit or hidden, that underlie cutting-edge technology platforms. This education should be coupled with ethical or, more accurately, bioethical literacy.

The European Union's Artificial Intelligence Act (AI Act) aims to regulate the development of a "risk-based" artificial intelligence framework, mandating the protection of human rights, democracy, environmental sustainability, and innovation. The practical implementation and effectiveness of this regulatory framework remain to be observed.

REFERENCES

Anastasiadou, S. (2013). Developing and evaluating a structural equation model measuring leadership changes in a lifelong learning world. *The International Journal of Educational Organization and Leadership*, *19*(2), 1–17. DOI: 10.18848/2329-1656/CGP/v19i02/48520

Anastasiadou, S. (2016). *Evaluation of the implementation of TQM principles in tertiary education using the EFQM excellence model: Research in educational departments of Greek universities* [Master's thesis, Greek Open University]. Patra, Greece.

Anastasiadou, S. (2019). Comparison of contemporary advanced statistical methods regarding construct validity evaluation of TEIque-SF instrument: Statistical implicative analysis vs. principal components analysis. In *9ème Colloque International sur Analyse Statistique Implicative (ASI 10)* (pp. 148–163). Belfort, France.

Anastasiadou, S. & Dimitriadou C. (2011). What does Critical Thinking mean? A statistical data analysis of pre-service teachers' defining statements. *International Journal of Humanities and Social Science* (IJHSS) Vol 1, No 7, Special Issue, pp 73-83.

Anastasiadou, S., & Karakos, A. (2011). The beliefs of electrical and computer engineering students regarding computer programming. *International Journal of Technology, Knowledge and Society*, *7*(1), 37–52. DOI: 10.18848/1832-3669/CGP/v07i01/56170

Anastasiadou, S., Masouras, A., & Papademetriou, C. (2023). *Attitudes toward Reproductive Tourism and Cross border reproductive care (CBRC): Legal, Economic, Ethical issues and dilemmas, possibilities and limitations. Proceedings of the 6th International Conference on Tourism Research, ICTR 2023*, Pafos, Cyprus, pp. 17-23. DOI: 10.34190/ictr.6.1.1189

Anastasiadou, S., Papalexandris, S. & Konteos, G. (2022). *The Effect of Emotional Intelligence on Perceived e-Service Quality, Consumers' Perceived Value, Purchase, Loyalty Intentions and Satisfaction*, Haicta. CEUR Workshop Proceedings (CEUR-WS.org), 151-161.

Anastasiadou, S., & Taraza, E. (2019). Total quality management: Implementation of the Six Sigma methodology for improving quality in higher education. In *ICERI2019: 12th Annual International Conference of Education, Research and Innovation* (pp. 9533–9537). Seville, Spain. DOI: 10.21125/iceri.2019.2318

Anastasiadou, S., & Tiliakou, C. (2016). Evaluation of the discriminating ability of the Greek state foreign language exam (KPG) for English by means of factorial analysis of correspondences. *International Journal of Learning and Teaching*, *8*(2), 129–140. DOI: 10.18844/ijlt.v8i2.518

Anastasiadou, S. D., Batiou, V., & Balkanos, E. (2014). Occupation mobility in Greece: A completely new area. *Global Journal of Business. Economics and Management*, *4*(2), 73–80.

Benzécri, J. P. (1973). *Data Analysis*. Analyse des Données.

Fornell, C., & Larcker, D. F. (1981). Evaluating structural equation models with unobservable variables and measurement error. *JMR, Journal of Marketing Research*, *18*(1), 39–50. DOI: 10.1177/002224378101800104

Fotiadis, Th., & Anastasiadou, S. (2018a). Contemporary advanced statistical methods for the science of marketing: Implicative Statistical Analysis vs Principal Components Analysis. Proceeding of the ICME 2018- 1st International Conference on Marketing and Entrepreneuship. Cyprus. https://hephaestus.nup.ac.cy/handle/11728/11393

Fotiadis, Th., & Anastasiadou, S. (2018b). Contemporary advanced statistical methods for the science of marketing: Principal Components Analysis vs Analyse Factorielle des Correspondances. In *Proceedings of the ICME 2018 – 1st International Conference on Marketing and Entrepreneurship*. Cyprus. Retrieved from https://hephaestus.nup.ac.cy/handle/11728/11392

Girasa, R. (2020). *Artificial Intelligence as a Disruptive Technology*. Palgrave, Mac Millan. DOI: 10.1007/978-3-030-35975-1

Kanellos, L. I. (2021). *Artificial Intelligence Applications in Law and Judicial Practice*. Nomiki Vivliothiki.

Kanellos, L. I. (2024). *The European Regulation on Artificial Intelligence and Its Implications for Greece*. Nomiki Vivliothiki., Retrieved from https://daily.nb.org/arthrografia/o-evropaikos-kanonismos-gia-tin-techniti-noimosyni-kai-oi-epiptoseis-tou-gia-tin-ellada/

Karapistolis, D. (2015). *Multivariate Statistical Analysis*. Athanasios Altintzis Editions.

Kousouni-Pantazopoulou, A. (2022). *Cloud Computing & Legal Issues*. Nomiki Vivliothiki.

Nunnally, J. C. (1978). An Overview of Psychological Measurement. In Wolman, B. B. (Ed.), *Clinical Diagnosis of Mental Disorders*. Springer. DOI: 10.1007/978-1-4684-2490-4_4

O'Neill, O. (2011). *Autonomy and Trust in Bioethics* (Dritsas, T., Trans.). Arsenidi Publishing.

Panagopoulou-Koutnatzi, F. (2023). *Artificial Intelligence: The Road to Digital Constitutionalism*. Papazisis Publishing.

Papadaki, Z. E., & Anastasiadou, S. D. (2019). Evaluating perception, expectation of consumers, and service quality gap in Greek banking in a period of financial crisis and capital controls. In N. Sykianakis, P. Polychronidou, & A. Karasavvoglou (Eds.), *Economic and Financial Challenges for Eastern Europe* (pp. 67–80). Springer Proceedings in Business and Economics. Springer, Cham. DOI: 10.1007/978-3-030-12169-3_5

Papadimitriou, I. (2007). *Data Analysis*. Tipothito Editions.

Roumeliotis, P. (2021). *Rupture: The Artificial Intelligence War*. A.A. Livanis Publishing.

Rueda, J. (2023). The ethics of doing human enhancement ethics. *Futures*, *153*, 103236. DOI: 10.1016/j.futures.2023.103236

Sindermann, C., Yang, H., Elhai, J. D., Yang, S., Quan, L., Li, M., & Montag, C. (2022). Quan4, L., Li, M., Montag, C. (2022). Acceptance and Fear of Artificial Intelligence: Associations with personality in a German and a Chinese sample. *Discover Psychology*, *2*(1), 8. DOI: 10.1007/s44202-022-00020-y

Souravlas, S., & Anastasiadou, S. (2020). Pipelined dynamic scheduling of big data streams. *Applied Sciences (Basel, Switzerland)*, *10*(14), 4796. DOI: 10.3390/app10144796

Souravlas, S., & Anastasiadou, S. (2022). On implementing social community clouds based on Markov models. *IEEE Transactions on Computational Social Systems*, *2022*. DOI: 10.1109/TCSS.2022.3213273

Souravlas, S., Anastasiadou, S., Economides, Th., & Katsavounis, S. (2023). Probabilistic community detection in social networks. *IEEE Access : Practical Innovations, Open Solutions*, *11*, 25629–25641. DOI: 10.1109/ACCESS.2023.3257021

Souravlas, S., Anastasiadou, S., & Katsavounis, S. (2021b). A survey on the recent advances of deep community detection. *Applied Sciences (Basel, Switzerland)*, *11*(16), 7179. DOI: 10.3390/app11167179

Souravlas, S., Anastasiadou, S., & Katsavounis, S. (2021c). More on pipelined dynamic scheduling of big data streams. *Applied Sciences (Basel, Switzerland), 11*(1), 61. DOI: 10.3390/app11010061

Souravlas, S., Anastasiadou, S., & Kostoglou, I. (2022b). A novel method for general hierarchical system modeling via colored Petri nets based on transition extractions from real data sets. *Applied Sciences (Basel, Switzerland), 13*(1), 339. DOI: 10.3390/app13010339

Souravlas, S., Anastasiadou, S., Tantalaki, N., & Katsavounis, S. (2021a). A dynamic load balanced task distribution strategy for heterogeneous cloud platforms based on Markov process modeling. *IEEE Access : Practical Innovations, Open Solutions, 10*, 2169–3536. https://ieeexplore.ieee.org/abstract/document/9729723

Souravlas, S., Anastasiadou, S. D., Tantalaki, N., & Katsavounis, S. (2022a). A fair, dynamic load balanced task distribution strategy for heterogeneous cloud platforms based on Markov process modeling. *IEEE Access : Practical Innovations, Open Solutions, 10*, 26149–26162. DOI: 10.1109/ACCESS.2022.3157435

Souravlas, S., Katsavounis, S., & Anastasiadou, S. (2020). On modeling and simulation of resource allocation policies in cloud computing using colored Petri nets. *Applied Sciences (Basel, Switzerland), 10*(16), 5644. DOI: 10.3390/app10165644

Taraza, E., & Anastasiadou, S. (2019). EFQM Excellence Model in Vocational Lyceum: Reliability and validity of the EFQM instrument. In *Proceedings of the 13th Annual International Technology, Education and Development Conference (INTED2019) (pp.* 2273–2285*)*. Valencia, Spain. DOI: 10.21125/inted.2019.0637

About the Contributors

Sofia D. Anastasiadou, is a Professor of Statistics and Research Methodology, Biostatistics Biomedical Research Methodology and at the School Health Sciences of the University of Western Macedonia (UOWM). She is the Dean of the Health Sciences of UOWM, Senate Member, and Member of post graduate studies committee. She is also a Member of the Senate and a MODIP Member of the same University. Her teaching focuses in the field of cognitive Research Methodology (Quantitative and Qualitative), Educational Research and Statistical Data Analysis, Big Data. She has written three books on Statistics and Research Methodology and has published over 300 articles in International and Greek Journals as well as in International and Greek Conferences.. She has organized and co-organized Greek and International Scientific Conferences. She is also a Member of a number of Scientific Committees of International Conferences and is an Associate Editor in a number of International Journals.

Theodore Chadjipadelis is a Professor of Applied Statistics at the Department of Political Science in Aristotle University of Thessaloniki. He is the Head of Laboratory of Applied Political Research and also Director of the MSc programs Political Analysis and Governance-Regional Development. He is also the Head of the Research Centre "Observatory for Democracy in Balkans, Black Sea and Eastern Mediterranean". He taught Statistics, Mathematics, Electoral Analysis, Governance, Local and Regional Government, Decision Theory in the Aristotle University as well as in the University of Thessaly and the Hellenic Open University. He has published more than 100 papers in scientific journals and conference proceedings in Greece and abroad relating to Electoral Analysis, Public Opinion analysis, laboratory planning, Applied Statistics, Civil and Regional Planning. He is the Head of the Committee for Control of Public Surveys (established following agreement of the Committee with the Ministry of Media). From April 2010 till 2012 he was Board Member of "Information Society SA" and since 2011 till 2015 he was President of the

Managing Committee at the University of Western Macedonia. He was a member of the Special Scientific Committee of the Ministry of Interior Decentralization and E-government for the new architecture of Administration and Regional Administration as well as the Special Committee for the electoral law. He has been member of the National Committee for Administrative Reform of the Ministry of Interior, Public Administration and Decentralization (as a representative of Aristotle University). He has been also a member of the scientific Committee of the "Institute of Local Government" (K.E.Δ.K.E-T.E.Δ.K.N.A.). He has also been representative of the Hellenic Ministry of Education in EUROSTAT and national representative at FP6 Initiative 7 (Citizens and Governance in Knowledge Society).

Nikolaos Koutsoupias is an Associate Professor of Applied Informatics & Data Analysis in the Department of International & European Studies of the School of Social Sciences, Humanities & Arts, University of Macedonia. He graduated from the Department of Computer Science, North Carolina State University (NCSU), Raleigh NC, USA and a PhD holder from the Department of Applied Informatics of the University of Macedonia. He worked as a Computing Lab Assistant at the Department of Computer Science and the Faculty of Social Sciences & Humanities, as a Teaching Assistant at the School of Electrical & Computer Engineering (NCSU) and as a Database Programmer of an International Insurance Group (ERGO). He taught IT & Data Analysis courses at the School of NCO Administration of the Hellenic Air Force and at the Departments of General Studies & Marketing of TEI of Thessaloniki. As Lecturer/pd407, Assistant and Associate Professor taught at the Aristotle University and later at the University of W.Macedonia. He is married with five children.

Angelos Markos is a Professor of Data Analysis in the Social Sciences at the Democritus University of Thrace, School of Education (Greece), board member of the Greek Society of Data Analysis (GSDA) and additional member of the International Federation of Classification Societies (IFCS) Council. In 2017, he was Visiting Scholar at the Department of Statistics, Stanford University. He hold a PhD degree from the University of Macedonia, Greece (2007). His areas of interest include multivariate analysis (especially dimension reduction and clustering), psychological testing and measurement in the social sciences and statistics education. I am creator and maintainer of the R packages clustrd and idm.

George C. Menexes is a professor of 'Biometry and Agricultural Experimentation' at the School of Agriculture of the Aristotle University of Thessaloniki. His research interests focused on the scientific research methodology, the development and application of statistical methods for the analysis of data resulted from experiments in the field of biological sciences and the methods of multivariate analysis of quantitative and categorical data. He has more than 240 publications (full articles) in scientific journals and conference proceedings and he has participated in 40 research programs. He has announced and/or presented more than 260 papers at National and International Conferences.

***

Thomas Alexandridis is an Assistant Professor in the Lab of Remote Sensing and GIS, Faculty of Agriculture, AUTH, Greece. He is an agronomist with a specialization in Applied Remote Sensing that started with a Masters' Degree in Cranfield University (United Kingdom) in 1996 and continued with a PhD from the School of Agriculture (AUTH) in 2003 in close collaboration with the International Water Management Institute (Sri Lanka and Pakistan). His research interests include digital image processing techniques and geographic analysis of spatial data for mapping agricultural resources (vegetation, soil and water), emphasizing on multiple scale issues in remote sensing and GIS. He has extensive experience in mapping vegetation types and monitoring vegetation parameters at wide scales (national and regional) using low resolution satellite images in several European countries, Pakistan and Mozambique. Fields of application include mapping crops, monitoring stress and land degradation using proximal and remote sensing data from satellites and UAVs in precision farming applications. He has authored and co-authored more than 80 journal and conference papers, and has participated in more than 20 research projects funded by the EU, European Space Agency, and various national sources, in several of them as Principal Investigator.

Vasiliki Bouranta holds a PhD in Political Science specializing in electoral systems and political and electoral competition, with an emphasis on the Balkan countries. Since 2021, she teaches as an adjunct lecturer in various academic institutions (Democritus University of Thrace, Aristotle University of Thessaloniki, Hellenic Open University) focusing on political analysis, political systems and state organization, political communication and marketing. She has participated in international and national research projects as a researcher and data analyst, as well as in international and national scientific conferences, workshops and seminars with announcements around her research interests. Her research interests and scientific activities focus on the analysis of electoral and political competition, with emphasis on the Balkan countries, the evolution of political communication and political marketing at national and European level, the course and development of multilevel governance and democratic institutions at central and local level in European countries. She is a member of the Hellenic Society of Political Science

(HPSA), the International Federation of Classification Societies (IFCS) and the Hellenic Society for Data Analysis (GSDA).

Zacharias Dermatis is an Assistant Professor at the Department of Management Science and Technology, University of the Peloponnese, specializing in Innovative and Digital Technologies for Quality of Life. He holds a PhD in Informatics, focusing on the use of GIS to analyze socioeconomic factors affecting elderly health, and a Master's degree in Computer Science and Technology. He has taught courses such as Management Information Systems, E-Government, Health Informatics, Strategic Management, and Geographic Information Systems (GIS). He is an active member of research laboratories and the iLEADS Research and Training Center. His research interests include Quality of Life, GIS, Databases, Data Digitization, and tools such as Python, SPSS, Jamovi, and Promethee. He has published extensively and participated in numerous conferences and collective scientific volumes.

Christos Dordas is a Professor of Agronomy and since 2015 he is the Head of the Laboratory of Agronomy. He was the Deputy Head of the School of Agriculture (2013–2017) and Head of the School of Agriculture (2017–2022). His research work addresses several issues in the field crops and plant physiology. He participated in numerous competitive projects funded from EU and national agencies. He supervised 15 master's students, five PhD candidates and postdoctoral researchers in Greece and is the author of more than 200 scientific publications, including more than 82 papers in peer-reviewed journals (> 4,600 citations, h-index: 33).

Lefkothea Karapetsi PhD candidate at the Department of Agriculture Crop Production and Rural Development, of University of Thessaly with PhD thesis title: "Comparative study at the molecular level of the interactions of parasitic organisms and plants using omics technologies». Graduated agronomist from the Department of Agriculture of the Aristotle University of Thessaloniki and holder of a master's degree in Plant Protection Sciences from the same department specializing in Phytopathology - Plant Virology. My internships were carried out at the Benaki Phytopathological Institute and the Valencian Institute of Agricultural Research (IVIA) in Spain. Current research interests focus on the use of modern molecular methods, such as High Resolution Melting Curve Technology (HRM) in the identification of plant species as well as Next Generation Sequencing (NGS) technologies, such as whole genome sequencing of the genome (WGS), the analysis of the transcriptome (RNA-Seq) in the study of plants and their interactions with plants and pathogenic microorganisms, with the aim of developing new molecular markers, the study of the microbiome through amplicon sequencing of the regions 16S and ITS in environmental samples, as well as the analysis of data obtained through these procedures. I have successfully attended the training seminars on "Meta-analysis of OMICs Data" from the ELIXIR Training network, "Basic laboratory techniques in Synthetic Biology" from the OMICEngine Research Infrastructure as well as the "Introduction to Galaxy and NGS analysis, RNA-Seq analysis with Galaxy and R and 16S Microbial Analysis" from the "Galaxy Training Network".

Efstratios Kiranas is an Emeritus Professor at the Department of Nutritional Sciences and Dietetics at the International Hellenic University. He is a Chemist-Technologist of Foods, specializing in the development of devices and methods of analyzing foods and biological liquids. His research interests focus on the preparation and the quality control of new food products, as well as on the study of nutritional habits and nutritional intake of groups of the population. He deals with the creation of databases, with informations for the nutritional habits of Greek adolescents, as well as the relationship of consumers with food additives, nutritional supplements and nutritional labeling. He is Director of the Research Laboratory "Investigation, Analysis and Evaluation of Foods and Nutritional Habits". He has been a member of national and European science committees (IDF-"Main Components in Milk", EFSA-"European Food Consumption Database", EFET- "Nutrition"), Researcher and Scientific supervisor of research programs. He is author of scientific books, publications in international journals, as well as author of announcements in international and Greek conferences. He has served as a member of Directors' Board in A.T.E.I. of Thessaloniki, Director of Technology of Foods and Nutrition Faculty, Vice President of Nutrition and Dietetics Department, Director of Foods and Nutrition Sector, Director of Research Laboratories, Vice President of Environmental Committee of ATEI of Thessaloniki and EMAS Environmental Management System manager.

Thomas Koutsos is Teaching Staff (E.DI.P. Special Laboratory Teaching Staff) in the Laboratory of Remote Sensing, Spectroscopy and GIS, Department of Hydraulics, Soil science and Agricultural Engineering, School of Agriculture, Faculty of Agriculture, Forestry and National Environment, Aristotle University of Thessaloniki. He holds a B.Sc. in Geology, M.Sc. in Applied and Environmental Geology, and a Ph.D. in Geo-Informatics, Environmental Geology, and Geographical Information Systems. His scientific interests lay in the fields of Geo-Informatics, Spatial Analysis, Geographical Information Systems, protection of the environment, management of environmental impacts and wastes, Data Visualization, and Spatial Statistics. He has also a Microsoft Certified Systems Engineer Certification (M.S.C.E. + Internet) and expertise in SQL Programming, Crystal Reporting and Database Management. Dr. Thomas Koutsos is scientific reviewer in more than 35 scientific Journals most of them with high Impact Factor. He has also more than 40 scientific publications, such as research articles in international journals of SCI, chapters in scientific books and series, posters and articles published at conference proceedings and workshops and he has been involved to at least 10 European co-funded projects.

Zacharenia Kyrana is a PhD candidate in Statistics at the School of Agriculture, Aristotle University of Thessaloniki. She holds an undergraduate degree in Statistics and Actuarial-Financial Mathematics and a postgraduate degree in Statistics and Data Analysis, from the School of Sciences, University of the Aegean. Her research interests include biometry, multivariate and multidimensional statistical-data analysis methods, and dimensionality reduction and clustering methods in data cubes and big data using various statistical software packages. She is the author of several scientific publications in peer-reviewed international journals, as well as in international and national conference proceedings.

Panagiotis Liargovas received his BA degree from the University of Athens and his MA and Ph.D degrees in Economics from Clark University, Worcester MA. He has been a Fulbright Scholar, a Bakalas Foundation Scholar, a Teaching Assistant Scholar (Clark University) and a European Commission trainee (Brussels). He has taught in many Universities including Clark University, Bologna University, and University of Strasbourg. He has served as economic advisor to the Greek Ministry of Finance, the Greek Ministry of Foreign Affairs and the Municipality of Athens in many occasions. For the period 2013-18 he was the Head of the Parliamentary Budget Office at the Hellenic Parliament. He is currently Chairman of the Board of the Centre of Planning and Economic Research (KEPE) and of the National Productivity Board. He is also Professor of Economics at the University of Peloponnese. He specializes on International and European Economics, Regional Economics and Economic Development and Policy. He has made more than 100 publications in academic journals and has published more than 20 books. His latest book (Pataki publishers, in Greek) was published in 2020 and was entitled "Ten years of crisis, three memoranda and a pandemic: Greece in search for a way out".

Alexandra-Maria Michaelidou specializes in Food Science and Nutrition at the Department of Food Science and Technology, School of Agriculture, Faculty of Agriculture, Forestry, and Natural Environment, Aristotle University of Thessaloniki (AUTh). She previously served as the Director of the MSc Program "Food Science and Nutrition" (2019–2023). Prof. Michaelidou has an extensive track record in research and education, having participated in 17 research and educational projects, with a leading role as coordinator or principal investigator in 9 of them.

She has supervised 4 doctoral dissertations (3 completed), 22 master's theses (18 completed), and over 100 undergraduate theses. Her academic contributions include more than 115 publications: 34 peer-reviewed articles in international journals (over 1,207 citations, h-index = 17, Scopus), 6 book chapters, and 67 conference papers presented at international and national peer-reviewed conferences. Prof. Michaelidou's research focuses on analyzing the nutritional properties of foods and evaluating their effects on metabolic health markers. Her expertise extends to the design and implementation of both in vitro and in vivo studies, particularly in assessing food digestibility and investigating the biofunctional properties of innovative food products.

Georgia Panagiotidou holds a PhD in Political Sciences with a specialization in applied political analysis. She has been teaching since 2016 at the School of Political Sciences of Aristotle University of Thessaloniki, as well as in the postgraduate program "Governance and Regional Development" as an academic scholar and adjunct lecturer. Additionally, she teaches as Collaborating Teaching Staff in the postgraduate program "Public Administration and Electronic Governance" at HOU and is a postdoctoral researcher in the Horizon Europe – AI4GOV program on artificial intelligence in governance. Her research interests and scientific activity focus on the analysis of political competition, parties, and electoral behaviour, and the identification of typologies of voters, candidates, parties, and issues using comparative data analysis methods under the "Candidates-Parties- Issues" scheme. She participates in research projects at the Laboratory of Applied Political Analysis of AUTH, she was member of the research team for Greece's first electronic Voting Advisor Application (VAA) in 2010 (helpmevote.gr) and participated in the COST True European Voter research project. Furthermore, she has conducted extensive empirical comparative research on political and electoral behaviour in Greece and at a pan-European level on behalf of European Union entities. She systematically participates in international scientific conferences with presentations of her research work and has published her scientific works in scientific journals and collective volumes. She is a member of the Hellenic, American, and International Political Science Associations (HPSA, APSA, IPSA), the International Federation of Classification Societies (IFCS), and the Greek Society for Data Analysis (GSDA).

Nikolaos Papafilippou is a PhD candidate in Data Analysis at the School of Agriculture, Aristotle University of Thessaloniki. He holds a Bachelor's degree in Mathematics and Economics and a Postgraduate degree in Education Studies and Mathematics Didactic. His research interests are primarily centered on multivariate and multidimensional data analysis, with a particular emphasis on dimension reduction on big datasets with categorical data from the field of biological and education sciences. He is the author of several scientific publications in international and national conferences. He is a member of the Greek Statistical Institute and the Greek Society of Data Analysis.

Panayiotis Papageorgiou is a maritime security specialist with experience in contract auditing and coordination in the shipping industry. He holds an MPhil in Maritime Security and an MSc in Maritime Operations Management from Liverpool John Moores University, as well as a Bachelor's degree in Business Administration from the Technological Educational Institute of Crete. He has worked as Contracts Auditor at Grimaldi Liverpool Agency Ltd, where he was involved in contract management, audits, tariff filing, and process optimization. He has also served as Financial Auditor and Chair of the Financial Audit Team of the Erasmus Student Network in Brussels, and as Office Assistant at the Greek Embassy in London. His research focuses on maritime security assessment methodologies, and he has published work on the use of the Bow-Tie diagram for risk evaluation. He is fluent in Greek and English, with conversational Spanish, and has strong skills in auditing, communication, and information systems. He has held multiple leadership roles in international volunteer organizations and has participated in numerous training events and conferences across Europe.

Emmanouil D. Pratsinakis is an PhD candidate at the School of Agriculture, Aristotle University of Thessaloniki (AUTh). Graduated from the School of Agriculture of AUTh with Postgraduate studies in the School of Agriculture of AUTh. His current research interests focused on Multivariate Data Analysis of mixed type data. In addition, his research utilises statistical and bioinformatics tools. He has published 10 publications in scientific journals and conference proceedings with full articles. He is a member of the Greek Society of Data Analysis.

Dimitrios Vagianos is a graduate of the School of Electrical and Computer Engineering of the Aristotle University of Thessaloniki. He holds a Master of Science degree with distinction in Mobile Communication Systems from the University of Surrey, UK, and a PhD degree from the Department of International and European Studies of the University of Macedonia in the field of Big Data Mining and Analysis of the Social Web. He is a member of the Laboratory Teaching Personnel at the University of Macedonia where he teaches in Greek and English stand-alone undergraduate and postgraduate courses. His scientific work has been published as peer-reviewed articles in noumerous scientific journals and as chapters in edited volumes of international publishers.

Theodora Vainterli is a highly qualified professional with a strong academic foundation in Political Science, Pedagogical Studies, and Counseling. She possesses substantial experience in journalism, project management within the Erasmus+ framework, and career guidance, and is currently a doctoral candidate at the Democritus University of Thrace.

Index

N

O

P

Q

R

S

T

U

V

W

Y